ADVANCES IN CHEMICAL ENGINEERING

Volume I

Advances in
CHEMICAL ENGINEERING

Edited by

THOMAS B. DREW
*Department of Chemical Engineering, Columbia University
New York, N. Y.*

JOHN W. HOOPES, JR.
*Atlas Powder Company
Wilmington, Delaware*

VOLUME I

1956

ACADEMIC PRESS INC. · PUBLISHERS · NEW YORK

TP
145
.D7
1956
v.1

Copyright © 1956, by
ACADEMIC PRESS INC.
125 EAST 23RD STREET
NEW YORK 10, N. Y.

ALL RIGHTS RESERVED
NO PART OF THIS BOOK MAY BE REPRODUCED IN ANY FORM,
BY PHOTOSTAT, MICROFILM, OR ANY OTHER MEANS,
WITHOUT WRITTEN PERMISSION FROM THE PUBLISHERS.

Library of Congress Catalog Card Number: 56-6600

PRINTED IN THE UNITED STATES OF AMERICA

CONTRIBUTORS TO VOLUME I

NATHANIEL F. BARR, *Chemistry Department, Brookhaven National Laboratory, Upton, New York*

R. BYRON BIRD, *Department of Chemical Engineering, University of Wisconsin, Madison, Wisconsin*

ERNEST J. HENLEY, *Department of Chemical Engineering, Columbia University, New York, New York*

A. B. METZNER, *Department of Chemical Engineering, University of Delaware, Newark, Delaware*

J. B. OPFELL, *California Institute of Technology, Pasadena, California*[1]

B. H. SAGE, *California Institute of Technology, Pasadena, California*

ROBERT W. SCHRAGE, *Esso Standard Oil Company, Linden, New Jersey*

ROBERT E. TREYBAL, *Department of Chemical Engineering, New York University, New York, New York*

J. W. WESTWATER, *Department of Chemistry and Chemical Engineering, University of Illinois, Urbana, Illinois*

[1] Present address: Cutter Laboratories, Berkeley, California.

PREFACE

The chemical engineer ministers to an industry of far flung interests. Its products range from soap to plutonium, from gasoline to paper, from antibiotics to cement. It flourishes on change: new products, processes, methods, and applications; new needs created and foreseen. Versatile men with breadth of interest in science and commerce have been demanded and the challenge of the field has found for it such men. The industry, in consequence, has grown apace—especially since 1900—and with it has proliferated a literature of truly vast proportions and variety. The end *Advances in Chemical Engineering* seeks to serve is to provide the engineer with critical running summaries of recent work: some that bring standard topics up to date; others that gather and examine the results of new or newly utilized techniques and methods of seeming promise in the field. Thereby we hope to help practitioners of the chemical engineering art to keep abreast the flood of information they have been creating.

The subjects covered will range widely because the field is broad. The unit operations will of course be covered but they are not the whole of chemical engineering. Topics from chemical technology, commercial development, and pertinent fields of applied science will, for example, also find a place. The authors have been asked to furnish definitive and critical reports giving their own analyses of their subjects, not merely annotated bibliographies. The editors have endeavored to invite as authors qualified scholars well versed in their fields and have given them free rein for their opinions.

The editors salute and thank the authors who have braved the vicissitudes of this new venture, and commend their efforts to their colleagues: the engineers of chemical industry.

New York, New York and THOMAS BRADFORD DREW
Wilmington, Delaware JOHN WALKER HOOPES, JR.
April, 1956

CONTENTS

CONTRIBUTORS TO VOLUME I v

PREFACE vii

Boiling of Liquids
J. W. WESTWATER, *Department of Chemistry and Chemical Engineering, University of Illinois, Urbana, Illinois*

I. Introduction 2
II. Nucleate Boiling and the Critical Temperature Difference 12
   Nomenclature 71
   References 73

Non-Newtonian Technology: Fluid Mechanics, Mixing, and Heat Transfer
A. B. METZNER, *Department of Chemical Engineering, University of Delaware, Newark, Delaware*

I. Fluid Classifications 79
II. Mechanics of Flow in Round Pipes 90
III. Miscellaneous Flow Problems 112
IV. Mixing of Non-Newtonian Fluids 119
V. Heat Transfer 121
VI. Rheology and Viscometry 138
   Nomenclature 148
   References 150

Theory of Diffusion
R. BYRON BIRD, *Department of Chemical Engineering, University of Wisconsin, Madison, Wisconsin*

I. Introduction 156
II. The Fluid Mechanical Basis for Diffusion 159
III. Calculation and Estimation of Diffusion Coefficients 182
IV. Solutions of the Diffusion Equations of Interest in Chemical Engineering . 198
V. Conclusions 228
   Nomenclature 231
   References 234

Turbulence in Thermal and Material Transport
J. B. OPFELL AND B. H. SAGE, *California Institute of Technology, Pasadena, California*

I. Introduction 242
II. Nature of Turbulence 242
III. Macroscopic Aspects of Steady, Uniform, Turbulent Flow 247
IV. Thermal Transport in Turbulent Flow 255

V. Material Transport in Turbulent Flow 267
VI. Combined Thermal and Material Transport in Turbulent Flow 278
VII. Summary 281
   Nomenclature 283
   References 285

Mechanically Aided Liquid Extraction
ROBERT E. TREYBAL, *Department of Chemical Engineering, New York University, New York, New York*

I. Introduction 290
II. Agitated Vessels 291
III. Agitated, Multistage Countercurrent Columns 310
IV. Pulsed Columns 317
V. Centrifugal Extractors 323
   Nomenclature 327
   References 328

The Automatic Computer in the Control and Planning of Manufacturing Operations
ROBERT W. SCHRAGE, *Esso Standard Oil Company, Linden, New Jersey*

I. Introduction 331
II. Computing Equipment 332
III. Control of Manufacturing Operations 341
IV. Statistical and Numerical Analysis 345
V. Mathematical Models 348
VI. Optimization Studies 356
   References 366

Ionizing Radiation Applied to Chemical Processes and to Food and Drug Processing
ERNEST J. HENLEY, *Department of Chemical Engineering, Columbia University, New York, New York,* AND NATHANIEL F. BARR, *Chemistry Department, Brookhaven National Laboratory, Upton, New York*

I. Introduction 370
II. Statement of Problem 371
III. Experimental Work on Initiation of Chemical Reactions 374
IV. Radiation Sterilization 397
   References 420

AUTHOR INDEX 427

SUBJECT INDEX 441

Boiling of Liquids

J. W. WESTWATER

Department of Chemistry and Chemical Engineering
University of Illinois, Urbana, Illinois

I. Introduction	2
A. Historical	2
1. Quenching	2
2. The Leidenfrost Phenomenon	3
B. Nukiyama and the Boiling Curve	3
C. Types of Boiling	4
1. Nucleate, Transition, and Film Boiling	4
2. Surface Boiling of a Subcooled Liquid	6
3. Bulk Boiling of a Volume-Heated Liquid	6
D. General Descriptions	6
1. Photographic Studies	7
2. The Sound of Boiling	11
II. Nucleate Boiling and the Critical Temperature Difference	12
A. Description from Photographic Studies	12
B. Theoretical Treatment of Nucleate Boiling	13
1. Equation of Rohsenow	13
2. Forster-Zuber Equation	16
3. Empirical Correlations	21
C. Theoretical Treatment of Nucleation	22
1. Classical Rate Theory	23
a. Homogeneous Case	24
b. Heterogeneous Case	34
2. Statistical Fluctuation Theory	39
D. Critical Temperature Difference	42
1. Maximum Efficiency	42
2. The Burnout Problem	42
a. Role of Heat Source	43
3. Physical Interpretations of the Critical ΔT	43
a. Thermodynamic Equation of State	44
b. Kinetic Viewpoint	48
E. Experimental Values for Nucleate Boiling	50
1. Effect of Type of Liquid	51
a. Water	51
b. Other Common Liquids	52
c. Liquid Metals	53
2. Effect of Type of Hot Solid	54
a. Typical Results	54
b. Measurement of Surface Temperature	55
3. Effect of Surface Texture	56
4. Effect of Geometric Arrangement	58

5. Effect of Pressure.. 60
   6. Effect of Surface Tension.. 60
   7. Effect of Agitation.. 62
   8. Effect of Impurities... 64
   9. Effect of Short-Wave Irradiation................................... 66
  F. Growth of Bubbles... 67
  Acknowledgment.. 71
  Nomenclature.. 71
  References.. 73

I. Introduction

The study of boiling liquids received little attention from chemical engineers, or other engineers for that matter, until very recently. The development of machines that evolve tremendous quantities of heat per unit volume has led to a sudden and great interest in boiling. Atomic reactors, jet engines, and rocket engines are the principal high-heat flux devices. Boiling happens to be a mode of heat transfer that can accommodate heat fluxes in the order of millions of B.t.u./(hr.)(ft.^{2}). The heat transfer rates possible with nonboiling fluids are far less.

Because boiling is a new science and data are rather scarce, much research is directed toward obtaining numerical values for specific test conditions. Representative data are presented in this chapter, but no pretense is made that the bibliography is complete; for example, the numerous tests made with evaporators are not included here.

The goal of this chapter is twofold. One object is to show the present state of knowledge of boiling liquids. In particular an effort is made to call attention to some of the experiments and thoughts on boiling by physicists, chemists, aeronautical engineers, mechanical engineers, and other persons not identified as chemical engineers. The second object is to indicate the particular parts of the subject which are in greatest need of more work.

A. Historical

It is curious that some of the most familiar phenomena in nature are among the least understood. Consider the boiling of liquids. Man's acquaintance with this must be as old as the discovery of fire, yet until our lifetime no serious study of boiling was attempted. The first few experiments showed that boiling is far more complex than anyone had expected.

1. *Quenching*

Blacksmiths and metallurgists have long known that hot metals seem to cool in two steps when quenched. At first the rate of cooling is very slow, in spite of the high metal temperature. Later the rate is rapid, although the metal temperature is much lower. The explanation of these facts was not possible until Nukiyama carried out his classical experiments, described later.

2. *The Leidenfrost Phenomenon*

The Leidenfrost phenomenon was first discussed in 1756 (L1). This phenomenon is the occurrence of a "repulsion" between a liquid and a very hot solid. For example, a drop of water on a hot plate will dance around noisily for some time before evaporating. On a moderately warm plate the phenomenon does not appear, and evaporation is very rapid. Nukiyama's test shed some light on this mystery.

In 1926 Mosciki and Broder (M10) made some studies of electrically heated, vertical wires submerged in cold water and in heated water. They showed that "subcooled" boiling results in greater heat fluxes than can be obtained with the liquid at its boiling point. These tests anticipated in part the tests made soon after by Nukiyama.

B. NUKIYAMA AND THE BOILING CURVE

In 1934 Nukiyama (N2) carried out a simple experiment which resulted in a great advance in the science of boiling. He submerged a thin platinum wire in water at 212° F. and heated the wire electrically to produce boiling. He discovered that the rate of heat transfer from the wire to the water increased steadily as the wire temperature was increased until the wire temperature reached about 300° F. At this temperature an unexpected thing happened; the wire temperature jumped suddenly to about 1800° F. A further increase in the wire temperature resulted in a smooth increase in the heat transfer rate.

The sudden temperature jump for the platinum wire was puzzling, and so Nukiyama tried wires of nickel and alloys having melting points lower than platinum has. When these wires experienced the temperature jump they melted. This was the phenomenon now known as *burnout*.

Nukiyama next investigated the in-between region of 300° to 1800° F. He used platinum, allowed the jump to occur, and then lowered the metal temperature gradually. It was possible to cool the wire to about 570° F. A smooth decrease in the heat flux occurred as the temperature was decreased. At 570° F. another temperature jump occurred, to less than 300° F.

Nukiyama's boiling curve is shown in Fig. 1. He concluded that at least two types of boiling occur: for water, one was below 300° F. and one above 570° F. He postulated that a third type might exist, represented by the dotted line on the figure. It was obvious that if this third type existed it would have a peculiar characteristic: any increase in the metal temperature would cause a decrease in the heat transfer rate. Such a happening is contrary to intuition.

C. Types of Boiling

1. *Nucleate, Transition, and Film Boiling*

Subsequent evidence proved that Nukiyama was right. At least three types of boiling exist. In Fig. 3, which shows some recent data (W3) for methanol boiling on a horizontal copper tube, the portion of the curve

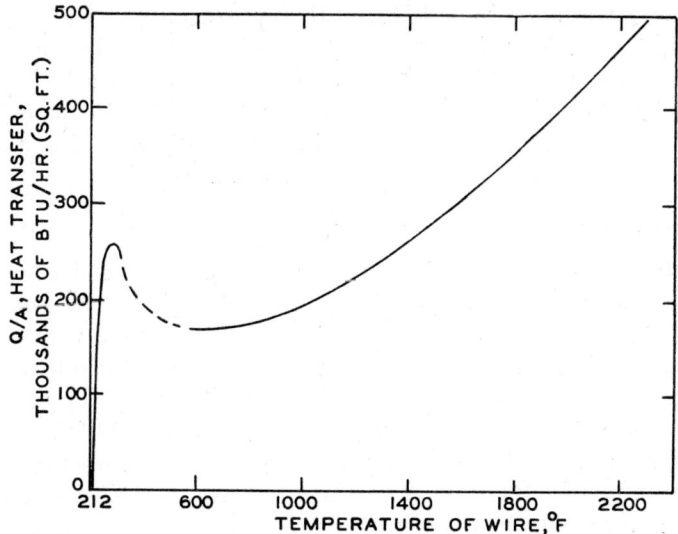

Fig. 1. The first boiling curve. Nukiyama's results for water boiling on a platinum wire are shown (N2).

AB corresponds to *nucleate* boiling. Unfortunately, the terminology used in boiling is somewhat confused—a characteristic common to most new sciences. Some writers call this *pool* boiling, *surface* boiling, and other names as well. *Nucleate* boiling seems preferable, because the outstanding feature of this type of boiling is that bubbles form at specific, preferred points on the hot surface. In other words, active nuclei exist on the solid. No one has proved what an active point is, but photographic evidence that something is special about certain points on the solid is convincing.

The region *BCD* is the *transition* region of boiling. No active centers exist. The heat flux from the hot solid to the boiling liquid decreases continuously as the temperature-difference driving force is increased.

This type of boiling has been called *metastable* by at least one writer. The word *transition* is preferred inasmuch as the word is well established for the in-between region encountered in fluid-flow studies.

From about E, in Fig. 3, on beyond indefinitely to increasing temperature differences, *film* boiling exists. This terminology is good, because a real film of vapor coats the hot solid during this type of boiling.

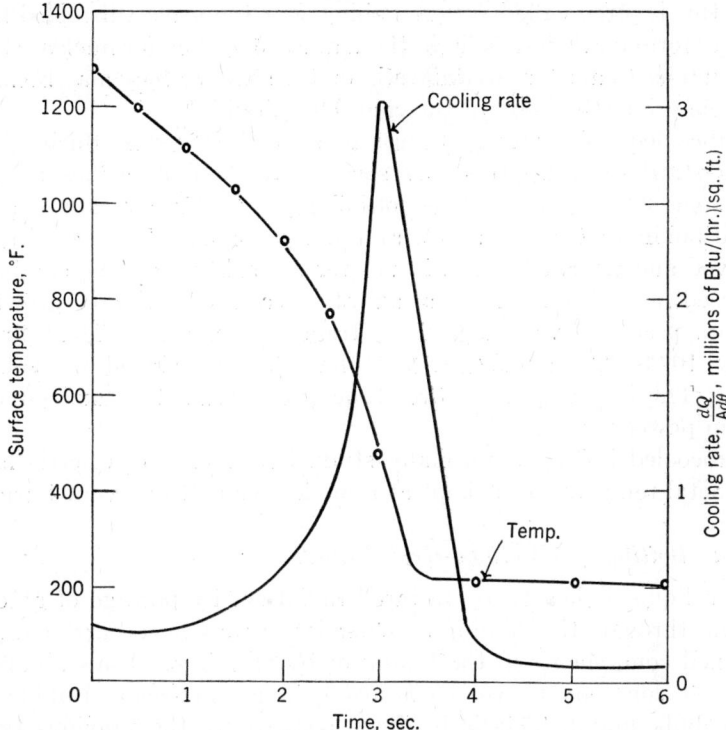

Fig. 2. Quenching curves for a 0.84-in.-diam. steel rod in city water (Y1).

A consideration of these three types of boiling results in an explanation for the distinct shape of quenching curves, temperature vs. time. Figure 2 shows some quenching data for a steel rod in water. The temperature decreases slowly at first, then very rapidly, and then slowly again. The explanation is that film boiling, transition boiling, then nucleate boiling, and finally free convection take place. The peak in the rate of cooling corresponds to the peak heat flux in the ordinary boiling curve (such as point B in Fig. 3). The S-shaped temperature-vs.-time curve in Fig. 2 is typical of the quenching phenomenon. By contrast, if a hot metal rod is cooled in a nonboiling liquid, the temperature will decrease asymptotically with time, according to a decay function, and no inflection point results.

2. Surface Boiling of a Subcooled Liquid

If a cold liquid (below the boiling point) is caused to flow past a hot solid, *subcooled* boiling results. Because the boiling exists on the hot solid only, this is often called *surface* boiling. For the usual subcooled boiling, bubbles form on the hot solid in the manner observed for nucleate boiling. The bubbles then collapse while still on the solid, or they may break loose and collapse in the body of the cold, bulk liquid.

Subcooled boiling often occurs in a nonflow system during the unsteady-state warm-up time. As soon as the liquid reaches its boiling point, one of the ordinary types of boiling results. If a steady-state, subcooled boiling is desired, a flow system is necessary. The liquid must be removed and replaced or it will not remain subcooled. The combination of forced flow plus the natural agitation caused by bubble growth and collapse produces very high heat transfer coefficients. Heat fluxes of nearly 10,000,000 B.t.u./(hr.)(ft.^{2}) have been reported for subcooled boiling. This is perhaps a hundred times greater than the fluxes commonly used in power plants.

Subcooled boiling is normally studied by means of electric heaters. The metal temperature is kept at a modest value to prevent burnout.

3. Bulk Boiling of a Volume-Heated Liquid

If a liquid is heated by infrared radiation, by passage of an electric current through the liquid, by chemical reactions occurring between dissolved components in the liquid, or by atomic reactions occurring in dissolved components, *volume-heated* boiling can occur. Bubbles grow in the bulk liquid. This is in sharp contrast to the previous types of boiling, all of which are concerned with bubbles growing on a hot solid. Volume-heated boiling and subcooled boiling may be special cases of nucleate, transition, and film boiling. Much more information is needed before this is certain.

D. General Descriptions

The most useful descriptions of the various types of boiling would be mathematical expressions that fit the physical observations. The available meager theories are described in the sections which follow. This section is concerned with two supplementary methods for describing boiling: the visual appearance of boiling and the sound of boiling.

1. *Photographic Studies*

Seeing a phenomenon stopped or slowed down is often a beginning to its understanding. Since 1936 a number of researchers have used photography to study boiling. A steady improvement in techniques has occurred since that date, and photography now is a powerful research tool. Some of the early still photographs were taken by Jakob (J1) and Drew and Mueller (D4) using ordinary exposure times. Their results proved that very short exposures were needed. Sauer (S1) and recently Corty and

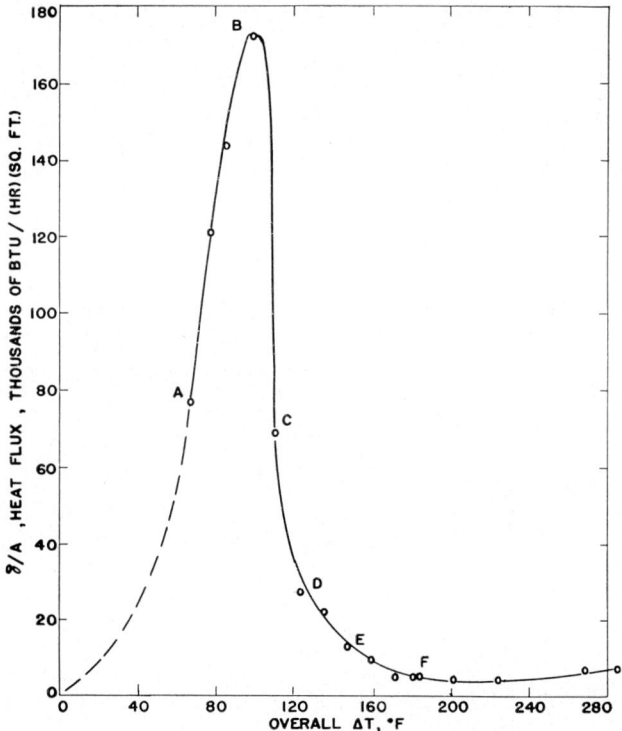

Fig. 3. Boiling curve for methanol. The liquid was outside a ⅜-in., steam-heated, horizontal copper tube, at 1 atm. The letters represent the conditions for the photographs, Figs. 4 to 9 (W3).

Foust (C4) obtained shots at 10^{-5}-sec. exposure. A set of photographs corresponding to the complete boiling curve for methanol has been obtained by Westwater and Santangelo (W3). Six of their photographs, taken at 10^{-6}-sec. exposure are included (Figs. 4 to 9).

Still photography is most suited to static conditions, but motion pictures are more valuable for dynamic conditions. Motion pictures of liquids boiling at a hot surface have been taken by Jakob (J1), by Corty and Foust (C5), and by others at the normal speed of 16 frames/sec. Such film shows blurred action in every frame. Drew and Mueller took

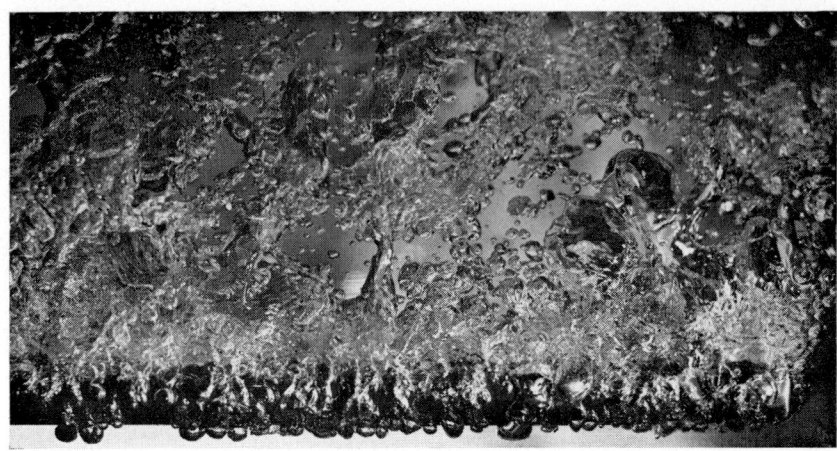

Fig. 4. Nucleate boiling of methanol. This is point A on the boiling curve, Fig. 3. Over-all $\Delta T = 67°$ F. Heat flux = 76,900 B.t.u./(hr.)(ft.^{2})(W3).

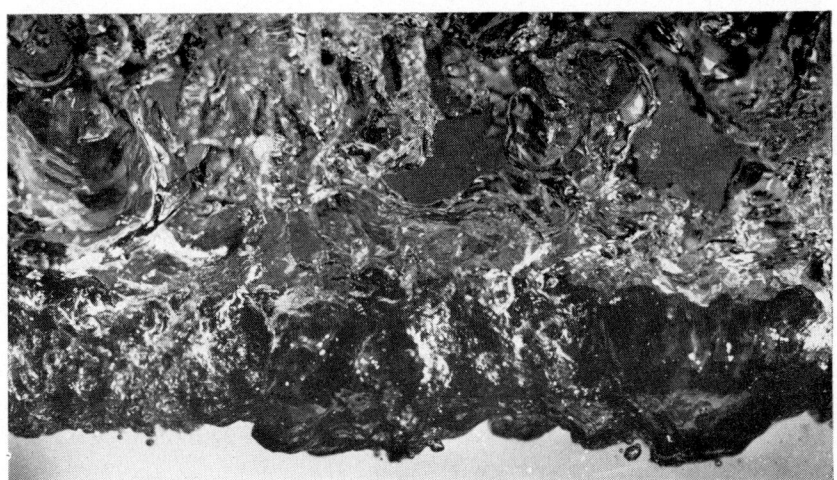

Fig. 5. Critical boiling of methanol. This is point B on the boiling curve. Over-all $\Delta T = 99.5°$ F. The heat flux is at a maximum = 172,000 B.t.u./(hr.)(ft.^{2})(W3).

higher speed motion pictures at 64 frames/sec. (D5). Jakob (J1, J2) achieved approximately 500 frames/sec., a speed at which some blurring still exists. By this time it was quite apparent that boiling is a high-speed phenomenon. Dew (D2) took motion pictures at 2000 frames/sec.,

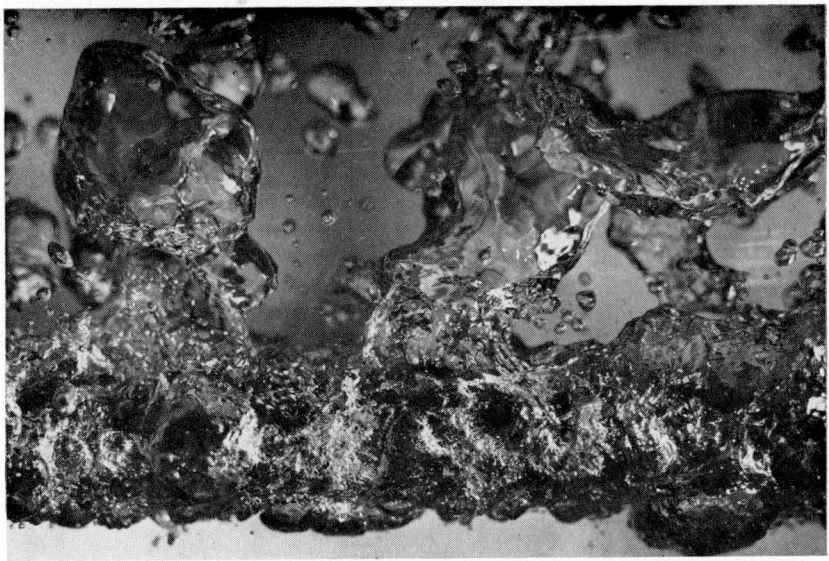

Fig. 6. Transition boiling of methanol. This is point C on the boiling curve. Over-all $\Delta T = 112°$ F. Heat flux = 69,000 B.t.u./(hr.)(ft.^{2})(W3).

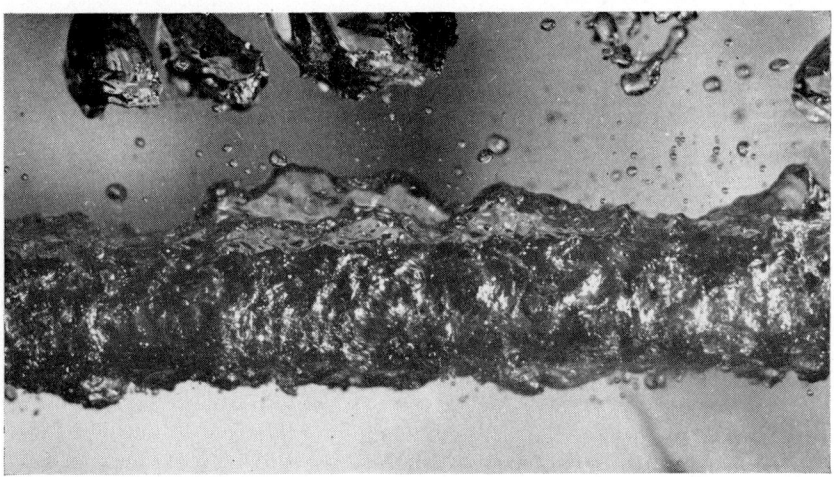

Fig. 7. Transition boiling of methanol. This is point D on the boiling curve. Over-all $\Delta T = 124°$ F. Heat flux = 27,200 B.t.u./(hr.)(ft.^{2})(W3).

and Westwater and Santangelo used 4000 frames/sec. (W4). Gunther (G7) obtained a speed of nearly 20,000 frames/sec., but the field of view was necessarily limited to a very small area.

The growth of bubbles in a volume-heated liquid has been determined by Dergarabedian (D1) using 1000 frames/sec. The bursting of bubbles

Fig. 8. Film boiling of methanol. This is point E on the boiling curve. Over-all $\Delta T = 148°$ F. Heat flux = 12,970 B.t.u./(hr.)(ft.^{2})(W3).

Fig. 9. Film boiling of methanol. This is point F on the boiling curve. Over-all $\Delta T = 181°$ F. Heat flux = 5,470 B.t.u./(hr.)(ft.^{2})(W3).

at a liquid-vapor interface has been studied by Aspden (A2) using a "speed" of 150,000 exposures/sec., but for three exposures only.

The visual appearance of the boiling action, given in Sec. IIA, IIIA, IVA, and VA, is based on the preceding works. Greatest emphasis has been given to the motion pictures taken at 4000 frames/sec. (W4).

2. The Sound of Boiling

Not only do the types of boiling look different, they sound different as well. Technical operators in charge of evaporators, reboilers, and other boiling equipment sometimes judge the operation of their equipment from the noise emitted.

The sound emitted from methanol boiling on a copper tube at 1 atm. has been measured (W2). No forced convection was used. Figure 10 shows how the sound, in total decibels for the frequency range of 25 to 7500 cycles/sec., varies with the temperature difference. Nucleate boiling

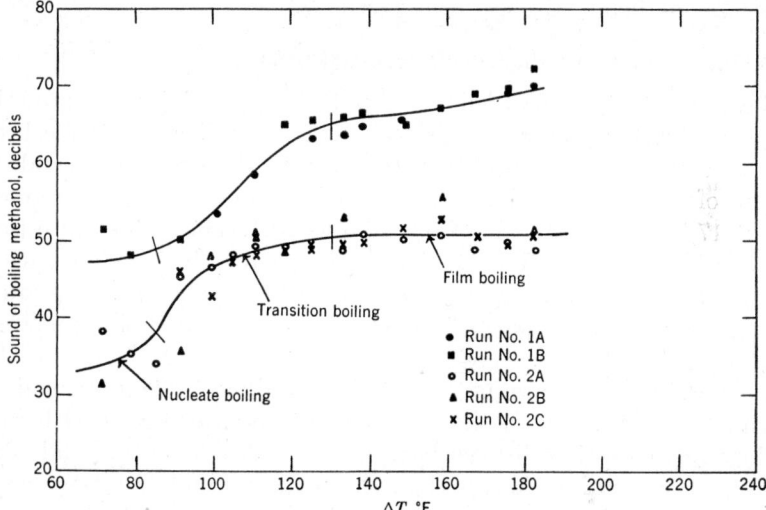

Fig. 10. The sound of boiling. Methanol was boiled on a ⅜-in., horizontal, steam-heated, copper tube at 1 atm. (W2).

is the most quiet, and film boiling is the noisiest of the three types of boiling studied. Runs 1A and 1B were taken with a thoroughly degassed liquid and a very clean tube. Runs 2A, 2B, and 2C were taken with a poorly degassed liquid and a slightly oxidized tube. It is seen that some of the variables known to affect boiling-heat transfer also affect boiling noise.

The sound of forced-flow, subcooled boiling was reported by Goldman (G6) under the attractive title "Boiling Songs." He noted that the noise increased as the heat flux increased toward the burnout condition. Under certain conditions of velocity, heat flux, liquid purity, and temperature difference, the "singing" became a "high frequency scream" or a "wailing banshee."

The sound emitted from hot metals as they are quenched has been noticed (M1): as the temperature decreases, the sound intensity and also the dominant pitch change.

II. Nucleate Boiling and the Critical-Temperature Difference

The nucleate type of boiling is far more common than any other type. This form usually occurs in commercial evaporators, still pots, and similar boiling equipment. Much attention has been given to nucleate boiling and a good many data are available. Because the critical temperature difference is the upper limit to nucleate boiling, it too has received notice. The number of data for it is far from ample, because of the burnout difficulty discussed in Sec. IID.

A. Description from Photographic Studies

The nucleate type of boiling (Fig. 4) is typified by the growth of bubbles at specific points on the hot solid. As soon as a bubble breaks loose, a new one begins to form at the identical spot on the solid. The emission of bubbles is rapid. For the conditions of Fig. 4, the frequency is about 17/sec., with a range from 15.5 to 20.1. The bubbles are rather small at the instant of release. The average for the above-mentioned case is 0.17 in., with a range from 0.10 to 0.20 in. The natural agitation is very good and must be part of the reason that nucleate boiling can result in such large heat fluxes.

The active points are not located according to any recognizable pattern. Some are so close that bubble interference occurs. Others are so widely scattered that distinct bare areas are visible on the solid. These bare areas are certainly hotter than the boiling liquid, yet they remain bare. In Fig. 4 the bubble-to-bubble spacing ranges from 0.058 to 0.46 in., averaging 0.103 in. An active point is suspected to be a tiny pit or scratch in the solid. However no proof exists, and it may be a tiny sharp point, a bit of impurity, or a boundary between metal crystals.

As the temperature difference increases, new active points appear. Finally the number becomes so large that no bare areas can be seen on the metal. This is the condition of maximum heat transfer illustrated in Fig. 5. The temperature difference is the critical ΔT.

It is strongly believed by many workers that the path of heat flow in nucleate boiling is from the solid to the liquid and hence to the growing bubbles. Considering the large bare areas in Fig. 4, it is inevitable that the heat flow from the solid to the liquid must be very important at modest ΔT values. Considering the visual evidence only, Fig. 5 for the critical ΔT shows no bare areas for the expected flow path. Motion pictures at the critical ΔT show violent motion but seem to indicate no liquid-solid contact. Figure 5 is the only published photograph of boiling at the critical ΔT. Photographs from other experimenters are needed to shed light on this provocative problem.

B. Theoretical Treatment of Nucleate Boiling

Inasmuch as nucleate boiling is the most common type of boiling, it has received much attention in the past. Until about 1950 the usual method of study was to attempt empirical correlations with the scanty data on boiling. None of the resulting equations was successful enough to warrant widespread adoption.

Since 1950 two new approaches have been published. Each of these is theoretical in part and empirical in part. Each begins by giving a physical picture of the mechanism of heat transfer during nucleate boiling and then proceeds to build an equation consistent with the described mechanism. Both attacks assume that during normal nucleate boiling heat flows from the hot solid into the liquid and from thence to the vapor bubbles. The controlling resistance to heat transfer is taken to be a "stagnant" film around the vapor bubbles. The final equations are those of Rohsenow, which have met with moderate success, and those of Forster and Zuber, which are promising but are too recent to have received thorough evaluation.

1. *Equation of Rohsenow*

Rohsenow and coworkers (R8, R9, R11, R12) use an attack in which the key feature is an assumption that the movement of bubbles at the instant of breaking away from the hot surface is of prime importance. The release of bubbles into the surrounding liquid is presumed to give good agitation, improving the convection currents and decreasing the thickness of the stagnant film through which the heat must flow.

Forster and Zuber (F6) on the other hand use a different approach. They assume that the movement of well-developed (large) bubbles is of little importance during nucleate boiling. The important movement is imagined to be that of the vapor-liquid interface of a growing bubble while the bubble is still attached to the surface. The linear velocity of this interface is much greater than the velocity described by Rohsenow; typical comparison values are 10 ft./sec. vs. 0.3 ft./sec. The Forster-Zuber derivation is concerned with small bubbles, on the average, compared with those treated by Rohsenow. Both approaches yield equations which are claimed to describe the same nucleate boiling. The form of the Rohsenow equation was influenced undoubtedly by the traditional equations for nonboiling liquids. For the case of heat transfer from a solid to a surrounding nonboiling liquid, it is known that the most reliable correlations are in the form of

$$\text{Nu} = (\text{Const.})(\text{Re})^a(\text{Pr})^s \tag{1}$$

This type of equation comes from dimensional analysis. The coefficient and exponents are found by experiment. If forced convection is used, the Reynolds number has the conventional meaning of $DV\rho/\mu$. If free convection is used, the Reynolds number is replaced by the Grashof number, which can be shown to have a meaning of a Reynolds number, owing to free convection (B8).

The Rohsenow approach is to use Eq. (1) modified for boiling. The correct Reynolds number is taken to be the Reynolds number of a bubble just after it breaks loose from the hot solid. Previous workers (J2) found experimentally that the velocity of a released bubble is constant for a short time. In addition it was shown that the diameters of the released bubbles of water or carbon tetrachloride are inversely proportional to the frequency of emission of bubbles that size, or

$$f \cdot D_b = C_1 \tag{2}$$

This relationship was verified at small heat fluxes only. Also a correlation for the average diameter of a bubble at the instant of break-off is available (F8) from experimental evidence for bubbles of steam and of hydrogen.

$$D_b = C_2 \beta \left[\frac{2g_0\sigma}{g(\rho_L - \rho_V)} \right]^{1/2} \tag{3}$$

If the bubbles are assumed to be spheres and if they leave the solid from n points per unit area, the estimated bubble velocity V becomes

$$V = fn\pi D_b^3/6 \tag{4}$$

The bubble Reynolds number then is

$$\text{Re} = \frac{D_b V \rho_L}{\mu_L} = \frac{C_3 \beta f n \rho_L D_b^3}{\mu_L} \left[\frac{2g_0\sigma}{g(\rho_L - \rho_V)} \right]^{1/2} \tag{5}$$

The bubble diameter can be eliminated because the product of the mass of vapor formed per unit time and the latent heat of vaporization must equal the heat transfer rate $q = w\lambda$.

$$\text{Re} = \frac{C_4 \beta q/A}{\mu_L \lambda} \left[\frac{g_0\sigma}{g(\rho_L - \rho_V)} \right]^{1/2} \tag{6}$$

The Nusselt number, hD_b/k_L, can be combined with Eq. (3) to give

$$\text{Nu} = \frac{h\beta}{k_L} \left[\frac{g_0\sigma}{g(\rho_L - \rho_V)} \right]^{1/2} \tag{7}$$

The Prandtl number is based on liquid properties on the grounds that the film resistance is probably a liquid film.

$$\mathrm{Pr} = \frac{C_L \mu_L}{k_L} \tag{8}$$

The final correlation becomes

$$\mathrm{Nu} = (\mathrm{Const.})(\mathrm{Re})^{2/3}(\mathrm{Pr})^{-0.7} \tag{9}$$

or $\dfrac{h\beta}{k_L}\left[\dfrac{g_0\sigma}{g(\rho_L-\rho_V)}\right]^{1/2} = (\mathrm{Const.})\left[\left(\dfrac{\beta q/A}{\mu_L \lambda}\right)\left(\dfrac{g_0\sigma}{g(\rho_L-\rho_V)}\right)^{1/2}\right]^{2/3}\left[\dfrac{C_L\mu_L}{k_L}\right]^{-0.7}$

(Rohsenow equation) (10)

In order to obtain the coefficient and the two exponents (2/3 and −0.7), Rohsenow used experimental data for five systems: water boiling on a 0.024-in.-diam. platinum wire (A1), water on a 1.5-in.-diam. horizontal tube (C6), and benzene, ethyl alcohol, and n-pentane on a chromium-plated horizontal surface (C2). The exponents for the Reynolds and Prandtl numbers were constant at 2/3 and −0.7 as shown. Unfortunately the coefficient varied from 0.006 to 0.015 from system to system.

It is not surprising that the foregoing equation is imperfect and contains a coefficient which is constant for one liquid-solid system only. Rather it is surprising that an equation based on the idea of a controlling macroconvection and on empirical relationships such as Eqs. (2) and (3) should fit any real system. An example of the suitable agreement between Addoms' data for water on a platinum wire, Fig. 11, and Rohsenow's equation is shown in Fig. 12. A few algebraic manipulations are required to obtain the coordinates of Fig. 12 from Eq. (10). By further algebra it is possible to show that the equation predicts that the heat transfer coefficient, h, is dependent on the temperature driving force ΔT (temperature of the solid minus the liquid

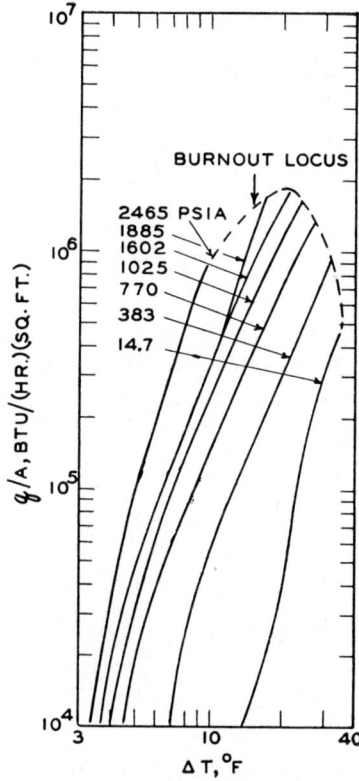

Fig. 11. Nucleate boiling of water. The liquid was in contact with a 0.024-in. electrically heated, platinum wire (A1).

boiling point) and other variables in the manner below.

$$h = (\text{Const.}) \frac{(\Delta T)^2 k_L^{5.1} g^{1/2} (\rho_L - \rho_V)^{1/2}}{\beta \sigma^{1/2} C_L^{2.1} \lambda^2 \mu_L^{4.1}} \tag{11}$$

(Rohsenow)

2. *Forster-Zuber Equation*

The starting point for the Forster-Zuber theory (F4, F5, F6) is the Rayleigh equation (R1) for a bubble growing in a liquid medium. In this

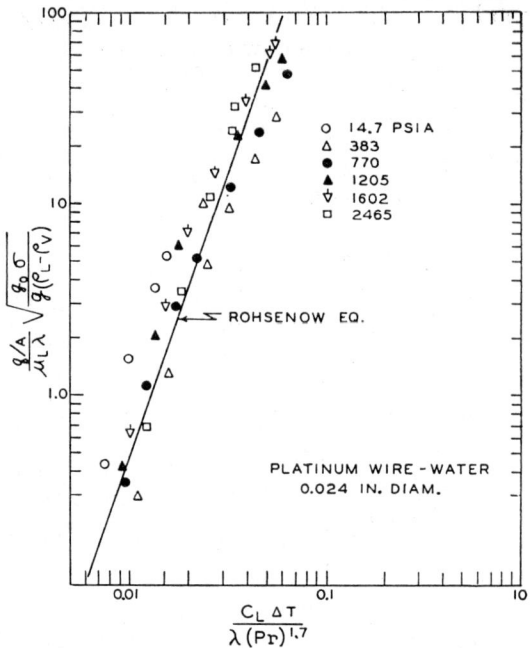

Fig. 12. Rohsenow's correlation for nucleate boiling of water. These are Addoms' data from Fig. 11 (R8).

development the bubble is not in contact with any solid surface.

$$R \frac{d^2 R}{d\theta^2} + \frac{3}{2} \left(\frac{dR}{d\theta} \right)^2 + \frac{2\sigma}{\rho_L R} = \frac{p_V - p_\infty}{\rho_L} \tag{12}$$

It is assumed that the bubble is spherical, that the liquid is incompressible, that viscosity effects may be neglected, and that the usual Thompson equation is applicable during the growth period (time from the appearance of a nucleus of size R_0, $R_0 = 2\sigma/(p_V - p_\infty)$, to some later time). By the use of these assumptions it is possible to treat the mathematical case of a moving, spherical heat transfer boundary (L2), to combine this with the

Clausius-Clapeyron equation, and to obtain a theoretical expression for a vapor bubble growing in a superheated liquid. The equation (F5, F6) is a second-order differential equation which is so complex as to be of limited usefulness without serious modification. Fortunately, the equation becomes enormously simpler if the inertia of the liquid can be ignored during bubble growth. Forster and Zuber give a careful discussion of the physical requirements for neglecting inertia of the liquid. These are that either the bubble must be very small or the temperature of the bubble

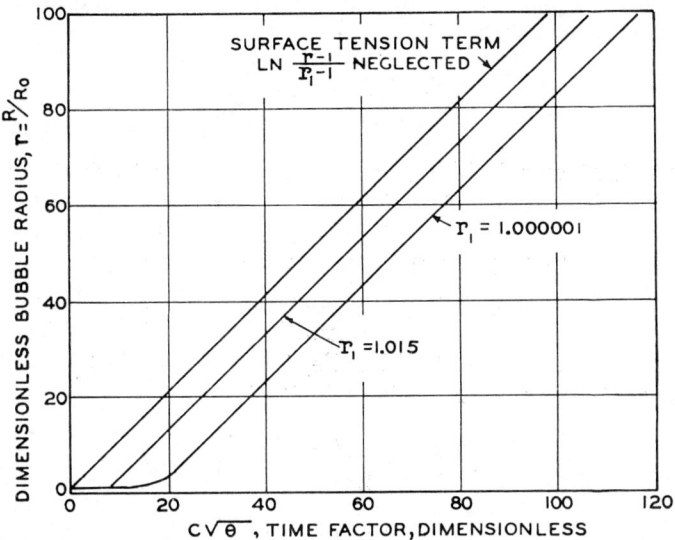

FIG. 13. Theoretical rate of growth for vapor bubbles. Two arbitrary starting radii are shown (F6).

must be nearly equal to the saturation temperature. The first requirement is met when a bubble is first growing from its nucleus, and the second is met when the bubble becomes "large." For all bubbles of intermediate size it is assumed that both the requirements are met in part, and so liquid inertia is negligible for them also. The growing-bubble equation becomes

$$r + \ln\left(\frac{r-1}{r_1 - 1}\right) = \frac{\Delta T C_{LP_L} \sqrt{\pi \alpha_L \theta}}{\lambda \rho_V R_0} \tag{13}$$

Here θ is the time for the bubble to grow from radius R_1 to R_2. The symbol r is a generalized radius R/R_0. Figure 13 shows the form of Eq. (13). The coordinates are general, dimensionless values. Arbitrary starting sizes are chosen as parameters varying from a bubble 0.0001% bigger in radius than a nucleus to one which is 1.5% bigger. The lines are straight when the time is large; at very small values of time, the lines are curved. This means that the logarithmic term in Eq. (13) is important only for bubbles which are just slightly larger than a nucleus. The logarithmic term entered the equation because of surface-tension considerations. The rate of growth of bubbles comparable to nuclei in size is therefore seen to be strongly dependent on surface tension, while the growth rate of larger bubbles is much less influenced.

The temperature of a bubble growing in a superheated liquid changes with the bubble size. If liquid inertia is negligible, the Forster-Zuber derivation gives the expression

$$\frac{T_0 - T_V}{T_0 - T_\infty} = 1 - \frac{R_0}{R} \tag{14}$$

which says that as a bubble grows from a nucleus to double the radius of a nucleus it will lose half its temperature of superheat. After the bubble has grown by several orders of magnitude, it will have practically no superheat and will be at the saturation temperature.

The velocity of the wall of a growing bubble can be obtained by differentiation of Eq. (13). Since a growing bubble is considerably larger than a nucleus during most of its life, the logarithmic term in Eq. (13) can be neglected. The wall radial velocity is

$$\frac{dR}{d\theta} = \frac{\Delta T C_L \rho_L}{2\lambda \rho_V} \left(\frac{\pi \alpha_L}{\theta}\right)^{1/2} \tag{15}$$

The radial velocity (bubble rate of growth) will decrease with increasing pressure. This has nothing to do with rate of nucleation. The rate of nucleation may increase or decrease with pressure as far as this equation is concerned.

The radial velocity is seen to be directly proportional to the temperature driving force and inversely proportional to the square root of the elapsed time of growth. On the other hand the bubble radius is proportional to the temperature driving force and to the direct square root of the time. The interesting result is that the product of the bubble radius and its radial velocity is independent of time.

$$R\frac{dR}{d\theta} = \left[\frac{C_L \rho_L \Delta T \sqrt{\pi \alpha_L}}{\lambda \rho_V}\right]^2 \tag{16}$$

The physical meaning of Eq. (16) is that small bubbles grow rapidly and large ones grow slowly but that the agitation in the surrounding liquid caused by the growth remains uniform.

The Reynolds number of a growing bubble can be defined (by means of the bubble radius instead of diameter) as

$$\text{Re} = \frac{R(dR/d\theta)\rho_L}{\mu_L} = \frac{\rho_L}{\mu_L}\left[\frac{C_L\rho_L\Delta T\sqrt{\pi\alpha_L}}{\lambda\rho_V}\right]^2 \qquad (17)$$

Forster and Zuber desired a correlation in the form of Eq. (1). A Nusselt number was therefore defined as

$$\text{Nu} = \frac{Rq/A}{k\Delta T} \qquad (18)$$

The bubble radius will not be known in the usual applications of boiling. However this radius may be eliminated by solving for its value in the equation describing bubble dynamics (F6):

$$R = \left(\frac{C_L\rho_L\Delta T\sqrt{\pi\alpha_L}}{\lambda\rho_V}\right)\left(\frac{2\sigma}{p_V - p_\infty}\right)^{1/2}\left(\frac{\rho_L}{p_V - p_\infty}\right)^{1/4} \qquad (19)$$

The final correlation is

$$\text{Nu} = 0.0015(\text{Re})^{0.62}(\text{Pr})^{1/3} \qquad (20)$$

or

$$\left(\frac{C_L\rho_L\sqrt{\pi\alpha_L}\,q/A}{k_L\lambda\rho_V}\right)\left(\frac{2\sigma}{\Delta p}\right)^{1/2}\left(\frac{\rho_L}{\Delta p}\right)^{1/4}$$
$$= 0.0015\left[\frac{\rho_L}{\mu_L}\left(\frac{C_L\rho_L\Delta T\sqrt{\pi\alpha_L}}{\lambda\rho_V}\right)^2\right]^{0.62}\left[\frac{C_L\mu_L}{k_L}\right]^{1/3} \qquad (21)$$

(Forster-Zuber equation)

A few manipulations show the dependence of the heat transfer coefficient on the other variables.

$$h = 0.0012\,\frac{(\Delta T)^{0.24}(p_V - p_\infty)^{0.75}k_L^{0.79}C_L^{0.45}\rho_L^{0.49}}{\sigma^{0.5}\lambda^{0.24}\mu_L^{0.29}\rho_V^{0.24}} \qquad (22)$$

(Forster-Zuber)

Although the Reynolds number occurs to about the same exponent in Eqs. (9) and (20), the meaning and numerical values of these Reynolds numbers are quite different. In addition the two equations disagree completely on the effect of the Prandtl number. Comparing Eqs. (11) and (22) shows that the two semitheoretical equations have little in common in respect to prediction of effects of variables on h. However, any change in ΔT in Eq. (21) must be accompanied by a change in $p_V - p_\infty$; so a comparison of the effects of ΔT is not so simple as first appears.

Verification of the Forster-Zuber formulas has been of two kinds, checks of Eq. (15) and of Eq. (21). Ellion (E1) conducted a photographic study to determine the rate of growth of bubbles of steam and of carbon tetrachloride on a horizontal heating strip. A typical radical velocity for steam growing from a radius of 0.001 to 0.01 in. was 10 ft./sec. at a heat flux of 1,150,000 B.t.u./(hr.)(sq. ft.). These measurements were made by use of subcooled liquids at atmospheric pressure. For the steam bubble mentioned, the wall was 48° F. hotter than the boiling point; the bulk liquid was 77° F. colder than the boiling point. In spite of the fact that Eq. (15) was derived for a bubble completely surrounded by superheated liquid, the equation gives a value of 11.2 ft./sec. for the predicted radial

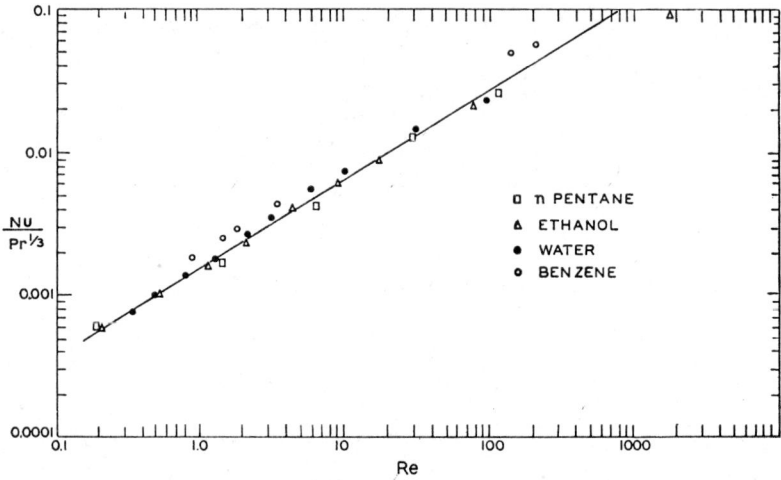

FIG. 14. The Forster-Zuber correlation for nucleate boiling of four liquids. Only data near the critical ΔT are represented (F6).

velocity under Ellion's conditions. Similar agreement between the equation and observed rates of growth is claimed for the photographic data of Gunther and Kreith (G8). Thus Eq. (15) gives good predictions for growth rates of vapor bubbles on solid surfaces.

A test of Eq. (21), or (20), is shown in Fig. 14. The same maximum heat-flux data of Cichelli and Bonilla (C2) for three organic liquids which produced three separate lines by Rohsenow's correlation give a single line on the Forster-Zuber plot. The data of Kazakova (K1) for water boiling at pressures varying from 10 to 80% of the critical pressure also fit the line nicely.

In one respect Eq. (20) is satisfying. The exponents on the Reynolds and Prandtl numbers are roughly the same as those used for ordinary forced-convection heat transfer. The negative sign on the Prandtl-number exponent in Rohsenow's equation has seemed "illogical" to many scientists. The "logical" exponents found by Forster and Zuber are interpreted to mean that the movement of the liquid-vapor interface during bubble growth is the correct velocity for incorporation in the Reynolds number. On the other hand Eq. (21) predicts the same heat transfer coefficient for a single liquid boiling on any hot surface (all heterogeneous cases) or boiling in the bulk (the homogeneous case). The validity of the same equation for all hot solids as well as for no hot bounding surface is open to serious question. Rohsenow's equation was developed for and is applied to the heterogeneous case only.

One objection to a Forster and Zuber assumption has been given by Zwick (Z1). Forster and Zuber state that the principal mechanism for heat transfer to a growing bubble is conduction across the film resistance. Zwick points out that heat can also flow by mass transport and that this convection should be included in the equations.

3. Empirical Correlations

In addition to the equations of Rohsenow and of Forster and Zuber, which are the most theoretical equations now available, there exist many equations which are simply empirical correlations. Little pretense is made that these additional equations are defensible on theoretical grounds beyond simple dimensional analysis. A few of these are given below.

1. Cryder and Gilliland (C7) observed nucleate boiling at 1 atm. for water, salt solutions, and five organic liquids on the outside of a 1.04-in. brass tube, range of $\Delta T = 2.5°$ to $25°$ F. (Note: $D =$ tube diameter.)

$$h = (\text{const.}) \frac{(\Delta T)^{2.39} k_L^{2.97} C_L^{0.43} \rho_L^{3.1} D^{2.1}}{\sigma^{1.65} \mu_L^{3.45}} \tag{23}$$

2. Cryder and Finalborgo (C6) observed nucleate boiling at various pressures for water, salt solutions, and four organic liquids on the outside of a horizontal brass tube.

$$\log h = 2.5 \log \Delta T + 0.014 T_s' - \text{Const.} \tag{24}$$

where T_s' is the boiling point, at the pressure used, in °F.; the constant varies from 2.05 to 5.15.

3. Jakob and Linke (J5) observed nucleate boiling at low bubble populations for water and carbon tetrachloride from flat vertical and horizontal chromium surfaces.

$$h = (\text{const.}) k_L^5 \left(\frac{\rho_L}{\sigma}\right)^{2.5} \left(\frac{\Delta T}{\rho_V \lambda C_1}\right)^4 \tag{25}$$

where C_1 is the same as in Eq. (2).

4. Insinger and Bliss (I1) observed nucleate boiling for water, sugar solution, and a variety of organic liquids from a vertical chromium tube.

The equation best fitting the data is

$$\log Y = 0.363 + 0.923 \log X - 0.047 (\log X)^2 \tag{26}$$

where X and Y are complex combinations of the variables considered. This equation is very well approximated by a much simpler relationship pointed out by Jakob (J2):

$$h = (\text{const.}) \left(\frac{k_L C_L}{\sigma \rho_V}\right)^{1.6} \frac{\rho_L^{2.6} (\Delta T)^{2.1}}{\lambda^{0.85}} \tag{27}$$

5. Bonilla and Perry (B7) observed nucleate boiling for water and numerous organic liquids on horizontal plates.

$$h = (\text{const.}) \frac{(\Delta T)^{2.7} k_L^{1.85} C_L^{1.85} \sigma^{0.85} \mu_{L,a}^{3.7} \rho_{L,a}^{2.7}}{\mu_L^{1.85} \rho_L^{0.85} \sigma_a^{2.7} \rho_{V,a}^{2.7} \lambda_a^{2.7} C_1^{2.7}} \tag{28}$$

where the a subscripts refer to physical properties at atmospheric pressure and C_1 is the same as in Eq. (2).

These five nontheoretical correlations are sufficient to show the difficulty of selecting a satisfactory empirical correlation. The various authors do not agree on the quantitative effects of such important variables as the temperature difference and surface tension; they fail even to agree on the qualitative effects of variables such as the liquid viscosity and the vapor density.

C. Theoretical Treatment of Nucleation

The traditional method of studying nucleate boiling by straightforward dimensional analysis has met with poor success. The methods of Rohsenow and of Forster and Zuber are an improvement. These workers also are forced ultimately to the use of dimensionless groups with empirical coefficients and empirical exponents. The real difficulty is that at least two separate and important processes are involved during nucleate boiling: the formation of bubbles which can grow and the subsequent growth of these bubbles. There is no reason to believe that the important variables for these two processes are the same or that either one of the two processes will be rate controlling under all conditions. If both processes are important and if even one key variable is important for one process and not important for the other, dimensional analysis for the combined processes is doomed to failure.

The problem of nucleation (birth of an interface capable of growth) followed by growth is a very common one in chemical engineering. Not only does it arise for nucleate boiling, but also for condensation, crystallization from solutions, and freezing. The rupture of glass subjected to hydrostatic tension is treated as nucleation (formation of first tiny cracks) followed by interface growth (F3). Even crushing and grinding are treated as nucleation and interface growth (S3) in an interesting attempt to explain why so much energy must be expended to get appreciable size reduction.

Nucleation during boiling and the growth of bubbles are considered separately in this chapter. Boiling nucleation was treated in 1926 by Volmer, who considered nucleation as a thermodynamic equilibrium state followed by a rate process. Other workers have followed Volmer's general scheme. All along, however, some workers have considered nucleation as a statistical density fluctuation followed by a rate process. The final forms of the resulting equations are quite similar. Both theories are outlined briefly in the following sections.

1. *Classical Rate Theory*

A few of the many contributors to the classical rate theory of boiling nucleation are Volmer (V1), Becker and Döring (B2), Frenkel (F7), Fisher (F3), and Bernath (B4). All agree that a prime requirement for nucleation to occur in a liquid is that the liquid must be superheated. The bubbles formed are cooler than the liquid; therefore nucleation is strictly irreversible. Because of the superheat, a temperature driving force exists between liquid and bubble. However, because surface tension forces are immense for tiny bubbles, a collapsing tendency exists which may counteract the tendency of a bubble to grow by absorbing heat. One problem faced by any theory of nucleation is to explain the formation of a bubble which will not collapse.

The kinetic theory assumes that some molecules in a liquid have energies far greater than the average energy of the remaining molecules. Their extra energy is called *activation* energy, if the magnitude is great enough to cause important results. The energy needed for "important results" depends on the process considered. For flow to result, the energy is rather small; for ultimate bubble growth, the energy is probably larger, perhaps being comparable to the heat of vaporization. If an activated molecule collides properly with a normal (average) molecule, the two may cling together to give an activated dimer. The activated dimer may grow to an activated trimer, and so on. The process is treated by the same rate-process scheme which is applied to polymerization. An activated cluster of x molecules behaves as a single large molecule. It is always capable of collecting one more molecule.

$$A_x + A_1 \rightarrow A_{x+1} \tag{29}$$

Here A_x signifies an activated cluster of x monomer units (single molecules), A_1 is a single simple molecule (unactivated), and A_{x+1} is an activated cluster of $x + 1$ monomer units. Of course, other modes of growth could be imagined, such as the collision of two dimers, but these are ruled out as being either very improbable or very difficult to treat mathematically. Only the simple growth, one molecule at a time, is considered.

If x becomes sufficiently large, the cluster is a vapor bubble. But since the process in Eq. (29) is imagined to be reversible, clusters can degrade to single molecules, step by step. It is never certain that a particular growing cluster will grow large enough to be a bubble. Probability theory is introduced at this point to estimate the distribution of clusters of various sizes, which is necessary to the development of a rate expression. It is imagined that clusters of all possible sizes exist and that the number of any one size depends on the "difficulty" of growth to that size in a step-by-step process. Thus

$$n_x = N \cdot e^{-\frac{W_x}{kT}} \tag{30}$$

Here n_x is the number of clusters containing x monomers, N is the total number of molecules (counting each cluster as a single molecule), and W_x is the work necessary to create a cluster of size x. The product kT will be recognized as the translational energy of an average liquid molecule; therefore the exponent gives the "surplus" energy of a cluster expressed as a fraction. As an excellent approximation, N is taken as being the total number of molecules before any clusters form. The number of clusters is usually a very small fraction of the total number of molecules. Equation (30) predicts that the number of clusters of size x decreases exponentially with the work of forming this size.

The work W_x of creating a cluster of x monomers will now be considered. This involves changing x superheated liquid molecules into a vapor cluster and will depend on whether the cluster forms within the liquid mass (homogeneous case) or whether it forms on the surface of a wall or foreign inclusion (heterogeneous case).

a. Homogeneous Case. A cluster is assumed to be spherical; the dividing boundary between the vapor and liquid is assumed to be distinct. No preferred structure or organization of the molecules at the interface is considered. The work, W, will be of two kinds: the work of shoving back the liquid and the work of creating an interfacial area. If the cluster has a radius R, volume v, and surface area A, then

$$W = -v\Delta p + \sigma A \tag{31}$$

The net pressure-volume work is

$$v\Delta p = v(p_V - p_L) \tag{32}$$

where p_V is the pressure of the vapor inside the sphere and p_L is the external pressure on the liquid. The famous equation of Laplace relates the pressure difference Δp to the radius of a bubble and the surface tension.

$$\Delta p = p_V - p_L = \frac{2\sigma}{R} \tag{33}$$

$v = \frac{4}{3}\pi R^3$ and $A = 4\pi R^2$, and so Eq. (31) becomes

$$W = -\frac{8\pi R^2 \sigma}{3} + 4\pi R^2 \sigma = \frac{4}{3}\pi R^2 \sigma \tag{34}$$

The work of forming a cluster is determined by the radius and surface tension. Equation (34) can be put in an alternate form by combining it with Eq. (33),

$$W = \frac{16\pi\sigma^3}{3(p_V - p_L)^2} \tag{35}$$

A third form, preferred by many writers, for the work of forming a cluster can be obtained by substituting Poynting's equation for a perfect gas

$$v_L(p_V - p_L) = kT \ln \frac{p_\infty}{p_V} \tag{36}$$

in Eq. (35).

$$W = \frac{16\pi\sigma^3 v_L^2}{3\left(kT \ln \frac{p_\infty}{p_V}\right)^2} \tag{37}$$

The volume of one molecule of liquid, v_L, is often represented as

$$v_L = M/(\rho_L N')$$

Equations (34), (35), and (37) are equivalent.

There is nothing in the work equations to indicate the existence of a critical size for a cluster. However, if resort is made to thermodynamics, the free energy of the entire system may be found and this value will yield helpful information. The system is imagined to be subjected to a small change of cluster radius and the laws of equilibrium are applied to the system during this imposed change.

Before a cluster is generated, the total free energy of the superheated liquid is

$$F_1 = (n_L + n_V)\eta_L \tag{38}$$

where $n_L + n_V$ is the total number of molecules, and η_L is the chemical potential of an average liquid molecule. The chemical potential is defined as $(\partial F/\partial n)_{T,P}$, the partial free energy of one molecule. After nucleation occurs, some molecules exist as a cluster with a different chemical potential η_V. The chemical potential does not account for surface effects, and so a term of this type must be included in a statement of the total free energy of the liquid plus one vapor cluster.

$$F_2 = n_L \eta_L + n_V \eta_V + 4\pi R^2 \sigma \tag{39}$$

Note that some "liquid energy" has been converted as work utilized to create a surface. Although the formation of a nucleus is not truly reversible, the reversible case can be handled mathematically, and so it is assumed to be a satisfactory approximation. Thus

$$\Delta F = n_V(\eta_V - \eta_L) + 4\pi R^2 \sigma \tag{40}$$

It is not obvious how ΔF varies with the size of a cluster, because η_V depends on the size, but an indirect scheme is available for determining the desired information. For one particular size, R_0, there is assumed to be a value of ΔF for a condition of stability. This means that for a superheated liquid at a stated temperature and pressure, one and only one cluster size is capable of existence for long. This cluster is called a *nucleus*. A stable cluster is really in a metastable state, as discussed later. However, for any degree of equilibrium, ΔF must be unaffected by infinitesimal changes in the cluster size. So $d(\Delta F)/dR = 0$. If v_V is defined as the volume occupied by one vapor molecule, then $n_V = 4\pi R_0^3/(3v_V)$. These two manipulations produce a solution for the quantity $\eta_V - \eta_L$ in Eq. (40)

$$\eta_V - \eta_L = -\frac{2\sigma v_V}{R_0} \tag{41}$$

If R_0 (the radius of a nucleus) is very large, the chemical potential of a bubble is the same as for the surrounding liquid. If R_0 is very small, the bubble potential is small, because of the great crushing effect from surface tension.

The chemical potential is a function of temperature and pressure, and so $(d\eta)_T = v\,dp$. Then from Eq. (41), considering a change in R being brought about by a change in pressure,

$$(v_L - v_V)dp = 2\sigma v_V d\left(\frac{1}{R_0}\right) \tag{42}$$

The volume of one vapor molecule may be computed from the gas law. $pv_V = kT$, and the volume v_L may be neglected compared with v_V. The substitution and integration give the Thomson Equation (T2)

$$\ln \frac{p_V}{p_\infty} = \frac{2\sigma v_V}{kR_0 T} \tag{43}$$

where p_V is the vapor pressure in a stable cluster of radius R_0, and p_∞ is the saturation pressure (at $R = \infty$).

Combining Eq. (41) with (40) and using $n_V = 4\pi R^3/(3v_V)$ give the desired relationship between ΔF and R for clusters of any size.

$$\Delta F = 4\pi\sigma \left(R^2 - \frac{2R^3}{3R_0} \right) \quad (44)$$

A graph of Eq. (44) is shown in Fig. 15. The peak is of great interest, for this represents the metastable condition. Clusters of this size, R_0, are stable as long as they do not change in radius. However if the radius were to decrease slightly, say because of the loss of one molecule, the cluster would collapse with an accompanying decrease in free energy.

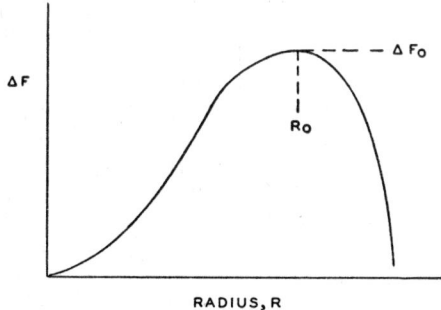

FIG. 15. Change in free energy for creation of vapor clusters in a liquid, Eq. (44).

If the nucleus were to gain one molecule, the cluster would then grow spontaneously while releasing free energy. Thus a nucleus is a metastable cluster which can disappear or can grow to a visible bubble. Both possibilities are equally probable.

The actual size of nuclei is of great interest. They are too small to be observed by use of microscopes. The few experimental values have been obtained by indirect means. Values cover the range of about 10 to several hundred molecules per nucleus. The values are not accurate, but they are believed to have the correct order of magnitude. Some typical calculated sizes are given in Table I.

The rate of appearance of nuclei in a superheated liquid determines the rate of macroscopic bubble formation and therefore must be of importance for nucleate boiling. This rate is found from a knowledge of the free-energy change necessary to form a nucleus. This is found by differentiation of Eq. (44), $d(\Delta F)/dR = 0$, which leads to $R = R_0$, and

$$\Delta F_0 = \tfrac{4}{3}\pi R_0^2 \sigma \quad (45)$$

This equation is equivalent to Eq. (34), therefore the free energy to form a nucleus is the same as the work to form a nucleus. This equation is not of great use at present because neither ΔF_0 nor R_0 are known with precision. If one were known, the equation would be useful for finding the other.

TABLE I
Liquids Superheated in Open, Glass Capillary Tubes at 1 Atm. (K2)

	B. P. (°F.)	Superheat		Estimated nucleus diam.[a] (Å.)
		Beyond B.P. (°F.)	Vapor pressure for saturation at max. temp. (atm.)	
SO_2	14	108	8.3	
Ethyl ether	95	195	15.1	90
CS_2	115	220	15.4	
Acetone	133	213	19.0	
Chloroform	142	202	14.5	
Methyl alcohol	151	205	26.5	94
Ethyl alcohol	172	221	29.9	50
Benzene	174	230	14.7	184
Water	212	306	54.2	146
C_6H_5Cl	270	212	10.9	334
Xylol	279	176	—	
C_6H_5Br	313	189	8.0	
Aniline	361	142	—	

[a] Calculated from the Laplace equation, Eq. (33).

Eyring and co-workers (G5) and others have shown that rate processes are determined by the activation energy (free-energy change needed to initiate the process) and by the temperature. For nucleation the expression becomes

$$\text{Rate} = C \cdot e^{-\frac{\Delta F_0}{kT}} \quad (46)$$

Thus the higher the temperature, the faster the rate becomes; the greater the activation energy, the slower the rate becomes.

For an exact consideration of the coefficient C, the references should be consulted; however a rough idea of the meaning follows, but this simple explanation is not rigorous. The coefficient C is a frequency factor which is proportional to the number of molecules present and also to the group kT/h. In addition it is likely that a molecule which moves from the liquid into the vapor experiences some resistance such as viscosity. Therefore the activation energy ΔF_0 should be increased by some amount

$\Delta F'$ which is the energy needed to overcome friction (or generally speaking $\Delta F'$ is the activation energy to cross a barrier). Combining this information leads to

$$\frac{dn_0'}{d\theta} = \frac{NkT}{h} \cdot e^{-\left[\frac{\Delta F'}{kT} + \frac{16\pi\sigma^3 v_V^2}{3(kT)^3\left(\ln\frac{p_\infty}{p_V}\right)^2}\right]} \qquad (47)$$

(Nucleation in a liquid)

Equation (47) gives the number of nuclei appearing per unit time in an assembly of N molecules at temperature T. The general form of the equation is well established. The many modifications that have been presented come from considerations of the "real" meanings of $\Delta F'$ and the coefficient C or from considerations of the relative magnitudes of the two terms in the exponent.

TABLE II
Cavitation Pressure of Water at 81° F. (F3)

Waiting time (sec.)	Fracture pressure, billions of dynes/cm.²		
	$\Delta F' = 0$	($\Delta F' = 5000$ cal./mole)	($\Delta F' = 10{,}000$ cal./mole)
10^{-15}	-1.74	-1.91	-2.14
10^{-6}	-1.46	-1.56	-1.68
1	-1.34	-1.41	-1.50
10^6	-1.24	-1.30	-1.36
10^{18}	-1.10	-1.13	-1.18

Fisher (F3), for example, argues that for cavitation[1] phenomena, $\Delta F'$ ought to be in the order of 5000 cal./mole. This is deduced from an analogy between viscosity and nucleation. In both cases a molecule escapes from its neighbors and relocates on the other side of a "boundary." Also for cavitation, p_L is a huge negative number. Thus it becomes convenient to express the work term in the exponent by Eq. (35) so that p_V may be neglected compared with p_L. It was shown by Fisher that compared with the work term the $\Delta F'$ term is not important for cavitation. This is true because the huge negative pressure dominates the exponent. Cavitation occurs if one bubble forms in a reasonable time. The reasonable time usually is taken to be 1 sec., although it can be shown that the choice of 1 sec. is not critical. In Table II the reasonable time for 1 mole of liquid is shown in a range from 10^{-15} sec. (time for sound to move from one molecule to the next) up to 10^{18} sec. (billions of years). In the

---

[1] The experimental method for studying cavitation is described in Sec. IID3a.

same table $\Delta F'$ is allowed three separate values. The fracture pressure for water at 81° F. is seen to be nearly independent of $\Delta F'$ and the waiting time. The fracture value is, however, highly dependent on temperature and surface tension. From the foregoing information Fisher presents a simple approximate expression for the fracture pressure.

$$\text{Fracture pressure} = -\left[\frac{16\pi\sigma^3}{3kT\ln\left(\frac{N'kT}{h}\right)}\right]^{1/2} \quad (48)$$

A sample of values predicted by this equation is given in Table III. Additional values are given in the literature (F3). The values are much closer to experimental values than are predictions based on Van der Waals' equation (discussed in Sec. IID3). However, Eq. (48) must fail if the temperature becomes very small or if the viscosity becomes very large.

TABLE III
Comparison of Observed and Theoretical (Kinetic Approach) Limiting Fracture Pressures (B12, B13, B15)

Description of liquid			Limiting cavitation pressures at approx. 81° F. (in atmospheres)				
Normal B. P. (°F.)	Critical state		Experimental, Briggs	Rate theory,		Eq.-of-state theory, Van der Waals
	T_c (°F.)	P_c (atm.)		Fisher	Bernath	
Water 212	705	218	−281	−1320	−1120	−16,800
Benzene 186	556	50	−152	−338	−284	−2340
Acetic acid 244	612	57	−288	−322	−262	−5420

The Volmer (V1) equation for liquids in tension has been modified by Bernath (B4). The principal alteration was the designation of the molecular heat of vaporization λ' as being equivalent to $\Delta F'$.

$$\frac{dn_0'}{d\theta} = N\omega A_0 B_2 \cdot e^{-\left[\frac{\lambda'}{kT} + \frac{W_0}{kT}\left(1 + \frac{2}{3}B_2\right)\right]} \quad (49)$$

(Nucleation in a liquid in tension: Bernath)

Here ω is the flow rate for molecules streaming across a boundary, given for the idealized case of molecules streaming from a region of pressure p_V to a region of zero pressure as

$$\omega = \frac{p_V}{\sqrt{2\pi mkT}} \quad (50)$$

The other symbols are

A_0 = surface area of a nucleus = $4\pi R_0^2$

B_2 = defined as $\ln\left(\dfrac{b-1}{2}\right)$

b = defined as $\dfrac{p_V - p_L}{p_V}$ or $\dfrac{2\sigma}{p_V R_0}$

The critical radius R_0 for cavitation is found by considering the gas-law expression for a bubble surrounded by a liquid in tension. From $p_V = \dfrac{2\sigma}{R} - p_L$,

$$\left(\dfrac{2\sigma}{R} - p_L\right)\left(\dfrac{4\pi R^3}{3}\right) = n_V k T \tag{51}$$

This equation is plotted in Fig. 16. The peak of the curve corresponds to R_0. Bubbles larger than this can grow without accumulating addi-

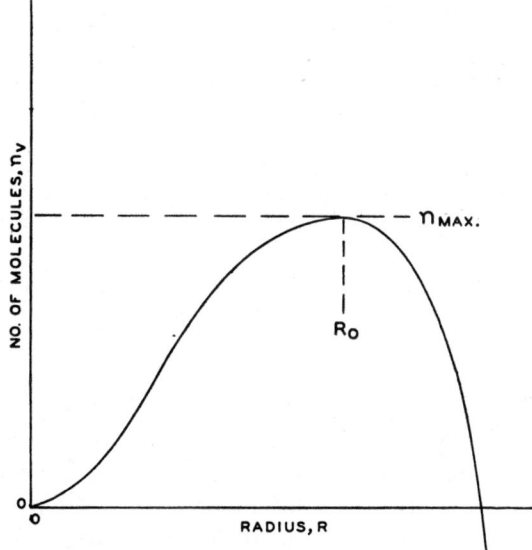

FIG. 16. Number of molecules for a vapor cluster of radius R in a liquid in tension.

tional molecules; therefore the liquid will rupture. The critical bubble size gives

$$R_0 = \dfrac{4\sigma}{3p_L}$$

$$n_0 = \dfrac{128\pi\sigma^3}{81kTp_L^2}$$

$$p_V = \tfrac{1}{2}p_L \text{ (at fracture)}$$

Equation (49) as given does not permit the pressure ratio to be equal to 3 or less, because B_2 then becomes zero or negative. The rate of nucleation thus would be predicted to be zero at a pressure ratio of 3; this is not in agreement with logical expectations or with experimental data. A clever method for approximating a reasonable value of B_2 is given by Volmer.

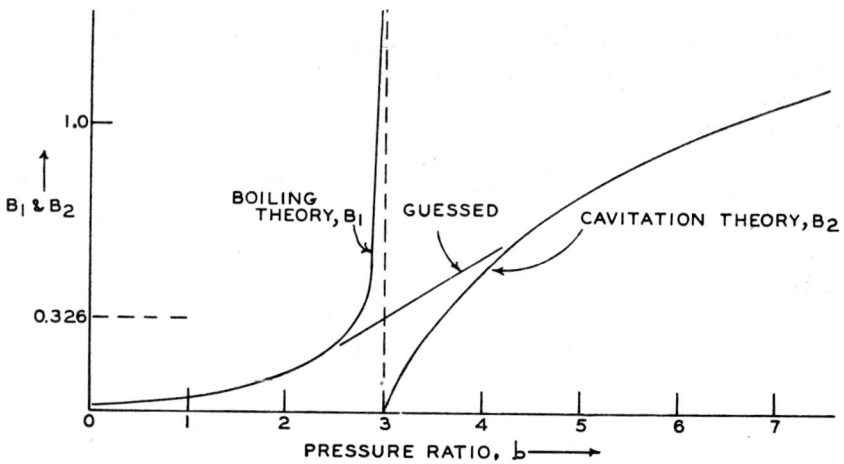

FIG. 17. Volmer's method for selecting the proper value for the factor B (V1).

This can be explained by considering Volmer's version of Eq. (47) for a superheated liquid at a positive pressure.

$$\frac{dn_0'}{d\theta} = NB_1 \cdot e^{-\left[\frac{\lambda'}{kT}+\frac{W_0}{kT}\right]} \tag{52}$$

(Nucleation in a boiling liquid: Volmer)

Here $B_1 = \sqrt{6\sigma/[\pi m(3-b)]}$. Although the cavitation equation, Eq. (49), does not permit $b < 3$, the boiling equation does not permit $b > 3$. When b becomes 3, the rate becomes infinite in Eq. (52). This is not sensible, because pressure ratios of more than 3 have been observed. The peculiar behavior at $b = 3$ is really an outcome of assumptions made during the derivations and need not have special physical meanings.

Boiling and cavitation are intimately related. Volmer argued that the two types of rate equations, such as Eqs. (49) and (52), should merge or agree at some point. The equivalent pressure-sensitive groups, B_1 and B_2, are plotted[2] in Fig. 17. The estimated form which B_1 and B_2

---

[2] Takagi (T1) shows that five cases can be considered, depending on whether b is negative, between 0 and 1, between 1 and 3, exactly 3, or greater than 3. The simpler, Volmer approach is used in this chapter.

should have near $b = 3$ is indicated. At $b = 3$, $B_1 = B_2 = 0.326$. Volmer then chose 0.326 as being very reasonable for either nucleate boiling or cavitation. The value chosen need not have a high precision, because the rate equations are not very sensitive to B, compared with their sensitivity to temperature. A few values calculated using this B_2 in the cavitation equation, Eq. (49), are shown in Table III. Other values are given by Bernath (B4). The agreement with experiment is surprisingly good. A test of the boiling equation, Eq. (52), is provided by Wismer's data (W5) for superheated ether at atmospheric pressure. Figure 18

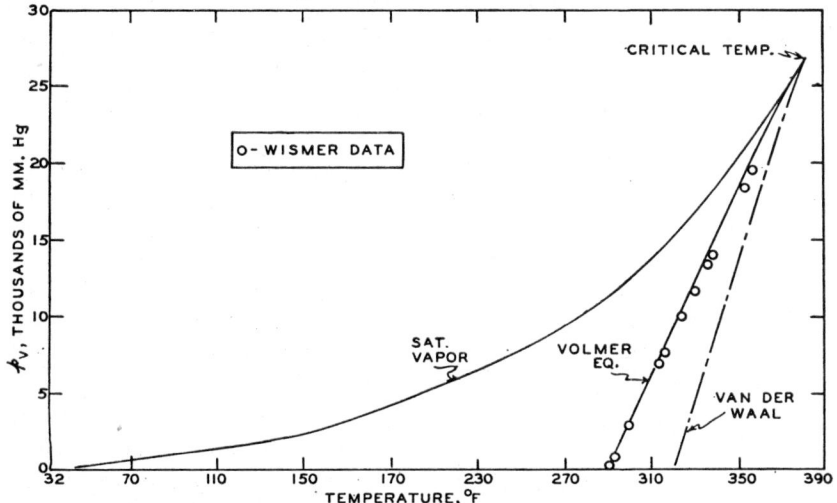

FIG. 18. Application of rate theory and equation-of-state theory to Wismer's data for ether superheated in glass. Horizontal displacements represent superheating; vertical displacements represent the liquid in tension. Wismer's original Van der Waals plot was different; above is the "corrected" form as given by Volmer (V1).

shows that the agreement between experiment and this set of observations is excellent. The rate theory of nucleation is therefore of real value in work dealing with boiling liquids.

A number of assumptions are involved in the derivation of the mathematical expressions for nucleation. First, the change of one phase into another is really a nonequilibrium process. There is no guarantee that equations derived for thermodynamic equilibrium will be valid when applied to nonequilibrium. For example, the relationship between surface tension and radius applies for a static bubble, but does it apply to a bubble which is changing in size? Second, it is imagined that any cluster which grows to a nucleus is bodily removed from the liquid. Thus nuclei cannot accumulate. This idea leads to a "chopped" distribution curve which is probably not correct. Third, it is not true that the boundary between a liquid phase and a vapor phase is a plane of zero thickness. Instead there exists a zone in which the density undergoes a steady change from the liquid value to the vapor value. Since a nucleus is extremely small, a sizable share of its molecules may be in the transition zone. This leads to inaccuracy in simple calculations of bubble volumes, surface forces, etc. Last, properties determined for large bodies (surface tension for example) may change if the body becomes very small. The Gibbs-Tolman-Koenig (H5) relationship shows how surface tension changes with radius of curvature. The exact expression is complex, but a simple first approximation is

$$\frac{\sigma}{\sigma_\infty} = 1 - \frac{z}{R} \tag{53}$$

where σ is the surface tension for a sphere of radius R, σ_∞ is the value for a flat plane, and z is the thickness of the boundary (transition zone). A calculation for liquid water drops at 212° F. gives z as about 4 molecules thick. For a drop having 100 molecules on a diameter, $\frac{\sigma}{\sigma_\infty} = 0.92$, or the surface tension has decreased by 8%. For a drop having a diameter of 8 molecules, the entire drop is a transition zone and the equation unfortunately becomes meaningless.

The first objection, that of nonequilibrium, has received a partial rebuttal from Rodebush (R5, R6, R7). The motions of translation and rotation tend to stabilize a cluster, as can be shown by considerations of the entropy of these two effects. It is also pointed out that water is an unusual material because the liquid molecules in the bulk material have a tetrahedral arrangement. Thus a tiny bubble will be surrounded by unsatisfied hydrogen bonds in the curved liquid surface. The effect on entropy of the interfacial organization probably means that the use of a constant molecular heat of vaporization λ' as used by Bernath and others is in error.

In spite of these difficulties, thermodynamics and the reaction-rate theory give a picture of nucleation which is reasonably consistent with experimental evidence. Researchers studying crystallization, condensation, and other nucleation phenomena have accumulated experimental values that show that this theoretical approach is a defensible one. The application of this theory to boiling has received scant attention; it is clear that the science of boiling will progress rapidly as the attention to nucleation theory expands.

b. Heterogeneous Case. The preceding section dealt with homogeneous nucleation, the formation of nuclei in the inner bulk of a pure liquid.

In practice, container walls are usually present, and the liquid is rarely free from dust particles, adsorbed gas, absorbed gas, and foreign ions. Heterogeneous nucleation refers to the formation of nuclei on a foreign object.

The theory of heterogeneous nucleation is the same as the theory for homogeneous nucleation except for two items: consideration is given to the possibilities that the work of forming a nucleus may be altered by the foreign object and that the number of sites for nucleation may be altered. It is very difficult to discover the number of sites available for heterogeneous nucleation except for the simple situations such as boiling on a heating tube or at the walls of a container. Usually the number of possible sites will be fewer for the heterogeneous case than for the homogeneous case. Consider a closed cubical container filled with N molecules. The possible homogeneous sites are N in number. The maximum heterogeneous sites amount only to $6N^{2/3}$. Since N is a huge number, $N > 6N^{2/3}$.

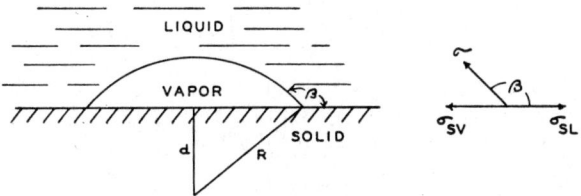

Fig. 19. Vapor bubble on a flat surface. The foregoing shape exists if the liquid wets the solid poorly.

As for the work of forming a nucleus, consider the simple case of the effect of a foreign solid surface. If the vapor and the liquid both wet the foreign surface, the work will be less than for the homogeneous case. This decrease will result in an increased rate of nucleation per nucleation site at the same superheat.

Assume that a bubble forming on a solid is a segment of a sphere, as shown in Fig. 19. This shape corresponds to a volume of minimum free energy. The contact angle β is greater than 90° for a system such as water on wax. It is usually less than 90° for systems such as water on metal. The reversible work of formation (F3) is

$$W = 2\pi R^2(1-m)\sigma + \pi R^2(1-m^2)(\sigma_{SV} - \sigma_{SL})$$
$$- \frac{\pi R^3}{3}(2 - 3m + m^3)p_V \quad (54)$$

where $m = \dfrac{d}{R} = \dfrac{\sigma_{SL} - \sigma_{SV}}{\sigma}$. If $\beta < 90°$, m is a negative number. Note that the area of the vapor-liquid interface is $2\pi R^2(1-m)$ that of the solid and liquid is $\pi R^2(1 - m^2)$, and the volume of the vapor is

$$\frac{\pi R^3}{3}(2 - 3m + m^3)$$

The work of forming a bubble of critical size, a nucleus, is

$$W_0 = \frac{16\pi\sigma^3\phi}{3(p_V - p_L)^2} \tag{55}$$

where $\phi = (2 + m)(1 - m)^2/4$. This work equation is identical with Eq. (35) except for the inclusion of ϕ. Equation (45) takes the form

$$\Delta F_0 = \tfrac{4}{3}\pi\sigma R_0^2\phi \tag{56}$$

Equation (56) states that if the liquid completely wets the solid, then $\beta = 0$, $m = -1$, and $\phi = 1$. In this case the heterogeneous situation becomes identical with the homogeneous one. However, if the liquid and vapor are equally attracted to the solid, $\beta = 90°$, $m = 0$, and $\phi = \tfrac{1}{2}$. In this case the free energy of activation and the work of forming a nucleus are each reduced to half. If the vapor wets the solid completely, $\beta = 180°$, $m = +1$, and $\phi = 0$. Then ΔF_0 (or W_0) is reduced to zero; superheating should be impossible.

The rate of nucleation in the presence of a foreign surface can be written readily by the use of the correct energy terms. The expression may be put in a convenient form for comparison with the homogeneous case.

$$\text{Hetero. rate} = \text{Homo. rate} \cdot \left(\frac{N_{HT}}{N_{HM}}\right) \cdot e^{1-\phi} \tag{57}$$

The rates here are number of nuclei appearing per unit time. The symbol N_{HT} refers to the number of sites available for heterogeneous nucleation and N_{HM} is the number for homogeneous nucleation. The first is proportional to solid surface area, the second to liquid volume. The value ϕ depends only on the contact angle for wetting.

Equation (57) predicts that the results to be expected by a change from a very poor catalyst for nucleation (a material giving a contact angle of near zero) to the best catalyst for nucleation (contact angle of 180°) is an increase in the rate of nucleation at a site by a factor, e. To get an increase greater than this would mean either $\beta > 180°$ (an interesting theoretical possibility which is discussed by Frenkel) or the presence of gas-containing cavities on the solid.

Good data for testing Eq. (57) are not available. Perhaps the data used for testing the homogeneous case should be considered as applicable

TABLE IV
Effect of Contact Angle on Cavitation
Temp. 81° F. (F3)

		Contact angle (°)	Cavitation pressure (atm.)
Homogeneous	Water	0	−1320
	Mercury	0	−22300
Heterogeneous	Water-paraffin wax	105	−820
	Water-montan wax	140	−290
	Mercury-glass	140	−4800
	Mercury-steel	154	−2100

to the heterogeneous case, because boundary walls are always present when data are obtained.

Fisher derives an equation for cavitation in the presence of solid surfaces. Assuming that the liquid tensile pressure is huge compared with

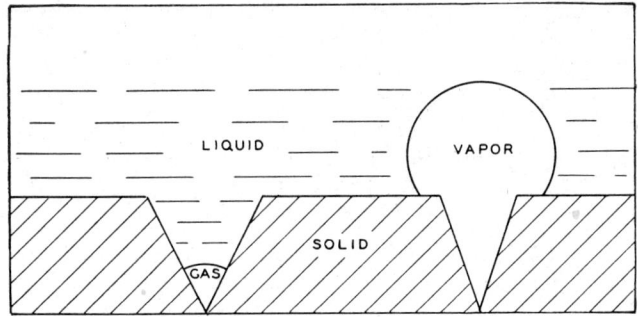

FIG. 20. Nucleation from pits on a solid surface. The foregoing bubble shapes exist if the liquid has good wetting action on the solid.

the vapor pressure in a bubble and that the liquid is in a cubical container, the expression is

$$\text{Fracture pressure} = -\left[\frac{16\pi\sigma^3\phi}{3kT \ln\left(\frac{6(N')^{2/3}kT}{h}\right)}\right]^{1/2} \quad (58)$$

The symbols have the same meaning as in Eq. (48). A comparison of theoretical fracture pressures from Eqs. (48) and (58) is shown in Table IV.

If a gas-filled cavity has a sharp apex such as shown in Fig. 20, the resulting gas body can serve as a nucleation site. The gas can gain vapor easily and grow to a bubble of critical size. The work of forming such a bubble will be much less than for homogeneous nucleation. A starting cluster is already in existence, and there is no need to overcome the enormous crushing force of surface tension which is so important for clusters which must start from zero size. Even if the gas molecules (usually air) which are originally in a conical cavity are gradually removed in escaping bubbles, the pit will remain an active site. The escaping gas molecules will be replaced by vapor molecules. A conical pit cannot fill with liquid without the expenditure of a huge amount of work, because the surface-tension forces will oppose the movement of liquid into the final tip. On the other hand if a pit has a rounded bottom, it may serve

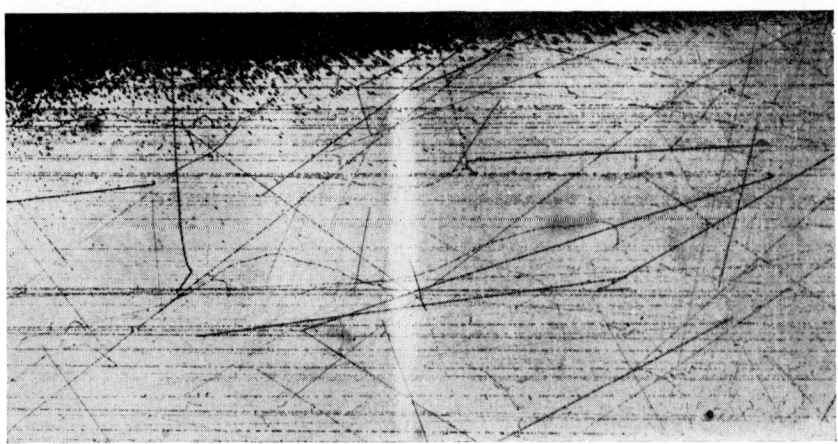

Fig. 21. Bubble-chamber photograph. Nucleation of superheated ether, at 1 atm., from a Cosmotron proton beam (D. A. Glaser).

as a nucleating site as long as some inert gas is present. When the inert gas is gone, the pit can fill with liquid and nucleation can cease at this site.

Nucleation from ions formed by radiation has been observed recently (G1, G2, G3, G4). The laboratory technique involves the use of a bubble chamber which operates much like the Wilson cloud chamber. High energy radiation from such sources as radioactive cobalt or a synchro-cyclotron is allowed to pass through superheated liquid ether, pentane, or hydrogen. Distinct bubble tracks appear, such as shown in Fig. 21, and then the liquid explodes. The bubble population on a track is roughly 100/cm. The present theory is that a high-energy particle passing through a vapor cluster produces numbers of ions of a common sign. These ions repel one another and separate, enlarging the cluster to a nucleus of critical size. Thus electrostatic forces overcome surface-tension forces. If this theory is valid, such nucleation should not be possible with liquids which are good electrical conductors.

Glaser (G2) presents an equation for the necessary conditions to achieve nucleation by irradiation.

$$p_\infty - p_L = \frac{3}{2}\left(\frac{4\pi\epsilon\sigma^4}{n_i e^2}\right)^{1/3} \qquad (59)$$

Here p_∞ is the saturation pressure at the temperature used, p_L is the external pressure on the liquid, n_i is the number of ions in the nucleus, e is the electrical charge per ion, and ϵ is the dielectric constant. The agreement between predicted critical temperatures of superheat and measured values is good within 15° or 20° F.

Nucleation from suspended inert gas particles is a familiar occurrence. The inert gas could originate as gas absorbed on a solid, as a gas trapped in pits on the solid, or as gas which was dissolved in the liquid and which came out of solution as the temperature was increased.

Nuclei formed from inert gas particles require less superheat in the liquid than do homogeneous nuclei. A metastable nucleus is one with a radius described by the modified Laplace equation.

$$(p_g + p_V) - p_L = \frac{2\sigma}{R} \qquad (60)$$

The partial pressure of the inert gas in the nucleus is p_g.

By use of Eq. (60) it is straightforward to derive the corresponding rate of nucleation. Two interesting facts appear. One is that the rate of nucleation must gradually decrease as boiling proceeds. Each escaping bubble will carry some of the inert gas with it, and the supply of inert gas will deplete. Thus unless gas is introduced by a chemical reaction or by some mechanical means, nucleation initiated by an inert gas is an unsteady-state process.

The other interesting fact is that stable gas plus vapor bubbles can exist in a liquid which is below its boiling point. Thus in Eq. (60) for given values of p_L and R there exist values of p_g which are great enough so that p_V may be quite small. Thus bubbles of air plus steam appear in water "boiling" at temperatures below 212° F. at atmospheric pressure. This occurrence is well known to observers who heat liquids in open-top or transparent vessels.

2. *Statistical Fluctuation Theory*

The fluctuation theory has received attention because it avoids some of the serious assumptions involved in the rate theory. The beginnings of fluctuation theory were presented by Einstein. Various workers since have attempted to refine the theory to an easily used form. The principal proponent at present is Reiss (R2, R3).

The advantages of the fluctuation theory are that it does not require that clusters be spheres, they need not have sharply defined bounding surfaces, nor is an equilibrium between phases assumed. The disadvantage is a practical one: how can the work term (defined later) be evaluated?

The density, temperature, energy, and other properties of a given mass of liquid are statistical averages. For a tiny volume within the liquid the properties are not constant. The density, for example, fluctuates rapidly about the mean value.

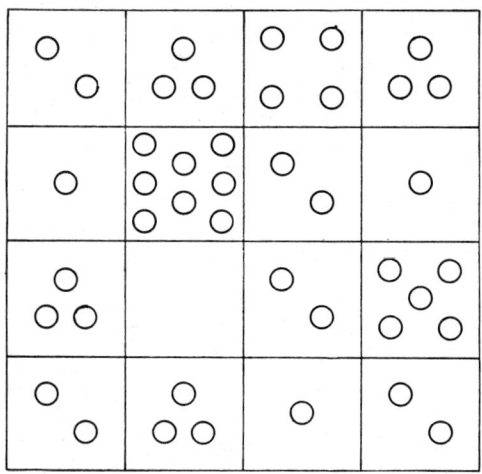

FIG. 22. Fluctuation picture of the liquid state.

Imagine the liquid to be divided into microscopic cells as shown in Fig. 22. The volume of a cell must be chosen carefully to have a particular, special value, described later. A cell is small enough so that occasionally it may be empty. At other times one or many molecules may occupy the cell. Consider all possible configurations of all the molecules in all the cells. The most probable configuration will exist for a large mass for a long time. In this respect the most probable distribution is a description of the steady-state condition. The problem is to calculate the most likely distribution. Once this is known, the rate of "formation" of any density in a cell can be calculated.

If the liquid molecules follow the gas law, a "configurational integral" solution of the possible configurations can be deduced (R2). A highly condensed summary of the procedure follows.

$$\text{One configuration} = \frac{d^N \Omega}{d\omega^N} = V \cdot e^{-\frac{W_\omega}{kT}} \tag{61}$$

$$\text{Configuration integral} = \Omega = V^N \tag{62}$$

The symbol ω is the volume occupied by a molecule in translational motion, and W_ω is the total potential energy due to the configuration of the total N molecules occupying the total volume V. The integral Ω may be thought of as a probability term. If V and N are large, the most probable configuration becomes very probable indeed. If these terms become very small, the most probable configuration is not very likely, and the fluctuation theory becomes unreliable.

From the configuration integral it is possible to derive an expression showing the most probable distribution of molecules among the cells. An approximate derivation can be used to arrive at the same distribution and to give some idea of the meaning of the terms involved. Let g_1 be the most probable number of molecules in a cell. Then the volume of a cell is $g_1 v_L$, and the total number of cells is $\dfrac{N}{g_1}$. What is the work of "forming" a cell with x molecules? This can be calculated by forming the cell in four steps.

1. First, the force field around x molecules must be discharged. Call this work w_1. The discharge must be done reversibly.

2. Now the x molecules are ideal and will follow the gas law. Expand these molecules from original volume xv_L to a final volume $g_1 v_L$. This work must be done isothermally and reversibly.

$$w_2 = -xkT \ln \frac{g_1 v_L}{xv_L} \tag{63}$$

3. The remaining $N - x$ molecules are compressed isothermally and reversibly from original volume $V - xv_L$ to a final volume $V - g_1 v_L$. Thus the total volume remains constant. This work is approximately

$$w_3 = (g_1 - x)kT \tag{64}$$

4. Last, the force field is recharged. The work required is w_4. The total work of forming a cell of x molecules is the sum of the four processes, or

$$W_x = W_F + xkT \ln \frac{x}{g_1} + (g_1 - x)kT \tag{65}$$

Here W_F is $w_1 + w_4$, a term which accounts for interaction among the x molecules in a cell and the molecules in surrounding cells. The number of cells with x molecules is

$$n_x' = (\text{No. of cells}) \cdot e^{-\frac{W_x}{kT}} \tag{66}$$

or

$$n_x' = \frac{N}{g_1} \cdot e^{-\left[\frac{kT\left(g_1 - x + x \ln \frac{x}{g_1}\right)}{kT} + \frac{W_F}{kT}\right]} \tag{67}$$

It is obvious that Eqs. (30) and (67) are very similar. In the first case a sphere of variable radius is considered; in the second a fixed cell of variable contents is considered. Once the distribution such as Eq. (67) is available, the conventional rate theory is used to obtain the rate of formation of critical cells.

As pointed out, the value g_1, must be selected properly. Roughly speaking it will have a value such that the density of a cell when one molecule is in the cell will be equal to the vapor density. In any case it seems to be possible to select this value so that the distribution will predict the existence of nuclei, that is, cells which have the proper density and energy to cause spontaneous growth of a new phase. The evaluation of the interaction term, W_F, is unsatisfactory. However the fluctuation theory cannot be dismissed. Light scattering measurements are strong proof that the assumed fluctuations are very real.

D. Critical Temperature Difference

Nucleate boiling from a solid surface is possible within a certain temperature-difference range only. If the temperature difference between the hot solid and the boiling liquid is very small, heat transfer to the liquid will occur by free convection and no bubbles will be created. If the temperature difference is increased, nucleate boiling occurs. As the ΔT is increased, the heat transfer rate increases, up to a point. The existence of an upper limit to nucleate boiling is of extreme importance to engineers.

1. *Maximum Efficiency*

Inasmuch as the critical ΔT corresponds to a maximum heat flux, this ΔT will permit a maximum duty for an evaporator, reboiler, or other boiling equipment. In practice, industrial equipment is designed to operate at slightly less than the critical ΔT. This gives a performance which is somewhat less than the optimum, but it provides insurance against exceeding the critical ΔT.

If the critical ΔT is exceeded, three possible events can result. Film boiling may occur, and the heat flux will remain nearly constant; transition boiling may occur, with an accompanying sharp decrease in the heat flux; or the heat source may melt and *burnout* result. Which of these events occurs depends on the source of heat and the materials of construction.

2. *The Burnout Problem*

The occurrence of burnout is a serious inconvenience, for the boiling process must then come to a halt. Burnout during boiling is the result of an unsteady-state process. For a short period of time the heat supplied to the heater exceeds the heat transferred to the boiling liquid. The accumulation of heat causes a rapid rise in the temperature of the heater. The origin for the unbalance can be understood by reference to Fig. 1. Suppose a boiler tube is operating smoothly at the critical ΔT. If for any reason the ΔT should increase slightly, the transition type of boiling will begin and h (the heat transfer coefficient for the boiling liquid) will decrease. Unless there is a decrease in the heat generated by the heat source, the steady state is destroyed.

Most researchers use "critical ΔT" to correspond to either of two conditions, maximum heat flux or maximum heat transfer coefficient. The two values are nearly identical for many systems. However, in some cases (boiling mercury, for example) the two are distinctly different. Evidence is needed to show which of these is more significant, i.e., which corresponds to burnout.

a. Role of Heat Source. The type of heat source plays a dominant role in determining the result of exceeding the critical ΔT. If heat is supplied from a condensing vapor or from a hot fluid, the unsteady state is self-regulating. Any decrease in h causes an increase in the wall temperature and a decrease in the driving force from hot fluid to the solid. The heat flux decreases and a new steady state is established. The result is a more or less smooth operation in the transition region of boiling. The apparatus will not be destroyed, but the operating conditions will be inefficient.

If an electric heating element is used, or if the heat is generated by chemical or nuclear reactions, the results are entirely different. The unsteady-state condition resulting from a slight increase beyond the critical ΔT is not self-regulating. If h decreases, the metal temperature will increase, but this will have no effect on the heat generation. Heat will accumulate in the metal until the metal melts or until the value of $h\Delta T$ increases sufficiently to reestablish a new steady state. If the metal is a common structural material such as steel or copper, burnout occurs. If platinum or some other high-melting-point metal is used, stable film boiling can occur. In the latter case the new heat flux will be slightly greater than that at the critical ΔT, but the operating ΔT will be enormously greater. For example, Nukiyama's results show that if water is boiled on a platinum wire, the critical ΔT is about 120° F. and the equivalent film-boiling ΔT is in the order of 1800° F.

3. *Physical Interpretations of the Critical ΔT*

It is always desirable to have physical interpretations of unusual phenomena. The existence of a critical temperature difference for nucleate boiling has challenged many thinkers. The easiest explanation occurs in numerous publications: the number of bubbles on a hot solid is known to increase with the temperature driving force; thus it can be argued that at some ΔT the bubbles become so prolific that they merge into a continuous vapor film.

This explanation does not permit predictions of the critical ΔT for various liquids. It does not tell what the quantitative effects of changing the pressure, the type of hot solid, the geometry of the system, etc.,

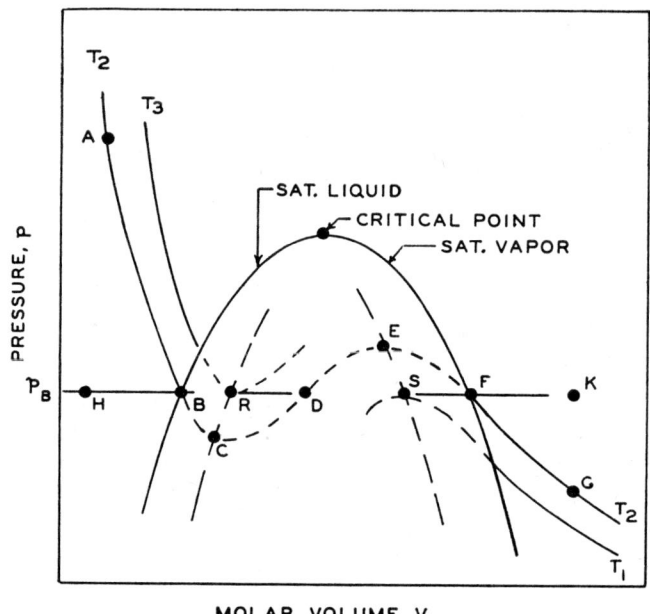

Fig. 23. Graph of equation of state, such as Van der Waals' expression, showing metastable conditions.

should be. Thus it is desirable to have a different physical explanation, one that involves mathematical expressions which could lead to new information.

Two theoretical explanations follow. Although they are not perfect, they are encouraging attempts to go beyond the superficial, easy explanation.

a. Thermodynamic Equation of State. A conventional p-V-T diagram for a single pure substance is shown in Fig. 23. The line AB represents a typical isotherm for the liquid state, and FG is an isotherm for the vapor at the same temperature. At pressure p_B, liquid at temperature T_2 and vapor at T_2 would be in equilibrium; or we say that T_2 is the usual boiling point of the liquid at pressure p_B, and p_B is the usual vapor pressure of the liquid at temperature T_2.

If a subcooled (below the boiling point) liquid at point A is expanded isothermally, it will boil at point B ordinarily. However, if the liquid is of good purity and its container is not contaminated, it is easy to proceed along the line toward C. Such a liquid is then under tension; that is, an external, mechanical pulling force is being resisted by the liquid. Numerous investigators have demonstrated the existence of liquids in

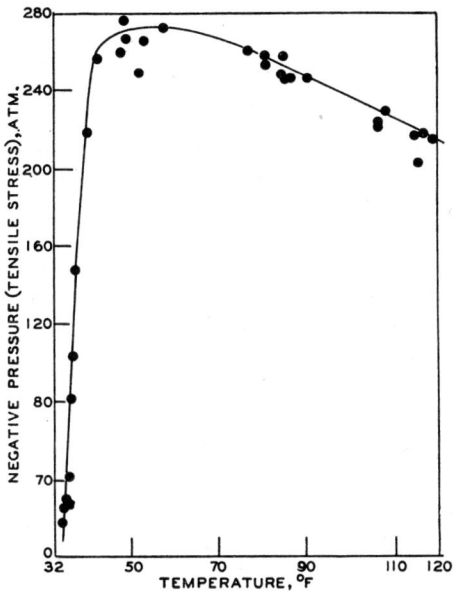

Fig. 24. Tensile strength (maximum negative pressure) of pure water in glass (B12).

great tension, and the colorful term *negative pressure* has been used to describe the condition of the liquids when sustaining stresses greater than 1 atm.

Briggs (B12) was able to subject water at room temperature to a negative pressure of nearly 270 atm. The experimental technique consisted of spinning a horizontal glass tube ("scrupulous cleanliness is necessary") about a vertical axis located at its center. The tubing contained the liquid and was open at both ends. The centrifugal force needed to break the liquid column was observed. The experimental results are shown in Fig. 24.

When a liquid in tension finally breaks, a cavity appears. Cavitation is known to occur with some pumps, ship screws, and other hydraulic equipment. The sudden movement of the liquid phase can cause severe damage to such machinery. A discussion of cavitation and the accompanying damage is given by Plesset and Ellis (P2).

Briggs' results, as well as those of others, prove that the dotted line BC in Fig. 23 represents attainable physical conditions. Similarly the line FE is real, for it represents the condition of a vapor compressed isothermally beyond its vapor pressure (saturation pressure). Van der Waals' equation[3] and certain other equations of state have the mathematical form described by the line $ABCDEFG$. The portions BC and EF are called *metastable*. The question of importance in a discussion of boiling is does the portion CDE have any physical significance?

Hirschfelder and co-workers (H6) give a detailed discussion of this question. It is proved that a liquid in tension such as that at temperature T_2, located on the line BC is unstable to macroscopic fluctuations in density but is stable (thermodynamically) to tiny fluctuations. Portion EF represents supersaturated vapor which also is stable (thermodynamically) to tiny fluctuations. The line CDE is proved to correspond to liquid or vapor which is thermodynamically unstable to density fluctuations of any magnitude whatever. This means that states corresponding to CDE are "completely unattainable."

To return to boiling, if a cold liquid is heated to boiling at constant pressure, the path on Fig. 23 will be from H to B. It is possible to superheat the liquid beyond B. From the standpoint of a "correct" equation of state, the liquid could be superheated to point R (temperature T_3) but not beyond this. Thus a maximum possible superheat, $T_3 - T_2$, is predictable theoretically. The dashed line which includes points C and R gives the boundary for possible liquid superheat. Similarly, the line through E and S is the boundary for possible vapor supercooling. Wismer (W5) calculated the maximum permissible temperature to which three different liquids could be superheated (at atmospheric pressure) and tested the results experimentally. The agreement between experiment and theory was rather encouraging. For example, liquid ethyl ether was heated to 289° F. under pressure and was then observed to remain a liquid for at least 1 sec. when the pressure was reduced to 1 atm. Van der Waals' equation gives a predicted value of 253° F. for ether at 1 atm. The normal boiling point of ether is 95° F. Other results are shown in Table V. The column of pressures represents the pressure which the liquid would exert if it were boiling at the maximum observed superheat. The last column gives the maximum temperature calculated from Van

---

[3] Van der Waals' equation is $(p + a/V^2)(V - b) = RT$, where p is the pressure, V the molar volume, T the absolute temperature, a the Van der Waals constant to account for attractive forces between molecules, and b the Van der Waals constant to account for the finite volume of molecules.

der Waals' equation. Of course neither Van der Waals' equation nor any other existing equation of state is regarded as being precise. The peak in the curve of Fig. 24, for example, is not predictable from present equations.

TABLE V
COMPARISON OF OBSERVED AND THEORETICAL MAXIMUM SUPERHEATING OF LIQUIDS

	Boiling point (°F. at 1 atm.)	(W5) Observed max. temp. (°F.)	Vapor pressure corresponding to observed temp. (atm.)	Max. temp. predicted by Van der Waals' Eq.[a] (°F.)
Ether	95	289	15.4	253
Isopentane	84	277	14.3	246
Ethyl chloride	54	259	20.0	240

[a] Using Van der Waals' $a = 9RT_cV_c/8$, $b = V_c/3 = RT_c/(8p_c)$, and values of T_c and p_c from Stull, D. R., *Ind. Eng. Chem.* **39**, 517 (1947).

During ordinary nucleate boiling, the liquid is in definite contact with a hot solid. The solid must be hotter than the boiling point of the liquid, or vapor bubbles will not appear. A temperature gradient is set up in the liquid in the region close to the solid. For example Jakob and Fritz (J3) obtained evidence in a typical experiment that the liquid temperature changed by over 18° F. in a distance of a few millimeters. Presumably the liquid molecules touching the solid are at the same temperature as the solid surface. Thus a heated, boiling liquid has amounts of superheat varying from the temperature of the bulk liquid to the solid-surface temperature.

According to the preceding argument concerning the limit to the metastable liquid state, a liquid cannot be hotter than the theoretical temperature given by the trough in a "correct" equation of state. The critical temperature difference from solid to bulk liquid then is the ΔT between the saturated-phase boundary and the trough boundary on a p-V-T diagram. With reference again to Fig. 23 at pressure p_B, the critical ΔT is represented as the temperature at R minus the temperature at B. This is $T_3 - T_2$ in the sketch.

What happens if the ΔT from solid to bulk liquid is made greater than the critical ΔT? From the thermodynamic viewpoint based on equations of state, the liquid cannot possibly exist at the temperature of this solid. However there is no reason why vapor cannot exist at this elevated temperature. Therefore vapor will form at the heating surface and will keep the surface coated continuously. If any vapor escapes from the surface, new vapor must form instantly to take its place.

A desirable feature of the foregoing theory is that it explains the existence of a critical ΔT for nucleate boiling. Unfortunately it gives no indication of the existence of the transition type of boiling. The predicted critical ΔT represents both the end of nucleate boiling and the beginning of film boiling. In reality these values are not the same. The theory suffers a practical objection also; there are no equations of state reliable enough to permit highly accurate predictions of the value of the critical ΔT. The theory predicts that the critical ΔT of a liquid is determined by the type of liquid and the external pressure applied to the liquid. At low external pressures the critical ΔT should be large; at the critical external pressure (peak of the phase diagram in Fig. 23) the ΔT should be zero. A ΔT found in practice may be less than the theoretical limiting ΔT given by a correct equation of state (owing to rate considerations discussed later) but no reason can be given for a too-high value in practice. Theoretically, agitation or forced convection should have no effect on the critical ΔT. Neither should the geometric arrangement (vertical vs. horizontal, plates vs. tubes, etc.) have any effect on the critical ΔT. The type of hot solid, its surface texture, the presence of trace impurities, and surface active agents should have no effect unless they can speed nucleation. In this case the critical ΔT will be decreased.

If the liquid is subcooled, the highest solid temperature which will allow nucleate boiling is the same as if the liquid were at its boiling point. The ΔT between the solid and the bulk liquid is of far less significance than the temperature of the solid minus the liquid boiling point. The critical value expressed this latter way should be exactly the same as for a liquid actually at its boiling point.

b. Kinetic Viewpoint. The kinetic viewpoint offers a second interpretation of the critical temperature difference for nucleate boiling and leads to somewhat different results.

The rate of nuclei formation in a liquid is strongly dependent on temperature, as was discussed in Sec. IIC. When the liquid is below its boiling point, the net rate of nuclei formation is zero, because all bubbles formed, of any size whatever, collapse. As the liquid is heated slightly above its boiling point, the rate of nuclei formation becomes real and positive. At small values of superheat the rate is extremely small, but as ΔT is increased to a certain value, the rate becomes extremely large. According to the simplified equations of Fisher, the rate becomes effectively infinite at a definite superheat. Figure 25 gives a pictorial representation of the rate of nuclei formation as a function of the superheat (expressed as a "superpressure").

The critical ΔT for ordinary nucleate boiling exists when the liquid portion immediately in contact with the heat source is hot enough to nucleate (form bubbles) at an "infinite" rate. Fisher (F3) and Bernath (B4) calculated the limiting vapor pressures corresponding to effectively infinite nucleation rates for several substances. As pointed out previously, their equations are somewhat different and lead to slightly different results. Their results plus recent experimental values obtained by Briggs (B12, B13, B14, B15) are shown in Table III.

Briggs' experimental procedure involved the use of a centrifuge as described before. His observations consisted of the ambient temperature and the force needed to rupture the liquid. Table III presents these

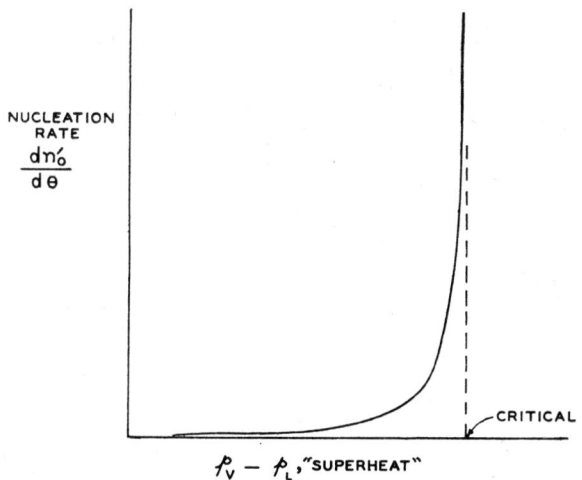

Fig. 25. Effect of superheat on the rate of formation of nuclei. Equation (47) gives this form.

cavitation measurements in terms of tensile (fracture, rupture, or cavitation) pressures.

Included in Table III are limiting pressures as calculated from Van der Waals' equation. The kinetic theory gives fair agreement with the experiments for benzene and acetic acid. Neither this theory nor the equation-of-state theory gives good results for water, presumably because water forms aggregates in the liquid state. None of the experimental values are closer than an order of magnitude to the equation-of-state prediction. This means that Van der Waals' equation is not exact, especially in the region of negative pressures.

All these fracture pressures are large negative pressures, the liquid being under great tension. The values show that the kinetic approach can be used to predict limiting values for this condition. If a substance were tested under conditions such that the limiting fracture pressures were not negative, these values could be used to predict maximum superheats obtainable at real, positive pressures. Ordinary liquids at temperatures well above room temperature should give the desired kind of cavitation pressures. In Fig. 24 the limiting negative pressure is seen to decrease at the higher temperatures. Briggs remarks that the cavitation pressure should be zero at the critical temperature (705° F. for water). This is true if the liquid is subjected to an ambient pressure equal to its vapor pressure. But if the ambient pressure is kept atmospheric, the rupture pressure should be zero at a temperature above 212° F. (for water) but appreciably below the critical temperature.

The kinetic approach, like the equation-of-state approach, is satisfying in that it predicts a limiting temperature beyond which a metastable liquid cannot exist (for a stated pressure). This critical ΔT can be a function of the type of hot solid, traces of impurities, and in short, anything which could affect the process of nucleation. The slowest nucleation and therefore the highest attainable superheats will occur when a pure liquid is heated internally, with no contact existing between the liquid and any solid. The presence of retaining walls, dust, gases, etc., will provide surfaces which can act as catalysts. The process of nucleation is easily catalyzed, and catalysis will reduce the limiting superheat. If small quantities of a negative catalyst (inhibitor) were introduced into a pure liquid, the attainable superheat should be unaffected. In this case the liquid will boil as soon as any tiny portion of the pure liquid reaches the temperature of infinite nucleation rate.

The rate of nucleation and therefore the critical ΔT will not be affected by the general geometric setup or by subcooling the bulk liquid. The critical ΔT will be independent of agitation or forced convection unless the agitation is so tremendous that it can lead to activation of molecules. This does not seem a likely occurrence.

If a hot solid is heated above the temperature which corresponds to the critical ΔT for a liquid in contact with the solid, vapor should form at effectively infinite speed. This vapor will coat the solid continuously. As quickly as any vapor is removed, it will be replaced with new vapor which forms instantly as the inrushing liquid touches the solid.

E. EXPERIMENTAL VALUES FOR NUCLEATE BOILING

Unless otherwise specified, the nucleate-boiling values presented in this section refer to liquids boiling on hot solid surfaces. The liquids are not subcooled and the agitation is caused by natural convection only.

The number of variables which affect boiling is surprisingly large. The following discussion considers the ones which are of particular interest. Certain properties of the liquid (viscosity, specific heat, density, thermal conductivity, and latent heat of vaporization) are undoubtedly variables, but these items are fixed usually by the choice of liquid. They are not given special consideration here. The effect of liquid level has been demonstrated for evaporators, and so this item is omitted also.

A study of the published information on boiling reveals numerous disagreements among observers. In some cases the lack of agreement is caused by real differences between the test conditions such as smoothness of the solid, purity of the liquid, etc. However great quantities of boiling data are of unknown accuracy. Faulty temperature-measurement methods, such as discussed later, are used with surprising frequency. Erroneous heat transfer rates must be common also. The best method of proving the correctness of instrumentation for a heat transfer experiment is to obtain agreement between measured input and output of heat. Yet probably nine tenths of published tests on boiling involve no heat balances whatever. The only proof of accuracy is the confidence and enthusiasm of these particular experimenters.

1. Effect of Type of Liquid

a. Water. For water at 1 atm. many observations are available. The data of Nukiyama (Fig. 1) and Rinaldo (Fig. 32) are typical. Other observers include Castles (C1), Miller (M8), Jakob (J2), Braunlich (M2), Cryder and Finalborgo (C6), Addoms (A1), Kazakova (K1), Farber, and Scorah (F1), Lyon and co-workers (L3), and Moscicki and Broder (M10).

The numerical observations vary widely. General agreement exists that the maximum heat flux is about 300,000 or 400,000 B.t.u./(hr.)(ft.^{2}). This flux is greater than the maximum for organic liquids by a factor of about three to eight times. The value is dependent on the metal surface, its roughness, and other factors discussed later.

The reported values for the critical ΔT vary from about 40° to 90° F., with most observations being close to 45° to 50° F. The critical ΔT is dependent also on some of the factors discussed later.

Addoms obtained a reasonable correlation for the maximum heat flux for water and five organic liquids by using a log-log plot of

$$(q/A)_{max.} \cdot (\lambda \rho_V)^{-1} \cdot [(\rho_L C_L)/(gk_L)]^{1/3} \text{ vs. } (\rho_L - \rho_V)/\rho_V$$

The result was a smooth curve. Rohsenow and Griffith (R12) have correlated the same data to obtain a straight line.

$$\frac{(q/A)_{max.}}{\lambda \rho_V} = 143 \left(\frac{\rho_L - \rho_V}{\rho_V}\right)^{0.6} \tag{68}$$

These correlations are presumed to include the effect of pressure, because the density group is essentially a pressure factor. The correlations are probably not general, because they omit variables such as surface tension, the type of solid, and the geometric arrangement. Another weakness is that the burnout data for the five organic liquids are estimates (and reported so) by Cichelli and Bonilla (C2). These researchers did not heat their surface to destruction. Thus the accuracy of the maximum q/A values is unknown.

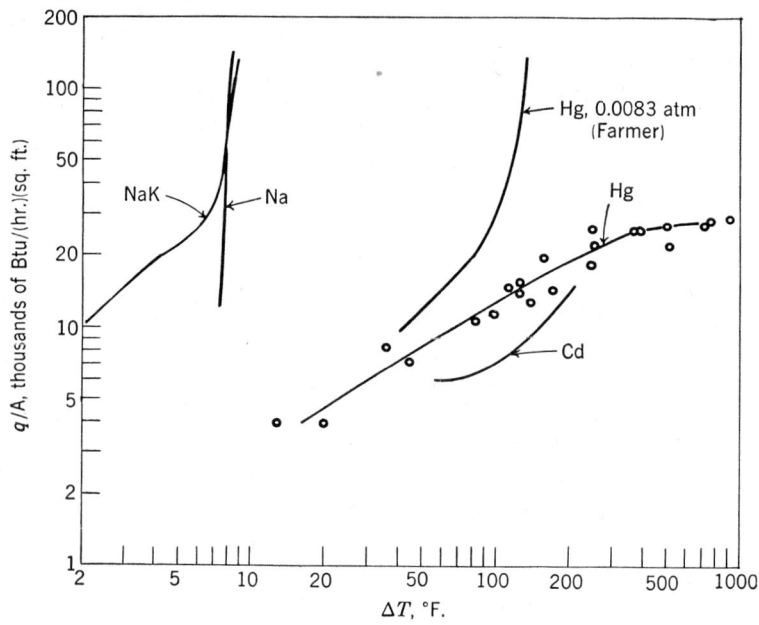

Fig. 26. Liquid metals boiling on a horizontal stainless steel tube at 1 atm. Data points are shown for mercury only (L3).

b. Other Common Liquids. The nucleate-boiling behavior of other common liquids is much like that of water. Typical boiling curves for this region at atmospheric pressure are included for ethanol (Fig. 29), isopropanol (Fig. 41), methanol (Fig. 3), ether (Fig. 30), and also benzene, toluene, trichlorethylene, and carbon tetrachloride (Fig. 39). Curves are available in the literature for many other liquids including *n*-butanol, and kerosene (C7), *n*-propane, *n*-pentane, *n*-heptane, and benzene (C2), isobutanol (M2), ethyl acetate (S1), Freon 113 (C4), and also Freon 12, methyl chloride, sulfur dioxide, and *n*-butane (M7). This list is certainly not complete.

The noticeable feature for these liquids is that they give maximum heat fluxes which are small compared with the heat flux of water. The peak values are in the order of 50,000 to 150,000 B.t.u./(hr.)(ft.^{2}). The critical temperature difference seems to be about 40° to 90° F., or essentially the same as for water.

c. Liquid Metals. Liquid metals have some desirable characteristics as heat transfer media. They have very low viscosities and high thermal conductivities. Thus circulation can be obtained with low power requirements, and temperature gradients through the liquid are rather flat.

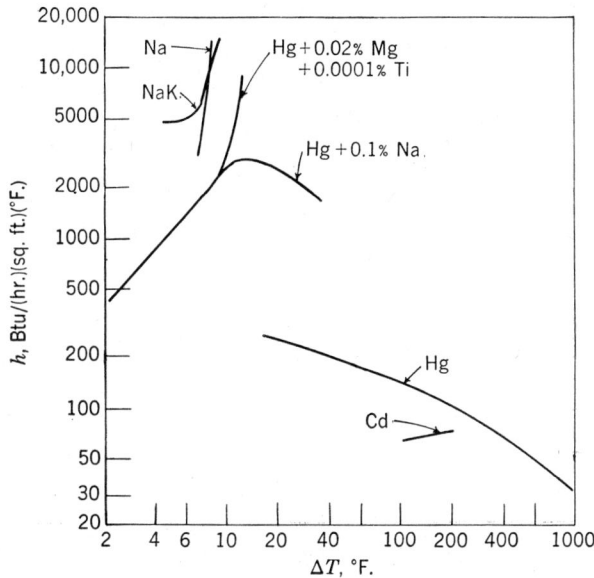

FIG. 27. Individual heat transfer coefficients for boiling metals. Data from Fig. 26 are included (L3).

The high boiling points mean that liquid metals are usable at atmospheric pressure at high temperatures. Other liquids require pressure application if they are to be used at high temperatures. Data (L3) for four liquid metals are shown in Figs. 26 and 27. The data of Farmer (F2) for mercury at vacuum are described as "preliminary" and may be subject to revision.

For the region of low ΔT (say below 25° F.) the liquid metals give high values of h. For example at a ΔT of 10° F. water gave $h = 500$ B.t.u./(hr.)(ft.^{2})(F.), mercury plus a wetting agent gave 3000, and NaK gave an extrapolated h of about 20,000, all tested in the same equipment (L3).

Thus liquid metals in nucleate boiling are excellent heat transfer agents, being superior even to water. Water in turn is definitely superior to the other common nonmetallic liquids.

2. Effect of Type of Hot Solid

Much confusion has existed concerning the possible effects of changing the type of solid from which nucleate boiling is taking place. If nucleation rates are important, the type of solid will be important. If the principal growth of bubbles occurs on the solid surface, the type again will be important. The solid should be unimportant only for the case of nucleation and growth occurring in the bulk liquid away from the solid surface.

a. Typical Results. Mead *et al.* (M6) found that the maximum superheat attainable in nonboiling water was about 10° F. higher in a stainless

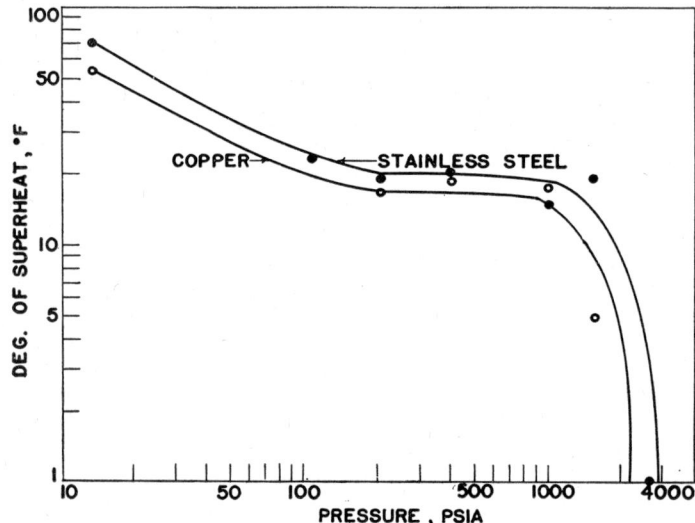

FIG. 28. Effect of pressure on maximum experimental superheat for nonboiling water (M6).

steel vessel than for a copper vessel, as shown in Fig. 28. Years ago Wismer and co-workers (K2) found that the maximum temperature to which liquid ether could be heated at atmospheric pressure was 259° F. for glass, 248° F. for silver, and 140° F. for brass and steel. For boiling water the h vs. ΔT curves obtained by Mead were distinctly different for copper and stainless steel. Similar results have been obtained for the boiling curves h vs. ΔT of: methanol with copper and stainless steel (M8); Freon 113, n-pentane, and ether with copper and nickel (C4); and for ethanol with copper, gold, and chromium (B7). Curves for the latter are shown in Fig. 29. In all these cases copper gave a greater heat transfer coefficient, at a fixed ΔT, than did the other metals.

The critical ΔT and the maximum heat flux are also dependent on the metal used. The data in Fig. 29 for ethanol on three metals at 1 atm.

show that copper gives the smallest ΔT_c, the highest $h_{max.}$, and the highest $(q/A)_{max.}$. Castle's results for water at 1 atm. are typical of the effect of the choice of metal:

Metal	Max. B.t.u./hr.	ΔT_c, °F.
Aluminum	418,000	46
Nickel	295,000	32
Platinum	296,000	42

The usual correlations proposed for the critical temperature difference or for the maximum heat flux do not specify the metal (B5, R12). Obviously this contributes to the observed deviations between the cor. relations and experimental data. Castle's table above indicates how great the deviations may be.

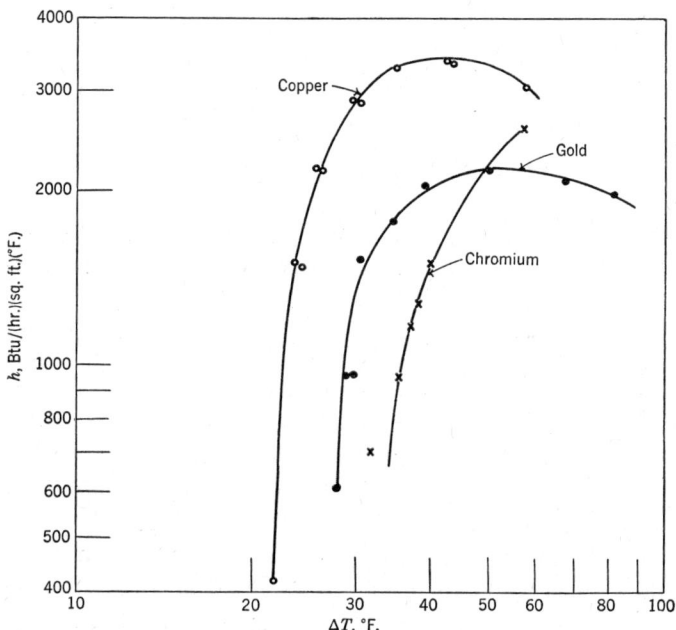

Fig. 29. Effect of type of metal on nucleate boiling. Ethanol was boiled at 1 atm. on polished surfaces (B7).

b. Measurement of Surface Temperature. An accurate determination of the critical temperature difference depends on a good measurement of the surface temperature of the hot solid. Because the peak heat transfer coefficients for boiling are quite large, special difficulties arise. A large number of publications show a lack of appreciation of the problems. For example, if a thermocouple is attached to the hot surface and the couple wires are extended into the boiling liquid, the indicated temperature will not be dependable. For one thing the wires will act as fins conducting heat away from the junction, thus decreasing the junction temperature. Second, the change in temperature at the junction will alter the boiling process near by. Some workers have used thermocouples welded to hot wires which were being used as the heat sources for boiling (F1). The resulting errors in the indicated temperatures were probably large.

If the thermocouple wires are located in a hole or groove in a metal tube or plate, the fin effect will be remedied, but the heat flow pattern through the solid will be altered. The correct surface temperature can be computed by the relaxation method. This "corrected" method has been used for boiling studies (S2), but many workers have made no correction for embedded wires.

One worker (B9) employed a tube of carbon as the heat source. Electric current was passed through this tube, and boiling occurred outside. A thermocouple was in contact with the inside surface. This is a reliable method of temperature measurement, for the air around the wires is at the same temperature as the inner surface.

Some researchers, aware of the temperature problem, elect to use electrically heated wires for heat sources. The same wires can be used as resistance thermometers with satisfactory accuracy. A drawback, however, is that tests cannot be made in the transition region of boiling, because of instability. In addition, unless the wires are quite small the currents needed become very great. For example, if a $\frac{1}{2}$-in.-diam. copper tube with a 0.03-in. wall is intended for use with water near the critical ΔT, a current of about 8,000 amp. is required.

A method of employing a tube carrying a modest electric current as a resistance thermometer, while the heat source is a condensing vapor, has been reported for boiling work (M8). The method was originally developed as a scheme for avoiding temperature-measurement errors with condenser tubes (J6).

3. *Effect of Surface Texture*

When nucleate boiling takes place on a hot solid, the surface texture of that solid becomes an important variable. A thorough study was made by Corty and Foust (C4), who boiled ether, pentane, and Freon-113 from horizontal, electrically heated copper and nickel surfaces. The surface texture was altered at will by use of different grades of emery polishing paper.

Figures 30 and 31 are representative of the results. It is obvious that a rough, clean surface is better for boiling heat transfer than is a smooth, clean one. In fact, for a given ΔT the heat transfer rate shown for the coarse polish is well over double that for the smooth metal.

The slopes of the boiling curves in Fig. 30 increase with increasing roughness. For the smooth metal the approximate relation is $h = C(\Delta T)^{12}$. For the roughest surface the exponent is 24. In no case did these experi-

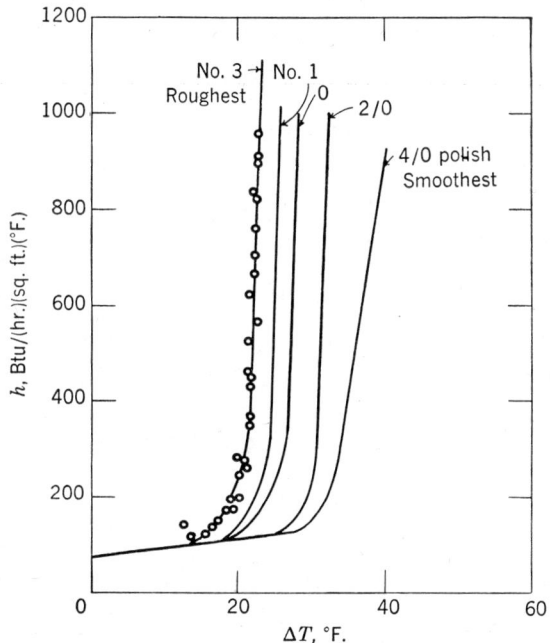

Fig. 30. Effect of surface texture on nucleate boiling. Ether was boiled on a nickel plate polished with different grades of emery paper. The pressure was atmospheric. Typical data are given for one curve only (C4).

menters get the smaller exponents commonly found by others. The tests were not extended to the burnout point.

The authors explain their results in terms of gas-filled pits on the metal surfaces. It is argued that a fine polish results in pits of a narrow size range. The sizes are presumed to be such that a rather high superheat is required to get nucleation. The coarse finishes are presumed to give a wide range of pit sizes. Some pits will have the correct size to permit nucleation at a low superheat.

Of course some increase in heat transfer should be expected with an increase in surface roughness just because of the surface-area increase.

However, the increase in the total heat transfer is greater than can be accounted for by area considerations. Other observers report similar results.

4. *Effect of Geometric Arrangement*

Inasmuch as a difference in testing techniques exists between different observers, a comparison of data for geometric arrangements from different laboratories is unwise. Only data from one laboratory for each case will be included in this section.

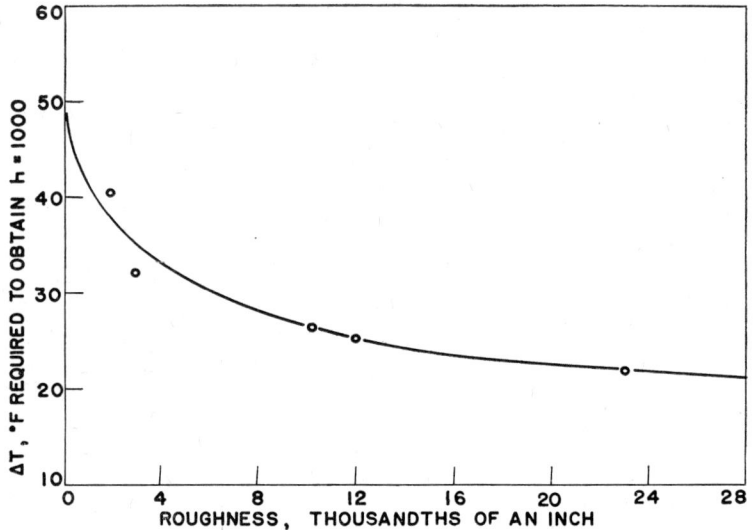

Fig. 31. Effect of surface texture. Roughness values were determined with a diamond-tipped profilometer. The data are from Fig. 30 (C4).

Both horizontal and vertical plates were used by Jakob and Linke (J4) for nucleate boiling of water at low heat fluxes. The results were nearly identical for these two cases. No other data are available for comparing orientation of plates.

Several comparisons of the performance of one tube vs. a bundle of tubes have been made. McAdams (M2) shows that a single, horizontal, 0.5-in. tube gave nearly the same performance as the average tube of a bundle of 60 tubes, through the nucleate region up to a ΔT of about 50° F. Above 50° F. the curves first diverged, then crossed at $\Delta T = 75$° F. and diverged again. Meyers and Katz (M7) found for boiling refrigerants that the performance of one horizontal tube in a bundle was quite different from that of the other tubes in the same bundle. The bottom tube in a vertical row of four gave heat transfer rates which were from 15 to 50% smaller than the rates for the higher tubes.

The effect of diameter has been tested for horizontal hot wires. Rinaldo (M3) used diameters of 0.024 in. and less. The data are shown in Fig. 32. The values at the lowest temperature differences are due probably to free convection rather than to boiling. For the range of

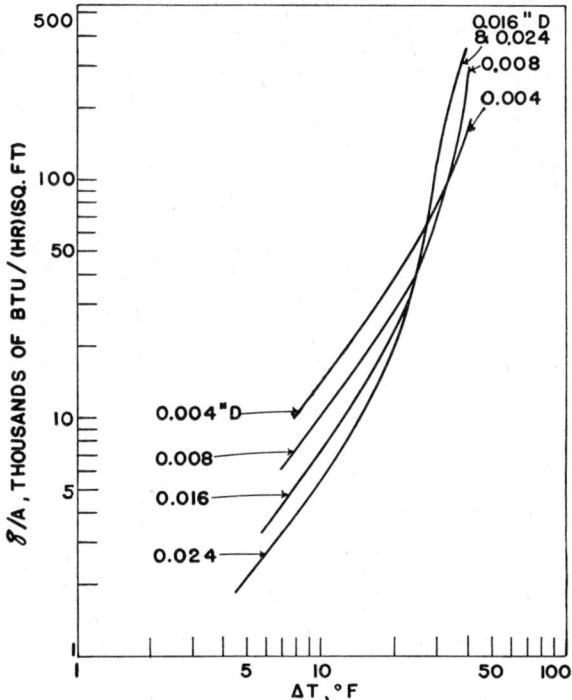

FIG. 32. Effect of wire diameter. Nucleate boiling of water at atmospheric pressure on horizontal platinum wires (M3).

well-developed boiling, the larger diameters seem to produce higher heat fluxes. The burnout values of the heat flux in Fig. 32 are dependent on diameter. The largest diameter gives almost twice the heat flux of the smallest diameter. The critical ΔT is not dependent on diameter.

Data from a single laboratory for tubes of different diameters are needed. Tubes are thicker than the bubbles produced, but the reverse is true for wires. The diameter effects may not be the same in the two cases. The equations of Rohsenow and of Forster and Zuber predict that the geometric arrangement is of no consequence. The prediction is not proved at present.

5. Effect of Pressure

Pressure is a vital factor for nucleate boiling. Figure 11 shows the results for water boiling on a 0.024-in. platinum wire at various pressures. At a given ΔT an increase in pressure causes an increase in heat flux. Similar results occurred when water and five organic liquids were boiled on a horizontal chromium plate (C2).

The maximum heat flux and the critical-temperature difference are functions of pressure also. Figure 33 shows the data of Cichelli and

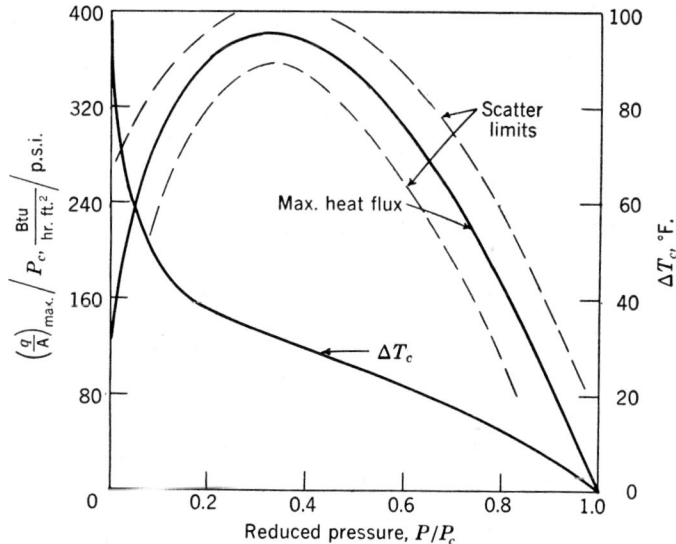

FIG. 33. Effect of pressure on the peak heat flux and critical ΔT. The data include water, ethanol, benzene, propane, pentane, and hexane on a horizontal chromium plate (C2).

Bonilla for these critical values plotted in such a way as to obtain a fairly good correlation. The optimum pressure for these six liquids is at about 35% of their critical pressures.

The maximum superheat which can be achieved with a nonboiling liquid is definitely pressure sensitive. This is evident in Fig. 18 and also in Fig. 28. Both plots show that the possible superheat (and therefore the possible values of ΔT_c) decreases to zero as the critical pressure is approached. This is in agreement with the equation-of-state theory and also the nucleation-rate theory.

6. Effect of Surface Tension

From a theoretical standpoint surface tension is an important variable. First consider nucleation. The rate of nuclei formation is proportional to

$e^{-\sigma^3}$; thus small decreases in σ should cause large increases in the number of nuclei. For liquids in tension, the rupture pressure is proportional to $\sigma^{3/2}$. Thus liquids with large surface tensions should be difficult to fracture. On the other hand, considering the equation of state only, no reason is apparent why a surface-active agent should affect the metastable state. On theoretical grounds only, therefore, two possibilities exist. A decrease in σ may have no effect or it may increase the rate of heat transfer. It cannot decrease the heat flow. Similarly, the critical ΔT should either remain constant or decrease as σ decreases.

The semitheoretical approaches which are concerned with bubbles already in existence predict that σ is an important variable also. Rohsenow and also Forster and Zuber give $h = C\sigma^{-0.5}$. This agreement is noteworthy.

Two experimental approaches have been used. The results for different pure liquids have been examined in attempts to detect the effect of σ, and surface-active agents have been added to pure liquids artificially to change σ. The results by either method are contradictory. Experimentally the effect of σ is still unproved.

If the nucleate-boiling region is represented by an equation such as $h = \text{(const.)} \sigma^z$, the results of a few workers will serve to show the present state of knowledge.

$z = -2.5$ Jakob and Linke (J5)
$z = -2$ Stoebe, Baker, and Badger (S4)
$z = -1$ McNelly (M5)
$z = 0.25$ Kutateladze (K5)
$z = 1.275$ Nakagawa and Yoshida (N1)

Several factors may be contributing to the difficulty. If a surface-active agent is added to a pure boiling liquid, the agent could dissolve and then migrate to the liquid-solid interface or to the liquid-vapor interface, or it can remain dispersed. A knowledge of the liquid-vapor interfacial tension tells nothing of the other interfacial tensions. If a surface-active agent is present in a liquid in the form of suspended particles, these could act as catalysts for nucleation. Last, surface tension is determined and defined under static conditions. It has been shown that σ measured under dynamic conditions is sometimes quite different (M9). These factors must be considered by any experimenter attempting to demonstrate the effect of surface tension.

One set of observations exists for the effect of surface-active agents on ΔT_c and $h_{\max}$. The values are somewhat incomplete because the ΔT_c for pure water could not be obtained with the equipment used. The accompanying table summarizes the results (M9). The results indicate that a decrease in σ causes a decrease in the maximum heat flux. Additional data are needed, because a general belief among those who quench metals is that surface-active agents increase the rate of cooling.

Liquid	ΔT_c, °F.	h_{max}, B.t.u./(hr.)(ft.^{2})(°F.)	$(q/A)_{max}$, B.t.u./(hr.)(ft.^{2})
Water	>100	>1830	>183,000
Water + 0.1% Drene	55	2090	115,000
Water + 1% Drene	36	1830	66,000
Water + 1% $C_{12}H_{25}SO_3Na$	69	2140	148,000

7. *Effect of Agitation*

As would be expected, agitation increases the heat transfer rate for nucleate boiling. Schweppe and Foust (S2) pumped saturated water through a heated tube to obtain the boiling curves in Fig. 34. Note the

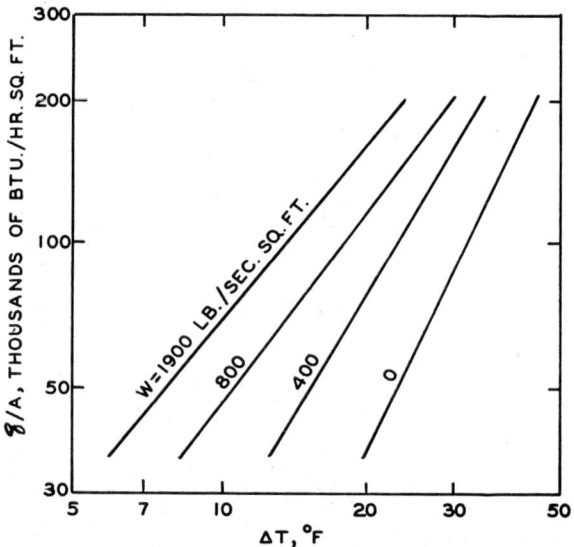

Fig. 34. Effect of agitation on nucleate boiling. Saturated water was pumped through a heated vertical tube, 0.62 to 1.05 in. in diameter, at the indicated flow rates. Pressure = 15 to 36 lb./sq. in. abs. (S2).

convergence of the lines as the ΔT is increased. Robinson and Katz (R4) injected Freon vapor into saturated liquid Freon boiling on the outside of tubes with results which were similar except that the lines actually coincided as the critical ΔT was approached. Beecher (B3) and others have obtained similar coincidence of boiling curves.

The data of Pramuk and Westwater (P5) show that the boiling curves for various degrees of agitation do not merge at the critical ΔT, although the approach is very close. Figure 35 shows the observations for methanol boiling while agitated with a three-bladed propeller. The increase in the peak heat flux is obvious. It is interesting also that the ΔT_c (corresponding

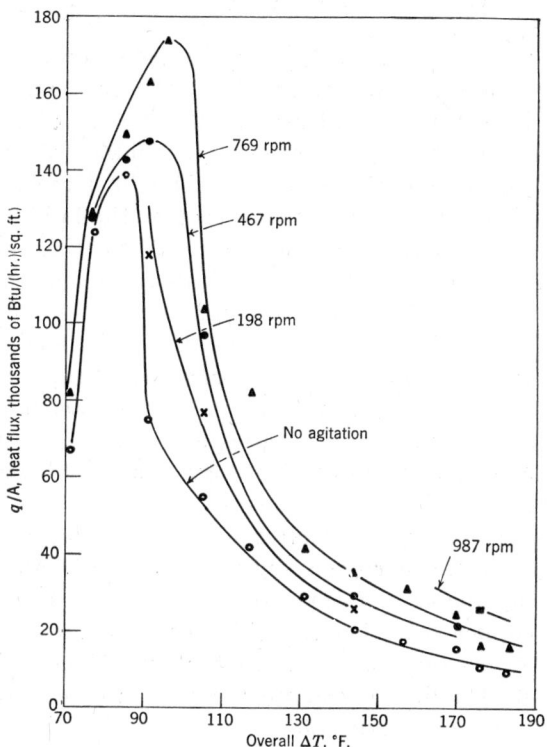

FIG. 35. Effect of agitation. Methanol was agitated with a three-blade propeller while boiling at 1 atm. on a ⅜-in. horizontal, steam-heated, copper tube. The shaft speed is indicated (P5).

to the maximum h) for the boiling-side coefficient is nearly constant at about 52° F. This estimate is obtained from the over-all ΔT by use of a calculated steam-side coefficient of 3000 B.t.u./(hr.)(ft.²)(°F.).

The constancy of the critical ΔT agrees with the equation-of-state explanation of superheated liquids and with the reaction-rate view of nucleation. It means there is an upper temperature limit above which a superheated liquid cannot exist (at a stated pressure) regardless of agitation.

The fact that an ordinary degree of agitation causes only a negligible change in the maximum heat flux, while a violent agitation causes an observable effect, can be explained. The natural agitation resulting from the boiling action itself results in a high degree of turbulence. Unless the added turbulence is of the same order of magnitude, it will scarcely affect the heat transfer coefficient.

8. *Effect of Impurities*

Nucleation rates are sensitive to the presence of foreign solid particles, because these objects may act as catalysts. If a nucleus is created on a solid particle, it will remain attached during part of the subsequent growth process. The growth equations for bubbles attached to solids have not been worked out mathematically, but it is rather obvious that interfacial tensions will be important as long as the bubbles are small.

The well-known "aging" phenomenon for heating surfaces is one aspect of the effect of impurities. If nucleate boiling is caused to occur on a freshly polished, clean surface, the boiling will not come to a steady-state condition for a period of at least a sizable fraction of an hour (sometimes much longer). During this time a degassing of the solid takes place. Chemical actions, which gradually diminish, may occur at the surface at the same time. After the proper aging, a heat transfer surface remains reasonably stable. Many researchers have adopted standard aging techniques for preparation of their heaters.

Corrosion of a surface may alter nucleate-boiling heat transfer in either direction. If the corroded surface is a good catalyst, the heat transfer rates can increase. Copper oxide is a better boiling surface for water than is copper (S2). On the other hand, oxidized stainless steel is sometimes better, but often poorer than unattacked stainless steel with water (B16). Oxidized chromium is a poor boiling surface for water compared with fresh chromium (J3).

If fouling (deposition of a solid from the liquid) occurs on a boiling surface, the effect on heat transfer will depend on whether the deposit is a catalyst and whether it is very bulky. If the deposit is thin, the heat transfer may increase. This is an unnatural occurrence to observers not familiar with boiling—a little fouling is sometimes a good thing!

Impurities can be present in the boiling liquid itself. The evidence is strong that traces of *dissolved* material (excluding surface-active agents of course) have but a small effect on boiling, whereas suspended material is much more important. Wismer's important experiments on superheated liquids show a distinct difference between the maximum superheat attainable with water at atmospheric pressure in the presence of dissolved material and the values resulting for suspended matter. Some of the results are given in Table VI.

Additional evidence of the small importance of dissolved impurities lies in the observations for water solutions. For glycerine in water, a smooth shift occurs in the values for ΔT_c and h_{max} as the concentration is increased from zero (M10). Similar results occur with sugar in water. Boiling curves for several concentrated solutions of salts in water have been reported (C6). Although these show that an appreciable shift results in the relationship between h and ΔT (compared with that for pure water) the effect does not seem to be unexpected or unusual.

TABLE VI
Effect of Impurities on Maximum Temperature for Superheated Water at 1 Atm. (K2)

	Max. temp. (°F.)
Pure water in glass	464
Solutions: 0.6% H_2SO_4	415
6.5% H_2SO_4	461
2.7% amyl alcohol	461
4% aniline	451
Colloids: 0.1% argyrol	336
3% starch	307
0.02% colloid Pt	275

The result to be expected when an immiscible liquid is added to a boiling liquid is uncertain. The addition of water to boiling butadiene has a negligible effect, whereas the addition of water to boiling styrene has a pronounced effect (B6). The unusual behavior of water-styrene mixtures is shown in Fig. 36. Note that all mixtures of the two components give poorer boiling coefficients than do either pure component.

Suspended inert gas bubbles, too small to be seen, seem to be present in some liquids in an as-received condition. These motes can also come into being if a liquid containing dissolved gases is heated. The evidence is clear that such motes have a strong effect on the nucleate-boiling curve. Pike and co-workers (P1) boiled tap water, deaerated water, and water which had been brought to equilibrium with bubbling air at various temperatures. It was found that an increase in air content causes an increase in h. Similar results are reported by McAdams (M4).

Very few tests have been performed with liquids and solids that were both truly degassed. Wismer and co-workers (K2) found that it was necessary to heat their glass tubes to a high temperature, near the softening point, in a good vacuum before they could get reproducible superheat data. Mead (M6) tried solvent cleaning, using successive washes with chromic acid, water, carbon tetrachloride, acetone, 5% HCl, and water. Although the glass surfaces seemed free of other impurities, gases were still present. Mead then used high pressure to force the motes into solution. The water to be tested was pressurized *in situ* for 15 min. at 20,000 p.s.i. Presumably the gases present on the glass and in the water passed into solution, because Mead found that the water-in-glass system could then be superheated to a maximum temperature of 358° F. (at 1 atm.). If no pressure pretreatment was used, the maximum possible temperature was 326° F. One objection to the pressure scheme is that the resulting

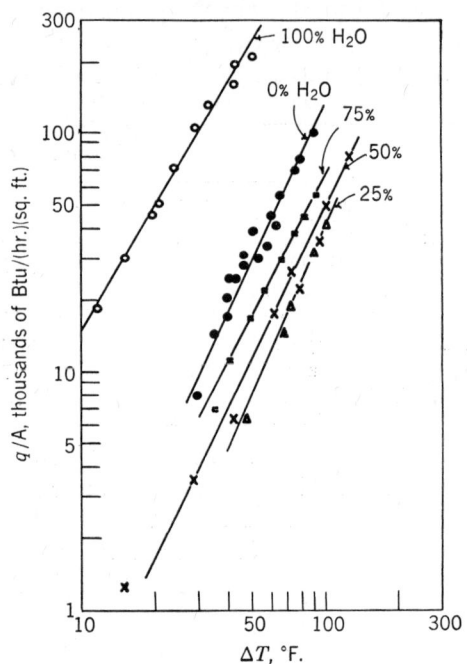

Fig. 36. Nucleate boiling of immiscible liquid mixtures. Water and styrene mixtures were boiled on a horizontal chromium plate at 0.053 atm., abs. (B6).

liquid may become supersaturated (with gas) when the pressure is released. It is doubtful that the gases will remain in solution indefinitely when heat is applied.

Another way of explaining why tiny suspended particles should be important for boiling is that these motes represent cracks or imperfections in the liquid structure. The tensile strength of the liquid will be reduced because of these flaws.

9. *Effect of Short-Wave Irradiation*

Knowledge of the effect of high-energy irradiation on nucleate boiling is not available. Such information is necessary for the design of boiling-cooling systems for atomic reactors. Glaser's observations with the bubble chamber, described in Sec. IIC1b, can be used for a guess. In order for the rate of nucleation to be affected strongly, the liquid must be easy to ionize and must be a poor conductor. Thus ordinary water should show little change in its boiling behavior during irradiation. Such materials as liquid hydrogen and the low-molecular-weight paraffins should be affected. Molecules containing heavy atoms should result in much coulomb scattering, poor ionization, and little effect on boiling.

F. Growth of Bubbles

As indicated in Sec. IIB, ordinary nucleate boiling is a two-step process. First, nuclei must appear. Second, the nuclei must grow into bubbles large enough to move away from the nucleation sites. The rate of heat absorption by the liquid may be controlled by either one or both of these two processes. The growth of a nucleus (tiny bubble) into ordinary bubbles has received attention recently. The theoretical attack of Forster and Zuber was discussed in Sec. IIB2. Inasmuch as the theory of Zwick and Plesset (P3, P4, Z1, Z2) represents another attempt to obtain exact expressions for bubble growth, and since the theory fits well with the few data for steam bubbles in superheated water, their theoretical method is summarized below.

If a metastable nucleus of critical size R_0 happens to gain a molecule, the metastable condition is unbalanced and spontaneous bubble growth begins. As the bubble grows, the surface tension forces rapidly become negligible. For example, when the radius becomes $2R_0$, the corresponding surface-tension "pressure," $2\sigma/R$ will have decreased to half its original value (for a pure, gas-free liquid). The relaxation of surface forces would cause a steady increase in the rate of bubble growth were it not for a braking action which limits the process.

For a bubble to grow, vapor must pass from the superheated liquid into the bubble. Thus latent heat of vaporization is removed from the surrounding liquid, and the liquid cools. The drop in liquid temperature near the bubble means a decrease in the driving force between liquid and bubble. This temperature drop strongly affects the bubble rate of growth. The rate can be shown to approach asymptotically a condition whereby the radius increases according to the square root of time.

The growth involves not only heat transfer but also mass transfer. Because of this, and also because growth is a dynamic process, the mathematical problem is quite complex. In order to describe the process in mathematical terms, a few reasonable assumptions can be used. The sole resistance to heat transfer is assumed to lie in a thin liquid film surrounding the bubble. The vapor in the bubble is assumed to be uniform. The bubble is imagined to be a sphere and to possess no rotation. The spherical condition means that buoyancy is being neglected. This seems to agree with observations for bubbles with radii less than 1 mm.

It is not necessary to assume the liquid film to be completely stagnant. Radial motion can be allowed for, but with some difficulty. It was noted in Sec. IIB2 that Forster and Zuber state that conduction is the chief mode of heat transfer (compared with convection due to radial motion). Eddies or motions of the liquid tangent to the bubble are neglected. The Zwick-Plesset theory likewise excludes eddies. The derivation is lengthy; therefore the final typical equations are presented here without proof.

The bubble radius at any time is given by

$$R\frac{d^2R}{d\theta^2} + \frac{3}{2}\left(\frac{dR}{d\theta}\right)^2 + \frac{2\sigma}{R\rho_L} = \frac{p_V - p_\infty}{\rho_L} \tag{12}$$

The integrated expression relating radius to time and temperature is a nonlinear, integrodifferential equation. It can be put into tractable forms by imagining four overlapping growth periods.

The first growth corresponds to a "relaxation period." The bubble grows from the nucleus radius R_0 to $\left(1 + \dfrac{1}{e}\right)R_0$, where e is the Napierian logarithm base. During this time growth is given by

$$R = R_0(1 + e^{H(\theta-\theta_0)}) \tag{69}$$

The time θ is measured from the instant the nucleus becomes unstable. The value H can be given a physical meaning. At first the growth rate is practically nil; then it begins to accelerate. Consider the total time required to grow from R_0 to $2R_0$; the "active" part of the period is equivalent to $1/H$ sec. The value of H is dependent on temperature. For water at 223° F. and 1 atm., $1/H$ is 1.5×10^{-6} sec. The second growth is called the "early phase." The third is the "intermediate phase." These are expressed as infinite series. The final growth is the "asymptotic phase." During this time the bubble temperature steadily approaches the boiling point of the liquid. The radius increases with the square root of time.

$$R = 2\left(\frac{3\theta}{\pi\alpha_L}\right)^{1/2}\left(\frac{k_L(T_L - T_V)}{\lambda\rho_V}\right) \tag{70}$$

During the final growth many complicating effects disappear and a simple heat balance can be used to show the radius-temperature-time relation.

$$\lambda \rho_V dv = k_L(4\pi R^2) \frac{(T_L - T_V)d\theta}{(\pi \alpha_L \theta/3)^{1/2}} \tag{71}$$

The left side is the heat gained by the bubble; the right is the heat transferred across the liquid film. The thickness of this film, $\sqrt{\pi \alpha_L \theta/3}$, increases steadily.

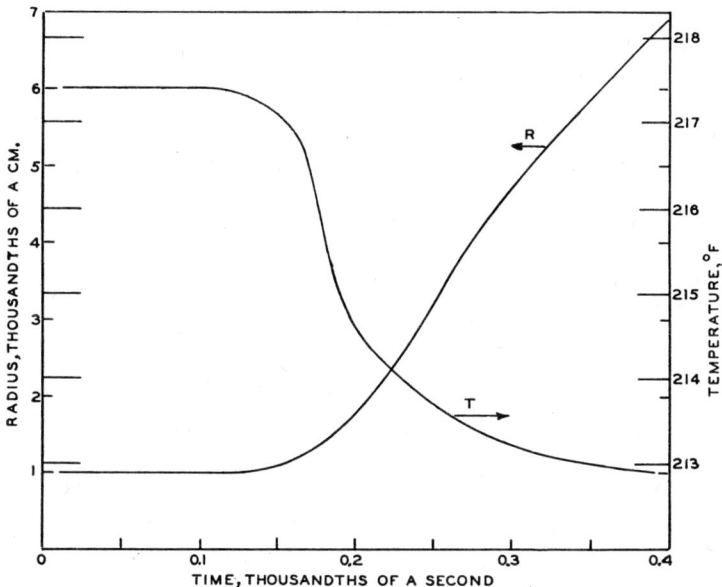

FIG. 37. Theoretical growth of a steam bubble in water (Z1).

A graph of the theoretical growth of a bubble of steam from a radius of 0.001 to 0.007 cm. in water at 217.4° F. and 1 atm. is shown in Fig. 37. The corresponding temperature of the bubble (and bubble wall) is included. Note that this growth occurs in only 0.0004 sec.

Laboratory measurements of bubble growth are rare. Dergarabedian (D1) used the high-speed-camera technique to observe the growth of steam bubbles in the observable (asymptotic) period. A sample of his results is shown in Fig. 38. The agreement between the experiments and the theoretical line (for asymptotic growth) is excellent. The instant of growth initiation is not known; so the time origins for the data and for the theoretical curve are arbitrary. The bubble-growth equations of Forster and Zuber also give good agreement with the data of Dergarabedian. Data are needed for other systems and particularly for microscopic bubble sizes to show whether the Forster-Zuber method is a satisfactory approximation in general.

After a bubble has become large enough for buoyant forces to be significant (R = about 1 mm.) the bubble will rise through the liquid and experience frictional drag on its surface. Haberman and Morton (H1) have studied the behavior of macroscopic bubbles in the absence

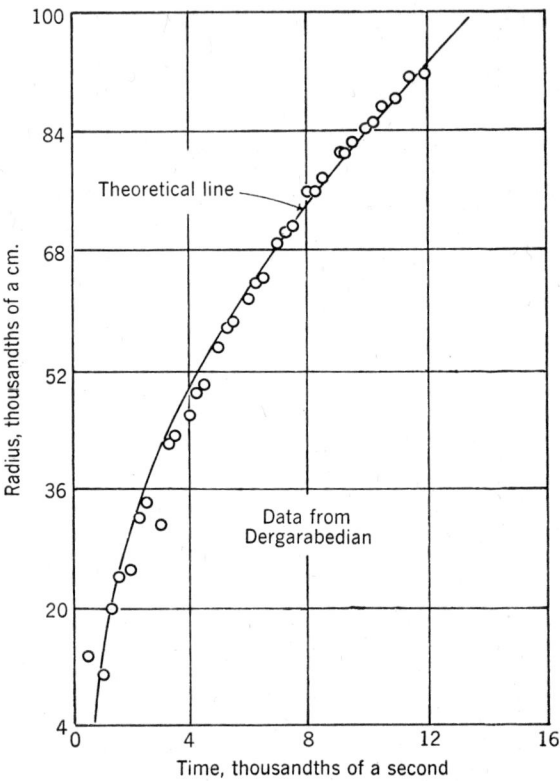

FIG. 38. Asymptotic growth of a steam bubble in water at 220.1° F., 1 atm. The line is a graph of Zwick's theoretical prediction (Z1).

of heat transfer. They bubbled air through eight liquids. Stokes's law fits the observations for velocity of rise for bubbles which are spherical (R = about 3 mm. or less). Bubbles of about 3- to 5-mm. radius are elliptical in shape, and larger ones are umbrella-shaped. The smallest bubbles move in straight lines unless disturbed. Bubbles large enough to have Reynolds numbers of 300 to 3000 move in spiral paths, and the largest bubbles move with a rocking motion. Some of these facts are evident in motion pictures of boiling. Figures 4 to 9 show various types of macroscopic bubbles. It is obvious that many bubbles formed during boiling are globs with no recognizable shapes.

Heat transfer to macroscopic bubbles has received but little attention. Jakob (J2) presents the growth curves for several steam bubbles up to a radius of 3.6 mm. The main experimental obstacle is that of determining the true volume of a nonspherical bubble.

Acknowledgment

Encouragement and assistance for the preparation of this chapter was furnished by the National Science Foundation. Helpful suggestions were given by L. Bernath, H. B. Clark, N. Zuber, and S. A. Zwick.

The following topics in "Boiling of Liquids" will be reviewed in a second installment to be published in Volume II of this series: III—Transition Boiling, IV—Film Boiling, V—Boiling in Subcooled Liquids, VI—Bumping during Boiling.

Nomenclature

a Arbitrary exponent.
A Surface area, sq. ft.
A_0 Surface area of a nucleus, sq. ft.
A_x Cluster of x molecules.
b Defined as $(p_V - p_L)/p_V$ or $2\sigma/(p_V R_0)$, dimensionless.
B_1 Defined as $[6\sigma/(\pi m)(3 - b)]^{1/2}$, dimensionless.
B_2 Defined as $\ln[(b - 1)/2]$, dimensionless.
$C, C_1, C_2, etc.$ Constants.
C_L, C_V Heat capacity of liquid, vapor, B.t.u./lb. (°F.).
d Bubble parameter, distance from solid to center of curvature of bubble, ft.
D Diameter of tube or wire, ft.
D' Heated perimeter divided by π, ft.
D_b Bubble diameter, ft.
D_e Equivalent diameter of flow passage, ft.
e Electrical charge on an ion.
e Base of Naperian logarithms, 2.718.

f Frequency, hr.$^{-1}$
F_1 Free energy of a liquid system, ft.-lb.
F_2 Free energy of a liquid containing a vapor cluster, ft.-lb.
ΔF Difference in free energy, ft.-lb.
ΔF_0 Free-energy change to produce a nucleus, ft.-lb.
F_ϵ Radiation emissivity factor, dimensionless.
g Acceleration of gravity, ft./hr.^{2}
g_0 Conversion factor, lb. mass × ft./lb. force × (hr.).^{2}
g_1 Most probable number of molecules per cell.
G Mass velocity, lb./hr. sq. ft.
h Individual heat transfer coefficient, B.t.u./(hr.)(sq. ft.)(°F.).
h Planck's constant, energy × time.
h' Heat transfer coefficient, neglecting radiation, B.t.u./(hr.)(ft.^{2})(°F.).

h_R	Heat transfer coefficient for radiation, B.t.u./(hr.)(ft.^{2})(°F.).	q	Heat transfer rate, B.t.u./hr.
H	Exponent in equation for bubble growth, hr.$^{-1}$ or sec.$^{-1}$	Q	Heat transfer, B.t.u.
		r	Radius ratio, R/R_0, dimensionless.
k	Boltzmann constant, energy/(degree temp.)(molecule).	r_1	Radius ratio, R_1/R_0, dimensionless.
		R	Radius of a bubble, ft.
k_L, k_V	Thermal conductivity of liquid, vapor, B.t.u./(hr.)(ft.)(°F.).	R_1	Arbitrary bubble radius, ft.
		R_0	Radius of a nucleus, ft.
L	Length of tube, ft.	Re	Reynolds number, $DV\rho/\mu$, dimensionless.
m	Weight of one molecule, lb.	s	Arbitrary exponent.
		T	Temperature, °F. or °R.
m	Contact angle parameter, equal to d/R or $-\cos\beta$.	ΔT	Temperature driving force, $T_L - T_V$ for homogeneous case, $T_s - T_\infty$ for heterogeneous, °F.
M	Molecular weight.		
n	Number of nucleating points per unit area, ft.$^{-2}$	T_{BP}	Boiling point of the bulk liquid at the existing pressure, T_∞, °F.
n_i	Number of ions per nucleus.		
n_0	Number of molecules in a nucleus.	T_L	Temperature of the bulk liquid, °F.
n_0'	Number of nuclei.	T_0	Temperature of superheated liquid, °F.
n_L, n_V	Number of molecules of liquid, vapor.	T_s	Temperature of a solid surface, °F.
n_x'	Number of clusters containing x molecules each.	T_V	Temperature inside a bubble, °F.
N	Total number of molecules in a system.	T_∞	Saturation temperature of a liquid flat surface at the existing pressure, °F.
N'	Avogadro number, molecules/mole.		
Nu	Nusselt number, hD/k_L or hD/k_V, dimensionless.	v	Volume of a bubble or cluster, cu. ft.
p	Pressure, lb./sq. ft.	v_L, v_V	Effective volume of one molecule of liquid, vapor, cu. ft.
Δp	Pressure difference, lb./sq. ft.		
p_g	Partial pressure of inert gas in a bubble, lb./sq. ft.	V	Volume of entire system, cu. ft.
p_L	Pressure imposed on a liquid, lb./sq. ft.	V	Velocity, usually ft./hr.
		w	Mass flow rate, lb./hr.
p_V	Pressure of the vapor in a bubble, lb./sq. ft.	w_1, w_2, w_3, w_4	Work terms in the fluctuation theory, ft. lb.
p_∞	Saturation pressure of a flat liquid surface at its existing temperature, lb./sq. ft.	W_F	Work to overcome molecular interactions, ft. lb.
		W_0	Work of forming a nucleus, ft. lb.
Pr	Prandtl number, $C_L\mu_L/k_L$ or $C_V\mu_V/k_V$, dimensionless.	W_x	Work of forming a cluster of x molecules, ft. lb.

W_ω Potential energy of a configuration, ft. lb.
x Number of molecules.
z Empirical exponent.
z Thickness of boundary zone between liquid and vapor, ft.
α_L Thermal diffusivity, $k_L/(\rho_L C_L)$.
β Contact angle measured through the liquid.
ϵ Dielectric constant, dimensionless.
η_L, η_V Partial free energy for a liquid molecule, a vapor molecule, $(\partial F/\partial n_L)_{T,p}$ or $(\partial F/\partial n_V)_{T,p}$.
θ Time, hr.
λ Latent heat of vaporization, B.t.u./lb.
λ' Molecular heat of vaporization, B.t.u./molecule.
μ_L, μ_V Viscosity of liquid, vapor, lb./(ft.)(hr.).
ρ_L, ρ_V Density of liquid, vapor, lb./cu. ft.
σ Surface tension, liquid-vapor interface, lb./ft.
σ_{sL} Interfacial tension, solid-liquid, lb./ft.
σ_{sV} Interfacial tension, solid-vapor, lb./ft.
σ_∞ Surface tension at a flat liquid-vapor interface, lb./ft.
σ' Stefan-Boltzmann radiation constant, B.t.u./(hr.)(ft.^{2})(°R.).^{4}
ϕ Contact angle parameter, $(2+m)(1-m)^2/4$, dimensionless.
ω Molecular volume of translational motion.
ω Flow rate of molecules streaming across a boundary, molecules/(ft.^{2})(hr.).
Ω Configuration integral used in fluctuation theory.

References

A1. Addoms, J. N., Doctor Science Thesis in Chemical Engineering, Massachusetts Institute of Technology, Cambridge, 1948.
A2. Aspden, R. L., *Natl. Advisory Comm. Aeronaut. Tech. Notes* **141**, N-34055 (1954).
B1. Banchero, J. T., Barker, G. E., and Boll, R. H., *Ann. Meeting Am. Inst. Chem. Engrs. St. Louis*, Preprint No. 3 (1953).
B2. Becker, R., and Döring, W., *Ann. Physik* **24**, 719 (1935).
B3. Beecher, N., M. S. Thesis in Chem. Engr., Massachusetts Institute of Technology, Cambridge, 1948.
B4. Bernath, L., *Ind. Eng. Chem.* **44**, 1310 (1952).
B5. Bernath, L., *Ann. Meeting Am. Inst. Chem. Engrs. Louisville, Ky.*, Preprint No. 8 (1955).
B6. Bonilla, C. F., and Eisenberg, A. A., *Ind. Eng. Chem.* **40**, 1113 (1948).
B7. Bonilla, C. F., and Perry, C. W., *Trans. Am. Inst. Chem. Engrs.* **37**, 685 (1941).
B8. Bosworth, R. C. L., "Heat Transfer Phenomena," p. 108. Wiley, New York, 1952.
B9. Bromley, L. A., *Chem. Eng. Progr.* **46**, 221 (1950).
B10. Bromley, L. A., Brodkey, R. S., and Fishman, N., *Ind. Eng. Chem.* **44**, 2966 (1952).
B11. Bromley, L. A., LeRoy, N. R., and Robbers, J. A., *Ind. Eng. Chem.* **45**, 2639 (1953).
B12. Briggs, L. J., *J. Appl. Phys.* **21**, 721 (1950).
B13. Briggs, L. J., *Science* **112**, 427 (1950).

B14. Briggs, L. J., *J. Chem. Phys.* **19**, 970 (1951).
B15. Briggs, L. J., *Science* **113**, 483 (1951).
B16. Buchberg, H., Romie, F., Lipkis, R., and Greenfield, M., "Heat Transfer and Fluid Mechanics Institute," p. 177. Stanford U. P., Stanford, 1951.
C1. Castles, J. T., M. S. Thesis in Chemical Engineering, Massachusetts Institute of Technology, Cambridge, 1947.
C2. Cichelli, M. T., and Bonilla, C. F., *Trans. Am. Inst. Chem. Engrs.* **41,** 755 (1945).
C3. Colburn, A. P., referred to by Castles (C1); also in unpublished U. of Delaware lecture material, 1941.
C4. Corty, C., and Foust, A. S., *Ann. Meeting Am. Inst. Chem. Engrs. St. Louis,* Prep. int No. 1 (1953).
C5. Corty, C., and Foust, A. S., "Surface Variables in Nucleate Boiling" (Motion Picture). Univ. of Michigan, Ann Arbor, 1953.
C6. Cryder, D. S., and Finalborgo, A. C., *Trans. Am. Inst. Chem. Engrs.* **33**, 346 (1937).
C7. Cryder, D. S., and Gilliland, E. R., *Ind. Eng. Chem.* **24**, 1382 (1932).
D1. Dergarabedian, P., *J. Appl. Mech.* **20**, 537 (1953).
D2. Dew, J. E., M. S. Thesis in Chemical Engineering, Massachusetts Institute of Technology, Cambridge, 1948.
D3. Dougherty, E. L., M. S. Thesis in Chemical Engineering, University of Illinois, Urbana, 1951.
D4. Drew, T. B., and Mueller, A. C., *Trans. Am. Inst. Chem. Engrs.* **33**, 449 (1937).
D5. Drew, T. B., and Mueller, A. C., "Boiling" (Motion Picture). E. I. du Pont de Nemours, Wilmington, Del., 1937.
E1. Ellion, M. E., Jet Propulsion Lab., Memo. 20–88, California Institute of Technology, Pasadena, 1954.
F1. Farber, E. A., and Scorah, R. L., *Trans. Am. Soc. Mech. Engrs.* **70**, 369 (1948).
F2. Farmer, W. S., *in* "Liquid Metals Handbook," 2nd ed., p. 205. Supt. of Documents, Washington, D. C., 1952.
F3. Fisher, J. C., *J. Appl. Phys.* **19**, 1062 (1948).
F4. Forster, H. K., *J. Appl. Phys.* **25**, 1067 (1954).
F5. Forster, H. K., and Zuber, N., *J. Appl. Phys.* **25**, 474 (1954).
F6. Forster, H. K., and Zuber, N., Conference on Nuclear Engineering, University of California, Los Angeles, 1955.
F7. Frenkel, J., "Kinetic Theory of Liquids," Chapter 8. Oxford U. P., New York, 1946.
F8. Fritz, W., *Physik. Z.* **36**, 379 (1935).
G1. Glaser, D. A., *Phys. Rev.* **87**, 665 (1952).
G2. Glaser, D. A., *Phys. Rev.* **91**, 762 (1953).
G3. Glaser, D. A., *Sci. American* **192**, No. 2, 46 (1955).
G4. Glaser, D. A., and Rahm, D. C., *Phys. Rev.* **97**, 474 (1955).
G5. Glasstone, S., Laidler, K. J., and Eyring, H., "Theory of Rate Processes." McGraw-Hill, New York, 1941.
G6. Goldman, K., Nuclear Development Associates, Report 10–68, White Plains, New York, 1953.
G7. Gunther, F. C., *Trans. Am. Soc. Mech. Engrs.* **73**, 115 (1951).
G8. Gunther, F. C., and Kreith, F., "Heat Transfer and Fluid Mechanics Institute," p. 113. Am. Soc. Mech. Engrs., New York, 1949.
H1. Haberman, W. L., and Morton, R. K., Report 802, AD 19377, Armed Services Tech. Information Agency, 1953.

H2. Hickman, K. C. D., *Science* **113**, 480 (1951).
H3. Hickman, K. C. D., *Ind. Eng. Chem.* **44**, 1892 (1952).
H4. Hickman, K. C. D., and Torpey, W. A., *Ind. Eng. Chem.* **46**, 1446 (1954).
H5. Hirschfelder, J. O., Curtiss, C. F., and Bird, R. B., "Molecular Theory of Gases and Liquids," p. 348. Wiley, New York, 1954.
H6. Hirschfelder, J. O., Curtiss, C. F., and Bird, R. B., "Molecular Theory of Gases and Liquids," p. 363. Wiley, New York, 1954.
I1. Insinger, T. H., and Bliss, H., *Trans. Am. Inst. Chem. Engrs.* **36**, 491 (1940).
J1. Jakob, M., *Mech. Eng.* **58**, 643 (1936).
J2. Jakob, M., "Heat Transfer," Vol. 1, Chapter 29. Wiley, New York, 1949.
J3. Jakob, M., and Fritz, W., *Forsch. Gebiete Ingenieurw.* **2**, 434 (1931).
J4. Jakob, M., and Linke, W., *Forsch. Gebiete Ingenieurw.* **4**, 75 (1933).
J5. Jakob, M., and Linke, W., *Physik. Z.* **36**, 267 (1935).
J6. Jeffrey, J. O., *Cornell Univ. Expt. Sta. Bull.* **21** (1936).
J7. Jens, W. H., *Mech. Eng.* **76**, 981 (1954).
J8. Jens, W. H., and Lottes, P. A., Report 4627, Argonne National Laboratory, 1951.
K1. Kazakova, E. A., *The Engineer's Digest* **12**, No. 3, 81 (1951).
K2. Kendrick, F. B., Gilbert, C. S., and Wismer, K. L., *J. Phys. Chem.* **28**, 1297 (1924).
K3. Kreith, F., and Foust, A. S., *Ann. Meeting Am. Soc. Mech. Engrs. New York*, Paper 54-A-146 (1954).
K4. Kreith, F., and Summerfield, M., "Heat Transfer and Fluid Mechanics Institute," p. 127. Am. Soc. Mech. Engrs., New York, 1949.
K5. Kutateladze, S. S., *Izvest. Akad. Nauk S.S.S.R. Otdel Tekh. Nauk*, p. 529 (1951).
L1. Leidenfrost, J. G., "De aquae communis nonnullis qualitatibus tractatus." Duisburg, 1756. (Discussed in reference D4.)
L2. Lipkis, R. P., Liu, C., Zuber, N., and Greenfield, M. L., Report 54-77, AECU 2950, University of California, Los Angeles, 1954.
L3. Lyon, R. E., Foust, A. S., and Katz, D. L., *Ann. Meeting Am. Inst. Chem. Engrs. St. Louis*, Preprint No. 6 (1953).
M1. Marx, J. W., and Davis, B. I., *J. Appl. Phys.* **23**, 1354 (1952).
M2. McAdams, W. H., "Heat Transmission," 3rd ed., Chapter 14. McGraw-Hill, New York, 1954.
M3. McAdams, W. H., Addoms, J. N., Rinaldo, P. M., and Day, R. S., *Chem. Eng. Progr.* **44**, 639 (1948).
M4. McAdams, W. H., Kennel, W. E., Minden, C. S., Carl, R., Picornell, P. M., and Dew, J. E., *Ind. Eng. Chem.* **41**, 1945 (1949).
M5. McNelly, M. J., *J. Imp. Coll. Chem. Eng. Soc.* **7**, 18 (1953).
M6. Mead, B. R., Romie, F. E., and Guibert, A. G., "Heat Transfer and Fluid Mechanics Institute," p. 209. Stanford U. P., Stanford, 1951.
M7. Meyers, J. E., and Katz, D. L., *Chem. Eng. Progr. Symposium Ser. No. 5*, **49**, 330 (1953).
M8. Miller, L. B., M. S. Thesis in Chemical Engineering, University of Illinois, Urbana, 1955.
M9. Morgan, A. I., Bromley, L. A., and Wilke, C. R., *Ind. Eng. Chem.* **41**, 2767 (1949).
M10. Moscicki, I., and Broder, J., *Roczniki Chem.* **6**, 319 (1926).
N1. Nakagawa, Y., and Yoshida, T., *Chem. Eng. (Japan)* **16**, No. 3, 6 (1952).
N2. Nukiyama, S., *Soc. Mech. Eng. Japan* **37**, No. 206, 267 (1934).

P1. Pike, F. R., Miller, P. D., Jr., and Beatty, K. O., Jr., *Ann. Meeting Am. Inst. Chem. Engrs. St. Louis*, Preprint No. 2 (1953).
P2. Plesset, M. S., and Ellis, A., *Trans. Am. Soc. Mech. Engrs.* **77**, 1055 (1955).
P3. Plesset, M. S., and Zwick, S. A., *J. Appl. Phys.* **23**, 95 (1952).
P4. Plesset, M. S., and Zwick, S. A., *J. Appl. Phys.* **25**, 493 (1954).
P5. Pramuk, F. S., and Westwater, J. W., *Ann. Meeting Am. Inst. Chem. Engrs. Louisville, Ky.*, Preprint No. 10 (1955).
R1. Rayleigh, Lord, *Phil. Mag.* **34**, 94 (1917).
R2. Reiss, H., *Ind. Eng. Chem.* **44**, 1284 (1952).
R3. Reiss, H., *J. Chem. Phys.* **20**, 1216 (1952).
R4. Robinson, D. B., and Katz, D. L., *Chem. Eng. Progr.* **47**, 317 (1951).
R5. Rodebush, W. H., *Chem. Revs.* **44**, 269 (1949).
R6. Rodebush, W. H., *Ind. Eng. Chem.* **44**, 1289 (1952).
R7. Rodebush, W. H., *Proc. Natl. Acad. Sci. (U. S.)* **40**, 789 (1954).
R8. Rohsenow, W. M., *Trans. Am. Soc. Mech. Engrs.* **74**, 969 (1952).
R9. Rohsenow, W. M., "Heat Transfer and Fluid Mechanics Institute," p. 123. Stanford U. P., Stanford, 1953.
R10. Rohsenow, W. M., and Clark, J. A., "Heat Transfer and Fluid Mechanics Institute," p. 193. Stanford U. P., Stanford, 1951.
R11. Rohsenow, W. M., and Clark, J. A., *Trans. Am. Soc. Mech. Engrs.* **73**, 609 (1951).
R12. Rohsenow, W. M., and Griffith, P., *Ann. Meeting Am. Inst. Chem. Engrs. Louisville, Ky.*, Preprint No. 9 (1955).
S1. Sauer, E. T., M. S. Thesis in Chemical Engineering, Massachusetts Institute of Technology, Cambridge, 1937.
S2. Schweppe, J. L., and Foust, A. S., *Chem. Eng. Progr. Symposium Ser. No. 5*, **49**, 77 (1953).
S3. Simmonds, W. H. C., *Ann. Meeting Am. Inst. Chem. Engrs. Houston*, Paper No. 43 (1955).
S4. Stoebe, G. W., Baker, E. M., and Badger, W. L., *Ind. Eng. Chem.* **31**, 200 (1939).
T1. Takagi, S., *J. Appl. Phys.* **24**, 1453 (1953).
T2. Thomson, Sir W. (Lord Kelvin), *Phil. Mag.* **42**, 448 (1871).
V1. Vomer, M., "Kinetik der Phasenbildung." Steinkopff, Dresden and Leipzig, 1939. Also reprinted by Edwards, Ann Arbor, 1945.
W1. Westwater, J. W., *Sci. American*, **190**, No. 6, 64 (1954).
W2. Westwater, J. W., Lowery, A. J., and Pramuk, F. S., *Science* **122**, 332 (1955).
W3. Westwater, J. W., and Santangelo, J. G., *Ind. Eng. Chem.* **47**, 1605 (1955).
W4. Westwater, J. W., and Santangelo, J. G., "A Photographic Study of Boiling" (Motion Picture). University of Illinois, Urbana, 1954.
W5. Wismer, K. L., *J. Phys. Chem.* **26**, 301 (1922).
Y1. Yoshida, T., *Trans. Japan Soc. Mech. Engrs.* **16**, No. 54, 32 (1950).
Z1. Zwick, S. A., Hydrodynamics Lab. Report 21–19, California Institute of Technology, Pasadena, 1954.
Z2. Zwick, S. A., and Plesset, M. S., Hydrodynamics Lab. Report 26–7, California Institute of Technology, Pasadena, 1954.

Non-Newtonian Technology: Fluid Mechanics, Mixing, and Heat Transfer

A. B. METZNER

Department of Chemical Engineering
University of Delaware, Newark, Delaware

Author's Note	78
I. Fluid Classifications	79
A. Fundamental Basis	79
B. Newtonian Fluids	80
1. Definition and Representation	80
2. Examples and Causes of Newtonian Behavior	82
C. Non-Newtonian Fluids	82
1. Examples and Causes of Non-Newtonian Behavior	82
2. Classical Methods of Classifying Non-Newtonian Fluids	83
a. Time-Independent Non-Newtonians	83
b. Time-Dependent Non-Newtonians	87
D. Recent Developments in the Engineering Classification of Fluid Behavior	88
E. Summary	89
II. Mechanics of Flow in Round Pipes	90
A. Bingham Plastics	90
B. General Discussion Applicable to All Fluids	95
C. Velocity Profiles	107
D. Effect of Surface Roughness	108
E. Effect of Temperature and Concentration on Flow Properties	109
F. Summary	110
1. Recommended Design Procedure	110
a. Laminar Flow	110
b. Transition and Turbulent Regions	111
2. Physical Properties	111
3. Future Work	111
III. Miscellaneous Flow Problems	112
A. Fluid Kinetic Energies	112
B. Flow Through Fittings and Entrance Losses	113
C. Flow in Annular Spaces	115
D. Two-Phase Flow	115
E. Nonisothermal Flow	116
F. Extrusion	117
G. Summary	118
IV. Mixing of Non-Newtonian Fluids	119
V. Heat Transfer	121
A. Heat Transfer to Dilute Suspensions	121
1. Evaluation of Thermal Conductivities	122
2. Evaluation of Viscosities	124

3. Design Equations for Use with Dilute Suspensions................. 125
4. Comparative Economics of Suspensions and Liquids as Heat Transfer
   Media.. 129
5. Summary.. 130
   a. Recommended Design Procedure.............................. 130
   b. Future Work.. 131
B. Heat Transfer to Highly Non-Newtonian Systems.................... 131
   1. Laminar-Flow Region.. 131
   2. Turbulent-Flow Region.. 132
      a. Bingham-Plastic Fluids... 132
      b. Pseudoplastic Fluids.. 134
   3. Summary... 136
      a. Recommended Design Procedure............................. 136
      b. Future Work.. 137
VI. Rheology and Viscometry.. 138
   A. Determination of Shear-stress–Shear-rate Relationships.............. 138
      1. Capillary-Tube Viscometers....................................... 138
      2. Rotational Viscometers.. 139
   B. Determination of the Relationship between $D\Delta P/4L$ and $8V/D$........ 141
      1. Capillary-Tube Viscometers....................................... 141
      2. Rotational Viscometers.. 141
   C. Determination of the Absence of Thixotropy and Rheopexy.......... 142
      1. Capillary Tubes... 142
      2. Rotational Viscometers.. 143
   D. Viscometer Design.. 143
      1. Capillary-Tube Viscometers....................................... 143
      2. Rotational Viscometers.. 146
      3. Miscellaneous and Control Instruments........................ 147
   E. Summary... 148
   Acknowledgment... 148
   Nomenclature... 148
   References... 150

Author's Note

The presentation in this chapter dwells rather heavily on the classification, measurement, and interpretation of non-Newtonian behavior. These rheological fundamentals have frequently been presented in literature which is unfamiliar to the engineer and have usually included much discussion of factors which at the present time are of minor engineering interest. Accordingly, it was felt that one of the primary needs in this field was a concise summary of these fundamentals and common definitions. It is hoped that thereby future developments may be undertaken in an orderly and rigorous manner, as contrasted to the relatively fruitless empiricism which has enveloped areas of this field in the past.

A discussion of the flow of non-Newtonian materials under certain complex industrial conditions such as in calendering and coating machines was not felt to be warranted at this time in view of the general dearth of published information in these areas. Occasionally statements or recommended design procedures are based on extremely limited information and may, perhaps, be proven partially incorrect by subsequent studies. The limitations of such conclusions have usually been indicated or are obvious from the preceding text. This procedure, while admittedly open to criticism, has been followed in the hope that such approximate indications will be more useful than the complete absence of any conclusions until further data become available.

I. Fluid Classifications

A. Fundamental Basis

Fluids are classified into the two main categories of Newtonian or non-Newtonian according to their behavior at constant temperature under imposed shearing forces. This behavior may be illustrated by the experiment shown in Fig. 1. Two parallel planes of area A, separated

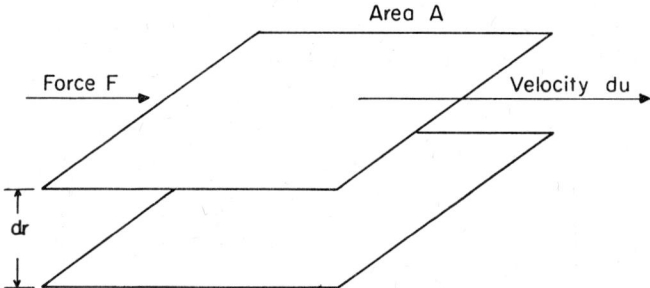

Fig. 1. Schematic illustration of response of fluids to an imposed shearing force.

by the differential distance dr, have the space between them filled with a fluid. Considering the lower plate fixed, a small force F applied to the upper plane will give it a velocity du in the direction of the force. Normally fluid velocities adjacent to a wall are zero; i.e., there is no slip between the wall and the fluid. Thus, the fluid adjacent to the upper wall will also have a velocity du in the direction of the applied force and the fluid next to the lower plane or wall will have a zero velocity in the same direction. In this manner a uniform velocity gradient of magnitude du/dr is set up in the fluid since the shearing force F is uniform across the distance dr. This is discussed, in more detail, for example, by Reiner (R4) and by Badger and McCabe (B1).

The velocity gradient du/dr is commonly referred to as the "shearing rate" or "rate of shear"; similarly, the shearing force per unit area (F/A) is called the "shearing stress" and is denoted by the symbol τ. Rheology is that branch of science which treats the relationships between an imposed shearing stress τ, the resultant shear rate du/dr, and any other variable that influences these relationships. As commonly used, it refers solely to the study of these problems in laminar flow.

It is obvious that the foregoing ideal system of two parallel planes must usually be modified to permit measurements in a practical manner. The various modifications available have led to an overwhelming confusion throughout the entire area of non-Newtonian technology. The need for a completely clear understanding of these experimental problems can accordingly not be overemphasized. They are, however, distinct from the theoretical understanding of non-Newtonian behavior and have accordingly been segregated in Section VI.

B. Newtonian Fluids

1. *Definition and Representation*

Newtonian fluids are those which exhibit a direct proportionality between shear stress and shear rate in the laminar-flow region. This is

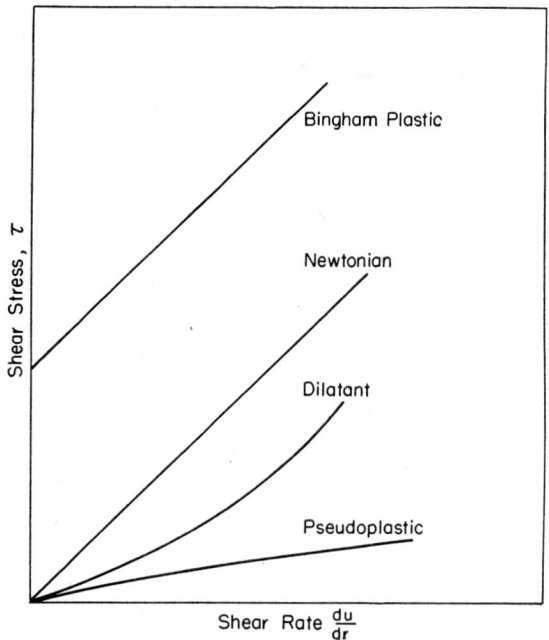

Fig. 2. Fluid-flow curves (arithmetic).

usually stated:

$$\tau = \frac{\mu}{g_c}(du/dr) \qquad (1)$$

where μ is the viscosity of the fluid. Graphically this relationship may be shown in either of two convenient manners.

  a. The classical method is to use an arithmetic plot of the flow curve, i.e., of the relationship between shear stress and shear rate. On arithmetic

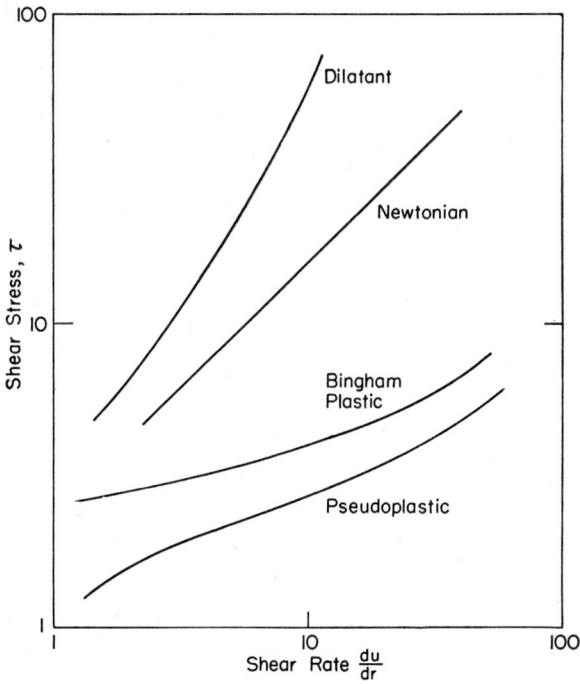

Fig. 3. Fluid-flow curves (logarithmic).

coordinates the Newtonian fluid is depicted by a straight line, through the origin, of slope μ/g_c. Such a diagram is shown in Fig. 2.

  b. A recent, and more useful, engineering approach is to use logarithmic coordinates as shown in Fig. 3. In this case all Newtonians have the same slope (unity) while μ/g_c is given by the intercept at a shear rate of unity. While this method of showing the flow curve has the usual advantages of logarithmic coordinates (a large range of both variables may be shown with the same percentage precision over the entire range), it is used primarily because the fluid properties can be most readily compared with those of non-Newtonians in this manner. The most important single factor to recognize at this point is that the flow curves of Newtonian fluids are completely characterized by a single physical property, viz., the viscosity μ. Accordingly one single experimental determination is sufficient to define completely the rheological or laminar-flow properties of the fluid. If a greater number of measurements are made at the same temperature (and pressure), they serve solely to confirm the experimental accuracy of the first one.

2. *Examples and Causes of Newtonian Behavior*

Newtonian behavior has been found to be common to the following systems: (*a*) all gases and (*b*) all liquids or solutions of low molecular weight (i.e., nonpolymeric) materials.

The common denominator of these Newtonian fluids is that the dissipation of viscous energy in them is due to the collision of reasonably small molecular species.

C. NON-NEWTONIAN FLUIDS

By definition the term "non-Newtonian" encompasses all materials which do not obey the direct proportionality between shear stress and shear rate depicted by Eq. (1).

1. *Examples and Causes of Non-Newtonian Behavior*

Non-Newtonian behavior is commonly observed in the following kinds of systems.

a. First, and most important, solutions or melts of high-molecular-weight polymeric materials are invariably non-Newtonian except when unusually dilute.

b. Suspensions of solids in liquids become increasingly non-Newtonian with increasing solids concentration and are particularly non-Newtonian if the solid tends to swell, solvate, or otherwise associate with the liquid phase.

The distinguishing feature of non-Newtonian systems is seen to be that the colloidal rather than the molecular properties are of significance. Philippoff (P4) has summarized the properties of the colloidal particles which are relevant in determining their rheological behavior as follows:

(1) Particle shape.
(2) Particle size.
(3) Particle flexibility and ease of deformation.
(4) The solvation of the particles by the continuous phase.
(5) The presence and magnitude of electrical charges on the particles.

Non-Newtonian behavior is most often pronounced at intermediate shear rates. At extraordinarily high or low shear rates many non-Newtonians approach Newtonian behavior.

2. *Classical Method of Classifying Non-Newtonian Fluids*

The completely general case is that of a fluid for which the relationship between shear stress and rate of shear is not linear and may also depend on both the duration of the shear and the extent of the deformation produced. Thus it is possible to divide non-Newtonian systems into three broad categories:

a. Fluids with properties independent of time or duration of shear.

b. More complex fluids for which the relationship between shear stress and shear rate depends upon the duration of shear.

c. Those systems which have many characteristics of a solid and primarily that of elastic recovery from the deformations which occur upon flow. This elastic recovery may be only a partial recovery if the deformations have been large.

Virtually all the engineering work to date has dealt with the first, and simplest, of these three subdivisions. In fact, publications of engineering interest concerning the last group of materials have been able to progress little beyond the point of drawing attention to the occurrence of such pseudosolid properties. The necessary limitation on the present compilation is therefore to exclude from consideration systems which possess elasticity. It would be expected that this limitation is of no consequence in situations in which the fluid is sheared continuously and extensively in a single direction, as in these instances any elastic properties would be manifested primarily as "end effects" in those regions where the shear begins and ends. For example, it has been shown (A1, M11, R10) that jellies such as Napalm—which exhibit appreciable elastic recovery—can be treated as though the elastic properties were negligible insofar as flow through a pipe of constant cross section is concerned. The influence of elastic recovery or similar properties on flow through fittings, across tubes, and in similar situations in which the end effects may be of primary importance has not yet been investigated.

a. Time-Independent Non-Newtonians. The following three materials fall into this category.

(1) BINGHAM PLASTIC OR PLASTIC FLUIDS. As shown in Fig. 2, this is the simplest of all non-Newtonian fluids in the sense that the relationship between shear stress and shear rate differs from that of a Newtonian fluid only by the fact that the linear relationship does not pass through the origin. Thus a finite shearing stress τ_y is necessary to initiate move-

ment. The equation of the flow curve, for stresses above τ_y, is

$$\tau - \tau_y = \frac{\eta}{g_c} (du/dr) \qquad (2)$$

where η, the "coefficient of rigidity" or "plastic viscosity," represents the slope of the flow curve. On logarithmic coordinates (Fig. 3) the flow curve becomes asymptotic to τ_y at low shear rates and approaches a slope of unity at very high shear rates.

The usual explanations of Bingham-plastic behavior assume that the fluid at rest contains a three-dimensional structure, of rigidity sufficiently great to resist the finite stress τ_y. When this stress is exceeded, the structure breaks down completely and the fluid displays the linear relationship between shear stress and shear rate common to Newtonian materials (H2). It has been suggested, for example, that the structure of a Bingham-plastic fluid at rest may be compared to a structure of building blocks: the structure is able to withstand a finite force, but once this force is exceeded the entire structure crumbles. One would expect from this analogy that slurries of nearly equidimensional particles in a liquid would be the most likely to exhibit Bingham-plastic behavior and that the yield value τ_y would increase as the tendency of the particles to adhere to one another increases. This is actually the case and the latter phenomenon has been vividly demonstrated (K2). Examples of fluids which have been stated to approximate Bingham-plastic behavior are drilling muds, suspensions of chalk, grains, rock, and sewage sludge (M10).

(2) PSEUDOPLASTIC FLUIDS. This most important classification is the one into which the majority of non-Newtonian fluids fall. It displays, on arithmetic coordinates, the concave-downward flow-curve relationship shown in Fig. 2. On logarithmic coordinates (Fig. 3) these materials exhibit flow curves having slopes between zero and unity. Sometimes the curve shows a point of inflection and the slope approaches a value of unity at extremely low as well as at very high shear rates (A3). Usually the flow curve is a straight line on logarithmic coordinates over rather wide (10 to 100-fold) ranges of shear rates (M11), sometimes having slopes of appreciably less than 0.10 (A1, M11). For any such straight-line region the flow curve may be defined by the equation

$$\tau = K(du/dr)^n \qquad (3)$$

where n has values of less than unity.

The viscosity of Newtonian fluids was defined as the ratio of shear stress to rate of shear $\tau/du/dr$ (the conversion factor g_c being temporarily omitted from discussion). For Bingham-plastic materials the corresponding viscosity is

$$\mu_a = \frac{\tau_y}{(du/dr)} + \eta \qquad (4)$$

For many pseudoplastic non-Newtonians,

$$\mu_a = \frac{K(du/dr)^n}{du/dr} = \frac{K}{(du/dr)^{1-n}} \qquad (5)$$

It is seen that for both these materials μ_a decreases as shear rate increases. The same conclusion may be reached qualitatively from consideration of Figs. 2 and 3, for doubling the shear rate does not double the shear stress, as is true for Newtonian fluids. For this reason μ_a is termed an "apparent viscosity"—the viscosity of a non-Newtonian fluid at a given shear rate. It is imperative to note that the term *viscosity* is meaningless for non-Newtonians unless the shear rate or shear stress to which this viscosity refers is stipulated.

The term *differential viscosity* is also in common use; it denotes the slope of the arithmetic shear-stress–shear-rate relationship at any given value of the shear rate; that is

$$\mu_d = \frac{d\tau}{d(du/dr)}$$

as compared with

$$\mu_a = \frac{\tau}{(du/dr)}$$

In view of the fact that the apparent viscosity gradually decreases with increased values of shear rate for pseudoplastic fluids, the physical explanation of pseudoplastic behavior is readily visualized; i.e., the intermolecular or interparticle interactions smoothly decrease with increasing rates of shear. This would be the case, for example, if the particles or molecules, initially randomly interspersed, were to align during shear so that their interactions were minimized. Such behavior could occur most readily with highly asymmetric particles, which, when the fluid is at rest, are randomly entangled in one another. Under shear the particles would tend to align themselves so that their major axes were parallel to the direction of motion. Since the random oscillations (Brownian motion) are relatively small for large molecules or suspended particles, the viscous dissipation of energy between such aligned particles is small. Each layer of aligned particles effectively moves parallel to the other with little interchange of the particles between adjacent layers. Since the disrupting influence of Brownian motion is constant at a fixed temperature, while the aligning forces of shear increase with increasing shear rate, the molecules become progressively more perfectly aligned at higher shear rates. The apparent viscosity therefore continues to decrease until extremely high shear rates are reached and no further increases in perfection of the alignment are possible. This explanation clearly shows why many pseudoplastic materials approach Newtonian behavior at high shear rates. Similarly at extremely low shear rates the balance between alignment and randomness is very strongly in the direction of the latter. Therefore, this balance again is not appreciably affected by small changes in shear rate; hence Newtonian behavior may be approached.

Undoubtedly other factors may promote pseudoplastic behavior. For example, a decrease in size of swollen, solvated, or even entangled molecules or particles may occur under the influence of shearing stresses, to give the same changes in fluid behavior as in particle alignment.

Examples of fluids which exhibit pseudoplastic behavior include most non-Newtonian fluid types: polymeric solutions or melts, as rubbers (B2, F2, M7), cellulose acetate (A3), and Napalm (A1); suspensions such as paints, mayonnaise, paper pulp, and detergent slurries (M10); and even dilute suspensions of inert, unsolvated solids (M13). Increasing pseudoplasticity of polymeric solutions may result from increases in concentration of the dissolved polymer as well as from increases in molecular weight of the polymer. As a result it is difficult to define the minimum molecular weight required to cause appreciable non-Newtonian behavior. Present indications, however, are that linear chains only a dozen atoms in length may be definitely non-Newtonian if they are appreciably solvated by the solvent used. It has been suggested that even aqueous glucose solutions may be slightly non-Newtonian at high concentrations (M6), although the glucose chain is only six carbon atoms long. On the other hand it should be noted that both high concentrations and high molecular weight are necessary to produce non-Newtonian behavior if the polymer does not associate strongly with the solvent.

Since the shear-stress–shear-rate properties of pseudoplastic materials are defined as independent of time of shear (at constant temperature), the alignment or decrease in particle size occurring when the shear rate is increased must be instantaneous. However, perfect instantaneousness is not always likely if the foregoing causes of pseudoplastic behavior are correct, as they are believed to be. Pseudoplastic fluids are therefore sometimes considered to be those materials for which the time dependency of properties is very small and may be neglected in most applications.

(3) DILATANT FLUIDS. Dilatant fluids display a rheological behavior opposite to that of pseudoplastics (Figs. 2 and 3) in that the apparent

viscosity *increases* with increasing shear rate. The best explanation for dilatant behavior (R4, S7) is still the original explanation given in 1888 by Osborne Reynolds. He assumed that these fluids, when at rest, consist of densely packed particles, in which the voids are small and perhaps at a minimum. Sufficient liquid is present only to fill these voids. The motion or shear of such fluids at low rates requires only small shearing stresses since the liquid lubricates the passage of one particle past another. At increasing rates of shear the dense packing of the solid particles is progressively broken up, and, since the packing initially involved a minimum of voids, there is now insufficient liquid present to enable the particles to flow smoothly past one another. Therefore the shear stress increases more than proportionately with shear rate.

The preceding explanation suggests that all suspensions of solids in liquids should exhibit dilatant behavior at high solids contents. Few data are available for evaluation of this conclusion, as the usual examples of dilatant behavior (starch, potassium silicate, and gum arabic in water) (A3, G3) are not true suspensions. The excellent studies of Daniel (D1) and Verway and De Boer (V3) have indicated under what conditions more dilute suspensions may also exhibit dilatancy. Some of these factors have been summarized by Pryce-Jones (P6). If Reynolds' explanation is a valid one, it should be possible to measure the expansion or dilation of the fluid with increases in shear rate. This has been done indirectly: Andrade and Fox (A5) measured the dilation of sand suspensions and arrays of cylinders upon the imposition of localized stresses.

b. Time-Dependent Non-Newtonians. This designation is meant to apply to those materials for which the shear stress, at a given rate of shear and temperature, is not a constant, as for the fluids discussed so far, but changes with the duration of the shear. Irreversible changes, i.e., those which might be due to actual destruction of the particles or molecules, are however excluded.

These time-of-shear–dependent non-Newtonians may be divided into two groups, depending on whether the shear stress increases or decreases with time of shear at a constant shearing rate. The former are termed *rheopectic* the latter *thixotropic* fluids (P3).

The causes for thixotropic and rheopectic behavior are possibly very similar to those for pseudoplasticity and dilatancy, respectively. The proposed causes of pseudoplasticity, i.e., the alignment of asymmetrical molecules and particles or the breakdown of solvated masses, could not always be expected to be instantaneous with respect to time. Therefore it seems that pseudoplastic behavior may simply be that form of thixotropy which has too small a time element to be measurable on most instruments in current use. Exactly the same argument may be applied to the relationship between rheopexy and dilatancy. Rheopexy has been observed in certain sols and in bentonitic clay suspensions (A3). Thixotropy is common to paints,[1] to ketchup and other foods, and to some polymeric solutions (A3). Thixotropy is thus of much greater practical importance than rheopectic behavior.

While rheological literature abounds with examples of thixotropic behavior, several leading authorities (L4, M12, W3) have shown that many of these "examples" are due to errors in experimental technique and not to the actual presence of thixotropy. In fact, the magnitude of the possible errors in most thixotropy studies to date as indicated by these references throws considerable doubt on the validity of most available quantitative conclusions concerning this phenomenon.

In conclusion, it should be noted that few of the comments in this section are original. These classifications have been presented by so many authors that any attempt to give detailed references would have been clumsy. Some of the best references, to which the reader is referred for further details, include the books by Reiner (R4), Philippoff (P4), Green (G3), and Scott-Blair (S7). Since all systems of classification are, to an extent, arbitrary, many others have been proposed. Reiner (R4) discusses several of these but none appear to possess obvious advantages over the more common system of classification discussed herein.

D. Recent Developments in the Engineering Classification of Fluid Behavior

It is of importance for the engineer to be familiar with the preceding classifications in order to understand the relative frequency of occurrence of various types of rheological behavior and to obtain a general understanding of the physical basis for these phenomena. On occasion, this understanding may result in far greater economic improvements than detailed design possibly can, since the possibility of modifying a fluid so that it exhibits a simpler or less viscous mode of behavior may open the door to a host of processing advantages. Unfortunately, too little is known of the complex factors responsible for changes in the flow behavior to enable such important changes in properties at will. For example, even the relatively simple problem of behavior of solid particles suspended in a liquid is still rather confused. In a recent issue of the *Journal of Colloid*

---

[1] Thixotropic behavior is responsible for the smooth flow of paints after the intense shear of a paint brush or spray. On standing without shear for a few moments, the fluid "recovers" its original viscous nature. The former behavior is desirable to permit the smooth flow necessary for ease of application and for removal of brush marks. The high consistency at rest is needed to prevent flow after application to vertical surfaces (L2).

Science the authors of one paper (O4) concluded that particle size is of no relevance in determining the flow behavior of suspensions, although the particle-size distribution is. The authors of the very next paper (Z1), on the other hand, documented the opposite conclusion. Until the reasons for such completely different types of behavior in apparently similar systems are understood, the role of the engineer is necessarily primarily one of evolving processes for fluids of predetermined behavior rather than one of questioning in detail how such behavior comes about or how it may be modified.

For engineering design work the classical rheological definitions are sometimes clumsy and inadequate. It has been pointed out (M10) that minor changes in temperature or composition of the fluid may completely alter its type of behavior. For example, some non-Newtonian fluids appear to exhibit Bingham-plastic behavior at temperatures up to 150°F., pseudoplastic behavior at slightly higher temperatures, and above about 210°F. they are substantially Newtonian (M9). Were the engineering research approach one of developing suitable design procedures for each of these three types of flow behavior, an individual faced with the simple problem of heating the fluid from 140° to 220°F. might have to employ a different design technique for each of the three different types of behavior encountered!

Accordingly pragmatic individuals have long been interested in the evolution of a universal explanation for fluid behavior of all kinds and the establishment of quantitative relationships with which to correlate and extrapolate the flow curves of all fluids, Newtonian and non-Newtonian alike. Progress toward the former of these two objectives appears to be too limited to warrant presentation at this stage, but recently a method has been proposed (F2, M11, W4) by means of which the flow behavior of all fluids which are not time dependent may be compared. Two indexes are necessary to accomplish this:

(1) The *flow-behavior index* n (or n'), which characterizes the extent of the deviation of a fluid from Newtonian behavior.

(2) The *consistency index* K (or K'), which characterizes the consistency or "thickness" of a fluid. It is analogous to the viscosity of a Newtonian fluid and similarly enables quantitative comparison of the consistency of fluids having identical flow-behavior indexes.

The precise definitions and utility of these indexes are discussed in Section IIB.

E. SUMMARY

Non-Newtonians of the pseudoplastic type are second in importance only to Newtonian fluids, while Bingham-plastic, dilatant, thixotropic, and rheopectic behavior are all of much less frequent occurrence. Further work is needed to help define the relationships, if any, between thixotropic and pseudoplastic behavior as well as between dilatancy and rheopexy. It would be desirable to be able eventually to relate fluid behavior of all types to the microstructure of the fluid and to use methods of approach which emphasize this factor theoretically. Unfortunately, work to date has not led to the development of such universal approaches, although Reiner's (R4) complex equations involving a "coefficient of structural stability" may be a fruitful start in this direction.

Until much more progress is made on these very fundamental approaches, greater accuracy is believed possible in engineering work that is based on the better developed discussion reviewed here. To this end a method has recently been proposed whereby the properties of all four time-independent fluids (Newtonian, pseudoplastic, dilatant, and Bingham plastic) may be quantitatively compared.

II. Mechanics of Flow in Round Pipes

A. Bingham Plastics

Since Bingham plastics represent the simplest non-Newtonian behavior it is logical that these fluids should have been the subject of the first and some of the most extensive studies. However, few, if any, actual materials appear to behave as true Bingham plastics. It has been pointed out (O6, R2) that the assumed linear relationship between shear stress and shear rate (Fig. 2) almost always breaks down if rheological measurements are made over a sufficiently wide range of shearing rates. Many real fluids, nevertheless, approach this behavior to a degree sufficient for most engineering applications. Therefore, for simple geometries in which the shear rates may be estimated, the assumption of Bingham-plastic behavior can frequently be justified. The most important situation for which this may be done is flow in round pipes. A review of this field is therefore warranted because the understanding of the behavior of materials which may be assumed to be Bingham plastics is considerably superior to the understanding of any other kind of non-Newtonian behavior.

Buckingham (B10) integrated the rheological equation [Eq. (2)] for the isothermal flow of Bingham plastics in round pipes. The familiar and important resulting equation:

$$\frac{8V}{D} = \frac{\tau_w g_c}{\eta} \left[1 - \frac{4}{3}\frac{\tau_y}{\tau_w} + \frac{1}{3}\left(\frac{\tau_y}{\tau_w}\right)^4 \right] \qquad (6)$$

enables one to calculate the relationship between pressure drop and flow rate provided only that the two fluid properties, τ_y and η, are known. It is not possible, however, to solve the equation for τ_w (or ΔP) directly.

Except for the experimental verification of the preceding equation, no significant development occurred until 1939–41, when Caldwell and Babbitt (C1) attempted to study the fluid mechanics of Bingham plastics in the turbulent region. Their publication contained many useful and fundamental ideas concerning the onset of turbulence, but the understanding of this and the allied problem of flow in the turbulent region was very diffuse until the work of Govier and Winning (G1) and Hedstrom (H4) became available. Most of the significant discoveries of Govier and Winning have been condensed in a recent NACA publication (W4). The understanding of non-Newtonian flow problems is significantly aided by these developments, which are accordingly reviewed in the following paragraphs.

In general, problems of turbulent flow have been too difficult to study mathematically, at least in their initial stages. If, therefore, one is to develop a relationship between pressure drop and flow rate which is valid for the turbulent- as well as the laminar-flow region, the approach must be empirical—and that of dimensional analysis would seem to be eminently suited to the problem at hand.

For Newtonian fluids, one may say

$$\Delta P = \theta'(D,L,V,g_c,\rho,\mu) \tag{7}$$

Similarly, for Bingham plastics,

$$\Delta P = \theta''(D,L,V,g_c,\rho,\eta,\tau_y) \tag{8}$$

where θ' and θ'' denote unknown functional relationships.

That is to say, the same variables are relevant in both problems with the exception of the rheological properties of the fluid. For Newtonian fluids, the viscosity μ defines these adequately; for Bingham plastics, the two parameters τ_y and η are required.

Application of the usual procedures of dimensionless analysis to Eq. (8) gives

$$\frac{D\Delta P g_c}{V^2 L \rho} = \theta'' \left[\frac{DV\rho}{\eta}, \frac{\tau_y D^2 \rho g_c}{\eta^2} \right] \tag{9}$$

For Newtonian fluids, the Fanning friction factor f is usually defined (M4):

$$\frac{D\Delta P g_c}{V^2 L \rho} = 2f \tag{10}$$

Arbitrarily using the same definition with Bingham plastics gives

$$2f = \theta'' \left[\frac{DV\rho}{\eta}, \frac{\tau_y D^2 \rho g_c}{\eta^2} \right] \tag{11}$$

Equation (11) states that the conventional Fanning friction factor, which may be used through Eq. (10) to calculate pipe-line pressure drops, is a unique function of two dimensionless groups for Bingham-plastic fluids. Newtonian fluids represent that special case for which τ_y, and hence the second dimensionless group, is equal to zero.

The actual choice of the form of dimensionless groups to be used for interpretation and correlation of data is somewhat arbitrary; those shown in Eq. (11) are believed to be especially suitable in view of the fact that the second is a constant in a given system. (It does not vary with flow rate.) In honor of the man who first proposed its use, it has been termed the Hedstrom number, N_{He}. The foregoing dimensional-analysis approach was apparently first developed by Govier and Winning, but their choice of dimensionless groups was slightly less convenient for design purposes.

The assumptions upon which these derivations were based were sufficiently general to assure the utility of these dimensionless groups in both laminar- and turbulent-flow problems. For the former case the results must be the same as are obtainable from the Buckingham relationship, Eq. (6). Perkins and Glick (P2) accordingly rearranged Eq. (6) in terms of these dimensionless groups to obtain

$$\frac{1}{N_{Re}} = \frac{f}{16} - \frac{1}{6} \frac{N_{He}}{(N_{Re})^2} + \frac{1}{3} \frac{N_{He}^4}{f^3 (N_{Re})^8} \tag{12}$$

It is thus possible to calculate theoretically the relationship between the friction factor and Reynolds and Hedstrom numbers in the laminar-flow region. Beyond the laminar-flow region the relationships between these three dimensionless groups must be determined experimentally.

At extremely high Reynolds numbers or, more correctly, at very high shear rates, the importance of fluid properties which are most significant at low shear rates must progressively decrease. The only variable in these three groups which has primary importance only at low shear rates is the yield value τ_y. Therefore the importance of the dimensionless group containing this parameter (N_{He}) must also decrease in importance at progressively higher Reynolds numbers in the turbulent-flow region.

These factors were clearly worked out by Winning and Govier as well as by Hedstrom; the resulting relationships are shown in Fig. 4. A major factor to note concerning this important accomplishment is that the transition from laminar to turbulent flow occurs at progressively higher Reynolds numbers as the Hedstrom number increases. Dashed lines have been shown in this region to indicate that the exact behavior is in doubt, although, as discussed in Section IIB, the bulk of the available evidence appears to favor the suggestion that the transition region begins at $f = 0.008$ and follows (approximately) the dashed lines. Correlation of the extensive data of Govier and Winning on this type of plot generally supports this conclusion. The data also indicate that the transition does not occur suddenly at the point of intersection of the lines for various values of Hedstrom number with the turbulent flow curve, as proposed by Hedstrom.

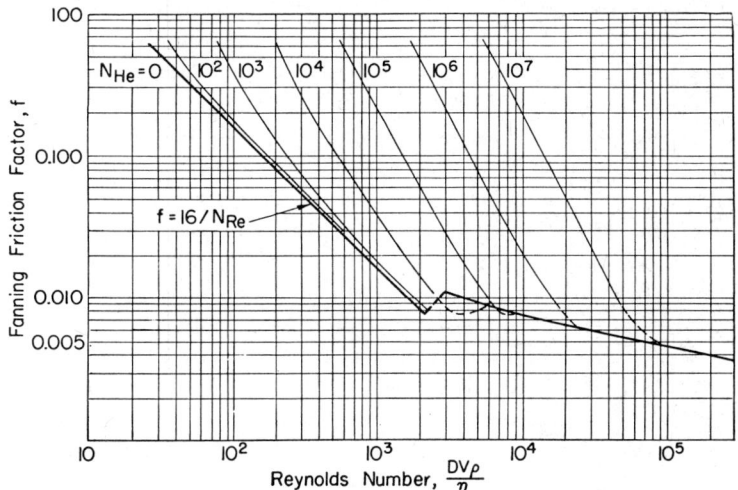

FIG. 4. Friction factor–Reynolds number diagram for Bingham plastics.

Reprinted with permission from Ind. Eng. Chem. **44**, 651 (1952).

In highly developed turbulent flow Govier and Winning (G1) found experimental data to fall within $\pm 10\%$ of the conventional Newtonian curve. This means that the effect of the Hedstrom number may be neglected beyond the end of the transition region. Their data were for five clay slurries in two pipe diameters (¾ and 1¼ in. I.P.S.) for Reynolds numbers to 200,000 and Hedstrom numbers to 560,000.

Since it is not likely that the viscous slurries which exhibit Bingham plastic behavior will frequently reach Reynolds numbers appreciably greater than 200,000 it is possible to conclude that Fig. 14 may be used to predict pressure drops accurately under all conditions of interest except in the transition regions. If a problem happens to fall into what may appear to be a transition region, use of Fig. 7 is recommended instead of Fig. 4.

The accuracy of pressure-drop predictions based on Fig. 4 depends primarily on the approach of real fluids to true Bingham-plastic behavior. For this reason it is imperative that any extrapolations of the rheological data from which the yield value τ_y and coefficient of rigidity η were calculated be very conservative. Within the range of wall shearing rates for which the given values of τ_y and η accurately represent the rheological data, however, predictions of excellent accuracy are possible. It is possible to check the wall shearing rate in a given problem by the methods discussed in Section IIB [Eq. (18)] in order to be certain that the rheological constants (τ_y and η) apply to the same region. Alternately, one may calculate the pressure drop with the assumed rheological data and then determine the shearing stress at the wall ($\tau_w = D\Delta P/4L$) to check whether the rheological data were in the region of this shearing stress. Caldwell and Babbitt (C1) present a good discussion of methods of obtaining τ_y and η directly from pipe-line or capillary-tube data; alternately one may calculate shearing stresses and shearing rates by the methods of Section VI for capillary-tube or pipe-line data as well as for rotational viscometric data. These shearing stresses may be plotted versus shearing rate to give the actual flow curve (Fig. 2) of the fluid. This latter procedure in effect differentiates the experimental data, thereby enabling one to see clearly the range of shearing stresses over which the linear Bingham-plastic relationship is valid. It is accordingly superior to the Reiner-Riwlin integrated equations (G3, R5, W4), which necessarily tend to obscure trends or deviations from the assumed behavior.

The importance of checking the identity of the shear-stress range in which τ_y and η were determined with that in which they are used is clearly illustrated by Weltmann (W4), who found errors of over 700-fold (i.e., over 70,000%) to occur in the calculated pressure drops when this was not done. The shearing stress τ on a cylindrical fluid element of radius r may readily be shown to be equal to $r\Delta P/2L$ [cf. reference (C1)], where $\Delta P/L$ is the pressure drop per unit length of pipe at the flow rate in question. At the wall of the pipe this reaches the maximum value, $\tau_w = D\Delta P/4L$. At the center line of the pipe it is zero. Since shear stresses vary from zero to τ_w in every pipe flow problem, it is theoretically necessary that the linear Bingham-plastic relationship be valid over this entire range. From a practical viewpoint its validity in the region near the wall is far more important than near the center line, but in most instances it will be found necessary that the linear Bingham-plastic relationship be closely obeyed over at least the upper half of the shear-stress range involved.

Other papers discussing the flow of Bingham plastics (A3, B5) suggest determination of the "effective viscosity" in turbulent flow by means of pressure-drop measurements in turbulent flow. This method is insensitive to viscosity, however, and since the effective viscosity of the plastics must be identical to the coefficient of rigidity, this method is of limited utility. The same would appear to be true of the dimensionless rearrangements of the Buckingham relation [Eq. (6)] made by McMillen (M5) for the laminar-flow region. McMillen's tentative criterion for the end of the laminar-flow region differs markedly from that discussed above and, in view of his limited supporting data, may be incorrect. Wilhelm and co-workers (W6) and Filatov (F1) achieved partial correlation of flow data by means of empirical procedures. Ooyama and Ito (O2) presented a complex form of the Reynolds number which is claimed to be applicable to Bingham plastics but which requires a trial-and-error calculation of the friction factor. In addition their criterion for the end of the laminar-flow region does not reduce to the common value for Newtonian fluids as the yield value approaches zero. It may be concluded that several of the papers reviewed in this paragraph had utility at the time of their publication, but none of these are believed to be so convenient or rigorous as is the procedure summarized by Fig. 4.

B. GENERAL DISCUSSION APPLICABLE TO ALL FLUIDS

The great practical importance of pseudoplastic behavior makes an understanding of these fluids of paramount importance. Their complexity is such, however, that an understanding of their behavior is frequently applicable to Newtonian, dilatant, and Bingham-plastic materials as well.

Only in recent years has progress toward an engineering understanding of pseudoplastic behavior been substantial. Most developments in this field stem from relationships developed by Rabinowitsch (R1) and exploited fully by Mooney (M15) in a classic paper that all engineers interested in non-Newtonian behavior must probably place at the top of their reading list.

Mooney clearly showed that the relationship between the shear stress at the wall of a pipe or tube, $D\Delta P/4L$, and the term $8V/D$ is independent of the diameter of the tube in laminar flow. This statement is rigorously true for any kind of flow behavior in which the shearing rate is only a function of the applied shearing stress.[2] This relationship between $D\Delta P/4L$ and $8V/D$ may be conveniently determined in a capillary-tube viscometer, for example. Once this has been done over the range of

[2] It is true, therefore, for Newtonian, Bingham-plastic, pseudoplastic and dilatant fluids. The same relationship can possibly be extended to thixotropic and rheopectic fluids by evaluating the shear stress at the wall over a differential length of tube, i.e., by replacing $D\Delta P/4L$ with $DdP/4dL$. This term will vary with distance along the pipe, however, and as no evident means of developing this relationship has been

$8V/D$ of interest, it may be used for the calculation of the relationship between pressure drop and flow rate in a pipe line of any size, provided only that the flow is laminar and that the laboratory data are at the correct temperature.

Alves, Boucher, and Pigford (A3) made use of the unique relationship between $D\Delta P/4L$ and $8V/D$ in correlating data on a wide variety of non-Newtonian suspensions and solutions. Since this method is essentially one of scaling up to the new conditions under consideration, it is very simple, direct, and accurate. Metzner (M10) pointed out that this scale up must obviously be at a constant value of V/D, not at a constant value of V, as has frequently been recommended. Furthermore, for all the fluids under consideration here a single experimental determination in a model of convenient size theoretically enables the prediction of the pressure drop in any larger pipe at the same value of V/D. Whenever measurements in several tubes of different diameters are possible, they should be carried out, however, to ensure the absence of complicating effects such as thixotropy. Schultz-Grunow (S6) recently described a method for making the plots of $D\Delta P/4L$ versus $8V/D$ dimensionless. Although this procedure proves the validity of the assumptions upon which it is based, it does not aid calculational procedures and hence is not generally recommended.

If only rotational viscometric data are available, the design of pipe lines is somewhat more complex. Such data are usually expressed in the form of a relationship between shear stress and shear rate. The shear stress on a cylindrical element of fluid of radius r flowing through a pipe in laminar motion is equal to $r\Delta P/2L$. If the corresponding shear rate $(-du/dr)$ can be expressed analytically, i.e., if the functional relationship

$$\frac{r\Delta P}{2L} = \varphi\left(\frac{-du}{dr}\right) \tag{13}$$

is known, Eq. (13) may be integrated to relate the point velocity u to radial position in the pipe. The total volumetric flow rate in a pipe may then be determined by a second integration:

$$V\frac{\pi D^2}{4} = \int_0^{D/2} 2\pi r u\, dr \tag{14}$$

---

published these latter two types of flow behavior must be excluded from discussion in this section.

Some fluids, as discussed in Section VI, may appear to "slip" at the wall. This factor may be accounted for quantitatively by methods given by Mooney (M15), but corrections for this factor are necessary before the proposed correlating methods may be applied.

If the relationship in Eq. (13) cannot be expressed analytically, the necessary double integration may be carried out graphically. Fortunately, this is usually not necessary, as several analytical relationships between shear stress and shear rate have been proposed as generally useful for these purposes:

1. The empirical power function

$$\tau = K(-du/dr)^n \tag{3}$$

probably represents the relationship between shearing stress and shear rate more closely for a very wide variety of non-Newtonian fluids than any other equation with only two constants. Scott-Blair (S7) has pointed out that this relationship was used as early as 1926, but from an engineering viewpoint the recent studies of Rankin (R2), Salt (S2), and Weltmann (W4) may be of greater interest. The resulting integrated equation (R2) may be expressed as follows:

$$\Delta P = 4K \left(\frac{6n+2}{n}\right)^n \frac{V^n L}{D^{n+1}} \tag{15}$$

Weltmann (W4) presented this relationship on a friction factor–Reynolds number diagram similar to Fig. 4 for Bingham-plastic fluids. Excellent agreement between predicted and measured results was found by Salt for two carboxymethylcellulose solutions; Weltmann shows no data to support her somewhat more useful rearrangement but cites three literature references for this purpose. Review of these shows that none dealt explicitly with this method of approach, as claimed.

Rheologists (e.g., Reiner (R4)) have long objected to the use of Eq. (3) on the basis that it is purely empirical, not being derived from any physical concepts. This criticism can hardly be considered valid in view of the fact that the "physical concepts" upon which theoretical equations are frequently based consist of purely mechanical analogs such as springs, dashpots, and blocks, which at best have only vague equivalents in any real system. It must be noted, however, that Reiner's objections that this equation does not correctly portray the behavior of many real fluids at extremely low and high shearing rates is certainly valid. Fortunately the error introduced by these deficiencies is not serious, at least for the problem of flow in round pipes. The region of low shear rates at the center of a pipe contributes only slightly to the over-all flow, and the behavior under very high shear rates may be correctly represented by fitting the equation to the experimental data in this region. The objections to use of Eqs. (3) and (15) may therefore be considered to be of minor engineering interest at the present time.

2. Investigators at the University of Utah (R10, S3, S10) have used the Eyring-Powell equation

$$\tau = \mu \frac{du}{dr} + \frac{1}{B}\sinh^{-1}\left(\frac{du/dr}{C}\right) \quad (16)$$

which contains the three constants μ, B, and C. In a particularly extensive and interesting thesis Stevens (S10) has integrated this equation numerically and presented the resulting relationship between pressure drop and flow rate upon generalized charts with dimensionless parameters. These charts make Eq. (16) almost as easy to use as Eq. (15). Good agreement between predicted and calculated pressure drops was obtained for a great variety of non-Newtonian fluids. However, it has been stated (M3) that Eq. (16) usually does not fit the rheological shear-stress–shear-rate data as well as Eq. (3). This factor would rule against the use of Eq. (16) in instances where extrapolations of the data were required.

The foregoing two equations probably represent the most useful approaches of this type, although many other equations have been proposed. None of these have been shown to be directly applicable to flow outside the laminar region or to be useful for predicting the onset of turbulence and the end of stable laminar flow.

A somewhat more general approach which has been shown to be at least partially useful for the latter two problems as well as for correlation of laminar-flow data is that of Metzner and Reed (M11). Their work is based upon an equation developed by Rabinowitsch (R1) and Mooney (M15) for the calculation of shear rates at the wall of a tube or pipe:

$$(-du/dr)_w = \frac{3}{4}\left(\frac{8V}{D}\right) + \frac{1}{4}\frac{D\Delta P}{4L}\frac{d(8V/D)}{d(D\Delta P/4L)} \quad (17)$$

This equation was rearranged to give

$$(-du/dr)_w = \frac{3n'+1}{4n'}\frac{8V}{D} \quad (18)$$

where

$$n' = \frac{d\left(\ln\frac{D\Delta P}{4L}\right)}{d(\ln 8V/D)} \quad (19)$$

Since the relationship between $8V/D$ and $D\Delta P/4L$ is independent of pipe diameter, the same is true of Eqs. (17) to (19) inclusive. They are applicable to all four types of common flow behavior, i.e., to pseudoplastic, Newtonian, Bingham-plastic, and dilatant fluids.

The derivative of Eq. (19) represents the slope of a line; hence it is rigorously permissible to write

$$\frac{D\Delta P}{4L} = \tau_w = K' \left(\frac{8V}{D}\right)^{n'} \tag{20}$$

If the derivative in Eq. (19), i.e., n', is a constant, the mathematical relationship between the $D\Delta P/4L$ and $8V/D$ in Eq. (19) is given by Eq. (20). If n' is not a constant but varies with τ_w, Eq. (20) represents the tangent to the curve at any chosen value of $D\Delta P/4L$ or $8V/D$.

The similarity between Eqs. (3) and (20) is obvious; the only important difference is that Eq. (3) is not useful for relating pressure drop to flow rate until it has been integrated to give Eq. (15), which in turn requires that the exponent n be constant over the entire range of shear rates in the pipe [from zero to $(-du/dr)_w$], unless laborious averaging procedures are used. Methods based on Eq. (15) are therefore approximations—although frequently of excellent engineering accuracy—while Eqs. (19) and (20) already relate ΔP to V and hence require no integration; they therefore present a theoretical rigorous relationship between pressure drop and flow rate. If n' varies from one value of τ_w to another, it is simply necessary to evaluate it at the point appropriate for the problem at hand.

The parameters n' and n characterize the extent of non-Newtonian behavior. When they are equal to unity, the fluids are Newtonian; values of n' and n less than unity indicate pseudoplastic or Bingham-plastic behavior, and values greater than unity characterize dilatant behavior. These dimensionless quantities will accordingly be called *flow-behavior indexes*, a term which is perhaps more suitable than the analogous *structure number* of Weltmann (W4). The use of exponents such as n or n' to characterize the degree of non-Newtonian behavior was proposed approximately simultaneously by three independent investigators (F2, M11, W4).[3] Since to date few fluids have been found for which n' differs appreciably from n, the two terms are used interchangeably in this book. Methods of obtaining these flow behavior indexes as well as the constants K and K' are discussed in Section VI. Since K and K' vary directly with

---

[3] It seems almost inconceivable that earlier investigators would not have made the same recommendation, as these power-function equations have been in use for approximately thirty years. Nevertheless, no early reference was found which explicitly suggested the use of n or n' for such purposes, although it is highly possible that such references may exist.

the consistency of the fluid, they may be considered to be indexes of the "viscosity" or consistency of the fluid (M11).[4]

It is useful to regroup the usual definition of the Fanning friction factor, Eq. (10), as follows:

$$f = \frac{D\Delta P}{4L} \bigg/ \frac{\rho V^2}{2g_c} \qquad (21)$$

and to adopt the standard Newtonian relationship between this friction

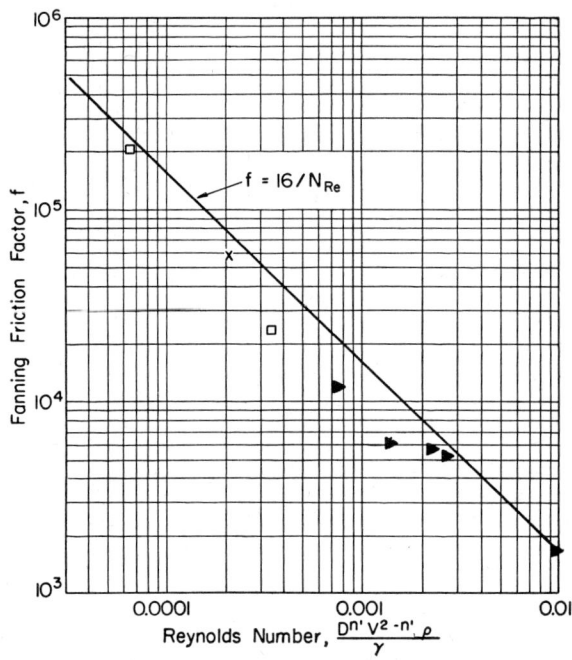

Fig. 5. Friction factor–Reynolds number diagram for non-Newtonians–low range. Taken from reference (M11) with permission.

factor and the Reynolds number (in laminar flow):

$$f = 16/N_{Re} \qquad (22)$$

Eqs. (20), (21), and (22) have been combined to solve for a generalized Reynolds number, which is applicable to Newtonian and non-Newtonian

---

[4] As pointed out in Section I, viscosities are really meaningful if compared only for Newtonian fluids or at specified shear rates for other materials. A similar limitation must be imposed on the consistency indexes K and K': values of either are comparable only for fluids with the same flow behavior indexes n or n'.

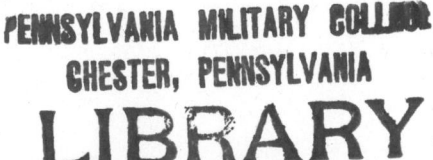

fluids alike (M11):

$$N_{Re} = \frac{D^{n'} V^{2-n'}}{\gamma} \rho \qquad (23)$$

where

$$\gamma = g_c K' 8^{n'-1} \qquad (24)$$

For the special case of Newtonian fluids, $n' = n = 1.00$, $K' = \mu/g_c$, and Eq. (23) becomes

$$N_{Re} = \frac{DV\rho}{\mu}$$

This equation shows that the conventional Reynolds number is simply a special case of the more general one defined by Eq. (23).

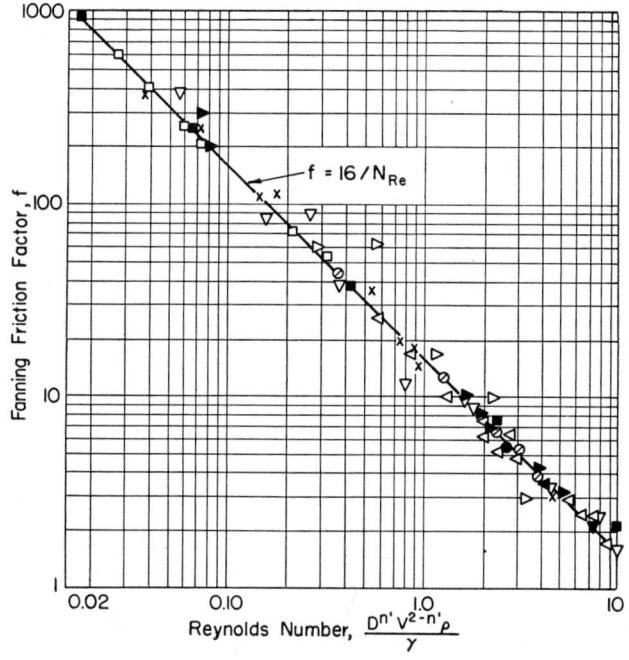

FIG. 6. Friction factor–Reynolds number diagram for non-Newtonians–medium range. Taken from reference (M11) with permission.

The utility of this generalized Reynolds number has been shown (M11) by the correlation of all available literature data on flow of non-Newtonian fluids on the conventional friction factor–Reynolds number diagram which is reproduced in Figs. 5, 6, and 7. The curves shown are not drawn through the data points but rather represent the conventional recommended relationship for Newtonian fluids. Table I summarizes the physical properties of the materials used and the sources of the data.

The rigor of this development requires that all fluid-flow data[5] follow the conventional $f = 16/N_{Re}$ relationship in the laminar-flow region. Accordingly, deviations from the theoretical curve in this region are due entirely to errors in measurement or calculation; the figures show that these errors were always low except at the lowest Reynolds numbers, where both the flow rates and pressure drops were extremely difficult to determine accurately.

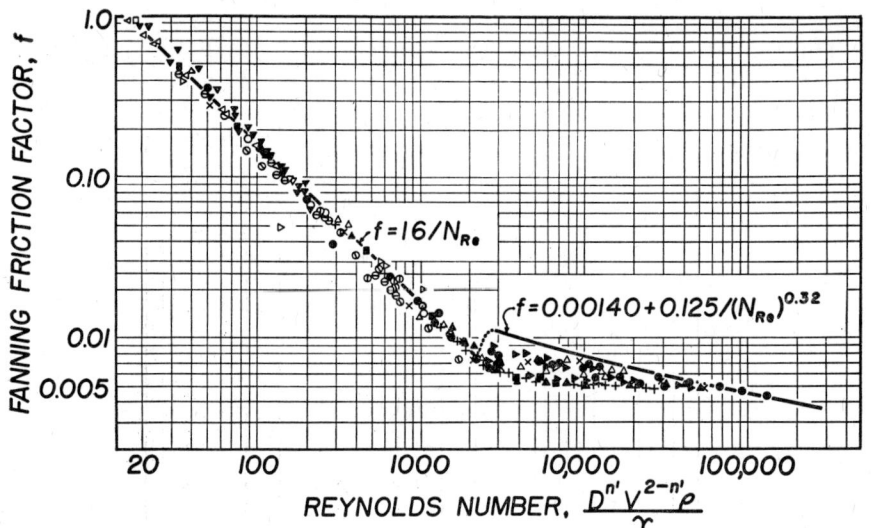

FIG. 7. Friction factor–Reynolds number diagram for non-Newtonians–high range. Taken from reference (M11) with permission.

The correlation of the Govier and Winning (G1) data by Weltmann (W4) on the assumption of Bingham-plastic behavior frequently showed order-of-magnitude deviations at low flow rates; in Figs. 5 to 7 only a negligible number of points deviate from the $f = 16/N_{Re}$ curve by more than ±25%. The data point which led to a 70,000% difference between calculated and observed friction factors in the former case showed only a 36% deviation in Fig. 6, for example. One may therefore conclude that in the laminar region use of Figs. 5 to 7 is much more accurate than prior-art procedures and have led to entirely acceptable pressure-drop predictions for all non-Newtonian fluids on which accurate data were available.

Eqs. (20) and (24) may be rearranged (M11) to give the generalized

---

[5] Exclusive of thixotropic and rheopectic fluids, and of the few materials (see Section VI) which may "slip" at the wall.

equivalent of Poiseuille's law for calculation of pressure drop in laminar flow:

$$\Delta P = \frac{32\gamma L V^{n'}}{g_c D^{n'+1}} \qquad (25)$$

which shows that the pressure drop is not very sensitive to either velocity or pipe diameter for highly non-Newtonian fluids with a flow-behavior index approaching zero. When n' is a constant, i.e., independent of the rate of shear, Eq. (25) reduces to Eq. (15).

TABLE I

Legend and Rheological Properties for Fluids Shown in Figures 5 through 7

Symbol used in figures	Nominal pipe size, in.	Composition of fluid	Rheological properties		Source of data
			n'	γ	
+	1	23.3% Illinois yellow clay in water	0.229	0.863	(C1)
⊕	⅞ and 1½	0.67% carboxy-methyl-cellulose (CMC) in water	0.716	0.121	(S2)
⊖	⅞ and 1½	1.5% CMC in water	0.554	0.920	(S2)
⊘	⅞ and 1½	3.0% CMC in water	0.566	2.80	(S10)
⊗	⅞, 1½, and 2	33% lime water	0.171	0.983	(S10)
◁	⅞ and 1½	10% napalm in kerosene	0.520	1.18	(S10)
▼	8, 10, and 12	4% paper pulp in water	0.575	6.13	(A2)
△	¾ and 1½	54.3% cement rock in water	0.153	0.331	(W6)
▲	4	18.6% solids, Mississippi clay in water	0.022	0.105	(G5)
●	¾ and 1¼	14.3% clay in water	0.350	0.0344	(G1)
▷	¾ and 1¼	21.2% clay in water	0.335	0.0855	(G1)
×	¾ and 1¼	25.0% clay in water	0.185	0.204	(G1)
▽	¾ and 1¼	31.9% clay in water	0.251	0.414	(G1)
□	¾ and 1¼	36.8% clay in water	0.176	1.07	(G1)
■	¾ and 1¼	40.4% clay in water	0.132	2.30	(G1)
▶	⅛, ¼, ½ and 2	23% lime in water	0.178	1.04	(A3)

The end of the stable laminar-flow region usually occurs near

$$N_{Re} = 2100$$

as for Newtonian fluids;[6] all fluids on which data were available entered the transition region below $N_{Re} = 2800$. The many diverse prior-art methods for predicting the end of the laminar-flow region for various

[6] This comment does not apply to the onset of turbulence in miscellaneous geometries, such as flow through short tubes or orifices as studied by Schnurmann (S4). This latter problem appears to be unsolved, partly because of the great experimental difficulties in techniques used to study this phenomenon to date.

kinds of non-Newtonian behavior were discussed (M11) and shown to reduce to this single criterion or in a few cases to be incorrect.

The transition region in Fig. 7 appears to extend to $N_{Re} = 70{,}000$ in some cases for highly pseudoplastic fluids. For these materials one would expect the necessity of high velocities (hence Reynolds numbers) for fully developed turbulence on physical grounds because the lower shear rates at the center of the eddies create regions of highly viscous behavior. Accordingly, unusually high shear rates would be necessary at the surfaces of the eddies in order to cause them to break up; this expected broadness of the transition region has been confirmed (T3) by dye-injection studies.

The importance of this physical picture, that the centers of eddies of pseudoplastic fluids are relatively motionless as compared with those of Newtonians, is difficult to overemphasize, as it illustrates the possible differences in the structure of turbulence between Newtonian and non-Newtonian fluids. One might expect on this basis that isothermal conditions for chemical reactions with appreciable heat effects may be progressively more difficult to obtain as the flow-behavior index n of a fluid decreases. In addition the energy losses in turbulent flow should be much less, since only the surfaces of the eddies undergo violent shearing rates. This effect is predicted by the exponent on the velocity term in the generalized Reynolds number, which has the effect of making pressure drop less sensitive to flow rate as n' approaches zero. The fact that a non-Newtonian solution may exhibit lower pressure drops in turbulent flow than the solvent used in preparing such a polymeric solution has been experimentally verified by two groups of investigators (A1, T2); however, two authors (O1, T2) interpreted the data in the second of the preceding references to be due to "slip" at the wall of tube. Although this complication cannot be ruled out definitely, it should be noted that no direct evidence is available in support of such a theory.

On the basis of this argument one would expect the width of the transition region to decrease as a fluid becomes more nearly Newtonian (i.e., as n or n' increase toward a value of unity). This is actually the case; the data on slightly non-Newtonian fluids ($n' > 0.72$) were omitted from Fig. 7 to avoid confusion, but some of these did indicate narrower transition regions. Further data are necessary to enable precise calculations of friction factors in the transition region according to the degree of non-Newtonian behavior (n' or n) and any other relevant variables, and to define precisely the width of the transition region, as that indicated by Fig. 7 would appear to be excessive. One other variable of importance may be pipe diameter, as data for different diameters do not appear to fall on exactly the same curves in the transition region of this diagram.

In view of the usually viscous nature of highly non-Newtonian materials it is not likely that Reynolds numbers appreciably greater than 70,000 will be very common, at least for some time to come. This fact places great importance on the region below $N_{Re} = 70,000$, and its detailed study would appear to be of primary importance. In well-developed turbulent flow, which apparently may be delayed to

$$N_{Re} = 70,000$$

the usual Newtonian relationship between friction factor and Reynolds number appears to be followed. This would be expected in view of the fact that the flow-behavior index n' or n approaches unity for many fluids at very high shear rates, as discussed in Section I. However, the data are much too meager in this region to enable more than tentative estimates of the flow behavior.

It should be noted that two constants are the minimum number which can be used to characterize any non-Newtonian fluid completely. Therefore, the form of the generalized Reynolds number is as simple as it can be made if it is to characterize all types of fluid behavior (M11).

The rigor and utility of the generalized Reynolds number [Eq. (23)] depends on the fact that n' and γ may vary with the imposed shearing stress. This in turn simply means that they must be evaluated at the proper value of the wall shearing stress $D\Delta P/4L$ or at the proper value of $8V/D$ in laminar flow. Outside the laminar-flow region the relationship between $D\Delta P/4L$ and $8V/D$ is no longer independent of pipe diameter (A3) but fortunately the variation of the flow-behavior index (n') with $D\Delta P/4L$ is very small. The tentatively recommended approach for the transition and turbulent regions is to calculate an approximate friction factor f, and hence ΔP and $D\Delta P/4L$, then to evaluate K' and n' at this value of $D\Delta P/4L$. If these indexes differ from those initially assumed to obtain a Reynolds number (hence f), one must recalculate the Reynolds number and friction factor. Usually once around this trial-and-error procedure should be sufficient since the friction factor is insensitive to small errors in N_{Re} in turbulent flow. For all the available data used in drawing up Figs. 5 to 7 there was no such appreciable change in fluid properties over wide ranges of the variables; hence this trial and error procedure has never been found necessary to date. An important by-product of this fact is that extrapolations of experimental data, when necessary, might best be carried out on the assumption of constancy of n' and n. However, extrapolation of rheological data to new shear rates is always a risky procedure for non-Newtonian fluids. Extreme caution is therefore recommended.

Other prior-art procedures for calculation of pressure drop in the turbulent-flow region suggest the use of apparent viscosities at infinite shear rate (W7) or experimentally determined "turbulent viscosities." The use of viscosities determined at "infinite" shear rates is obviously restricted to use at very high Reynolds numbers or to fluids which are nearly Newtonian in nature (so that viscosities over all shear-rate ranges are nearly identical). This means that the method of Winding *et al.* (W7) is not entirely general, but its simplicity makes it useful in industries dealing exclusively with only slightly non-Newtonian materials. The second procedure consists (A3, B5, W6) of experimentally determining pressure drops in turbulent flow, then using such data to calculate turbulent viscosities from the Newtonian friction factor–Reynolds number curve. These viscosities are then used for extrapolation to larger pipe diameters. Such a method should be excellent in view of the fact that it makes use of actual pilot plant data instead of physical properties determined on a laboratory scale. Obtaining such pilot plant data is more laborious, but the absence of fully developed design methods based on Fig. 7 makes the use of experimentally determined turbulent viscosities attractive where more precise pressure-drop predictions are necessary. Only small extrapolations may be made if such turbulent viscosities are to be used in precise work, however, as Ward (W1) has shown in an extensive and detailed study that these empirical viscosities may vary appreciably with both pipe diameter and flow rate.

Ibrahin (I1) has published an addition to the Navier-Stokes equations which was intended to modify them for use with non-Newtonian fluids. The modification was only for the purpose of taking fluid elasticity into account, a factor which does not appear to be necessary for the majority of materials showing non-Newtonian behavior. Instead, a redevelopment of the original equations to allow for variations in viscosity with shear rate is required, an approach that would appear to be very complex but perhaps rewarding.

Problems involving non-Newtonian flow in the paper industry are very difficult to treat experimentally because of the large size of some of the flowing particles. However, the laminar-flow methods discussed here have been shown to be as applicable to paper slurries as to any other non-Newtonian fluid (M11, R10); hence the originally novel methods of approach (D3, V1) may be no longer necessary. Certainly, the arbitrary substitution of four new flow regimes for the conventional laminar-, transition-, and turbulent-flow regions, as suggested by Van der Meer (V1), cannot be recommended. The chief need in this area would appear to be for methods of prediction of the flow properties n' and γ, so that pressure-drop estimates may be made with a minimum of experimental data. If the pipes used are small enough to modify the size or shape of the flowing flocs, these properties would be expected to change somewhat with pipe diameter.

Head (H3) has published a pertinent article concerning the metering of non-Newtonian fluids. He pointed out that pressure drop is insensitive to flow rate for fluids with a low flow-behavior index [cf. Eq. (25)]. Accordingly, metering methods which involve measurement of the pressure drop over any appreciable length of pipe are insensitive to changes in flow rate; i.e., any change in pressure drop at the metering element is diluted by the large, relatively invariant pressure drop along the pipe itself. He has concluded that sharp-edged orifices with *vena-contracta* taps are excellent metering devices. Similarly, sharp-edged floats were most suitable in rotameter-type flowmeters. Both orifice meters and rotameters were found suitable for metering of certain paper-pulp suspensions.

Spencer and Dillon (S9) have discussed the problem of elastic recovery of polymeric melts and solutions at the exit of a tube. For the polystyrenes studied, a criterion was developed for the point at which such recovery was sufficient to cause buckling of the filament leaving the tube.

C. Velocity Profiles

In some applications the actual shape of the velocity profile within the pipe may be of importance. In the turbulent-flow region there is little reason to believe that non-Newtonians have velocity profiles which differ appreciably from those of Newtonian fluids. In laminar flow the so-called "plug flow" of Bingham-plastic fluids in that region near the center line of the tube where the shearing stresses are less than τ_y has received wide attention and experimental verification (G2, G3, G4). It has been pointed out (M11) that the early verification of this phenomenon is not inconsistent with the view that only a few materials exhibit Bingham-plastic behavior, as highly non-Newtonian pseudoplastics (i.e., those with a flow behavior index approaching zero) also exhibit very flat velocity profiles, as shown in Fig. 8. Since fluids with $n' = n < 0.20$ are fairly common (Table I), it is easy to find materials which closely approach plug flow over an appreciable part of the tube diameter and which are pseudoplastic in behavior. In view of the common occurrence of pseudoplastic behavior and the frequent constancy of the flow-behavior indexes, integration of Eq. (3) is recommended to relate the local velocity to radial position in the pipe. The resulting equation, upon which the curves of Fig. 8 are based, is

$$\frac{u}{V} = \frac{1 + 3n}{1 + n} \left[1 - \left(\frac{r}{R}\right)^{\frac{n+1}{n}} \right] \qquad (26)$$

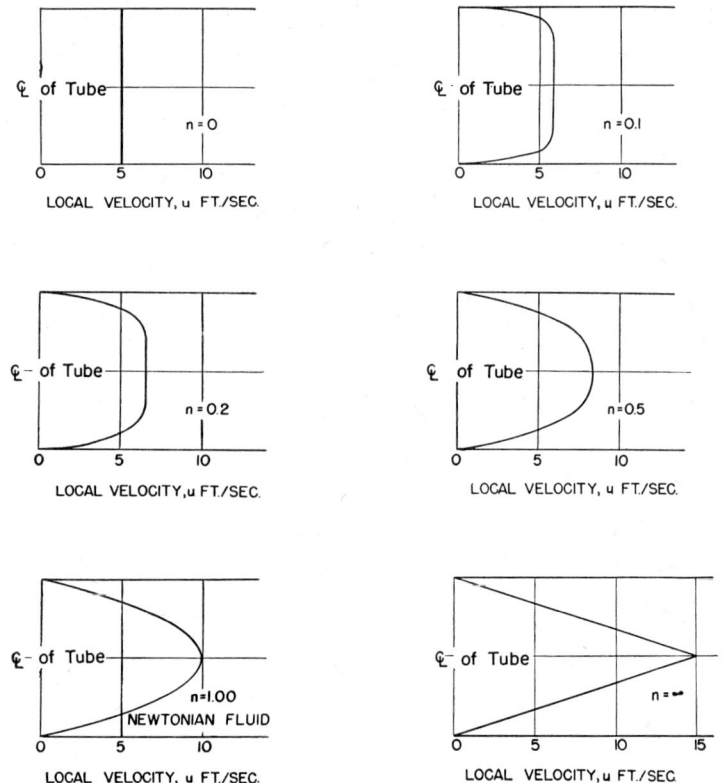

Fig. 8. Dependence of velocity profiles upon flow-behavior index. Average velocity of 5 ft./sec. in all cases.

D. Effect of Surface Roughness

Although the correlations of the prior sections did not include a discussion of pipe roughness, it would appear that some differences in pipe roughness may have occurred between different investigations. In laminar flow no effect of roughness would be expected in accordance with results for the laminar flow of Newtonian fluids. Outside the laminar region one might expect that pseudoplastic and Bingham-plastic non-Newtonian fluids would show a smaller dependency of pressure drop on pipe roughness than do Newtonians, since the shear rates of the fluid in the pits or depressions would be low, the viscosity high,[7] and the energy-

---

[7] The opposite conclusion would presumably apply to dilatant fluids and for these the pressure drop in a rough pipe may perhaps exceed that for a Newtonian fluid. A similar situation might also arise for Bingham-plastic and pseudoplastic materials which exhibit "elastic recovery" to a high degree.

dissipation rate therefore low. Until experimental results become available, the conventional relationships for Newtonian fluids may accordingly be recommended on the basis that, if in error at all, they are probably conservative.

As in the case of Newtonian fluids (R7) one would expect that the end of the stable laminar-flow region (generalized $N_{Re} = 2100$) should not be influenced by the roughness of the pipe. Until experimental data are available, this assumption is recommended although a contrary opinion has been published (W4).

E. Effect of Temperature and Concentration on Flow Properties

It has been pointed out by Weltmann (W4) that the complexity of some non-Newtonian systems leads to unusual changes in fluid properties with temperature. This may occur, for example, if solids tend to go in or out of solution or if the solids are more completely dispersed at the higher temperature. Most non-Newtonian fluids, however, do not show such unusual effects, and the changes in fluid properties with temperature and concentration of material in suspension or solution may be summarized as follows.

1. The flow-behavior index n' (or n) may be assumed independent of temperature and concentration for small changes in such conditions (R3). For larger changes, Vaughn (V2) has found the following generalizations to be valid for water-dispersible polymers:

a. The flow-behavior index increases toward unity as the fluid temperature increases; i.e., many pseudoplastic materials tend to approach Newtonian behavior at higher temperatures. This would be expected in view of the physical causes for such behavior discussed in Section I.

b. This index usually decreases as the concentration of dissolved or suspended solids increases in the range where these fluids are pseudoplastic in behavior. The flow-behavior index, however, does not change rapidly with either temperature or concentration of dissolved material.

2. The viscosity or consistency indexes K and K' frequently change as rapidly with temperature as the viscosity of the solvent or suspending medium (R3). For suspensions the ratio of K' to the viscosity of the suspending medium is frequently nearly constant for suspensions of a given concentration and independent of the viscosity of the suspending medium. For example, Reed (R3) found that this ratio changed only five fold for 150-fold changes in viscosity of the suspending medium; however, the consistency indexes increase rapidly with increases in concentration of a solution or suspension.

As indicated by the paragraph above, changes in K' with temperature are usually an order of magnitude greater than changes in n'. These generalizations cannot be considered to be rigorous in detail but they are believed to be sufficiently valid for any preliminary estimates which might precede experimental data on the fluid in question; usually they have been evaluated for slurries containing up to 25 vol. % of suspended solids or to about 3% dissolved polymeric material. The same conclusions are extended by recent more extensive data (B2, M7) on melts and solutions of rubbers in organic solvents. These show, as would be expected, that the flow-behavior index n changes rapidly near the limiting conditions of small quantities of polymeric material in the solvent but only very slightly over wide ranges of concentration above about 10% of dissolved polymer.

For Bingham plastics the changes in yield value τ_y with temperature have been reported generally to resemble changes of the flow-behavior index n'; similarly, the coefficient of rigidity η frequently changes as rapidly with temperature as the viscosity of the suspending medium (G1).

F. Summary

1. *Recommended Design Procedure*

a. Laminar Flow. If the viscometric data available are in the form of a curve of $D\Delta P/4L$ versus $8V/D$, this may be used directly for scale-up, or the constants K', (γ), and n' may be determined for use with the generalized Reynolds number $(D^{n'}V^{2-n'}\rho/\gamma)$ or the generalized Poiseuille relationship—Eq. (25). Use of the generalized Reynolds number–friction factor method is generally recommended on the basis that this number must usually be calculated in all scale-up procedures to ensure that laminar flow actually exists; once it has been calculated, the simplest procedure is to determine the friction factor, hence the pressure drop, from it.

If data from a rotational viscometer or other instrument are available in the form of curves of shearing stress versus shearing rate, they may be plotted logarithmically to determine the physical properties K and n. If, as is usually the case, the logarithmic plot is nearly a straight line over a five-to-ten fold range of shear rates, n may be taken as equal to n', and K' may be calculated from K [Eq. (54)]. The generalized Reynolds number is calculated to determine whether the flow is actually laminar under the proposed conditions. The actual pressure drop may then be calculated from the friction factor or from Eq. (15) or (25). When this has been done, the shear stress at the wall of the pipe $(D\Delta P/4L)$ should be checked to ensure that the chosen values of n and K are valid over approximately a twofold range of shear stresses up to the value $D\Delta P/4L$.

If the relationship between shearing stress and shearing rate is not nearly linear on logarithmic corrdinates, graphical integration of Eqs. (13) and (14) must be used, or an attempt may be made to fit other equations to the data [such as the Eyring-Powell relationship—Eq. (16)]. In most cases the predicted pressure drop will be accurate to within ±15%.

The assumption of Bingham-plastic behavior is not *generally* recommended in view of the rarity of close approach to such behavior, but the excellent design procedures available for these materials should be very useful to the few industries in which such fluids commonly occur.

b. Transition and Turbulent Regions. The end of the laminar-flow region may be assumed to occur at generalized Reynolds numbers between 2100 and 2800. Outside the laminar-flow region the use of the generalized Reynolds number and Fig. 7 is recommended; use of a mean curve through the data for fluids having a flow-behavior index n' of less than 0.50 and use of the Newtonian curves for $n' > 0.50$ should usually enable pressure-drop predictions within ±30%. If more accurate predictions are necessary, the experimental determination of turbulent viscosities according to the procedure of Alves *et al.* (A3) is recommended.

2. Physical Properties

Decreases in concentration or increases in temperature usually decrease the consistency indexes K and K' but leave the flow-behavior indexes n and n' relatively unaltered. The latter appear to be determined primarily by the components of the non-Newtonian fluid and increase only slightly with increases in temperatures or decreases in concentration for pseudoplastic materials.

3. Future Work

The orderly and complete understanding of flow behavior is an obvious prerequisite to progress in the fields of mixing and heat transfer and so is of primary importance to these fields as well as in its own. While the understanding of laminar flow in circular pipes is complete and detailed (except for thixotropic and rheopectic fluids), the same is not true of the transition and turbulent-flow regions. Since, as for Newtonian materials, reasonably high rates of mixing and heat transfer can be obtained only outside the laminar region, a study of the transition region is particularly necessary. The behavior of dilatant fluids under these conditions should be unusually interesting (M11). In view of the viscous nature of most non-Newtonians, generalized Reynolds numbers greater than 10^5 may not be of frequent practical interest, but study of this region might be useful to help to define flow problems at somewhat lower Reynolds numbers.

The work to date indicates that the generalized Reynolds number might be a good basis for the correlation and interpretation of such future fluid mechanics studies. However, this indication is to an extent due to the fact that no other approach has been given so adequate a trial. Use of other approaches is, therefore, definitely recommended if there is reason to believe that they may prove to be as valuable.

All pipe-line work to date has dealt with fluids which are not thixotropic and rheopectic. To an extent this may be justified because the limiting conditions (at startup—for thixotropic materials, and after long times of shear for rheopectic fluids) in pipe flow and some mixing problems are of primary importance. Design for these conditions would be similar to the techniques discussed herein for other fluids. This is not true of problems in heat transfer, however, and inception of work on the laminar flow of thixotropic fluids in round pipes would appear to be in order as a prerequisite to an understanding of such more complex nonisothermal problems.

III. Miscellaneous Flow Problems

A. Fluid Kinetic Energies

The preceding section dealt with that part of the pressure drop which was due to fluid friction; additional and possibly large energy requirements may be necessary, owing to changes in the kinetic energy of the moving stream. For Newtonian fluids it has been shown (M4) that the kinetic energy of the fluid in laminar flow is equal to V^2/g_c ft.-lb.$_F$/lb.$_M$; in turbulent flow the velocity profile is flat enough so that the kinetic-energy term may usually be evaluated simply by assuming plug flow; i.e., the kinetic energy is taken equal to $V^2/2g_c$ ft.-lb.$_F$/lb.$_M$. Since *non-Newtonians in turbulent flow* probably have velocity profiles very similar to those of Newtonian fluids, assumption of kinetic energies equal to $V^2/2g_c$ is recommended until velocity-profile data become available.

In laminar flow of Bingham-plastic types of materials the kinetic energy of the stream would be expected to vary from $V^2/2g_c$ at very low flow rates (when the fluid over the entire cross section of the pipe moves as a solid plug) to V^2/g_c at high flow rates when the "plug-flow" zone is of negligible breadth and the velocity profile parabolic as for the flow of Newtonian fluids. McMillen (M5) has solved the problem for intermediate flow rates, and for practical purposes one may conclude

$$\text{kinetic energy, ft. lb.}_F/\text{lb.}_M = \frac{V^2}{2g_c}(2 - \tau_y/\tau_w) \qquad (27)$$

This equation represents the complex theoretical kinetic-energy equation of McMillen (M5) within 2.5% for all values of τ_y/τ_w.

The kinetic energy of any moving stream in a round pipe is equal to (M4)

$$\text{K.E.} = \frac{1}{V\pi R^2 \rho} \int_0^R \frac{u^2}{2g_c} \cdot \rho \cdot 2\pi r u \, dr$$

$$= \frac{1}{VR^2} \int_0^R \frac{u^3 r \, dr}{g_c} \tag{28}$$

For non-Newtonians which obey the power-law relationship between shear stress and shear rate [Eq. (3)] Dodge has shown (D2) that Eq. (28) may be integrated to give

$$\text{K.E., ft. lb.}_F/\text{lb.}_M = \frac{V^2}{\alpha g_c} \tag{29}$$

where

$$\alpha = \frac{(4n + 2)(5n + 3)}{3(3n + 1)^2} \tag{30}$$

Equation (30) is shown graphically in Fig. 9.

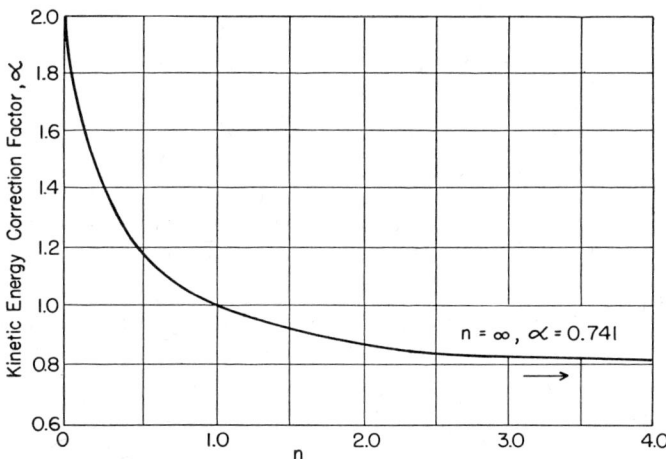

FIG. 9. Dependence of kinetic-energy correction factor upon flow-behavior index.

It may be concluded that kinetic-energy terms can be evaluated for all time-independent non-Newtonian fluids with an accuracy probably sufficient for all engineering work.

B. FLOW THROUGH FITTINGS AND ENTRANCE LOSSES

Very few data appear to be available in the literature on pressure losses due to flow of non-Newtonian fluids through fittings such as orifices, valves, and elbows. Preliminary work (H6) indicates that these appear to be of the same order of magnitude as pressure drops for Newtonian fluids at the same generalized Reynolds number in the laminar-flow region. At the time of this writing, the results are too tentative to enable any more detailed conclusions. Weltmann (W4) has pointed out that dilatant fluids should show large pressure losses on flowing through constrictions because of the increased consistency of these materials at high shear rates; this factor can perhaps be accounted for satisfactorily by use of the generalized Reynolds number. Mooney and Black (M16) studied the extrusion of rubber stocks through short tubes and orifices, but their results are difficult to compare with other work. Head's work (H3) has been briefly discussed in Section IIB.

McMillen (M5) has measured the pressure loss in the entrance region to a tube. Flow was in the laminar region in pipes ranging from $\frac{1}{8}$ to $1\frac{1}{2}$ in. in diameter (nominal). The total entrance loss for the napalm gels studied, including the kinetic-energy change of the fluid, could be expressed by the following dimensional equation:

$$\Delta P_C + \Delta P_{\text{K.E.}} = 0.12 V^{1.5} \tag{31}$$

The sum of ΔP_C and $\Delta P_{\text{K.E.}}$ include all losses in the entrance region except the loss due to the fluid friction along the length of pipe in this region. The fluid friction was evaluated from measurements downstream at a section well removed from the entrance. ΔP_C, the contraction loss itself, may be calculated as the difference between the total pressure loss given by Eq. (31) and the kinetic-energy pressure loss. The latter is the product of the kinetic energy as given by Eq. (27) and the fluid density.[8]

Comparison of McMillen's results with those for Newtonian fluids is instructive in two important respects. First, the non-Newtonian entrance loss was felt 40 diameters downstream from the entrance. Since the Reynolds number of the flow was only 50, the entrance loss for a Newtonian fluid (P3) would have only been felt downstream for a distance of 3 diameters. Second, comparison of the foregoing formula with the usually recommended procedure for Newtonian fluids (P3) indicates that the non-Newtonian pressure drop was approximately six times as great as that for Newtonian fluids under the same conditions. This figure was checked reasonably closely by the later work of Mooney and Black (M16), who found entrance losses of up to seven times those for comparable Newtonian fluids. Since this entrance loss (P3) is due to the energy required to set up the velocity profile, it might appear logical that

---

[8] The units must be converted, however. The units of V must be ft./sec. in Eq. (31), and ΔP_C and $\Delta P_{\text{K.E.}}$ must be in lb.$_F$/sq. in. The product of fluid density and Eq. (27) must therefore be multiplied by 144 before the result is substituted into Eq. (31).

non-Newtonians should exhibit lower entrance losses than Newtonian fluids. The reasons for this apparent discrepancy are not yet clear, although it has been suggested (W9) that the high non-Newtonian entrance losses are related to the energy requirements for alignment of the colloidal particles under the influence of the high shear rates in a contracted stream. Granted that the difference between Newtonian and non-Newtonian pressure losses through fittings is due to molecular or particle alignment, it would appear logical that the change in shear rates which the fluid undergoes on passing through fittings such as valves, orifices, abrupt contractions, etc., should determine the difference between the Newtonian and non-Newtonian pressure losses. Wood, Nissan, and Garner (W9) presented this theory and correlated a variety of orifice pressure drops on this basis. Unfortunately they used, as the Newtonian pressure drop, $1.14 \rho V^2/2g_c$—i.e., the pressure drop at an abrupt contraction, rather than in an orifice. This leaves their good correlation without any apparent physical basis, and its use cannot be recommended generally at this point.

The same authors presented data on pressure drop at abrupt contractions for which their method of correlation has a rational basis. In this work both the fluid used and the pipe diameter were changed simultaneously; hence the reasons for the absence of any correlation in this case cannot be specified. Their general method of approach appears to be a powerful and useful basis for further work but has not yet led to a recommended design procedure.

Calculation of entrance losses from experimental data is aided by the procedure of Hall and Fuoss (H1) in cases where direct measurements are not feasible.

C. Flow in Annular Spaces

Mori and Ototake (M17) have presented a mathematical analysis of the *laminar flow of Bingham-plastic materials* in the annulus between two concentric pipes. The complex results have been shown in convenient graphical form which enables one to solve for the flow rate corresponding to a given pressure gradient.

It would be most interesting to have this work extended to the more common power-function non-Newtonians.

D. Two-Phase Flow

Ward and DallaValle (W2) studied the two-phase cocurrent flow of air and four non-Newtonian fluids in three horizontal pipes (I.D. 0.82 to 1.60 in.). Flow rates were such that both phases were in turbulent motion. The flow-behavior index of the liquids used was varied from 0.31 to 1.00, a rather wide range. These authors found that the Lockhart-Martinelli (L3) correlation for the cocurrent flow of a gas and a Newtonian liquid,

$$\frac{(\Delta P/L)_{TP}}{(\Delta P/L)_{LP}} = \theta \left(\frac{(\Delta P/L)_{LP}}{(\Delta P/L)_{GP}} \right) \tag{32}$$

(where θ simply indicates a functional relationship) also gave good agreement with the non-Newtonian data. The mean deviation of the data from the single correlating line was about $\pm 10\%$ and all data fell within $\pm 20\%$ of the curve, an accuracy which is considerably superior to that reported for two-phase flow with a Newtonian liquid. It may be concluded that the Lockhart-Martinelli relationship is valid for non-Newtonian fluids over the entire range of variables covered. In view of the fact that methods are not yet available for calculation of the term $(\Delta P/L)_{LP}$—the liquid-phase (non-Newtonian) pressure drop—with this great an accuracy in the turbulent-flow region, it would appear to be unwise to attempt any improvement upon this correlation until the simpler problem of turbulent flow of Newtonian fluids is much better understood.

E. Nonisothermal Flow

No methods appear to be available for the precise prediction of pressure drop when a non-Newtonian fluid is being heated or cooled, but Vaughn (V2) has shown that the procedure recommended most recently by McAdams (M4, p. 149) for Newtonian fluids is slightly conservative when applied to pressure-drop data on the heating of non-Newtonian solutions in laminar flow. McAdams has suggested evaluation of the fluid properties at a film temperature t_f, defined as

$$t_f = \tfrac{3}{4} t_b + \tfrac{1}{4} t_w \tag{33}$$

Since Vaughn's work has indicated a stronger dependency of pressure drop upon wall temperature than is given by Eq. (33), it seems probable that use of this procedure would also be in error if applied to the cooling of non-Newtonian fluids, but in the direction of predicting pressure drops lower than those actually encountered. Until adequate design methods are developed, the following *conservative procedure is recommended.*

1. For *heating* of non-Newtonian fluids the pressure drop should be determined by evaluation of fluid properties at the film temperature as given by Eq. (33).

2. For *cooling* of non-Newtonian fluids, the pressure drop should be calculated by use of fluid properties at the wall temperature.

These recommendations assume that the consistency of the fluid decreases with increasing temperatures. For some complex non-Newtonian suspensions this may not be the case; accordingly, extreme care must be exercised to ensure the absence of such unusual complications. If such complications do occur, only order-of-magnitude estimates are possible at present.

F. Extrusion

An extrusion symposium (E1) contains papers which deal extensively with the mathematics of viscous flow in screw extruders but which are limited to Newtonian materials. An extension of this work to materials which may be assumed to be Bingham plastic in behavior has been reported in Japan (M18, M19). The first of these papers deals with a screw extruder with a uniform channel; the second with an extruder for which the depth of the channel decreases linearly with channel length. The mathematical results are shown graphically in terms of four dimensionless groups:

$$\frac{\eta v}{\tau_y g_c H}, \frac{2\tau_y/H}{dP/dL}, Q/vSH, 1 - H_0/H$$

where the symbols designate the following quantities *in this section only:*
  v = helical surface speed.
  H = depth of screw channel. H_0 refers to this depth at the point of maximum pressure.
  S = width of channel.
  Q = volumetric flow rate of polymer.

The other quantities have their usual significance. The first group is the product of the Reynolds and Hedstrom numbers; the second is analogous to the τ_y/τ_w ratio of Eq. (6). Experimental data were used to support the mathematical development only in the case of the uniform-channel extruder.

The generality of the assumption of Bingham-plastic behavior may be questioned, but the discussion above indicates a powerful method of approach to the problems in a field of major importance. Obviously much further work is needed.

An interesting study of the extrusion of lubricated polytetrafluoroethylene powders has been published by Lewis and Winchester (L1), who have concluded that much of the shearing takes place in the taper of the die before the "fluid" enters the tube, or "hole," of the die itself—i.e., that the end affects are of primary importance. Furthermore the data were interpreted to indicate an outstanding importance of the flow of the polymer in the material chamber well ahead of the die itself, an importance which is difficult to accept. Much further work is needed to demonstrate the generality of their conclusions, but this paper stands as a model on the basis that careful attempts were made to understand the actual behavior of the polymer in various parts of the system. The same is true of the empirical but erudite study of Beyer and Spencer (B3) of some factors determining the quality of injection-molded products. Although their article is entitled "Rheology in Molding," it was found necessary to assume Newtonian behavior of the melt: such a requirement was not necessary in work on extrusion through round tubes (S9).

Carley (C2) has used the power-function relation [Eq. (3)] in a mathematical development for slit-crosshead-die extruders. Experimental verification of what appears to be a most useful piece of work would appear to be desirable, but even without such data the qualitative trends discussed by Carley are of major significance.

Flow in rectangular channels has been treated by Beyer and Towsley (B4) and Rankin (R2). Rankin found it possible to simplify the complex relationship presented earlier by the other authors.

Meskat (M8) has presented a mathematical analysis of the effect of fluctuations in pressure and other variables on the comparative fluctuations in extrusion rates of Newtonian and non-Newtonian fluids. This work indicates the possibility of amplification of such fluctuations under certain circumstances with non-Newtonians rather than the uniform damping predicted for Newtonian behavior. If the validity of this analysis can be proved, it would warrant major attention being given to the problem of unsteady flow of non-Newtonian materials.

In general, most of the extrusion work has necessarily been of an empirical nature. Nevertheless, important practical conclusions have been drawn by each of the authors. The nature of the problem is such, however, that a more detailed review could still not encompass all the necessary trends; hence the interested reader is referred directly to the original sources for additional details.

G. Summary

With the exception of the discussion of fluid kinetic energies, for which well-developed design equations are available, the work in this section has served primarily to draw attention to the lack of information in this area.

The problem of flow through fittings and annular spaces has dealt only with the laminar region; the work on annuli was further restricted to the relatively unimportant case of Bingham-plastic behavior. The annular studies to date were quite well chosen in the sense that assumption of Bingham-plastic properties has led to a well-developed method of attack which may not have been possible if more complex non-Newtonian behavior had been considered. However, one can only conclude that much further work is needed to solve the industrially important problems of flow of all types of non-Newtonian fluids in ducts of miscellaneous shapes and through fittings. Such studies must eventually encompass the laminar-, transition-, and turbulent-flow regions. The same comments apply to prediction of pressure drop in pipes when the flow is not isothermal. A particularly important industrial problem is that of flow of non-Newtonian fluids through porous formations. This is of day-to-day importance in petroleum exploration work but not a single publication appears to be available in this field. Such studies would also be interesting as a particularly strong test of the actual generality and scope of the generalized Reynolds number.

Extrusion and extruder design are apparently almost entirely empirical at present. This field might well defy rigorous theoretical analyses until the simpler problems of flow in annular spaces and rectangular ducts and of nonisothermal fluid flow are understood.

Unsteady state phenomena have been stated to be of greater importance for non-Newtonian than for Newtonian materials and therefore warrant experimental investigation. The prediction of pressure drop for two-phase flow of a gas and a non-Newtonian fluid seems to be in a well-perfected state but requires extension to situations in which the liquid flow is laminar. Apparently no information is yet available on the problems of mixing, entrainment, and other similar relationships which are of importance if such contactors are to be designed for chemical rather than mechanical purposes.

IV. Mixing of Non-Newtonian Fluids

Magnusson (M1) reported a procedure for the calculation of the average apparent viscosity of a non-Newtonian fluid in agitated tanks but presented no method for using such results for equipment design. Schultz-Grunow (S5) studied the power requirements for agitation of slightly non-Newtonian fluids in the laminar-flow region. By applying dimensional analysis to the problem, he was able to evolve an entirely sound design procedure for those fluids with flow curves which may be approximated by the equation

$$\tau = E \text{ arc sin} \frac{du/dr}{C} \qquad (34)$$

Unfortunately, Schultz-Grunow states that the impellers which he used are of no particular interest industrially (S5). For this reason, and because the form of the correlation necessarily limits its utility to the laminar region where little actual mixing of non-Newtonians appears to take place (O6), this publication also leaves no recommended method of approach.

Brown and Petsiavas recently presented a paper (B9) in which the relationship between the power number $N_P = Pg_c/(\rho N^3 D^5)$; the Reynolds number, $N_{Re} = (D^2 N \rho)/\eta$; and a third dimensionless group was presented for fluids which may be assumed to approximate Bingham plastic behavior. In this work a simplification was made in interpretation of the viscometric data, as a result of which the last group was not actually dimensionless in the form used. Correction of this error is readily possible, however, to make this method of approach rigorous and somewhat more generally useful. Experimental data were presented to enable the necessary empirical positioning of the curves but additional data to prove the absence of scale effects are needed, as equipment of only one size was used. It is evident that Brown and Petsiavas have presented a good method of approach, although refinements are necessary to make it useful for equipment design generally.

Substitution of the generalized Reynolds number (Section IIB) for the simple Newtonian Reynolds number has been shown (O6) to enable approximate prediction of agitator power consumption for non-Newtonian fluids at low Reynolds numbers. The conventional Newtonian power number–Reynolds number charts which have been drawn up by Rushton et al. (R9) were shown to be applicable in the laminar region. This laminar region, however, appeared to extend to Reynolds numbers of 20 to 25, as compared with critical values of 8 to 10 for Newtonian liquids. Above Reynolds numbers of about 70 the conventional Newtonian curve again appeared to be followed.

The ranges of the variables covered by Otto and Metzner (O6) were as follows:

Impeller diameters: 2 to 8 in.
Tank diameters: 6 to 22 in.
Power input per unit fluid volume: 2 to 176 hp./1,000 gal.
Reynolds numbers: 0.8 to 165
Flow-behavior index (n): 0.25 to 1.00

(The standard flat-blade turbines of Mixing Equipment Company were used throughout this work.)

In summary, two of the principal approaches which were found useful for the calculation of pipe-line pressure drops have been extended to the problem of predicting power consumption for the agitation of non-Newtonian fluids. Extension of this work is required, but until further data become available, use of the standard power number–Reynolds number charts (with the generalized Reynolds number) is recommended. Between Reynolds numbers of 10 and 70 these charts will provide conservative design estimates for non-Newtonian fluids since the experimental data fall below the Newtonian curve in this region.

As in the case of agitation of Newtonian fluids, no over-all understanding is yet available for the prediction of mixing rates, uniformities, and efficiencies. Considerable need exists for an emphasis on these problems; the agitation number and "time-of-a-transfer-unit" concepts of Hixson *et al.* (H5) would appear to be recommended starting points in such studies. Rushton (R8) has presented a review of the basic problems in this field as well as results obtained to date with Newtonian fluids. Most of these appear to have emphasized the results of given mixing conditions, and much work is necessary to develop a quantitative understanding of the underlying reasons for the observed results and their various interrelationships.

V. Heat Transfer

A. Heat Transfer to Dilute Suspensions

It has been shown repeatedly (B6, M13, O5, S1, W8) that suspensions of "inert" solids in liquids are essentially Newtonian in behavior if the solids concentration is below about 10% by weight. "Inert" solids are those which may be assumed to undergo little or no physical change in the presence of the liquid. The investigators mentioned above have dealt with calcium carbonate, graphite, silica flour, glass beads, and powdered copper and powdered aluminum in water and with graphite-kerosene, graphite-ethylene glycol and aluminum-glycol suspensions. Other solids, such as the dilute clay suspensions used by Orr (O5), which are swelled or solvated appreciably by the liquid phase, may be expected to deviate from Newtonian behavior at far lower concentrations. However, as in the case of most of the inert solids the deviation from non-Newtonian behavior will always be small at sufficiently low concentrations. Therefore, in the absence of well-proved design concepts which may be universally applicable to Newtonian and non-Newtonian materials alike, an approach based on the assumption of Newtonian behavior may lead to estimates sufficiently accurate to be of value. It is with such suspensions that this section is primarily concerned.

In order to understand and correlate the heat transfer data, the relevant physical properties of the suspensions must be carefully evaluated. The experimental determination of heat capacity and density pose no particular problem. In many instances it is possible to estimate these values accurately by assuming them to be weight averages of those of the two components. In contrast, great difficulty is associated with the accurate determination of thermal conductivity and viscosity, largely owing to the fact that the solids tend to settle readily in any device where convection currents are eliminated, as they must be for these determinations. Thus the basic problem in this field may be stated to be the proper measurement and interpretation of thermal conductivity and viscosity data.

1. *Evaluation of Thermal Conductivity*

Orr and DallaValle (O5) made measurements on suspensions of settlable solids by adding to the suspension 2% by weight of the gel agar, which completely eliminated settling. By means of calibrating determinations it was ascertained that addition of the agar to water increased its thermal conductivity by 9%; as the net exponent on the thermal conductivity term in heat transfer correlations has a value of about two thirds, the maximum error incurred in such a correlation must be less than 6%.

Even more interesting and important than the foregoing unique method of measuring thermal conductivities of suspensions is the procedure used to calculate thermal conductivities theoretically. Orr and DallaValle noted that electrical and thermal fields are similar; hence the usual equation for calculation of electrical conductivity of a suspension should also be applicable to thermal conductivities. Their extensive tabulated results support this contention to within 3%. This equation is

$$k_s = k_l \left[\frac{2k_l + k_p - 2x_v(k_l - k_p)}{2k_l + k_p + x_v(k_l - k_p)} \right] \qquad (35)$$

where k_s represents the thermal conductivity of the suspension, k_l and k_p the conductivities of the liquid and solid particles, respectively, and x_v the volume fraction of solids in the suspension.

For those systems for which $k_p \gg k_l$, as, for example, suspensions of metallic powders in nonmetallic liquids, the preceding equation reduces to

$$k_s = k_l \frac{(1 + 2x_v)}{(1 - x_v)} \qquad (36)$$

That is, the ratio of the thermal conductivity of the suspension to that of the liquid is independent of the conductivity of the solid. Several values of k_s/k_l, calculated by means of Eq. (36), are shown in the following tabulation:

x_v	k_s/k_l
0.00	1.00
0.10	1.33
0.20	1.75
0.40	3.00
0.60	5.50

Since a 0.60-volume fraction of solids approaches the upper limit of solids which can be incorporated into a liquid, this tabulation clearly shows that it is impossible to increase the thermal conductivity of the liquid more than a few fold by addition of solids.[9]

Two other references (M13, S1) have discussed the problem of choice of correct thermal conductivity values. Miller (M13) worked with water-graphite and kerosene-graphite suspensions at concentrations of the solid between 4.7 and 13.5% by weight ($0.02 < x_v < 0.08$). This investigator noted the results of attempts to correlate the heat transfer data in two manners.

a. Evaluation of the thermal conductivity of the suspension by averaging the conductivities of the solids and the liquid according to the volume fraction of each phase present gave, as would be expected from the preceding discussion, predicted thermal conductivities which were erroneous by an order of magnitude.

b. Use of the liquid thermal conductivity (i.e., ignoring the thermal conductivity of the solids) worked acceptably well. It is to be noted that the thermal conductivities predicted by Eq. (36) would differ from those of the liquid phase by less than 25% at the solids concentrations used by Miller. Since the final correlating equation contained the thermal conductivity term raised to the 0.60 power, the error incurred by ignoring the effect of the solids would always be less than 14%, or almost within the scattering of the data. It may be concluded, therefore, that this work tends to support the conclusions reached by Orr and DallaValle.

Salamone and Newman (S1) recently studied heat transfer to suspensions of copper, carbon, silica, and chalk in water over the concentration range of 2.75 to 11.0% solids by weight. These authors calculated "effective" thermal conductivities from the heat transfer data and reached conclusions which not only contradicted Eqs. (35) and (36), but also indicated a large effect of particle size. However, if one compares the conductivities of their suspensions at a constant volume fraction of solids, the assumed importance of particle size is no longer present. It should also be noted that their calculational procedure was a difficult one in that it placed all undefined errors present in the heat transfer data into the thermal conductivity term. For example, six of the seven-

---

[9] The addition of this large a quantity of solids would probably not result in increased heat transfer rates as the adverse effect of increases in viscosity on addition of the solids would almost certainly be of greater magnitude than the improvement in thermal conductivity.

Equation (36) is accurate to within 2% if $k_P = 100 k_l$. For most suspensions of metallic powders the conductivity of the solid is at least a hundred times as great as that of the liquid.

teen available data points[10] on copper suspensions predict a conductivity less than that of water alone, and silica ($k = 0.20$) appears to increase the thermal conductivity of water ($k = 0.38$). However, all these variations were within the accuracy claimed for the data.

In summary it may be stated that all available data of known reliability support the Orr-DallaValle conclusion that thermal conductivities of suspensions may be calculated by means of Eq. (35). This does not prove that the other effects (such as particle size) do not in actuality influence the thermal conductivity of suspensions, but rather that within the ranges of particle sizes (2 to 260 μ) and Reynolds numbers (3,000 to 3×10^5) investigated, the effects were too small to be measurable. In view of the over-all conclusions (at the end of this section) in regard to the utility of further heat transfer studies in this field, it is not recommended that a more detailed investigation of the thermal conductivity of suspensions be undertaken at this time.

2. *Evaluation of Viscosities*

Basically, heat transfer investigators have used two distinct methods of viscosity evaluation.

a. Bonilla *et al.* (B6), Winding *et al.* (W8), Miller (M13), and Orr and DallaValle (O5) measured the actual relationship between shearing stress and rate of shear in a laboratory viscometer.

b. Miller (M13) and Salamone and Newman (S1) took pressure-drop measurements in a section of pipe (of the same diameter as the tube through which the heat was transferred) in the same apparatus in which the heat transfer rates were being measured.

The first of these two methods has the important advantage that measurements of the actual physical properties of the fluids are made and used to interpret the heat transfer data. It sometimes has the important practical disadvantage that such measurements on dilute suspensions are difficult to make in view of the settling tendency of the solid phase to leave nonuniform samples. This difficulty has perhaps been completely overcome in a device described by Orr and DallaValle (O5). Their viscometer is shown in Fig. 13 and described in Section VI.

The second approach to measurement of viscometric properties, that of using a pipe of the same diameter as the heat transfer test section, has the advantage that simple and direct measurements under conditions

---

[10] Salamone and Newman state that the constant (limiting) value of thermal conductivity is reached above Reynolds numbers of about 50,000. However, in the plots which were drawn up to show the effect of particle size only data at considerably higher Reynolds numbers appear to have been used. The identical procedure was followed in the calculations discussed here.

identical to those in the heat transfer equipment are possible. The ease of obtaining these data is especially noteworthy. Further, any problems as to the applicability of laboratory measurements of physical properties (under laminar-flow conditions) to the problem of heat transfer under turbulent conditions is avoided. This same factor may also prove to be a limitation, however, since turbulent-flow pressure-drop measurements are not very sensitive to viscosity, and, if laminar-flow physical properties do not correlate such data, the question arises as to what physical properties the turbulent-flow pressure drops represent.

This second method does not lend itself to the development of quantitative correlations which are based solely on true physical properties of the fluids and which, therefore, can be measured in the laboratory. The prediction of heat transfer coefficients for a new suspension, for example, might require pilot-plant-scale turbulent-flow viscosity measurements, which could just as easily be extended to include experimental measurement of the desired heat transfer coefficient directly. These remarks may best be summarized by saying that both types of measurements would have been desirable in some of the research work, in order to compare the results. For a significant number of suspensions (four) this has been done by Miller (M13), who found no difference between laboratory viscosities measured with a rotational viscometer and those obtained from turbulent-flow pressure-drop measurements, assuming, for suspensions, the validity of the conventional friction-factor—Reynolds-number plot.[11] It is accordingly concluded here that use of either type of measurement is satisfactory; use of a viscometer such as that described by Orr (O5) is recommended on the basis that fundamental fluid properties are more readily determined under laminar-flow conditions, and a means is provided whereby heat transfer characteristics of a new suspension may be predicted without pilot-plant-scale studies.

3. *Design Equations for Use with Dilute Suspensions*

The pioneering work in this field seems to be due to Winding, Dittman, and Kranich (W8), who investigated the heat transfer characteristics

[11] Two of Miller's suspensions were slightly non-Newtonian in behavior ($0.8 < n' < 1.00$). For these he determined the differential viscosities over the range of shear rates from 5.8 to 77 sec.$^{-1}$ with a MacMichael viscometer. They were constant over about the upper 70% of this shear-rate range and were found to be equal to the experimentally determined turbulent-flow viscosities in the heat exchanger.

The shear rates in the heat exchanger are believed to have been much higher than the maximum shear rates obtained in the viscometer; hence constancy of the differential viscosities at these higher shear rates is indicated. They should not, however, be considered equal to the apparent viscosities at infinite shear rate, as they were termed by Miller, as no data are available to support such a statement.

of slightly non-Newtonian synthetic rubber (GR-S) latices. These investigators used calorimetrically measured values of specific heat and thermal conductivity and, for the viscosity of the fluid, used the limiting value of the differential viscosity at an infinite shear rate, as determined with a capillary-tube viscometer. Their results may be summarized as follows.

a. The standard Dittus-Boelter equation, viz.,

$$\frac{hD}{k} = 0.023 \left(\frac{DG}{\mu}\right)^{0.8} \left(\frac{C_p\mu}{k}\right)^x \qquad (37)$$

was found to correlate the data in a tube with an actual inside diameter of 0.781 in. The value of x was taken as 0.30 for cooling and 0.40 for heating, in accordance with the usual recommendation (M4) for the use of this equation. The scattering of the data from the correlating line was appreciable; only about half the data points fell within $\pm 10\%$ and several deviated by as much as 35%. No consistent trend may be noted to the deviations, which may probably be ascribed to the absence of any term to account for the variation of physical properties with temperature, such as the Sieder and Tate μ/μ_w correction factor.

The foregoing data were for the cooling of five different latices covering a range of Reynolds numbers of $2200 < \frac{DG}{\mu} < 26{,}000$. It is at first surprising that the correlation does not break down as Reynolds numbers of 2100 are approached, but, as will be discussed later, this is possibly due to the particular choice of the viscosity term. The differences in fluid properties between the various latices may perhaps be best described by the values of the flow-behavior index n', which in turn may be estimated from the published data on the laminar flow of these materials (W7). The flow-behavior index n' varied between 0.824 (pseudoplastic) and 1.00 (Newtonian). Solids contents and conversions of monomeric material varied between 14 to 28 and 30 to 75%, respectively.

b. On the basis of six data points $\left(4300 < \frac{DG}{\mu} < 8200\right)$ it appears that the data for both heating and cooling in a ⅜-in. tube may also be represented by the preceding equation provided the constant (0.023) is reduced to a value of about 0.017.

The over-all conclusions to be drawn from this work are not entirely satisfactory in that one is left with the need, apparently, to use a slightly different equation for every single tube diameter. However, in view of the pioneering nature of this paper the degree of correlation achieved is noteworthy. The reasons for the uncorrelated effect of tube diameter probably lie with the chosen constant values of the viscosity. The accompanying fluid-flow work showed that these fluids in turbulent flow revealed slightly higher "apparent" viscosities in small pipes than in larger ones.

From a practical viewpoint it is noteworthy that the tube sizes used approximated industrial heat-exchanger tubes. Accordingly, it would appear to be a sound procedure to use the correlations proposed for fluids of approximately the same physical properties and near the ranges of conditions studied, which may be summarized as follows:

n': 0.82 to 1.00

μ_d: 2 to 7 cp.

D: 0.3 to 0.8 in. (actual inside diameter)

$\dfrac{C_p \mu}{k}$: 17 to 96

$\dfrac{DG}{\mu}$: 2200 to 26,000

Essentially the same correlating equation was used by Miller (M13), who independently explored the range of Reynolds numbers from about 15,000 to 200,000 and found good agreement between Eq. (37) and his results[12] over the entire range, for the following suspensions:

a. coal in water: 1.9 to 7.3 wt. % solids
b. calcium carbonate in water: 2.5 to 15.6 wt. % solids
c. graphite in water: 5.8 to 13.5 wt. % solids
d. graphite in kerosene: 4.7 wt. % solids

In this work several fluids were appreciably non-Newtonian. As discussed in the previous section, the shear rates at which the differential viscosities were taken do not seem to coincide with the values which were probably occurring in the heat exchanger. However, the fluids were not sufficiently non-Newtonian in nature to make the choice of the shear-rate range particularly critical. For such materials, accordingly, the recommended design procedure is to determine the approximate values of shearing stress at the wall from pressure-drop estimates (Sections IIB and IIIE) and then to calculate differential viscosities at the same value of shearing stress or the corresponding shearing rate. These, when

---

[12] Miller's recommended equation actually had a higher constant (0.029) substituted for the value of 0.023 shown in Eq. (37). This was found necessary to enable correlation of his data on both water and aqueous suspensions, i.e., he was able to conclude that the same correlating equation is applicable to pure liquids and suspensions alike, although the coefficients were high for both. For the former, Eq. (37) is usually suitable; hence it may be concluded that the consistently high coefficients reported by Miller were probably due to unusual experimental factors and are not of general interest.

used with Eq. (37), would probably predict heat transfer coefficients correct to within ±25 to 35%. However, care must be exercised to note that the problem at hand falls into the ranges of experimental conditions studied by these investigators, as otherwise the errors involved might be considerably greater.

Orr and DallaValle (O5), on the basis of their extensive experimental results, recommend use of the equation

$$\frac{hD}{k} = 0.027 \left(\frac{DV\rho}{\mu_s}\right)^{0.8} \left(\frac{C_p\mu_s}{k}\right)^{1/3} \left(\frac{\mu_l}{\mu_{lw}}\right)^{0.14} \qquad (38)$$

for correlation of heat transfer data to slurries over the range of Reynolds numbers from 10,000 to 300,000, although their own data verify the utility of the equation to Reynolds numbers as low as 3,000. The viscosity of the suspension was calculated by means of the relationship

$$\mu_s = \mu_l/(1 - x_v/x_{vb})^{1.8} \qquad (39)$$

This empirical equation was found to represent the viscosity data with the following accuracies:

x_v/x_{vb}	Accuracy, %
0.20	±15
0.40	±30
0.60	±85

The deviation of their own data from Eq. (38) was occasionally greater than ±30%, which represents a slightly lower accuracy than the correlations of Winding et al. (W8) or Miller (M13). While this might be attributed partly to the greater range of variables covered by Orr, it is also likely that much of the scatter is due to the complete absence of any term in Eq. (38) to account for the non-Newtonian behavior of many of the slurries used. Reference to Orr's original data (O3) shows that several of the fluids were quite highly non-Newtonian, with flow-behavior indexes as low as 0.65.

Only one other publication on the subject of heat transfer to suspensions is available. Salamone and Newman (S1) recommended the correlation

$$\frac{hD}{k_l} = 0.131 \left(\frac{DV\rho_s}{\mu_s}\right)^{0.62} \left[\frac{(C_p)_l\mu_s}{k_l}\right]^{0.72} \left(\frac{k_p}{k_l}\right)^{0.05} \left(\frac{D}{D_p}\right)^{0.05} \left[\frac{(C_p)_p}{(C_p)_l}\right]^{0.35} \qquad (40)$$

No term is included in this equation which might account for the relative quantities of the two phases present, except in the viscosity and density terms. Obviously the specific-heat and thermal conductivity terms must also depend on the relative quantities of the two phases present, as well as on the corresponding values for each of the phases individually. The limitations of this approach have placed an undue and perhaps unrealistic importance on the dimensionless groups containing density and viscosity terms. Equation (40) is untenable, therefore, in at least two respects:

  a. The unusually high value (0.72) of the exponent on the Prandtl number, which is indicated to be incorrect by all prior work on both suspensions and single-phase systems.

  b. The predicted net effect of viscosity on the heat transfer coefficient is $h \propto (\mu)^{0.10}$ as compared to the usual (M4, M13, O5, W8) $h \propto \left(\dfrac{1}{\mu}\right)^{0.4 \text{ to } 0.5}$.

Thus, according to Eq. (40) heat transfer coefficients to viscous fluids are higher than those to less viscous materials, in contradiction to the behavior usually observed. Although the experimental data upon which Eq. (40) was based appear to be precise, it may be concluded that the correlation itself is not generally useful.

4. *Comparative Economics of Suspensions and Liquids as Heat Transfer Media*

Since several of the foregoing investigations appear to have been motivated by the hope of increasing the heat transfer rates of liquids by the addition of small quantities of solids of high thermal conductivity, it is relevant to consider the results in this light. For this purpose the tabulated data of Salamone and Newman (S1) on aqueous suspensions of copper may be used and the following conclusions drawn.

  a. At a constant value of the mass flow rate G, the film coefficients increased, with increased amounts of solids added to the water, over the range 2 to 10 wt. % solids. The first small quantity of solids added reduced the film coefficients so markedly, however, that even at the highest concentration studied the film coefficients were still below those of water alone. For example, the film coefficients for water, 2.50% solids, and 10.6% solids, suspensions at identical temperatures and flow rates of 126 lb./min. (inside a ½-in. I.P.S. pipe) were 3330, 3100, and 3210 B.t.u./(hr.)(sq. ft.)(°F.) respectively.

It appears from these data that the suspension always compares unfavorably to the pure liquid on this basis unless high solids concentrations are used.

  b. On a basis of equal power requirements, the following conditions held.

  (1) The film coefficients increased with increasing solids content of the suspension, above 2.5% solids. The film coefficients of the 2.50% solids suspensions were well below those of water alone.

  (2) Below a solids concentration of about 6% by weight, the film

coefficients of water are higher than those of the copper suspension; above this concentration the suspension has progressively higher film coefficients than water.

c. On the above-mentioned basis of equal power requirements, the suspension is the better heat transfer medium (at higher solids concentrations) primarily because its higher density decreases the power requirement at a given pressure drop. This beneficial effect of density is considerably offset by the simultaneous viscosity increases.

This discussion indicates that only small beneficial effects are possible on addition of solids and that addition of heavy salts, which could increase density without an appreciable effect on viscosity, would appear more interesting heat transfer mediums than suspensions. It is possible that in these systems too, however, the effects would usually not be of sufficient magnitude to warrant the complications introduced.

Orr and DallaValle (O5) compared suspensions of approximately 50 wt. % copper with water. They also concluded that under these conditions the suspension is a slightly better heat transfer medium but that the improvement was too small to be of practical interest.

5. *Summary*

a. Recommended Design Procedure. (1) The conventional forms of the Dittus-Boelter (37) or Sieder-Tate (38) equations have been shown to be applicable to suspensions. The former,

$$\frac{hD}{k} = 0.023 \left(\frac{DG}{\mu}\right)^{0.8} \left(\frac{C_p\mu}{k}\right)^x$$

is to be slightly preferred since it has been shown that small deviations from Newtonian behavior may be taken into account by use of the differential viscosity (at the correct wall shear stress) in place of the Newtonian viscosity.

(2) The physical properties should be determined as follows:

Thermal conductivities may be calculated by means of Eqs. (35) or (36) if the solids are truly inert or experimentally determined if the solids tend to associate or solvate in the liquid. In experimental determinations the procedure of Orr and DallaValle (O5) is recommended for suspensions which tend to settle.

Viscosities must be determined experimentally. For settlable solids, the suspension viscometer of Orr and DallaValle (O5) is recommended. For other suspensions any good rotational or capillary-tube viscometer is suitable (see Section VI).

If the suspension is non-Newtonian, the viscosity used should be the differential viscosity at the shear stresses found in the heat exchanger.

These may be estimated by the methods of Sections IIB (Fig. 7) and IIIE.

Specific heats and densities may be assumed to be weight averages if the solids are truly inert; otherwise experimental determinations are necessary. The experimental work poses no unusual difficulties in this case.

Predictions accurate to within ±30% are believed probable over the following ranges of variables:

DG/μ: 10,000 to 200,000
$C_p\mu/k$: 1.4 to 100
n': 0.8 to 1.0
D: ⅜ to 1.5 in.
μ: 2 to 13 cp.

b. Future Work. Relatively few practical applications are to be found for dilute suspensions of inert solids in liquids, and for these the art appears to be adequately defined at present. Since higher concentrations of inert solids and most concentrations of solids which associate with the liquid are appreciably non-Newtonian in behavior, it would appear that future work might best be directed toward a study of highly non-Newtonian systems. An adequate understanding of that field should simultaneously provide what little additional information is needed in the field of heat transfer to slightly non-Newtonian suspensions.

B. Heat Transfer to Highly Non-Newtonian Systems

Two experimental publications (B6, C3) and one theoretical paper (P5) are available in the field of heat transfer to liquids which are appreciably non-Newtonian in behavior. The data of Bonilla *et al.* (B6) have frequently been used by workers in the field of heat transfer to dilute suspensions but may more properly be treated here along with other fluids whose heat transfer characteristics are appreciably influenced by their departure from Newtonian behavior.

1. *Laminar-Flow Region*

Pigford (P5) has stated that the heat transfer coefficients of *Bingham-plastic fluids* in laminar flow will be greater than those of Newtonian fluids by a factor of approximately $1 + (\frac{1}{9})(\tau_y/\tau_w)$. Furthermore, the heat transfer coefficients should increase less rapidly than with the ⅓ power of flow rate found for Newtonian fluids.

For pseudoplastic fluids the dependence of heat transfer coefficients upon flow rate should be the same as for Newtonians, but their numerical value should be greater at any flow rate by the factor $\sqrt[3]{(3n' + 1)/4n'}$.

This relationship appears to be valid for reasonably large values of n' (above about 0.2) but would indicate infinite heat transfer coefficients for the limiting case of n' equal to zero. Such a limiting case corresponds to perfect "plug flow" (Fig. 8), for which McAdams (M4, p. 238) has shown the heat transfer coefficient to be twice as great as for Newtonian fluids when $wC_p/kL = 200$ and only about 30% greater when $wC_p/kL = 10$. Accordingly, modification of the foregoing theoretical relationship for very highly pseudoplastic materials appears to be necessary. Until this has been done and the predictions for both types of non-Newtonian behavior have been verified experimentally, adoption of the conservative design procedure of using the usual Newtonian heat transfer coefficients is suggested. McAdams (M4) recommends an equation which reduces to

$$\frac{h_a D}{k}\left(\frac{\mu_w}{\mu}\right)^{0.14} = 1.75 \left(\frac{wC_p}{kL}\right)_b^{1/3} \qquad (41)$$

when natural-convection effects are negligible, as is the usual case with the viscous non-Newtonians processed in the laminar-flow region. Presumably the ratio γ_w/γ would be used to replace the Newtonian μ_w/μ term. Preliminary results indicate that Eq. (41) will then be accurate to within 40% for non-Newtonians with a flow behavior index above 0.10 (V2).

If, on the other hand, dilatant behavior is encountered, the cube-root relationship of Pigford indicates lower heat transfer coefficients than for Newtonian fluids. Its use is recommended in this case on the basis that it does not appear to approach any unusual limit as n' approaches infinity. It has been pointed out (V2) that the assumptions used by Pigford would not be expected to break down in the case of highly dilatant behavior as they do near the limiting condition of plug flow. Vaughn (V2) has been concerned with the appropriate modification of the foregoing theoretical design equations, but the work is too incomplete at the time of this writing to be presented here.

2. *Turbulent-Flow Region*

   a. *Bingham-Plastic Fluids.* Viscometric data taken by Bonilla et al. (B6) were interpreted and reported on the assumption that the 0 to 18 wt. % calcium carbonate slurries used exhibited Bingham-plastic behavior. These viscometric data were not, however, used in the empirical correlation developed by the authors. McAdams (M4, p. 223) has pointed out that the particular correlating equation chosen by these authors may lead to predictions of zero or even negative heat transfer coefficients within the range of the authors' data. Accordingly, their empirical equation cannot be recommended for design purposes, but the work is excellent from an over-all viewpoint because of its pioneering nature in a very difficult field.

By use of the data of Bonilla et al., the heat transfer results may be recorrelated in the following manner:

For Newtonian fluids in turbulent flow,

$$h = \theta(D,V,C_p,k,\rho,\mu,g_c) \tag{42}$$

For Bingham plastics, similarily,

$$h = \theta'(D,V,C_p,k,\rho,\tau_y,\eta,g_c) \tag{43}$$

when θ and θ' denote unspecified functional relationships.

Application of dimensional analysis to Eq. (42) leads to the dimensionless groups of the Dittus-Boelter equation; for Eq. (43), one obtains

$$\frac{hD}{k} = \theta' \left[\frac{DV\rho}{\eta}, \frac{C_p\eta}{k}, \frac{D^2\rho\tau_y g_c}{\eta^2} \right] \tag{44}$$

or

$$N_{Nu} = \theta'(N_{Re}, N_{Pr}, N_{He}) \tag{44a}$$

As it is permissible to recombine dimensionless groups at will, it is convenient to rearrange Eq. (44) to eliminate the coefficient of rigidity η from the Hedstrom number. The purpose of such a rearrangement is to reduce the temperature sensitivity of this group, as all the other physical properties are rather insensitive to temperature changes. Such rearrangement gives

$$\frac{hD}{k} = \theta'' \left[\frac{DV\rho}{\eta}, \frac{C_p\eta}{k}, \frac{D^2\tau_y \rho g_c C_p^2}{k^2} \right] \tag{45}$$

The available understanding of flow of Bingham-plastic fluids (Section IIA) indicates that the effect of the last dimensionless group should decrease progressively with increased Reynolds numbers. Furthermore, the relationship between the Nusselt, Reynolds, and Prandtl numbers should probably be very similar to that for Newtonian fluids, which merely represent the special case of a Bingham plastic for which the yield value, τ_y, approaches zero. An empirical correlation of the Bonilla data which meets these requirements has been developed (M9). Unfortunately, the resulting correlation cannot be considered valid as no change in heat transfer behavior appeared to occur as the Reynolds numbers decreased to values below 2100. The lack of such a change in mechanism was believed to be due to inaccuracies in the measurement of the rheological properties of the fluids used.[13] Accordingly, one must conclude that the functional relationship between the variables of Eq. (45) is still to be determined.

b. Pseudoplastic Fluids. The one publication in this field, by Chu, Brown, and Burridge (C3), is an outgrowth of the thesis work (B8) by the last two men, who recommended that the equation

$$J + 100 = 0.56 \left(\frac{DG}{\mu_a}\right)^{0.82} \tag{46}$$

be used for correlation of the data, where

$$J = \left[\left(\frac{hD}{k}\right)\left(\frac{C_p\mu_a}{k}\right)^{-0.37} + 20(1 - y^y)\right](L/D)^{0.63}$$

and

$$y = \frac{\mu_0 - \mu_a}{\mu_a}$$

Equation (46) may lead to predictions of zero or even negative heat transfer coefficients for the authors' fluids within the range of Reynolds numbers for which the equation is claimed to be applicable. The parameter y, which is included to account for the deviation from non-Newtonian behavior, has a value of zero for Newtonian fluids and increases gradually toward infinity as the non-Newtonian character increases in the direction of pseudoplasticity. However, the peculiar form of the chosen function of this parameter does not uniquely characterize non-Newtonian behavior; it has been shown by Branch (B7) that the term $1 - y^y$ first increases as μ_0/μ_a increases, then goes through a maximum, and finally decreases, reaching negative values for $\mu_0/\mu_a > 2.0$.

The rather extensive data of these investigators have been recorrelated by Branch, as shown in Figs. 10, 11, and 12. The physical basis of this empirical correlation was developed as follows.

The velocity profiles of pseudoplastic non-Newtonian fluids (Fig. 8) in laminar flow deviate from the Newtonian parabola in the same way as the velocity profile of Newtonian liquids changes when heat is being transferred to them (M4, p. 229), since in both cases the viscosity of the fluid is lower at the wall than at the center of the tube. For the Newtonian

---

[13] Three factors support the contention that the rheological properties η and τ_y may not be correct: (a) no temperature dependency measurements are available and the room-temperature data cannot possibly apply rigorously; (b) with the more dilute suspensions, Bonilla *et al.* (B6) encountered serious settling problems in the MacMichael viscometer used to measure these properties; (c) for all suspensions the reported values of τ_y are surprisingly low.

situation, the empirical correction factor of Sieder and Tate, $(\mu/\mu_w)^{0.14}$, has been found fairly satisfactory for correlation of data in both the laminar- and turbulent-flow regions. Similar factors, therefore, which might account for the change in non-Newtonian viscosities with shear rate (such as μ_∞/μ_0 or μ_∞/μ_a), should be useful in accounting for the difference between Newtonian and non-Newtonian data. As in the case of heat transfer to Bingham-plastic fluids, it may be expected that the magnitude

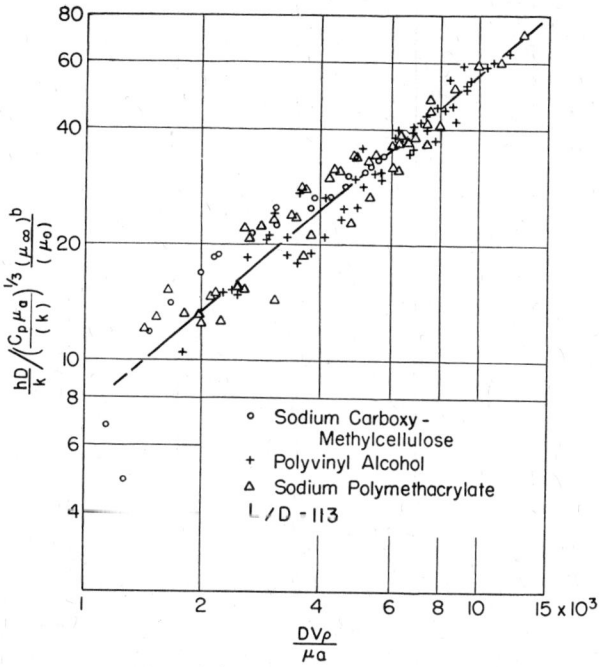

FIG. 10. Heat transfer correlation of Branch (B7) for pseudoplastic fluids.

of the correction should decrease as the Reynolds number increases (Fig. 11), as at high turbulence levels the shear rates should be reasonably high at all radii.

Examination of Fig. 10 shows that the majority of the data points are in the transition region where interpretation of data is difficult; hence it is impossible to recommend the correlation for Reynolds numbers outside the actual range covered. Since the data appear to fall about 50% above the conventional Newtonian curve at $N_{Re} = 10,000$ it must also be recommended that safety factors of at least this magnitude be applied to Fig. 10. The effect of L/D shown by Fig. 12 seems to extend to unusually high Reynolds numbers; in all cases D was constant at 0.433

in. The fact that there seems to be no break into laminar flow below $N_{Re} = 2100$ is inconclusive but would appear to be due to erroneous data since the same effect is noted both in the original (B8) and recorrelated (B7) work.[14]

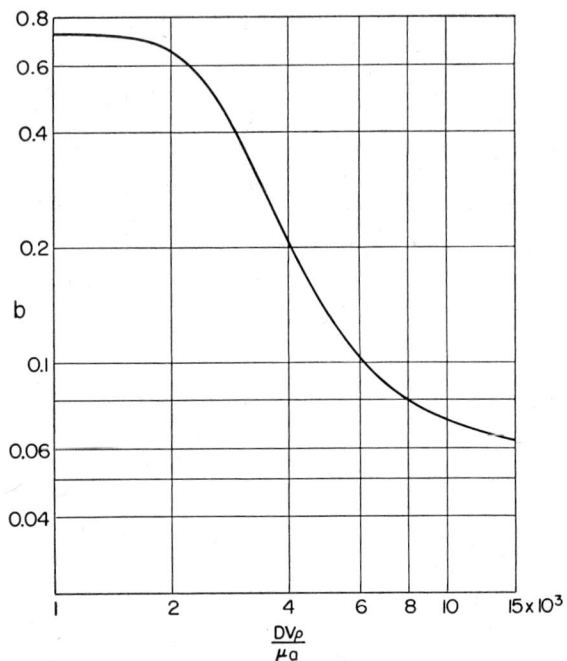

Fig. 11. Dependence of the parameter b (Figure 10) on Reynolds number.

All the heat transfer data were for cooling of the polymeric solutions. The solutions were dilute, with apparent viscosities of the order of 1 to 15 cp., but nevertheless the deviation from Newtonian behavior was very significant. The flow-behavior index n' varied from 0.65 to 1.00.

3. *Summary*

a. *Recommended Design Procedure.* In the *laminar region* a simplification of the conventional Newtonian heat transfer correlation [Eq. (41)] presented by McAdams is recommended for use with Bingham-plastic or

[14] Branch (B7) used as a measure of the apparent viscosity $\tau_w/(8V/D)$ instead of $\tau_w / \left(\frac{-du}{dr}\right)_w$. The former is not a true ratio of shear stress to shear rate for non-Newtonians, but its use is justifiable since it may be rigorously shown to be that apparent viscosity which must be used in order to define the pressure drop precisely in laminar flow. The difference between these two types of apparent viscosity is small for the work in question and, to avoid confusion, has not been considered in this presentation.

pseudoplastic fluids. This procedure will predict conservative values of the actual non-Newtonian heat transfer coefficients. For dilatant fluids ($n' > 1.00$) it is recommended that the heat transfer coefficients for Newtonian fluids be multiplied by the factor $\sqrt[3]{(3n' + 1)/4n'}$. Limited experimental data indicate an accuracy of $\pm 40\%$ for these procedures.

In the *turbulent-flow region* Figs. 10 to 12 are recommended for order-of-magnitude predictions. Two methods have been suggested for the correlation of accurate data when such become available.

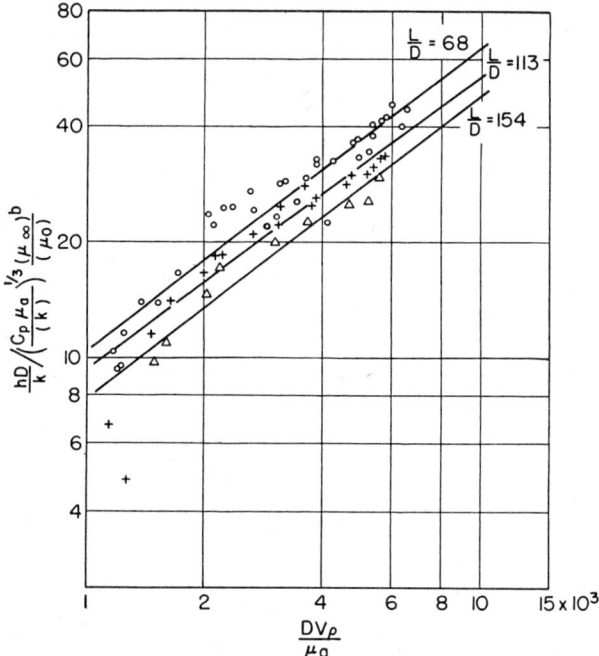

Fig. 12. Effect of L/D for pseudoplastic fluids. Correlation of Branch (B7).

b. Future Work. It might appear that future workers should concentrate first on the problem of heat transfer in the laminar region in order to benefit from the excellent available understanding of the flow under such conditions; logical starting points would appear to be the equations of Pigford (P5). As several groups of investigators are already actively pursuing this field, new workers might do well to attack the turbulent heat transfer problem directly. Two methods have been suggested for correlation of such data, but perhaps even more interesting would be the determination of the applicability of the generalized Reynolds number (Section IIB) to the problem of interpretation of heat transfer data. If this suggestion is adopted, experimental equipment should be constructed to enable the simultaneous study of the turbulent-flow problem.

This is suggested to confirm the applicability of this Reynolds number (or any other similar method of approach) to the correlation of turbulent-flow data prior to the complication of the fluid behavior by the imposition of temperature gradients. While the recommendation that extensive heat transfer work be delayed until the flow problem has been defined might initially appear to be unduly cautious, the work reviewed here indicates that such caution might be well worth while in the long run.

VI. Rheology and Viscometry

It is perhaps self-evident that a thorough knowledge of the methods of measuring flow properties must precede any progress in the development of engineering design methods and an understanding of flow through equipment. Nevertheless, almost every one of the numerous failures to reach a desired objective in this field may be attributed to an inadequate understanding on the part of the investigator of this particular problem. Accordingly, all newcomers as well as a substantial fraction of the experienced investigators in this field must take careful note of these factors. Otherwise, repetition of the many ludicrous statements in the recent engineering literature concerning the "problem" of correlating data from rotational instruments with data taken in a tube or pipe line may occur. In general, there is no problem in obtaining shearing-force–shearing-rate relationships from either type of instrument and in comparing data from the two; the alleged theoretical superiority of one type of instrument over the other is usually purely subjective. The sole exceptions to this statement occur in measurement of thixotropy and in the determination of the consistency and flow behavior indexes K' and n'.

The over-all problem of measurement of fluid properties can be subdivided into four categories:

A. Determination of the relationship between shearing stress and shearing rate, at constant temperature, over wide ranges of these variables.

B. Determination of the relationship between shear stress at the wall of a round tube or pipe, $D\Delta P/4L$, and the term $8V/D$.

C. Determination of the absence of thixotropy or rheopexy.

D. Viscometer design.

A. Determination of Shear-stress–Shear-rate Relationships

1. *Capillary-Tube Viscometers*

The flow rate of the fluid through a tube of known dimensions should be measured at several different pressure drops. From these original data the shear stress at the wall of the tube $\left(\dfrac{D\Delta P}{4L}\right)$ and the term $8V/D$ are calculated for each pressure drop and flow rate. A plot of $D\Delta P/4L$ versus $8V/D$ is conveniently made on logarithmic paper and its slope determined at arbitrary values of $D\Delta P/4L$ (or $8V/D$). Application of the Rabinowitsch-Mooney equation (Section IIB) gives

$$(-du/dr)_w = \frac{3n' + 1}{4n'} \frac{8V}{D} \tag{18}$$

where n' is the slope of the foregoing logarithmic plot at the corresponding value of $8V/D$. In this way the wall shear rate may be calculated at several values of wall shear stress and the results may be shown graphically as on Fig. 2. Alternatively, $D\Delta P/4L$ may be plotted logarithmically versus $(-du/dr)_w$; the slope of the resulting curve at any point is equal to the flow-behavior index n. Extension of the tangent to the curve (at the point where the slope is taken) to a shear rate of unity gives the corresponding numerical value of the consistency index K, since $\tau = K$ at this point.

2. Rotational Viscometers

With these instruments the torques measured at various rotational speeds of the cup or bob represent the original data. The shearing stress on the fluid at the wall of the cylindrical bob is given by

$$\begin{aligned}\tau_i &= \text{force/area} \\ &= \frac{2t}{D_i} \Big/ \pi D_i l \\ &= \frac{2t}{\pi D_i^2 l}\end{aligned} \tag{47}$$

An assumption implicit in the foregoing equation is that the shearing stresses are confined entirely to the cylindrical surface of the cup or bob, i.e., that there are no end effects due to the bottom. Many viscometers which minimize such end effects are available; accordingly the recommended procedure is to use such an instrument and to correct for the small remaining end effect by use of an equivalent bob length l, which differs slightly from the actual length. This equivalent bob length may be calculated by calibration of the viscometer with liquids of known viscosity.

The problem of calculation of the shearing rate at the surface of the bob is much more difficult but has recently been solved by Pawlowski (P1) and Krieger and Maron (K1). Their equations are entirely independent of the shear-stress–shear-rate relationship of the fluid. The Krieger and Maron equation consists of an infinite series, which, according to the authors, converges sufficiently (if the cup to bob diameter ratio is less than 1.2) to enable use of only three terms as follows:

$$(-du/dr)_i = \frac{4\pi N}{1 - 1/s^2}\left\{1 + k_1\left(\frac{1}{n''} - 1\right) + k_2\left[\left(\frac{1}{n''} - 1\right)^2 + \frac{d\left(\frac{1}{n''} - 1\right)}{d(\log t)}\right]\right\} \quad (48)$$

where k_1 and k_2 are constants of the instrument as follows:

$$k_1 = \frac{s^2 - 1}{2s^2}\left(1 + \frac{2}{3}\ln s\right) \quad (49)$$

$$k_2 = \frac{s^2 - 1}{6s^2}\ln s \quad (50)$$

For a given cup and bob combination, the term n'' represents the slope of a logarithmic plot of torque t versus rotational speed N at the particular value of N in question. Since the derivative in the last term of Eq. (48) is normally small, one may usually take as the final shear-rate equation:

$$(-du/dr)_i = \frac{4\pi N}{1 - 1/s^2}\left[1 + k_1\left(\frac{1}{n''} - 1\right) + k_2\left(\frac{1}{n''} - 1\right)^2\right] \quad (51)$$

The values of shear stress and shear rate calculated by means of Eqs. (47) and (51) may be used to draw a flow curve or to evaluate K and n, as in the case of capillary-tube data.

It may be noted that Eq. (51) may not be used to calculate rates of shear at the *cup* of a rotational viscometer. Therefore, regardless of the position at which the torque is measured, it is necessary to calculate shear stress at the bob by use of Eq. (47) in order to relate the shearing stress and rate at the same position in the viscometer.

It can be shown (V2) that Eq. (51) is identical to an equation presented earlier by Alves, Boucher, and Pigford (A3) for power-function non-Newtonians. This is due to the fact that neglecting the derivative in Eq. (48) implies assumption of this type of fluid behavior. Eq. (51) is to be slightly preferred to that of Alves *et al.* on the basis that its readily possible to take the slope of a semilogarithmic plot of the term $(1/n'') - 1$ versus the torque t in order to determine the error involved in neglecting the derivative of Eq. (48) for those fluids for which the term n'' is not independent of shear rate or torque.

For viscometers having a cup-to-bob diameter ratio which is essentially infinite, as for the case of the Brookfield Synchrolectric cylindrical-bob viscometers rotating in a large beaker or tank, the shear rate at the bob is given (A3, K1) by the relation

$$(-du/dr)_i = \frac{4\pi N}{n''} \qquad (52)$$

In this case Eq. (52) is rigorous for all fluids which are not time dependent.

The few viscometers which have a cup-to-bob diameter ratio of greater than 1.2 but for which this ratio cannot be considered to be infinite may be treated by use of more terms of the infinite series in Eq. (48). Such equations being of restricted interest, they are not tabulated here.

Since the data from both rotational viscometers and capillary tubes may be used to obtain the desired shear stress-rate of shear relationships, it may be concluded that properly designed viscometers of both types are theoretically of equal utility. The reader who may be concerned by the many invalid literature statements to the contrary should refer to some of the many references (A3, K1, O6, P4, R2, V2) where this has also been proved experimentally on a great variety of non-Newtonian materials.

B. Determination of the Relationship between $D\Delta P/4L$ and $8V/D$

1. *Capillary-Tube Viscometers*

With these instruments the relationship between $D\Delta P/4L$ and $8V/D$ is obtained directly. On a logarithmic plot of $D\Delta P/4L$ versus $8V/D$ the slope of the curve at any point is equal to the flow-behavior index n'; extension of the tangent to the curve at this point to a value of $8V/D$ of unity gives the corresponding value of the consistency index K'.

2. *Rotational Viscometers*

The problem of calculating the relationship between wall shear stress and $8V/D$ from shear-stress–shear-rate data depends on the assumption of an algebraic relationship between the shear stress and shear rate unless graphical procedures are employed. For the very important case of fluids which follow the power-law relationship between shear stress and shear rate, Eq. (3), it has been shown (M11) that

$$n' = n \qquad (53)$$

$$K' = K\left(\frac{3n'+1}{4n'}\right)^{n'} \qquad (54)$$

These relationships may be used with Eq. (20) to draw a curve of $D\Delta P/4L$ versus $8V/D$.

It is important to note that, although one may occasionally encounter fluids to which the power-law relationship does not apply over a sufficiently wide range to make Eqs. (53) and (54) valid for engineering purposes (say within 5%), only one such material, a concentrated Napalm gel, has come to the writer's attention to date (M3).

C. Determination of the Absence of Thixotropy and Rheopexy

The discussion in this chapter applies rigorously only to fluids which show negligible time dependence of fluid properties at constant temperature. It was pointed out in section IIB that pipe lines and mixers may frequently be designed for the limiting cases of flow properties at zero or infinite times of shear for thixotropic and rheopectic materials, respectively, but this is not true of heat exchange equipment, in which case the actual flow properties of the fluid in the exchanger are of importance. Accordingly, it is necessary, first, to determine the existence of thixotropy or rheopexy and, second, to develop design methods which are useful for such fluids. Unfortunately, no knowledge whatever is available on the second of these problems.

The existence of time dependence, as the study of other viscometric problems, may be determined with either capillary-tube or rotational viscometers.

1. *Capillary Tubes*

a. Thixotropic fluids show decreases in consistency with increasing times of shear at a given shearing stress ($D\Delta P/4L$). Therefore, the term $8V/D$ (or the flow rate—if diameter is held constant) will increase as the length of the tube is increased at constant shearing stress. Conversely, the term $8V/D$ will decrease with increasing tube diameter at constant values of tube length and $D\Delta P/4L$ (A3), as the velocity normally in-increases in this case, giving shorter times of shear and higher consistencies. Ambrose and Loomis (A4) have verified this effect of tube diameter experimentally.

b. Rheopectic fluids show exactly the opposite effects: at a given $D\Delta P/4L$ the value of $8V/D$ will increase with increasing tube diameter and decrease with increasing tube length.

Care must be taken to ensure the absence of significant amounts of shear before the fluid enters the tube in which its pressure drop and flow rate are determined. Otherwise, the material may have reached its limiting condition at "infinite" time of shear and the time effects might no longer be evident. This might occur, for example (A3) if the fluid passed through a pump before entering the test section.

2. *Rotational Viscometers*

In rotational viscometers thixotropy shows up as a progressive decrease in torque as time increases (until the "infinite-time" properties are reached) at constant rotational speed. Care must be taken, however, not to confuse the initial oscillation of a viscometer cup or bob, which is due to inertial forces, with thixotropy. Conversely, rheopectic behavior shows progressively increasing torques at constant rotational speeds.

Since fluid shear rates vary enormously across the radius of a capillary tube, this type of instrument is perhaps not well suited to the quantitative study of thixotropy. For this purpose, rotational instruments with a very small clearance between the cup and bob are usually excellent. They enable the determination of "hysteresis loops" on a shear-stress–shear-rate diagram, the shapes of which may be taken as quantitative measures of the degree of thixotropy (G3). Since the applicability of such loops to equipment design has not yet been shown, and since even their theoretical value is disputed by other rheologists (L4), they are not discussed here. These factors tend to indicate that the experimental study of flow of thixotropic materials in pipes might constitute the most direct approach to this problem, since theoretical work on thixotropy appears to be reasonably far from application. Preliminary estimates of the experimental approach may be taken from the one paper available on flow of thixotropic fluids in pipes (A4). In addition, a recent contribution by Schultz-Grunow (S6) has presented an empirical procedure for correlation of unsteady state flow phenomena in rotational viscometers which can perhaps be extended to this problem in pipe lines.

D. Viscometer Design

1. *Capillary-Tube Viscometers*

The essential features of capillary-tube viscometers may be summarized to be

  a. Equipment for the precise control and measurement of the applied pressure (to about 300 lb./sq. in.).

  b. A material storage chamber (volume of at least 250 ml.).

  c. A small tube or pipe of known dimensions, through which the measured flow occurs (approx. dimensions of 0.05 in. in diameter by 12 in. in length).

  d. Precise temperature control of the material in the storage chamber and capillary tube.

e. Flow-rate determination equipment (usually scales and a stopwatch).

The figures in parentheses indicate the scale of equipment which has been found to be very useful in work on viscous non-Newtonian slurries and solutions at the University of Delaware (R2). In some cases the measured pressure drop must be corrected for pressure loss due to kinetic energy imparted to the fluid, pressure losses at the entrance or exit of the tube (i.e., end effects), and occasionally the pressure due to the head of liquid in the tube and material storage chamber must be considered (H1). The first and last of these three effects require straightforward corrections; the end effects can usually be estimated in only one of two ways. First, measurements of pressure drop in tubes of various lengths and constant diameter, when extrapolated to zero tube length, give a precise but tedious evaluation of end effects. Second, calibration with Newtonian liquids of known viscosity and preferably with viscosities and densities in the same range as those of the unknown materials to be evaluated, provides an easy but only approximate method of correction. Since the latter correction will seldom be precise, it is important to check that it is small; normally all these corrections are small for viscous non-Newtonians and no particular difficulty is encountered with them. With low-consistency fluids, extreme difficulties have been reported (B8, C3).

Chemical engineers may be surprised to note that rheologists have long been concerned with "slip" of the fluid at the wall of a viscometer. There seems to be no sound reason to believe that true slip actually occurs, i.e., that the fluid velocity is not zero directly next to the wall, but an analogous phenomenon which may be treated in the same way has been observed (G3, R2, R4) in suspensions of solids in liquids. In such materials the solids sometimes appear to separate from the liquid next to the wall, leaving a thin film of the liquid, of lower consistency than the suspension, which shears easily and which may account for an appreciable part of the total velocity gradient within the fluid. Mooney (M15) has presented in detail a rigorous procedure for accounting for such slip which rheological measurements are made in capillary tubes. Slip is readily noticeable if tubes of several diameters are used in that the data of $D\Delta P/4L$ versus $8V/D$ no longer fall on the same curve for various tube diameters (i.e., it is similar to thixotropic behavior in this sense).

For rheological measurements the flow through the tube must be laminar; no orifices or other restrictions may be present (as they would induce turbulence), and the flow rate (or pressure drop) must be variable. Apparently very few commercially available instruments meet these requirements, but it is often readily possible to build such equipment. Severs (S8) has discussed these factors in some detail. A recent development is the varying-head viscometer (M2), which enables the rapid measurement of a complete flow curve with a single sample of fluid. Although it should be possible to correct this instrument for various end effects, this has not yet been done. Therefore at present this instrument is recommended for study of non-Newtonian fluids only if corrections

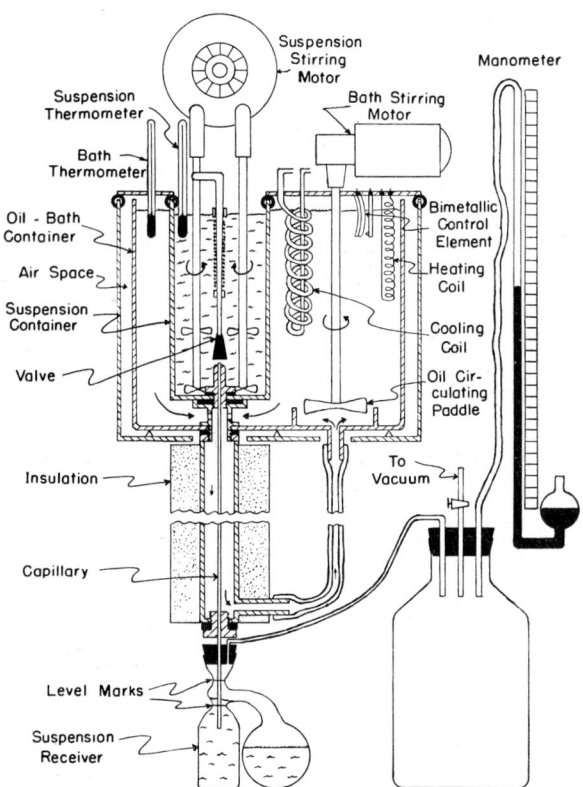

FIG. 13. Orr-DallaValle suspension viscometer. Reprinted with permission from *Chem. Eng. Progr. Symposium Ser. No. 9* **50**, 29 (1954).

for kinetic energy, end effects, etc., are unnecessary. This is usually the case with viscous materials, but insufficient pressure drop is available in this instrument to permit its use with extremely viscous fluids.

Orr and DallaValle (O5) have designed a capillary-tube viscometer which is especially suited for use with suspensions which tend to settle. This instrument, shown in Fig. 13, consists of the following four components:

(i) A well-agitated suspension container.

(ii) A capillary tube through which the suspension flows in laminar motion.

(iii) A suspension receiver fitted with a pressure controller and manometer.

(iv) A constant-temperature bath.

In this apparatus settling of the suspension cannot occur except in the capillary tube itself (where the holdup is of only a few seconds' duration and the settling therefore negligible in suspensions of normal densities and viscosities) or in the suspension receiver, where settling is irrelevant. By changing the diameter of the capillary tube and varying the vacuum to the suspension receiver, the applied shearing stress may be easily varied over a ten- to twentyfold range. This is frequently satisfactory, as suspensions which settle readily are often not highly non-Newtonian. Accordingly, for such materials it is not particularly important to obtain data over a greater range of shear rates. It may be noted, however, that this does represent a significantly greater range than is available on many commercial rotational viscometers.

2. *Rotational Viscometers*

These instruments are fairly common; examples are the Brookfield (Brookfield Engineering Laboratories, Stoughton, Mass.), the Precision-Interchemical (Precision Scientific Co., Chicago), the Hercules Hi-Shear (Martinson Machine Co., Kalamazoo, Mich.), the Stormer (Central Scientific Co., Chicago; A. H. Thomas Co., Philadelphia), the MacMichael (Fisher Scientific Co., New York), and the Fann V-G (Geophysical Machine Works, Houston), the Hagan (Hagan Corp., Pittsburgh) and the Drage (Drage Products, Inc., Union City, N. J.).[15]

As in the case of capillary-tube units, the shear rate (rotational speed) should be variable over wide ranges (10- to 1000-fold) and baffles or other obstructions which could interfere with the laminar-flow pattern must be absent. Since the fluid is sheared for long periods of time in these instruments, temperature control is much more critical, especially in the case of high-consistency materials, for which temperature rises of over 20°C. (W2) have been recorded. Weltmann and Kuhns (W5) subsequently presented an erudite mathematical analysis of the temperature distribution within the layers of sheared fluid.

For the investigation of thixotropy, it may be desirable that the range of shear rates in the viscometer be small (W4). This means that the

[15] No attempt has been made to list all viscometers or suppliers but simply to mention some of the better instruments available and representative sales agencies in the United States. A very detailed list, including many European viscometers, is given in the Proceedings of the First International Rheological Congress (1948).

ratio of cup-to-bob diameter should be very close to unity and both should be as large as conveniently possible. Under these conditions the width of the zone to which the fluid is confined is very small and approaches the ideal arrangement of two parallel flat plates. An instrument embodying these features to a great degree is the recently developed instrument of Merrill (M6), which has a clearance between the cup and bob of only 0.006 in. At the high shear rates for which this viscometer was designed temperature control is extremely difficult but good control can apparently be achieved by using runs of very short duration, although this requires limiting the number of readings taken per run and per hour. Naturally, such instruments can be used only if the particles (or flocs in a flocculated suspension) are small in size.

For practical reasons some rotational viscometers are fitted with bobs of complex geometry. For one of these an empirical method has been presented (M10) which enables use of the instrument for the sizing of pipe lines. It would appear likely that this same method could be extended to other geometries provided that the flow of the fluid around the viscometer bob is laminar.

A rather complete survey of the entire field of viscometry, including the mathematical relationships applicable to various types of instruments, has been made by Philippoff (P4). The problem of "slip" at the walls of rotational viscometers has been discussed by Mooney (M15) and Reiner (R4). Mori and Ototake (M17) presented the equations for calculation of the physical constants of Bingham-plastic materials from the relationship between an applied force and the rate of elongation of a rod of such a "fluid."

3. *Miscellaneous and Control Instruments*

The unsuitable nature of many commercial instruments which are in common use clearly illustrates the confusion prevalent in the field of viscometric measurements. Many instruments measure some combination of properties which depend only partly on the fluid consistency since the flow is not laminar. In others the shear rates are indeterminate and the data cannot be interpreted completely. Examples of such units include rotational viscometers with inserted baffles, as in the "modified" Stormer; instruments in which the fluid flows through an orifice, as in the Saybolt or Engler viscometers; instruments in which a ball, disk, or cylinder falls through the fluid, as in the Gardiner mobilometer. Recently even the use of a vibrating reed has been claimed to be useful for measurement of non-Newtonian viscosities (M14, W10), although theoretical studies (R6, W10) show that true physical properties are obviously not obtainable in these instruments for such fluids. These various instruments do have a restricted utility for control purposes since a deviation of any one of several physical properties from its fixed value would result in a changed instrument reading, but it must be obvious that they are of no significant utility for research or design purposes. Since the shear rates must be variable over wide ranges, many instruments used commonly by the petroleum industry in which the force producing flow cannot be varied are also unsatisfactory although the resistance to motion may be due to purely viscous forces. It is in a sense unfortunate that such instruments are in widespread industrial use as they eliminate potentially extensive sources of plant-scale physical property data; furthermore the adaptation of rheologically-useful instruments to control work would seem to require only minor ingenuity to match the convenience of most of these empirical instruments. The sad state of affairs in this area is clearly brought out by Thomas's (T1) review of continuous-recording viscometers: of the seven instruments discussed only three are both available commercially and capable of obtaining rheologically-useful data.

E. Summary

It may be concluded that well-designed rotational and capillary-tube viscometers are generally more useful than any other type. Accordingly, rheological equations for interpretation of data have been presented for these two types of instruments.

Viscometer design to fill simultaneously the needs of quality-control and production personnel as well as development engineers would appear to be a stimulating and rewarding field to which, evidently, little intelligent attention has been given to date.

Acknowledgment

D. W. Dodge, G. L. Houghton, R. E. Otto, R. L. Pigford, Sidney Rankin, and R. D. Vaughn contributed directly to the final manuscript. Many of the authors of work reviewed in this chapter supplied supplementary data, figures, and calculations to enable more detailed analysis and comparison of various sources of information than would otherwise have been possible.

Nomenclature

Note:—Most of the correlating equations are dimensionless; hence any consistent set of units may usually be used. The units in the following table represent those normally used by the author.

a Empirical exponent in Eq. (32), dimensionless
A Surface area, sq. ft.
b Empirical exponent in Figs. 10 to 12, dimensionless (see also under "Subscripts")

B, C Constants in the Eyring-Powell relationship, Eq. (16)
C, E Constants in Eq. (34)
C_p Specific heat at constant pressure, B.t.u./(lb.$_M$)(°F). $(C_p)_p$ refers to the specific heat of solid

particles in a suspension and $(C_p)_l$ to that of the liquid phase of a suspension

d Differential operator, dimensionless

du/dr Shear rate, sec.$^{-1}$ $(du/dr)_w$ and $(du/dr)_i$ refer to the shear rate at a wall and at the inner cylinder (cup) of a rotational viscometer, respectively

D Diameter of pipe or impeller, ft. D_p = particle diameter; D_i and D_o refer to the bob and cup diameters of a rotating cylindrical viscometer

$\dfrac{D \Delta P}{4L}$ Shear stress at the wall of a pipe, lb.$_F$/sq. ft.

f Fanning friction factor, dimensionless

F Shearing force, lb.$_F$

g_c Dimensional conversion factor, 32.2 ft. lb.$_M$/(sec.^{2})(lb.$_F$)

G Mass velocity, lb.$_M$/(sec.)(sq. ft.) or lb.$_M$/(hr.)(sq. ft.)

h Film coefficient of heat transfer, B.t.u./(hr.)(sq. ft.)(°F.). The coefficient h_a is based on an arithmetic-mean driving force

k Thermal conductivity, B.t.u./(hr.)(ft.)(°F.). k_s, k_l, and k_p refer to the thermal conductivities of a suspension, a liquid, and the solid particles of a suspension, respectively

k_1, k_2 Instrument constants defined by Eqs. (49) and (50), dimensionless

K Fluid consistency index defined by Eq. (3), (lb.$_F$)(sec.^{n})/sq. ft.

K' Fluid consistency index, defined by Eqs. (20) and (54), (lb.$_F$)(sec.$^{n'}$)/sq. ft.

K.E.   Fluid kinetic energy, (ft.)(lb.$_F$/lb.$_M$)

l Equivalent length of the bob of a rotational viscometer, ft.

L Length of a pipe or tube, ft.

n Flow behavior index, defined by Eq. (3), dimensionless

n' Flow behavior index defined by Eq. (19), dimensionless

n'' $\dfrac{d(\log t)}{d(\log N)}$, dimensionless

N Rotational speed, rev./min. or rev./sec.

N_{He} Hedstrom number, dimensionless [Eq. (11)]

N_{Nu} Nusselt number $\dfrac{hD}{k}$, dimensionless

N_P Power number $Pg_c/(\rho N^3 D^5)$, dimensionless

N_{Pr} Prandtl number, dimensionless, taken as $\dfrac{C_P \mu}{k}$ for a Newtonian fluid and as $\dfrac{C_P \eta}{k}$ for Bingham plastics

N_{Re} Reynolds number, dimensionless, taken as $DV\rho/\mu$ and $DV\rho/\eta$ for Newtonian and Bingham-plastic fluids respectively. The generalized Reynolds number $D^{n'} V^{2-n'} \rho / \gamma$ is applicable to all except thixotropic and rheopectic fluids

P Power, ft. lb.$_F$/sec.

ΔP Pressure drop, lb.$_F$/sq. ft. $(\Delta P/L)_{TP}$, $(\Delta P/L)_{LP}$, and $(\Delta P/L)_{GP}$ refer to the pressure drop per unit length of pipe for two-phase flow, flow of the liquid phase, and flow of the gas phase, respectively

ΔP_C Pressure drop at a contraction, lb./sq. in.

ΔP_{KE} Pressure drop due to the kinetic energy of the fluid, lb./sq. in.

r Lineal or radial distance, ft.

R Radius of a pipe (internal), ft.

s D_o/D_i, dimensionless

t Temperature, °F. t_b, t_w, and t_f are the bulk, wall, and film temperatures of the fluid, respectively

u Local velocity, ft./sec.

V Average or bulk velocity, ft./sec.

$8V/D$ Shear rate of a Newtonian fluid at the wall of a pipe (laminar flow), sec.$^{-1}$

w Mass flow rate, lb.$_M$/hr. or lb.$_M$/sec.

x Dimensionless exponent in the Dittus-Boelter equation [Eq. (37)]

x_v Volume fraction of solids in a suspension, dimensionless

x_{vb} Volume fraction of solids in a sedimented suspension, dimensionless

α Kinetic-energy correction factor defined by Eq. (30), dimensionless, or a symbol indicating a direct proportionality

γ Denominator of the generalized Reynolds number, defined by Eq. (24), $\text{lb.}_M/(\text{ft.})(\text{sec.}^{2-n'})$ γ_w refers to this term evaluated at the wall temperature

η Coefficient of rigidity of a Bingham-plastic fluid, $\text{lb.}_M/(\text{sec.})(\text{ft.})$ or $\text{lb.}_M/(\text{hr.})(\text{ft.})$

$\theta, \theta', \theta'', \theta'''$ Indicators of unspecified functional relationships

μ Viscosity of a Newtonian fluid, $\text{lb.}_M/(\text{sec.})(\text{ft.})$ or $\text{lb.}_M/(\text{hr.})(\text{ft.})$. μ_s refers to the viscosity of a suspension, μ_l to that of a liquid phase, and μ_w and μ_{lw} to the viscosities of a fluid and the liquid phase at the wall temperature, respectively

μ_a Apparent viscosity of a non-Newtonian fluid at some specified shear rate, $\text{lb.}_M/(\text{sec.})(\text{ft.})$ or $\text{lb.}_M/(\text{hr.})(\text{ft.})$. μ_0 and μ_∞ refer to the apparent viscosities of non-Newtonian fluids at zero and infinite shear rates, respectively

μ_d $\dfrac{d\tau}{d(du/dr)}$. Differential viscosity, $\text{lb.}_M/(\text{sec.})(\text{ft.})$ or $\text{lb.}_M/(\text{hr.})(\text{ft.})$

π 3.1415 . . .

ρ Fluid density, $\text{lb.}_M/\text{cu. ft.}$

τ Shear stress (F/A), $\text{lb.}_F/\text{sq. ft.}$ τ_w refers to the shear stress at the wall of a round pipe $(D\Delta P/4L)$ and τ_i to the shear stress at the wall of a viscometer bob

τ_y Yield value or yield stress of a Bingham-plastic fluid, $\text{lb.}_F/\text{sq. ft.}$

φ Indicator of an unspecified functional relationship

SUBSCRIPTS

b Bulk temperature
l Liquid
p Particle
s Suspension

REFERENCES

A1. Agoston, G. A., Harte, W. H., Hottel, H. C., Klemm, W. A., Mysels, K. J., Pomeroy, H. H., and Thompson, J. M., *Ind. Eng. Chem.* **46**, 1017 (1954).
A2. Allis-Chalmers Manufacturing Co., Bulletin No. 1649.
A3. Alves, G. E., Boucher, D. F., and Pigford, R. L., *Chem. Eng. Progr.* **48**, 385 (1952).
A4. Ambrose, H. A., and Loomis, A. G., *Physics* **4**, 265 (1933).
A5. da C. Andrade, E. N., and Fox, J. W., *Proc. Phys. Soc. (London)* **B62**, 483 (1949).
B1. Badger, W. L., and McCabe, W. L., "Elements of Chemical Engineering," 2nd ed., p. 31. McGraw-Hill, New York, 1936.
B2. Bestul, A. B., and Belcher, H. V., *J. Appl. Phys.* **24**, 696 (1953).
B3. Beyer, C. E., and Spencer, R. S., "Rheology in Molding." Paper presented at the Kansas City American Chemical Society Meeting, 1954.
B4. Beyer, C. E., and Towsley, F. E., *J. Colloid Sci.* **7**, 236 (1952).
B5. Binder, R. C., and Busher, J. E., *J. Appl. Mech.* **13**, A101 (1946).
B6. Bonilla, C. F., Cervi, A., Colven, T. J., and Wang, S. J., *Chem. Eng. Progr. Symposium Ser. No. 5* **49**, 127 (1953).
B7. Branch, R. E., Jr., Master of Chemical Engineering thesis, University of Delaware, Newark, 1954.

B8. Brown, F., and Burridge, K. G., Master of Chemical Engineering thesis, Polytechnic Institute of Brooklyn, 1952.
B9. Brown, G. A., and Petsiavas, D. N., Paper presented at the New York American Institute of Chemical Engineers Meeting (December 1954).
B10. Buckingham, E., *Proc. Am. Soc. Testing Materials* **21,** 1154–61 (1921).
C1. Caldwell, D. H., and Babbitt, H. E., *Trans. Am. Inst. Chem. Engrs.* **37,** 237 (1941).
C2. Carley, J. F., *J. Appl. Phys.* **25,** 1118 (1954).
C3. Chu, J. C., Brown, F., and Burridge, K. G., *Ind. Eng. Chem.* **45,** 1686 (1953).
D1. Daniel, F. K., *India Rubber World* **101,** 33 (1940).
D2. Dodge, D., Private communication, 1955.
D3. Durst, R. E., and Jenness, L. C., *TAPPI* **37,** 417 (1954).
E1. Extrusion Symposium, *Ind. Eng. Chem.* **45,** 969 (1953).
F1. Filatov, B. S., *Kolloid. Zhur.* **16,** 65 (1954).
F2. Fok, S. M., Ph.D. thesis in Chemical Engineering, Case Institute of Technology, Cleveland, 1954.
G1. Govier, G. W., and Winning, M. D., Paper presented at the Montreal Meeting, American Institute of Chemical Engineers (1948). Based on Master of Science thesis in Chemical Engineering by Winning, University of Alberta, Edmonton, 1948.
G2. Green, H., *Proc. Am. Soc. Testing Materials* **20,** 451 (1920).
G3. Green, H., "Industrial Rheology and Rheological Structures." Wiley, New York, 1949.
G4. Green, H., and Haslam, G., *Ind. Eng. Chem.* **17,** 726 (1925).
G5. Gregory, W. B., *Mech. Eng.* **49,** 609 (1927).
H1. Hall, H. T., and Fuoss, R. M., *J. Am. Chem. Soc.* **73,** 265 (1951).
H2. Harper, R. C., Jr., and Riseman, J., *J. Colloid Sci.* **9,** 81 (1954).
H3. Head, V. P., *TAPPI* **35,** 260 (1952).
H4. Hedstrom, B. O. A., *Ind. Eng. Chem.* **44,** 651 (1952).
H5. Hixson, A. W., Drew, T. B., and Knox, K. L., *Chem. Eng. Progr.* **50,** 592 (1954).
H6. Houghton, G. L., and Ching, G. P. K., B.Ch.E. thesis, University of Delaware, Newark, 1955.
I1. Ibrahin, A. K., *J. Chem. Phys.* **22,** 1274 (1954).
K1. Krieger, I. M., and Maron, S. H., *J. Appl. Phys.* **25,** 72 (1954).
K2. Kruyt, H. R., and van Selms, F. G., *Rec. trav. chim.* **62,** 415 (1943).
L1. Lewis, E. E., and Winchester, C. M., *Ind. Eng. Chem.* **45,** 1123 (1953).
L2. Lewis, W. K., Squires, L., and Broughton, G., "Industrial Chemistry of Colloidal and Amorphous Materials." Macmillan, New York, 1942.
L3. Lockhart, R. W., and Martinelli, R. C., *Chem. Eng. Progr.* **45,** 39 (1949).
L4. Lower, G. W., Walker, W. C., and Zettlemoyer, A. C., *J. Colloid Sci.* **8,** 116 (1953).
M1. Magnusson, K., *Ing. Vetenskaps Akad.* **23,** 86 (1952).
M2. Maron, S. H., Krieger, I. M., and Sisko, A. W., *J. Appl. Phys.* **25,** 971 (1954).
M3. Matthews, T. A., II., Private communication, 1954.
M4. McAdams, W. H., "Heat Transmission," 3rd ed. McGraw-Hill, New York, 1954.
M5. McMillen, E. L., *Chem. Eng. Progr.* **44,** 537 (1948).
M6. Merrill, E. W., *J. Colloid Sci.* **9,** 7 (1954).
M7. Merrill, E. W., *J. Colloid Sci.* **9,** 132 (1954).
M8. Meskat, W., *Chem. Ing. Tech.* **24,** 333 (1952).

M9. Metzner, A. B., Lecture Notes in Chemical Engineering 564 (Special Topics in Heat Transfer), University of Delaware, Newark, 1953.
M10. Metzner, A. B., *Chem. Eng. Progr.* **50**, 27 (1954).
M11. Metzner, A. B., and Reed, J. C., *A.I.Ch.E. Journal* **1**, 434 (1955); technical report to Office of Ordnance Research, U. S. Army, under Contract No. DA-36-034-ORD-1495 (August, 1954).
M12. Mill, C. C., *Chemistry & Industry*, p. 156 (1952).
M13. Miller, A. P., Jr., Ph.D. thesis in Chemical Engineering, University of Washington, Seattle, 1953.
M14. Minneapolis-Honeywell Instrumentation Data Sheet No. 10.13-2a; Minneapolis-Honeywell Regulator Co., Philadelphia, 1953.
M15. Mooney, M., *J. Rheol.* **2**, 210 (1931).
M16. Mooney, M., and Black, S. A., *J. Colloid Sci.* **7**, 204 (1952).
M17. Mori, Y., and Ototake, N., *Chem. Eng. (Japan)* **17**, 224 (1953).
M18. Mori, Y., and Ototake, N., *Chem. Eng. (Japan)* **18**, 221 (1954).
M19. Mori, Y., and Ototake, N., *Chem. Eng. (Japan)* **19**, 9 (1955).
O1. Oldroyd, J. G., *Proc. 1st Intern. Rheol. Congr.* **2**, 130 (1948).
O2. Ooyama, Y., and Ito, S., *Chem. Eng. (Japan)* **14**, 96 (1950).
O3. Orr, Clyde, Jr., Ph.D. thesis in Chemical Engineering, Georgia Institute of Technology, Atlanta, 1952.
O4. Orr, C., Jr., and Blocker, H. G., *J. Colloid Sci.* **10**, 24 (1955).
O5. Orr, C., Jr., and DallaValle, J. M., *Chem. Eng. Progr. Symposium Ser. No. 9* **50**, 29 (1954).
O6. Otto, R. E., and Metzner, A. B., "Agitation of non-Newtonian Fluids." Paper presented at the Seventh Annual Delaware Chemical Symposium (1955). Technical Report to Office of Ordnance Research, U. S. Army, under Contract No. DA-36-034-ORD-1495 (April, 1955).
P1. Pawlowski, J., *Kolloid-Z.* **130**, 129 (1953).
P2. Perkins, A., and Glick, J. J., B.Ch.E. thesis, University of Delaware, Newark, 1954.
P3. Perry, J. H., "Chemical Engineers Handbook," 3rd ed. McGraw-Hill, New York, 1950.
P4. Philippoff, W., "Viskosität Der Kolloide." (1942). Lithoprinted by J. W. Edwards, Ann Arbor, Mich. (1944).
P5. Pigford, R. L., *Chem. Eng. Progr. Symposium Ser. No. 17*, **51**, 79 (1955).
P6. Pryce-Jones, J., *Proc. Univ. Durham Phil. Soc.* **10**, 427 (1948).
R1. Rabinowitsch, B., *Z. physik. Chem.* **A145**, 1 (1929).
R2. Rankin, S., Master of Chemical Engineering thesis, University of Delaware, Newark, 1955.
R3. Reed, J. C., Master of Chemical Engineering thesis, University of Delaware, Newark, 1954.
R4. Reiner, M., "Deformation and Flow." Lewis, London, 1949.
R5. Reiner, M., and Riwlin R., *Kolloid-Z.* **43**, 1 (1927).
R6. Roth, W., and Rich, S. R., *J. Appl. Phys.* **24**, 940 (1953).
R7. Rouse, H., "Elementary Mechanics of Fluids," p. 205. Wiley, New York, 1946.
R8. Rushton, J. H., *Chem. Eng. Progr.* **50**, 587 (1954).
R9. Rushton, J. H., Costich, E. W., and Everett, H. J., *Chem. Eng. Progr.* **46**, 395, 467 (1950).
R10. Ryan, N. W., Stevens, W. E., and Christiansen, E. B., *A.I.Ch.E. Journal* **1**, 544 (1955).

S1. Salamone, J. J., and Newman, M., *Ind. Eng. Chem.* **47,** 283 (1955).
S2. Salt, D. L., M.S. thesis in Chemical Engineering, University of Utah, Salt Lake City, 1949.
S3. Salt, D. L., Ryan, N. W., and Christiansen, E. B., *J. Colloid Sci.* **6,** 146 (1951).
S4. Schnurmann, R., *Proc. 1st Intern. Rheol. Congr.* **2,** 142 (1948).
S5. Schultz-Grunow, F., *Chem. Ing. Tech.* **26,** 18 (1954).
S6. Schultz-Grunow, F., *Kolloid-Z.* **138,** 167 (1954).
S7. Scott Blair, G. W., "Introduction to Industrial Rheology." J. & A. Churchill, London, 1938.
S8. Severs, E. T., and Austin, J. M., *Ind. Eng. Chem.* **46,** 2369 (1954).
S9. Spencer, R. S., and Dillon, R. E., *J. Colloid Sci.* **3,** 163 (1948); **4,** 241 (1949).
S10. Stevens, W. E., Ph.D. thesis in Chemical Engineering, University of Utah, Salt Lake City, 1953.
T1. Thomas, B. W., *Ind. Eng. Chem.* **45,** No. 6, 87A (1953).
T2. Toms, B. A., *Proc. 1st Intern. Rheol. Congr.* **2,** 135 (1948).
T3. Toms, B. A., *J. Colloid Sci.* **4,** 511 (1949).
V1. Van der Meer, W., *TAPPI* **37,** 502 (1954).
V2. Vaughn, R. D., Ph.D. thesis in Chemical Engineering, University of Delaware, Newark, 1956.
V3. Verway, E. J. W., and DeBoer, J. H., *Rec. trav. chim.* **57,** 383 (1939).
W1. Ward, H. C., Ph.D. thesis in Chemical Engineering, Georgia Institute of Technology, Atlanta, 1952.
W2. Ward, H. C., and DallaValle, J. M., *Chem. Eng. Progr. Symposium Ser. No 10* **50,** 1 (1954).
W3. Weltmann, R. N., *Ind. Eng. Chem.* **40,** 272 (1948).
W4. Weltmann, R. N., *Natl. Advisory Comm. Aeronaut.* **Tech. Note 3397** (1955).
W5. Weltmann, R. N., and Kuhns, P. W., *J. Colloid Sci.* **7,** 218 (1952).
W6. Wilhelm, R. H., Wroughton, D. M., and Loeffel, W. F., *Ind. Eng. Chem.* **31,** 622 (1939).
W7. Winding, C. C., Baumann, G. P., and Kranich, W. L., *Chem. Eng. Progr.* **43,** 527, 613 (1947).
W8. Winding, C. C., Dittman, F. W., and Kranich, W. L. "Thermal Properties of Synthetic Rubber Latices," Report to Rubber Reserve Company. Cornell University, Ithaca, 1944.
W9. Wood, G. F., Nissan, A. H., and Garner, F. H., *J. Inst. Petroleum* **33,** 71 (1947).
W10. Woodward, J. G., *J. Colloid Sci.* **6,** 481 (1951).
Z1. Zettlemoyer, A. C., and Lower, G. W., *J. Colloid Sci.* **10,** 29 (1955).

Theory of Diffusion

R. Byron Bird

Department of Chemical Engineering
University of Wisconsin, Madison, Wisconsin

I. Introduction... 156
II. The Fluid Mechanical Basis for Diffusion............................. 159
    A. The Equations of Change for a Pure Fluid......................... 159
        1. The Equation of Continuity.................................... 160
        2. The Equation of Motion....................................... 161
        3. The Equation of Energy Balance................................ 162
    B. The Equations of Change for a Multicomponent Mixture............. 165
        1. The Equations of Change in Terms of the Fluxes 165
        2. The Fluxes in Terms of the Transport Coefficients................. 166
        3. Some Comments Concerning the Equations of Change............. 169
    C. Diffusion in Binary Systems..................................... 170
        1. Definitions of Concentrations, Velocities, and Fluxes.............. 170
        2. Ordinary Diffusion in Binary Systems........................... 171
        3. Thermal Diffusion in Binary Systems............................ 176
    D. Diffusion in More Complex Systems.............................. 177
        1. Diffusion in Multicomponent Systems........................... 177
        2. Diffusion in Turbulent Systems................................. 178
        3. Diffusion by Forced and Free Convection........................ 179
        4. Diffusion in Multiphase Systems................................ 180
III. Calculation and Estimation of Diffusion Coefficients..................... 182
    A. Diffusion Coefficients in Dilute Gases............................. 182
        1. Intermolecular Forces... 183
        2. Classical Calculations of the Diffusion Coefficients................. 186
        3. Quantum Calculations of the Diffusion Coefficients................ 190
        4. Empirical Formulas... 190
    B. Diffusion Coefficients in Dense Gases............................. 191
        1. Enskog's Rigid-Sphere Theory.................................. 191
        2. Applications of Enskog's Theory to Real Gases 192
    C. Diffusion Coefficients in Liquids................................. 195
        1. Hydrodynamical Theories...................................... 195
        2. Activated State Theories....................................... 196
        3. Empirical Relations... 197
IV. Solutions of the Diffusion Equations of Interest in Chemical Engineering.. 198
    A. Steady State Diffusion Problems in Nonflow Systems................ 199
        1. Diffusion Through a Stagnant Fluid............................. 199
        2. Diffusion Through a Stagnant Fluid (Spherical Coordinates)........ 200
        3. Diffusion Through a Nonisothermal Film........................ 201

        4. Diffusion Accompanied by Chemical Reaction..................... 202
        5. Diffusion with Thermal Diffusion............................... 203
        6. Diffusion with Pressure Diffusion.............................. 204
    B. Unsteady State Diffusion Problems in Nonflow Systems.............. 205
        1. Unbounded Equimolal Counterdiffusion.......................... 205
        2. Bounded Equimolal Counterdiffusion............................ 206
        3. Partially Bounded Equimolal Counterdiffusion.................. 207
        4. Diffusion in a Two-Phase System............................... 208
        5. Diffusion Accompanied by Rapid Chemical Reaction............... 209
        6. Diffusion Accompanied by a Slow Chemical Reaction.............. 210
    C. Steady-State Diffusion in Flow Systems............................ 211
        1. Diffusion into a Falling Film................................. 211
        2. Diffusion in Flow Through Circular Tubes...................... 216
        3. Diffusion from a Point Source into a Moving Fluid................ 218
        4. Diffusion in Flow Through a Tubular Reactor................... 219
    D. Diffusion in More Complex Systems................................. 220
        1. Unsteady State Diffusion in Evaporation....................... 220
        2. Theory of the Thermal Diffusion Column........................ 223
        3. Some Additional Solutions Discussed in the Literature............ 227
V. Conclusions.. 228
   Appendix... 229
   Nomenclature... 231
   References... 234

I. Introduction

In almost every branch of chemical engineering there occur problems of diffusion and mass transfer[1]—in applied kinetics, where the rate of a chemical reaction is frequently controlled by the diffusion of a reactant through a film or within a catalyst pellet; in mixing and blending, where uniformity of product may depend upon molecular or turbulent mass transfer; in applied electrochemistry, where diffusion may be the limiting factor in corrosion processes; and, above all, in the separation processes (absorption, adsorption, distillation, extraction, and ion exchange), where diffusion and interphase mass transfer play key roles. The design methods for most of these processes have been based upon experimental data, empirical procedures, and operating experience. Better understanding of the processes and improvement in design practice will be gained by the study of the various elementary transport processes involved.

There are numerous mechanisms by which mass transfer can occur: (1) *ordinary diffusion*, which results from a gradient in the concentration; (2) *thermal diffusion*, which results from a gradient in the temperature; (3) *pressure diffusion*, which results from a gradient in the hydrostatic

---

[1] The terms *diffusion* and *mass transfer* seem to be used more or less interchangeably in the literature. The tendency seems to be to use *mass transfer* as the general term embracing all mechanisms of transport of a chemical species and to reserve *diffusion* for mass transport by molecular motion within a single phase.

pressure; (4) *forced diffusion*, which results from different external forces acting upon the different species present; (5) *mass transfer by forced convection*, which results from the over-all motion of the fluid, this motion being produced by the expenditure of energy upon the fluid; (6) *mass transfer by free convection*, which results from the over-all motion of the fluid, the motion being produced by inequalities in the density of the fluid; (7) *turbulent mass transfer*, which results from the motion of eddies through the fluid; and (8) *interphase mass transfer*, which results from a nonequilibrium situation at an interface. The chemical engineer wants to know what the correct starting point is for the analytical description of these processes and then to what extent the theory can be used to aid in the understanding of the various basic mass transfer processes occurring in the chemical engineering operations. This review is intended to indicate the present status of these aspects of the theory of diffusion.

The development of the theory describing the various mechanisms of mass transfer in flow systems consists of the following steps. First, the *basic differential equations* for fluid systems with diffusion must be established. These are the so-called "equations of change," which comprise the equations of continuity for each chemical species, the equations of motion, and the equation of energy balance. These relations provide the starting point for study of diffusion in laminar- and turbulent-flow systems and for simultaneous heat and mass transfer. In their general form these equations are written in terms of the "fluxes" (that is, the mass-flux vector, the heat-flux vector, and the viscous-stress tensor). For mass transfer studies the equations of continuity are the most important. It is from these equations that one derives Fick's second law of diffusion, after making appropriate assumptions. Next the fluxes must be related to the driving forces which cause the flow of mass, momentum, and energy. In these relations certain fluid properties—the transport coefficients—are involved. Of primary interest in the theory of diffusion is, of course, the mass-flux vector and its relation to concentration gradients and diffusion coefficients. These *diffusion coefficients* have to be calculated or estimated in order to make calculations of practical interest. For dilute and moderately dense gases there are excellent theoretical means whereby all the transport coefficients, including the coefficients of diffusion, self-diffusion, and thermal diffusion, may be computed from information about intermolecular forces. In dense gases and liquids elegant theories are lacking, and recourse is usually taken to empiricism. Finally, *solutions of the diffusion equations* for systems of engineering interest have to be obtained. For simple systems analytical solutions may frequently be worked out. For somewhat more complex systems the basic differential equations may be solved by semianalytical approximation procedures or by numerical methods. And for very complex systems dimensional analysis, coupled with experimental data, has to be employed. Regardless of which of these three modes of solution is employed, the fundamental starting point for the problem is the set of equations of change.

In Sec. II of this review a discussion is given of the equations of change and the fluxes. In Sec. III the present status of calculation and estimation of diffusion coefficients is summarized. And in Sec. IV a survey is given of some of the solutions to diffusion problems of interest in chemical engineering. The greater part of the discussion given here is directed toward the study of diffusion in gaseous- and liquid-flow systems consisting of two components; the topics of multicomponent diffusion, turbulent diffusion, and interphase diffusion are considered only briefly, inasmuch as considerably less is known in these areas. Considerable emphasis is placed on the attack of diffusion problems by means of the equations of change describing the transport of mass, momentum, and energy in single-phase flow systems. One of the big obstacles in their use has been the differences in definitions and nomenclature by chemists, physicists, and engineers. A concerted effort is made here to interrelate various viewpoints, definitions, and systems of notation, in order to help bridge the gap between the pure scientist on the one hand and the engineer on the other.

There are a number of books and review articles concerning various aspects of diffusion, which should be consulted for additional sources of information and literature references. In the field of *kinetic theory* the monograph of Chapman and Cowling (C3) is the basic reference for the discussion of the Chapman-Enskog development. More recent calculations based on this theory as well as the additional developments due to Hirschfelder, Curtiss, and collaborators have been discussed in another book (H11) and in several review papers (B11, B12). In addition there is a review article by Schottky (S3) which gives a more complete survey of the German literature; and a review by de Boer (B15) discusses primarily the present status of intermolecular forces. Also extensive literature references are given in three review articles by E. F. Johnson (J5, J6, J7). An excellent summary of the nonequilibrium statistical mechanical approach to kinetic theory has been prepared by Montroll and Green (M10). For the *fluid mechanics and thermodynamics* of transport phenomena, there are the monographs of Prigogine (P8, P9) and de Groot (G12) which give an extensive treatment of the application of the Onsager relations to electrochemical, thermochemical, electrokinetic, and other phenomena. The applications to the transport phenomena of multicomponent fluid mixtures have been discussed by Curtiss (C12) and by Kirkwood and Crawford (K9). Some *solutions to the diffusion equations* are given in the books by Jost (J13) and by Barrer (B5) and in a review article by Fürth (F13). In this connection it should also be pointed out that the solutions to many diffusion problems are to be found in the excellent monograph of Carslaw and Jaeger (C2). Surveys of the *applications of diffusion theory to chemical engineering* are given by Sherwood and Pigford (S9), Colburn and Pigford (C5), Coulson and Richardson (C8), and Matz (M8). More recent summaries of the technical literature have been given by Pigford (P4, P5) and by Wilke (W10).

Several topics in diffusion have arbitrarily been excluded from the following discussion: diffusion in electrolytic solutions (F3, G9, H1, O2, R3, Y1), diffusion in ionized gases (K4), diffusion in macromolecular systems (W2), diffusion through membranes (F14), use of diffusional techniques in isotopic separations (S18), diffusion in metals (S8), and neutron diffusion (F2, G5, H15, W15).

II. The Fluid Mechanical Basis for Diffusion

First a derivative is given of the equations of change for a pure fluid. Then the equations of change for a multicomponent fluid mixture are given (without proof), and a discussion is given of the range of applicability of these equations. Next the basic equations for a multicomponent mixture are specialized for binary mixtures, which are then discussed in considerably more detail. Finally diffusion processes in multicomponent systems, turbulent systems, multiphase systems, and systems with convection are discussed briefly.

A. THE EQUATIONS OF CHANGE FOR A PURE FLUID

The equations of change are the differential statements of three laws of physics applied to fluid motion: the *equation of continuity* is a statement of the principle of conservation of mass in a fluid system; the *equation of motion*, the statement of Newton's Second Law of Motion as applied to an element of fluid moving with the mass average velocity of the fluid; the *equation of energy balance* is the statement of the principle of conservation of energy (or the first law of thermodynamics) to an element of fluid moving with the mass average velocity. These equations are derived here for a pure fluid, by assuming that the fluid is a continuum. The assumption of a continuum is not necessary, for the equations may also be derived by means of a molecular approach. This point is discussed further in Sec. II,B,3. The derivations presented here are somewhat simplified by the use of rectangular elements of volume $\Delta x \Delta y \Delta z$; the resulting equations are the same as those obtained by the use of more sophisticated developments employing arbitrary volume elements and Green's theorem.

1. *The Equation of Continuity*

Consider in the region where a fluid is flowing a small element of volume $\Delta x \Delta y \Delta z$ fixed in space (see Figure 1). The mass of fluid contained within that volume element at any time is then $\rho \Delta x \Delta y \Delta z$, and the time rate of increase of the mass contained therein is $(\partial \rho / \partial t) \Delta x \Delta y \Delta z$. This time rate of change results from the fact that in general the mass of fluid flowing into $\Delta x \Delta y \Delta z$ is not equal to that flowing outward; for example,

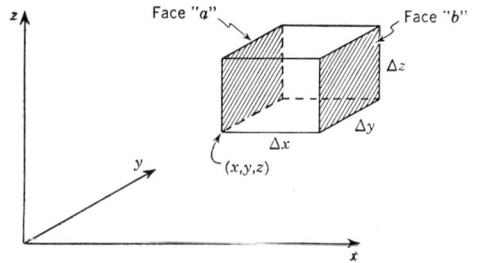

FIG. 1. Volume element used in making mass balance.

the net rate of increase of mass due to flow in through face a and out through b is $[\rho(x)v_x(x)\Delta y \Delta z - \rho(x + \Delta x)v_x(x + \Delta x)\Delta y \Delta z]$, and similar expressions can be written for the other faces of the elemental cube. Hence the mass balance on the element gives

$$(\partial \rho / \partial t)\Delta x \Delta y \Delta z = [\rho(x)v_x(x)\Delta y \Delta z - \rho(x + \Delta x)v_x(x + \Delta x)\Delta y \Delta z] + \text{(terms describing flow across faces perpendicular to the } y \text{ and } z \text{ axes)} \quad (1)$$

The entire equation is now divided through by $\Delta x \Delta y \Delta z$, and the limit is taken as the volume $\Delta x \Delta y \Delta z$ approaches zero. This limiting process gives

$$\frac{\partial \rho}{\partial t} = -\left(\frac{\partial}{\partial x}\rho v_x + \frac{\partial}{\partial y}\rho v_y + \frac{\partial}{\partial z}\rho v_z\right) \quad (2)$$

which is the *equation of continuity* for a pure fluid. It may be written in terms of vector notation thus:

$$\frac{\partial \rho}{\partial t} = -(\nabla \cdot \rho \mathbf{v}) \quad (3)$$

or in terms of the substantial derivative[2] thus:

$$\frac{D\rho}{Dt} = -\rho(\nabla \cdot \mathbf{v}) \qquad (4)$$

From either of these equations it may be seen that for an incompressible fluid $(\nabla \cdot \mathbf{v}) = 0$.

2. The Equation of Motion

Consider a fluid element of constant mass $\rho \Delta x \Delta y \Delta z$ moving along with the local fluid velocity $\mathbf{v}$. The x component of momentum of this fluid element is $\rho v_x \Delta x \Delta y \Delta z$. The momentum of the fluid element as it moves along with the local fluid velocity is a function of both space and time. The total derivative of the momentum of the fluid element with respect to time is then $\rho \Delta x \Delta y \Delta z (Dv_x/Dt)$. According to Newton's second law this quantity is to be equated to the forces acting on the element of mass: the net force in the x direction due to the difference in pressure on faces a and b, which is $[p(x)\Delta y \Delta z - p(x + \Delta x)\Delta y \Delta z]$, the net force in the x direction due to the difference in the viscous stresses,[3] which is

$$[\tau_{xx}(x)\Delta y \Delta z - \tau_{xx}(x + \Delta x)\Delta y \Delta z] + [\tau_{yx}(y)\Delta x \Delta z - \tau_{yx}(y + \Delta y)\Delta x \Delta z]$$
$$+ [\tau_{zx}(z)\Delta x \Delta y - \tau_{zx}(z + \Delta z)\Delta x \Delta y]$$

and finally the force to an external body force, which is $\rho \hat{F}_x \Delta x \Delta y \Delta z$. These forces are equated to the time rate of change of momentum, and the resulting equation is divided through by $\Delta x \Delta y \Delta z$; when the limit is taken as $\Delta x \Delta y \Delta z$ goes to zero, we get

$$\rho \frac{Dv_x}{Dt} = -\frac{\partial p}{\partial x} - \left(\frac{\partial \tau_{xx}}{\partial x} + \frac{\partial \tau_{yx}}{\partial y} + \frac{\partial \tau_{zx}}{\partial z} \right) + \rho \hat{F}_x \qquad (5)$$

Similar equations may also be written for the y and z components of the momentum. The three equations may be summarized in terms of one vector equation

$$\rho \frac{D\mathbf{v}}{Dt} = -\nabla p - (\nabla \cdot \boldsymbol{\tau}) + \rho \hat{\mathbf{F}} \qquad (6)$$

[2] The derivative Df/Dt is called the "substantial derivative" or the "derivative following the motion." It describes the time rate of change of a property $f(x,y,z,t)$ as would be noted by an observer floating along with the fluid with the local velocity $\mathbf{v}$. Analytically it is expressed as

$$\frac{Df}{Dt} = \frac{\partial f}{\partial t} + (\mathbf{v} \cdot \nabla)f = \frac{\partial f}{\partial t} + v_x \frac{\partial f}{\partial x} + v_y \frac{\partial f}{\partial y} + v_z \frac{\partial f}{\partial z}$$

[3] The element τ_{yx} of the second-order stress tensor $\boldsymbol{\tau}$ is the force-per-unit area acting in the x direction on an elemental fluid surface perpendicular to the y direction. The stress tensor is symmetric, and so $\tau_{yx} = \tau_{xy}$.

which is the *equation of motion*. This states that the acceleration of a fluid element results from the action of pressure forces, viscous forces, and external forces.

For most problems one needs to know how the elements of the second-order shear tensor τ are related to the velocity gradients and the coefficient of viscosity. It may be shown from the thermodynamics of irreversible processes (G12, C12, B11) that for a Newtonian fluid the diagonal and nondiagonal elements of τ have the form

$$\tau_{xx} = -2\mu \frac{\partial v_x}{\partial x} + \left(\frac{2}{3}\mu - \mu'\right)(\nabla \cdot \mathbf{v}) \tag{7}$$

$$\tau_{xy} = -\mu \left(\frac{\partial v_y}{\partial x} + \frac{\partial v_x}{\partial y}\right) = \tau_{yx} \tag{8}$$

in the Cartesian coordinate system. The quantities μ and μ' are the coefficients of shear viscosity and bulk viscosity[4] respectively. For non-Newtonian fluids these relations have to be replaced by the appropriate empirical relations used for the description of such substances (i.e., Bingham plastic, power-law model, etc.). When the density is assumed to be constant and the shear viscosity is assumed to be independent of position, the equation of motion then becomes

$$\rho \frac{D\mathbf{v}}{Dt} = -\nabla p + \mu \nabla^2 \mathbf{v} + \rho \hat{\mathbf{F}} \tag{9}$$

The components of this equation are sometimes referred to as the *Navier-Stokes equations;* when the viscosity is set equal to zero (inviscid fluid), then these equations reduce to the *Euler equations*.

3. The Equation of Energy Balance

Once again consider a fluid element of constant mass $\rho \Delta x \Delta y \Delta z$ moving along with the local fluid velocity $\mathbf{v}$. Let the total energy per unit mass of fluid be $\hat{E}$. This energy can arbitrarily be broken up into kinetic energy $\hat{K}$ and internal energy $\hat{U}$ (both per unit mass). Hence the first law of thermodynamics as applied to the fluid element under consideration is

$$\rho \frac{D\hat{E}}{Dt} = \rho \frac{D\hat{Q}}{Dt} - \rho \frac{D\hat{W}}{Dt} \tag{10}$$

or

$$\rho \frac{D\hat{U}}{Dt} = \rho \frac{D\hat{Q}}{Dt} - \rho \frac{D\hat{W}}{Dt} - \rho \frac{D\hat{K}}{Dt} \tag{11}$$

It is from this statement of the first law that the energy-balance equation is derived.

---

[4] The coefficient of bulk viscosity is of importance only in compressible flow, as may be seen from Eq. (7). This transport coefficient has been studied only slightly and is not further considered here; the subject has been reviewed by Karim and Rosenhead (K1) and has been considered further in a recent symposium under the leadership of Rosenhead (R5).

The first term on the right-hand side of Eq. (11) may be written in terms of the net influx of energy across the surfaces of the fluid element, thus

$$\rho \frac{D\hat{Q}}{Dt} \Delta x \Delta y \Delta z = [q_x(x)\Delta y \Delta z - q_x(x+\Delta x)\Delta y \Delta z]$$
$$+ \{\text{two additional terms containing } q_y \text{ and } q_z\} \quad (12)$$

in which q_x, q_y, q_z are the components of the energy flux vector $\mathbf{q}$ (energy per unit time per unit area). The work term consists of the work being done by the fluid element on its surroundings against the hydrostatic pressure, against the viscous shear forces, and against the external body forces; this term can be written thus

$$\rho \frac{D\hat{W}}{Dt} \Delta x \Delta y \Delta z = - [p(x)v_x(x)\Delta y \Delta z - p(x+\Delta x)v_x(x+\Delta x)\Delta y \Delta z]$$
$$- \{\text{two similar terms for } pv_y \text{ and } pv_z\}$$
$$- [\tau_{xx}(x)v_x(x)\Delta y \Delta x - \tau_{xx}(x+\Delta x)v_x(x+\Delta x)\Delta y \Delta z]$$
$$- [\tau_{yx}(y)v_x(y)\Delta x \Delta z - \tau_{yx}(y+\Delta y)v_x(y+\Delta y)\Delta x \Delta z]$$
$$- [\tau_{zx}(z)v_x(z)\Delta x \Delta y - \tau_{zx}(z+\Delta z)v_x(z+\Delta z)\Delta x \Delta y]$$
$$- \{\text{three similar terms in } \tau_{xy}v_y, \tau_{yy}v_y, \tau_{zy}v_y\}$$
$$- \{\text{three similar terms in } \tau_{xz}v_z, \tau_{yz}v_z, \tau_{zz}v_z\}$$
$$- \rho[v_x \hat{F}_x + v_y \hat{F}_y + v_z \hat{F}_z]\Delta x \Delta y \Delta z \quad (13)$$

The kinetic-energy term is given by

$$\rho \frac{D\hat{K}}{Dt} \Delta x \Delta y \Delta z = \rho \frac{D}{Dt}\left(\frac{1}{2}\mathbf{v}^2\right) \Delta x \Delta y \Delta z$$
$$= \rho\left(\mathbf{v} \cdot \frac{D\mathbf{v}}{Dt}\right) \Delta x \Delta y \Delta z$$
$$= [-(\mathbf{v} \cdot \nabla p) - (\mathbf{v} \cdot (\nabla \cdot \boldsymbol{\tau})) + (\mathbf{v} \cdot \rho \hat{\mathbf{F}})]\Delta x \Delta y \Delta z \quad (14)$$

In going from the second to the third line on the right hand side of this equation, use was made of Eq. (6) after it had been multiplied by the vector $\mathbf{v}$ (scalar product). Now when Eqs. (12), (13), and (14) are divided through by $\Delta x \Delta y \Delta z$ and the limit is taken as $\Delta x \Delta y \Delta z$ goes to zero, the following expressions are obtained:

$$\rho \frac{D\hat{Q}}{Dt} = -(\nabla \cdot \mathbf{q}) \quad (15)$$

$$\rho \frac{D\hat{W}}{Dt} = (\nabla \cdot p\mathbf{v}) + (\nabla \cdot (\boldsymbol{\tau} \cdot \mathbf{v})) - (\mathbf{v} \cdot \rho \hat{\mathbf{F}}) \quad (16)$$

$$\rho \frac{DK}{Dt} = -(\mathbf{v} \cdot \nabla p) - (\mathbf{v} \cdot (\nabla \cdot \boldsymbol{\tau})) + (\mathbf{v} \cdot \rho \hat{\mathbf{F}}) \quad (17)$$

Substitution of these three expressions into Eq. (11) gives

$$\rho \frac{D\hat{U}}{Dt} = -(\nabla \cdot \mathbf{q}) - p(\nabla \cdot \mathbf{v}) - (\boldsymbol{\tau} : \nabla \mathbf{v}) \qquad (18)$$

This is the *equation of energy balance*, which states that the internal energy of a fluid changes because of the flow of energy into it, because of the pressure-volume work done on the surroundings by the fluid element, and because of loss of energy by viscous dissipation.[5] The energy-balance equation may be written in an alternate form in terms of the temperature:

$$\rho \hat{C}_v \frac{DT}{Dt} = -(\nabla \cdot \mathbf{q}) - T \left(\frac{\partial p}{\partial T}\right)_\rho (\nabla \cdot \mathbf{v}) - (\boldsymbol{\tau} : \nabla \mathbf{v}) \qquad (19)$$

It should be noted that it is the heat capacity at constant volume which appears in this equation and not the heat capacity at constant pressure.

For most problems one needs to know how the components of the energy-flux vector are related to the space derivatives of the temperature. For conductive heat transfer[6] the necessary relations are of the form

$$q_x = -k(\partial T/\partial x) \qquad (20)$$

in which k is the coefficient of thermal conductivity. The expressions for the components of the energy-flux vector in Eq. (20) and the components of the stress tensor in Eqs. (7) and (8) may then be substituted into Eq. (19) to obtain the differential equation for the temperature distribution in terms of the transport coefficients μ and k. Several special cases of the energy-balance equation are of interest, in which it is assumed that k is independent of position and that there are no viscous dissipation effects:

Ideal Gas: $\qquad\qquad\qquad \rho \hat{C}_v \dfrac{DT}{Dt} = k\nabla^2 T - p(\nabla \cdot \mathbf{v}) \qquad (21)$

Ideal Gas at
Constant Pressure: $\qquad \rho \hat{C}_p \dfrac{DT}{Dt} = k\nabla^2 T \qquad\qquad\qquad (22)$

Incompressible Fluid: $\qquad\;\; \rho \hat{C} \dfrac{DT}{Dt} = k\nabla^2 T \qquad\qquad\qquad (23)$

in which $\hat{C}$ can be either $\hat{C}_p$ or $\hat{C}_v$ for an incompressible fluid.

In general the flow of a pure fluid is described by the equation of continuity, the three equations of motion, and the equation of energy balance. In addition, one has to specify boundary and initial conditions and also the dependence of ρ on p and T (the thermal equation of state) and the dependence of $\hat{C}_v$ or $\hat{U}$ on p and T (the caloric equation of state).

[5] The viscous dissipation term is normally not important. Its significance has been considered in connection with lubrication theory (V1), flow through tubes (B20), extrusion of plastics melts (B10), and viscometry in rotating-cylinder systems (W6).
[6] There is also an additional contribution to the energy flux vector describing energy transport by radiation. See discussion in connection with Eq. (29).

Further the pressure and temperature dependences of all the transport coefficients involved have to be specified. The solution of the equations of change consistent with this additional information then gives the pressure, velocity, and temperature distributions in the system. A number of solutions of idealized problems of interest to chemical engineers may be found in the work of Schlichting (S1); there viscous-flow problems, nonisothermal-flow problems, and boundary-layer problems are discussed.

B. The Equations of Change for a Multicomponent Mixture

In the preceding section it is shown that the equations of change for a pure fluid are expressed in terms of the momentum-flux (stress tensor) and the energy-flux vector. Expressions are there given which relate the fluxes to the transport coefficients. In this section there are presented, without derivation, the equations of change for a viscous, multicomponent fluid mixture in terms of the mass, momentum, and energy fluxes; then general expressions are given which relate the fluxes to the transport coefficients. In the following section the equations for two-component diffusing systems are dealt with in more detail.

1. *The Equations of Change in Terms of the Fluxes* (H11, C11, C13)

For a multicomponent fluid there is an equation of continuity for each chemical species, which can be derived by the methods of the preceding section. These equations when added together give the over-all equation of continuity [Eq. (4)]. The equation of motion for a mixture differs but slightly from that for a pure substance in that it must account for the fact that the various chemical species present may be acted upon by different external body forces (as would be the case when a solution containing ions of different charges is subjected to an electric field). The equation of energy balance differs only in the inclusion of an additional term describing the work done by the diffusion of the various species against the different external forces. The equations of change for a ν component mixture are

$$\frac{D\rho_i}{Dt} = -\rho_i(\nabla \cdot \mathbf{v}) - (\nabla \cdot \mathbf{j}_i) + k_i \qquad (i = 1,2,3 \ldots \nu) \qquad (24)$$

$$\rho \frac{D\mathbf{v}}{Dt} = -\nabla p - (\nabla \cdot \boldsymbol{\tau}) + \sum_{i=1}^{\nu} \rho_i \hat{\mathbf{F}}_i \qquad (25)$$

$$\rho \frac{D\hat{U}}{Dt} = -(\nabla \cdot \mathbf{q}) - p(\nabla \cdot \mathbf{v}) - (\boldsymbol{\tau} : \nabla \mathbf{v}) + \sum_{i=1}^{\nu} (\mathbf{j}_i \cdot \hat{\mathbf{F}}_i) \qquad (26)$$

In these equations ρ_i is the mass density (g. cm.$^{-3}$) of the ith chemical species, k_i is the rate of production of the ith chemical species by chemical reaction (g. cm.$^{-3}$ sec.$^{-1}$), and $\hat{F}_i$ is the external body force per unit mass acting on the ith species. The velocity $\mathbf{v}$ is the local mass average velocity (that velocity measured by a Pitot tube), ρ is the over-all density of the fluid, and $\hat{U}$ is the local thermodynamic internal energy (per unit mass) of the mixture. The $\mathbf{j}_i$ are the fluxes of the various chemical species in g. cm.$^{-2}$ sec.$^{-1}$ with respect to the local mass average velocity, $\mathbf{v}$. It should be noted that $\Sigma \mathbf{j}_i = 0$, $\Sigma k_i = 0$, and $\Sigma \rho_i = \rho$; these relations are used in deriving the over-all equation of continuity [Eq. (4)] by adding up the individual equations of continuity given in Eq. (24).

The energy-balance equation may also be written in terms of the temperature. If no assumptions are made, then Eq. (26) may be transformed into (B11):

$$\rho \hat{C}_v \frac{DT}{Dt} = -(\nabla \cdot \mathbf{q}) - p(\nabla \cdot \mathbf{v}) - (\boldsymbol{\tau} : \nabla \mathbf{v}) + \sum_{i=1}^{\nu} (\mathbf{j}_i \cdot \hat{\mathbf{F}}_i)$$
$$+ \left(p - T \frac{\partial p}{\partial T}\right)(\nabla \cdot \mathbf{v})$$
$$+ \sum_{i=1}^{\nu} \left[(\nabla \cdot \mathbf{j}_i) + k_i\right]\left[\frac{\bar{U}_i}{M_i} + \left(p - T \frac{\partial p}{\partial T}\right)\frac{\bar{V}_i}{M_i}\right] \quad (27)$$

in which the derivative $(\partial p/\partial T)$ is taken at constant composition and volume, and $\bar{U}_i$ and $\bar{V}_i$ are the partial molal internal energy and partial molal volume respectively. The fourth, fifth, and sixth terms on the right-hand side of Eq. (27) are probably not very important: the fourth term is identically zero if all species are under the influence of the same external force (such as gravity, for example); the fifth and sixth terms are identically equal to zero if the fluid is a perfect gas; and the second and fifth terms are zero for incompressible fluids.

2. *The Fluxes in Terms of the Transport Coefficients*

In order to make use of the equations just given, it is necessary to have expressions for the fluxes in terms of the driving forces and the transport coefficients. Such expressions can be obtained by the application of the Onsager relations in the thermodynamics of irreversible processes (G12, C11). In all the results thus obtained for the fluxes there is implied the assumption that the system under consideration is not far removed from equilibrium; the fluxes then are all proportional to the first power of the gradients in the physical properties. Very little has been done along the lines of examining the restrictions imposed by this assumption or in extending the theory to include higher derivatives of the physical properties.

The expression for the *stress tensor* $\boldsymbol{\tau}$ in a fluid mixture is the same as that for a pure substance. The tensor components given in Eqs. (7) and (8) may be summarized conveniently in tensor notation:

$$\boldsymbol{\tau} = -\mu[(\nabla \mathbf{v}) + (\nabla \mathbf{v})\dagger] + (\tfrac{2}{3}\mu - \mu')(\nabla \cdot \mathbf{v})\boldsymbol{\delta} \qquad (28)$$

in which $\boldsymbol{\delta}$ is the second-order unit tensor, and $(\nabla \mathbf{v})\dagger$ is the transpose of the dyadic $\nabla \mathbf{v}$. It should be borne in mind that Eq. (28) is valid for Newtonian fluids only.[7]

The *energy flux vector* for a fluid mixture has been shown to be the sum of four contributions: $\mathbf{q}^{(T)}$, the energy flux due to a temperature gradient (ordinary heat conduction); $\mathbf{q}^{(d)}$, the energy flux due to the fact that the diffusing species carry with them a certain amount of intrinsic energy; $\mathbf{q}^{(\rho)}$, the energy flux associated with a superimposed concentration gradient (a very small effect known as the "diffusion-thermo effect" or the "Dufour effect"); and finally, $\mathbf{q}^{(r)}$, the energy flux by radiation. Hence we make the following summary:

$$\mathbf{q} = \mathbf{q}^{(T)} + \mathbf{q}^{(d)} + \mathbf{q}^{(\rho)} + \mathbf{q}^{(r)} \qquad (29)$$

$$\mathbf{q}^{(T)} = -k\nabla T \qquad (30)$$

$$\mathbf{q}^{(d)} = + \sum_{i=1}^{\nu} (\bar{H}_i/M_i)\mathbf{j}_i \qquad (31)$$

in which $\bar{H}_i$ is the partial molal enthalpy.[8] We have given here the formulas for the energy flux due to conduction and to diffusion. The energy flux associated with the Dufour effect is quite small, and the general expression for it has never been developed; for dilute gases, however, the expression is known (H11, C11). No expression is given for the energy

---

[7] For non-Newtonian fluids the best that one can do at the present time is to make use of various empirical models for non-Newtonian flow. For example, for the incompressible Bingham plastic one can use in place of Eq. (28) the expression

$$\boldsymbol{\tau} - \boldsymbol{\tau}_0 = -\mu((\nabla \mathbf{v}) + (\nabla \mathbf{v})\dagger) \qquad (28a)$$

in which $\boldsymbol{\tau}_0$ is the yield stress. The substitution of Eq. (28a) into Eq. (6) leads to the Buckingham-Reiner equation for the flow of a Bingham plastic in circular tubes and to the Reiner-Riwlin equation for the tangential flow in a cylindrical annulus. Many substances obey a power law relationship of the form:

$$|\boldsymbol{\tau}| = [m(\nabla \mathbf{v}) + (\nabla \mathbf{v})\dagger]^{1/n} \qquad (28b)$$

in which m and n are constants describing the fluid. The various empirical models that can be used for describing non-Newtonian fluids are considered in detail by Reiner (R1).

[8] One might expect that the partial-molal energy (rather than enthalpy) should appear in the expression for the energy flux by diffusional processes. This point is discussed briefly for dilute-gas binary mixtures by Chapman and Cowling (C3, p. 145ff.).

flux due to radiation. If radiative energy transport is taking place, and if the radiation is in equilibrium with the matter, then the expression for the radiative energy flux is of the same form as that for the conduction term [Eq. (30)]. This condition is, however, ordinarily not satisfied, and the radiative energy flux depends in a detailed manner on the frequency and intensity of the radiation and the properties of the material (H11, pp. 720–727).

The *mass flux vector* is also the sum of four components: $\mathbf{j}_i^{(x)}$, the mass flux due to a concentration gradient (ordinary diffusion); $\mathbf{j}_i^{(p)}$, the mass flux associated with a gradient in the pressure (pressure diffusion); $\mathbf{j}_i^{(F)}$, the mass flux associated with differences in external forces (forced diffusion); and $\mathbf{j}_i^{(T)}$, the mass flux due to a temperature gradient (the thermal diffusion effect or the Soret effect). The mass flux contributions may then be summarized:

$$\mathbf{j}_i = \mathbf{j}_i^{(x)} + \mathbf{j}_i^{(p)} + \mathbf{j}_i^{(F)} + \mathbf{j}_i^{(T)} \tag{32}$$

$$\mathbf{j}_i^{(x)} = \frac{c^2}{\rho RT} \sum_{\substack{j=1 \\ k \neq j}}^{\nu} M_i M_j D_{ij} \left[x_j \sum_{\substack{k=1 \\ k \neq j}}^{\nu} \left(\frac{\partial \tilde{G}_j}{\partial x_k} \right)_{\substack{T,p,x_s \\ s \neq j,k}} \nabla x_k \right] \tag{33}$$

$$\mathbf{j}_i^{(p)} = \frac{c^2}{\rho RT} \sum_{j=1}^{\nu} M_i M_j D_{ij} \left[x_j M_j \left(\frac{\bar{V}_j}{M_j} - \frac{1}{\rho} \right) \nabla p \right] \tag{34}$$

$$\mathbf{j}_i^{(F)} = \frac{c^2}{\rho RT} \sum_{j=1}^{\nu} M_i M_j D_{ij} \left[x_j M_j \left(\hat{F}_j - \sum_{k=1}^{\nu} \frac{\rho_k}{\rho} \hat{F}_k \right) \right] \tag{35}$$

$$\mathbf{j}_i^{(T)} = -D_i^T \nabla \ln T \tag{36}$$

In these equations $\tilde{G}_i$ is the partial molal free energy (chemical potential) and $\bar{V}_j$ the partial molal volume. The M_j are the molecular weights, c is the concentration in moles per liter, ρ is the mass density, and x_i is the mole fraction of species i. The D_{ij} are the multicomponent diffusion coefficients, and the D_i^T are the multicomponent thermal diffusion coefficients. The first contribution to the mass flux—that due to the concentration gradients—is seen to depend in a complicated way on the chemical potentials of all the components present. It is shown in the next section how this expression reduces to the usual expressions for the mass flux in two-component systems. The pressure diffusion contribution to the mass flux is quite small and has thus far been studied only slightly; it is considered in Sec. IV,A,6. The forced diffusion term is important in ionic systems (C3, Chapter 18; K4); if gravity is the only external force, then this term vanishes identically. The thermal diffusion term is important only in cases where there are very steep temperature gradients; the existence of this term is the basis for the operation of the Clusius-Dickel column (F10, G11), which has been used for the separation of isotopes (F11) and for the rectification of complex liquid mixtures of organic compounds (J9, J10, J11).

It should be emphasized that the flux vectors for which expressions have been given in Eqs. (28) through (36) are all defined here as fluxes with respect to the mass average velocity. Not all authors use this convention, and considerable confusion has resulted in the definition of the energy flux and the mass flux. Mass fluxes with respect to molar average velocity, stationary coordinates, and the velocity of one component (such as the solvent, for example) are all to be found in the literature on diffusional processes. Research workers in the field of diffusion should be meticulous in specifying the frame of reference for fluxes used in writing up their research work. In the next section this important matter is considered in detail for two-component systems.

Several remarks need to be made concerning the definition of the diffusion coefficients in Eqs. (32) to (36) above. The multicomponent diffusion coefficients D_{ij} used here differ from those used in references (H11) and (B11). The latter must be multiplied by cRT/p to get the diffusion coefficients defined here. For perfect gases, of course, this difference in definition is unimportant since cRT/p is unity. For liquid, the definition used here is to be preferred, inasmuch as it is more in conformity with the customary definition used by experimentalists. The D_{ij} used here are defined in such a way that $D_{ij} \neq D_{ji}$ and $D_{ii} = 0$.

For a two-component mixture the multicomponent diffusion coefficients D_{ij} become the ordinary binary diffusion coefficients $\mathfrak{D}_{ij}$. For these quantities $\mathfrak{D}_{ij} = \mathfrak{D}_{ji}$ and $\mathfrak{D}_{ii} = 0$. For a three-component system the multicomponent diffusion coefficients are not equal to the ordinary binary diffusion coefficients. For example, it has been shown by Curtiss and Hirschfelder (C12) in their development of the kinetic theory of multicomponent gas mixtures that

$$D_{12} = \mathfrak{D}_{12} \left[1 + \frac{x_3 \left[\left(\frac{M_3}{M_2}\right) \mathfrak{D}_{13} - \mathfrak{D}_{12} \right]}{x_1 \mathfrak{D}_{23} + x_2 \mathfrak{D}_{13} + x_3 \mathfrak{D}_{12}} \right] \tag{37}$$

with similar relations for D_{13} and D_{23}.

3. *Some Comments Concerning the Equations of Change*

In Sec. II,B,1 the equations of change are given in terms of the fluxes, and in Sec. II,B,2 the fluxes are in turn given in terms of the gradients of the physical properties and the transport coefficients. The substitution of these expressions into the general equations of change then gives the general expressions for the basic differential equations in terms of the transport coefficients. Needless to say the equations are seldom used in their entirety or in complete generality. Nevertheless these general relationships are useful as an aid to systematizing the analytical results; one then uses these equations to solve problems by discarding all terms which are physically unimportant in the problem at hand.

In Sec. II,A the equations of change are derived by assuming that the fluid is a continuum. A physically more satisfying derivation may be performed in which one starts directly from considerations of the fundamental molecular-collision processes occurring in the fluid. For dilute monatomic gases and gas mixtures one can start with the Boltzmann equation and derive the equations of change as a consequence of the laws of conservation of mass, momentum, and energy in the collisions between two atoms (C3, H11); in this way it is not necessary to make the assumption of a continuum. The equations of change have also been obtained from the kinetic theory of dilute polyatomic gases (W4) and from the kinetic theory of dense gases both in classical theory (I1) and in quantum theory (I2). Hence the equations of change have certainly been adequately justified, and the engineer has in them a solid starting point for studies in fluid mechanics, mass transfer, and heat transfer.

The derivations from the molecular viewpoint emphasize the fact that the equations of change are useful only under such conditions that the local density, velocity, and temperature do not change over distances of the order of a mean free path. Hence the equations of change are not useful for describing a "Knudsen gas," in which the gas is so dilute that the dimensions of the containing vessel or immersed objects are of the same order of magnitude as the mean free path of the molecules. Nor are the equations of change of use in describing shock waves in which the macroscopic properties as functions of the distances undergo an abrupt change within a distance of a few mean free paths.

In turbulent flow the eddies, which are superimposed on the over-all flow pattern, have dimensions large compared with a mean free path. Hence turbulent motion is macroscopic rather than molecular and the equations of change do apply to turbulent flow. This subject is discussed further in Sec. II,D,2.

C. Diffusion in Binary Systems

In this section it is shown how the multicomponent equations just discussed can be applied to the discussion of binary systems. First a summary is given of the various notations used in discussing two-component systems. Then some important special results are given for the diffusion and thermal diffusion in binary systems. These equations are used as the starting point for the discussions in Sec. IV.

1. *Definitions of Concentrations, Velocities, and Fluxes*

In reading the literature on the fluid mechanics of diffusion, one encounters numerous difficulties because of the diversity of reference frames and definitions used by authors in various fields. Frequently more time is spent in "translating" from one system of notation to another than is spent in the actual study of the physics of the problem. It is hoped that the "glossaries" of terminology and symbols given here will be of use to those whose fields of research require a familiarity with the literature on diffusion. This exposition should emphasize the extreme importance of giving lucid definitions in any discussion of diffusion and mass transfer.

There are numerous ways of expressing concentration in diffusion problems, the most important for our purposes being mass density, molar density, mass fraction, and mole fraction. The chemical engineer and the chemist are familiar with the relationships between these quantities. Table I is given for the sake of summarizing the notation used here.

There are also many ways of expressing the velocity of a chemical species present in a flow system. We do not concern ourselves here with the instantaneous velocity of the individual molecules of a species, but rather with the average macroscopic velocities with which the species travel. These may be measured from a stationary coordinate system, but for flow systems the velocities of individual species are frequently measured from a coordinate frame moving with (a) the mass-average velocity of the stream, (b) the molar average velocity of the stream, or (c) the velocity of one particular component. The mass- and the molar-average velocity are defined in Table II and the notation for the various velocities of an individual species is given.

Because of the various definitions of concentrations and velocities, there are also various ways of defining the flux of any given species. The most important of these definitions are given in Table III. The flux $\mathbf{j}_A$ (with respect to the mass-average velocity) is that used in the basic flow equations in previous sections and for which general expressions are given in Eq. (32). All the definitions given in Table III may be found in the current literature; the fluxes $\mathbf{j}_A$ and $\mathbf{\phi}_A$ tend to be favored by physicists, whereas the fluxes in molar units $\mathbf{N}_A$ and $\mathbf{\Phi}_A$ seem to be generally employed by chemists and chemical engineers. The tabulation at the bottom of Table III should be useful for converting between the various types of notation.

To illustrate how the relations between the fluxes in Table III were obtained, the relation between $\mathbf{j}_A$ and $\mathbf{\Phi}_A$ is derived:

$$
\begin{aligned}
\mathbf{\Phi}_A &= c_A \mathbf{w}_A & \text{(by definition)} \\
&= c_A[\mathbf{v}_A + (\mathbf{v} - \mathbf{w})] & \text{(by relation between } \mathbf{v}_A \text{ and } \mathbf{w}_A\text{)} \\
&= c_A \left[\mathbf{v}_A - \frac{1}{c}(c_A \mathbf{v}_A + c_B \mathbf{v}_B) \right] & \text{(by expression for } \mathbf{w} - \mathbf{v} \text{ in Table II)} \\
&= c_A \mathbf{v}_A \left[1 - \frac{c_A}{c} - \frac{c_B}{c}\frac{\mathbf{v}_B}{\mathbf{v}_A} \right] \\
&= \mathbf{J}_A \left[1 - \frac{c_A}{c} + \frac{c_A}{c}\frac{M_A}{M_B} \right] & \text{(by the relation } \mathbf{j}_A + \mathbf{j}_B = 0\text{)} \\
&= \mathbf{J}_A \left[\frac{\rho}{cM_B} \right] \\
&= \mathbf{j}_A \left[\frac{\rho}{cM_A M_B} \right] & (38)
\end{aligned}
$$

2. Ordinary Diffusion in Binary Systems

The equation of continuity for component A in a two-component system is according to Eq. (24):

$$(\partial \rho_A/\partial t) + (\nabla \cdot \rho_A \mathbf{v}) + (\nabla \cdot \mathbf{j}_A) = k_A \tag{39a}$$

or

$$(\partial \rho_A/\partial t) + (\nabla \cdot \mathbf{n}_A) = k_A \tag{39b}$$

TABLE I
Notation for Concentrations in Binary Systems

$c = c_A + c_B$ = over-all concentration in g.-mole cm.$^{-3}$ (molar density)
$c_A = \rho_A/M_A$ = concentration of A in g.-mole cm.$^{-3}$
$x_A = c_A/c$ = concentration of A expressed as mole fraction

$\rho = \rho_A + \rho_B$ = over-all concentration in g. cm.$^{-3}$ (mass density)
$\rho_A = c_A M_A$ = concentration of A in g. cm.$^{-3}$
$\omega_A = \rho_A/\rho$ = concentration of A expressed as mass fraction

$$x_A = \frac{\omega_A/M_A}{\dfrac{\omega_A}{M_A} + \dfrac{\omega_B}{M_B}} \qquad \omega_A = \frac{x_A M_A}{x_A M_A + x_B M_B}$$

$$dx_A = \frac{d\omega_A}{M_A M_B \left(\dfrac{\omega_A}{M_A} + \dfrac{\omega_B}{M_B}\right)^2} \qquad d\omega_A = \frac{M_A M_B dx_A}{(x_A M_A + x_B M_B)^2}$$

TABLE II
Notation for Velocities in Binary Systems

$\mathbf{u}_A$ = average velocity of species A with respect to a stationary coordinate system
$\mathbf{v}_A$ = average velocity of species A with respect to the local mass-average velocity $\mathbf{v}$ (that is, $\mathbf{v}_A = \mathbf{u}_A - \mathbf{v}$)
$\mathbf{w}_A$ = average velocity of species A with respect to the local molar-average velocity $\mathbf{w}$ (that is, $\mathbf{w}_A = \mathbf{u}_A - \mathbf{w}$)

$\mathbf{v}$ = mass-average velocity = $(1/\rho)(\rho_A \mathbf{u}_A + \rho_B \mathbf{u}_B) = (\omega_A \mathbf{u}_A + \omega_B \mathbf{u}_B)$
$\mathbf{w}$ = molar-average velocity = $(1/c)(c_A \mathbf{u}_A + c_B \mathbf{u}_B) = (x_A \mathbf{u}_A + x_B \mathbf{u}_B)$

$\mathbf{v} - \mathbf{w} = (1/\rho)(\rho_A \mathbf{w}_A + \rho_B \mathbf{w}_B) = (\omega_A \mathbf{w}_A + \omega_B \mathbf{w}_B)$
$\mathbf{w} - \mathbf{v} = (1/c)(c_A \mathbf{v}_A + c_B \mathbf{v}_B) = (x_A \mathbf{v}_A + x_B \mathbf{v}_B)$

The definitions of concentrations, velocities, and fluxes in Tables I, II, III may be used to rewrite this equation in terms of molar concentrations and molar fluxes:

$$(\partial c_A/\partial t) + (\nabla \cdot c_A \mathbf{w}) + (\nabla \cdot \mathbf{\Phi}_A) = K_A \tag{40a}$$

or

$$(\partial c_A/\partial t) + (\nabla \cdot \mathbf{N}_A) = K_A \tag{40b}$$

the last equation being in the form more familiar to chemical engineers. In order to obtain the differential equations for diffusion, one has to substitute expressions for the mass fluxes into the foregoing equations of continuity. We now consider the expressions for these fluxes in two-component systems in which there is occurring only ordinary diffusion.

In the absence of pressure diffusion, forced diffusion, and thermal

TABLE III
Notation for Mass-Flux Vectors in Binary Systems

Quantity	With respect to stationary axes	With respect to v	With respect to w
Average velocity of species A	$\mathbf{u}_A$	$\mathbf{v}_A$	$\mathbf{w}_A$
Flux of A (g. cm.$^{-2}$ sec.$^{-1}$)	$\mathbf{n}_A = \rho_A \mathbf{u}_A$	$\mathbf{j}_A = \rho_A \mathbf{v}_A$	$\boldsymbol{\phi}_A = \rho_A \mathbf{w}_A$
Flux of A (mole cm.$^{-2}$ sec.$^{-1}$)	$\mathbf{N}_A = c_A \mathbf{u}_A$	$\mathbf{J}_A = c_A \mathbf{v}_A$	$\boldsymbol{\Phi}_A = c_A \mathbf{w}_A$
$\boldsymbol{\phi}_A = \left(\dfrac{\rho}{cM_B}\right)\mathbf{j}_A$	$\boldsymbol{\phi}_A = \left(\dfrac{\rho}{c}\dfrac{M_A}{M_B}\right)\mathbf{J}_A$	$\mathbf{j}_A + \mathbf{j}_A = 0$	
$\boldsymbol{\Phi}_A = \left(\dfrac{\rho}{cM_AM_B}\right)\mathbf{j}_A$	$\boldsymbol{\Phi}_A = \left(\dfrac{\rho}{c}\dfrac{1}{M_B}\right)\mathbf{J}_A$	$\boldsymbol{\Phi}_A + \boldsymbol{\Phi}_B = 0$	
$\mathbf{n}_A = \omega_A(\mathbf{n}_A + \mathbf{n}_A) + \mathbf{j}_A$	$\mathbf{n}_A = x_A\left(\mathbf{n}_A + \dfrac{M_A}{M_B}\mathbf{n}_B\right) + \boldsymbol{\phi}_A$		
$\mathbf{N}_A = \omega_A\left(\mathbf{N}_A + \dfrac{M_B}{M_A}\mathbf{N}_B\right) + \mathbf{J}_A$	$\mathbf{N}_A = x_A(\mathbf{N}_A + \mathbf{N}_B) + \boldsymbol{\Phi}_A$		

diffusion Eq. (33) becomes for a binary mixture of A and B:

$$\begin{aligned}\mathbf{j}_A &= (c^2/\rho RT)M_B M_B \mathfrak{D}_{AB}[x_B(\partial \bar{G}_B/\partial x_A)_{p,T}\nabla x_A] \\ &= -(c^2/\rho RT)M_A M_B \mathfrak{D}_{AB}[x_A(\partial \bar{G}_A/\partial x_A)_{p,T}\nabla x_A] \\ &= -(c^2/\rho)M_A M_B \mathfrak{D}_{AB}(\partial \ln a_A/\partial \ln x_A)\nabla x_A \end{aligned} \qquad (41)$$

The expression in the second line is obtained by use of the Gibbs-Duhem relation ($x_A d\bar{G}_A + x_B d\bar{G}_B = 0$), and the third line[9] by use of the relation $\bar{G}_A = G_A^0(T) + RT \ln a_A$.

For ideal gas mixtures and for dilute liquid solutions the activity is equal to the mole fraction, and then the various mass fluxes may be written:[10]

[9] In concentrated solutions one could rewrite the last line of Eq. (41) as

$$\mathbf{j}_A = -(c^2/\rho)M_A M_B \mathfrak{D}_{AB}'\nabla x_A \qquad (41a)$$

in which the diffusion coefficient $\mathfrak{D}_{AB}' = (\partial \ln a_A/\partial \ln x_A)\mathfrak{D}_{AB}$. Indeed this is universally done by the experimentalists who use concentrations rather than activities as the driving force in the analysis of their experiments. The quantity $\mathfrak{D}_{AB}'$ is for some systems considerably more dependent upon concentration than is the coefficient $\mathfrak{D}_{AB}$ (P6, S16, S17, H11, pp. 631–632). The necessity of introducing the factor ($\partial \ln a_A/\partial \ln x_A$) in solutions of electrolytes has also been discussed by Onsager and Fuoss (O2).

[10] For dilute gases Eq. (43) may be written as

$$\boldsymbol{\Phi}_A = -c\mathfrak{D}_{AB}\nabla x_A = -(p\mathfrak{D}_{AB}/RT)\nabla x_A \qquad (43a)$$

The appearance of the grouping $p\mathfrak{D}_{AB}/RT$ has led some authors (see, for example,

$$\mathbf{j}_A = -(c^2/\rho)M_A M_B \mathfrak{D}_{AB}\nabla x_A = -\rho\mathfrak{D}_{AB}\nabla\omega_A \tag{42}$$
$$\mathbf{\Phi}_A = -(\rho^2/cM_A M_B)\mathfrak{D}_{AB}\nabla\omega_A = -c\mathfrak{D}_{AB}\nabla x_A \tag{43}$$
$$\mathbf{n}_A = -\rho\mathfrak{D}_{AB}\nabla\omega_A + \omega_A(\mathbf{n}_A + \mathbf{n}_B) \tag{44}$$
$$\mathbf{N}_A = -c\mathfrak{D}_{AB}\nabla x_A + x_A(\mathbf{N}_A + \mathbf{N}_B) \tag{45}$$

For most diffusion calculations one makes use of special cases of the above expressions. For example, for systems of constant mass density ρ (dilute liquid solutions at constant temperature and pressure) and for systems of constant molar density c (ideal gases at constant temperature and pressure) Eqs. (42) and (43) may be simplified to:

$$\mathbf{j}_A = -\mathfrak{D}_{AB}\nabla\rho_A \qquad \rho = \text{constant} \tag{46}$$
$$\mathbf{\Phi}_A = -\mathfrak{D}_{AB}\nabla c_A \qquad c = \text{constant} \tag{47}$$

In chemical engineering calculations it has been customary to use Eq. (45) with c constant in one of two special forms:

 a. Equimolal counterdiffusion of A and B. When A and B are diffusing counter to one another in such a way that $N_A = -N_B$, then

$$N_A = -\mathfrak{D}_{AB}\nabla c_A \tag{48}$$

This is the correct expression for use in the analysis of closed diffusion-cell experiments for the measurement of diffusion coefficients. Equation (48) is known as *Fick's First Law of Diffusion*. Note that $\mathbf{N}_A = -\mathbf{N}_B$ corresponds to saying that $\mathbf{w} = 0$.

 b. Diffusion of A through stagnant B. When A is diffusing through B, which is not moving, then $\mathbf{N}_B = 0$ and

$$N_A = -\frac{\mathfrak{D}_{AB}}{1 - x_A}\nabla c_A \tag{49}$$

This expression is used in discussing the diffusion through films in mass transfer calculations. Note that in very dilute solutions of A the quantity $(1 - x_A) \approx 1$ and Eq. (49) then simplifies to Fick's first law.

Thus far the equation of continuity in terms of the mass flux has been discussed, and expressions for the mass flux in terms of the diffusion

---

p. 460 of W1) to define this group as the diffusion coefficient. It seems to this author that it is preferable to use one diffusion coefficient throughout the entire development. Other authors follow the scheme proposed by de Groot (G12, pp. 100–111), who prefers to write

$$(\text{flux}) = -\begin{pmatrix}\text{over-all}\\ \text{density}\end{pmatrix}\begin{pmatrix}\text{diffusion}\\ \text{coefficient}\end{pmatrix}\begin{pmatrix}\text{concentration}\\ \text{gradient}\end{pmatrix}$$

and let the expressions of this type define various diffusion coefficients; that is, for every choice of concentration and flux, the diffusion coefficient is different. This approach has led de Groot to the definition of several kinds of barycentric diffusion coefficients (fluxes with respect to the mass average velocity) and several kinds of molecular diffusion coefficients (fluxes with respect to the molar average velocity); de Groot (G12) points out that Prigogine has defined still other diffusion coefficients based on fluxes with respect to the mean volume velocity and the velocity of a single component in the mixture. The comments made in this and the preceding footnote should emphasize the importance of a clear specification of definitions of diffusion coefficients in publications on the subject.

coefficient have been given. These two results may now be combined to obtain the equation of continuity in terms of the diffusion coefficient. Substitution of Eq. (41) into Eq. (39a) gives the most general form of the diffusion equation. Ordinarily, however, one makes use of less general expressions. Two special cases are particularly important, since many problems are approximately described by these assumptions:

   a. Constant mass density ρ; $\mathfrak{D}_{AB}$ independent of position

Substitution of Eq. (46) into Eq. (39a) gives[11]

$$(\partial \rho_A/\partial t) + (\mathbf{v} \cdot \nabla)\rho_A = \mathfrak{D}_{AB}\nabla^2 \rho_A + k_A \tag{50}$$

This equation may be written in terms of c_A and K_A by dividing through by M_A. In the absence of chemical reactions $k_A = 0$, and this equation is of the same form as Eq. (22) for heat transfer. This similarity is of considerable help in connection with the study of analogous processes in heat and mass transfer.

   b. Constant molar density c; $\mathfrak{D}_{AB}$ independent of position

Substitution of Eq. (47) into Eq. (40a) gives[11]

$$(\partial c_A/\partial t) + (\mathbf{w} \cdot \nabla)c_A = \mathfrak{D}_{AB}\nabla^2 c_A + K_A - (c_A/c)(K_A + K_B) \tag{51}$$

When A is present in very small amount, the last term on the right may be neglected; and if in the reaction A $\rightarrow$ B one mole of A produces one mole of B, then the last term vanishes identically.

For diffusion in dilute non-reacting liquid systems which are not flowing ($\mathbf{v} = 0$) and for equimolal counterdiffusion in ideal non-reacting gases at constant temperature and pressure ($\mathbf{w} = 0$), Eqs. (50) and (51) simplify to:

$$(\partial c_A/\partial t) = \mathfrak{D}_{AB}\nabla^2 c_A \tag{52}$$

which is known as *Fick's Second Law of Diffusion*. Many authors leave their readers with the impression that Fick's first and second laws [that is, Eqs. (48) and (52)] are the fundamental starting point for the study of diffusion. It is true that for many physical situations these laws are valid or approximately valid. Several discussions concerning the applicability of Fick's laws have been given by Babbitt (B2, B3). An important class of problems not covered by Fick's laws are those involving evaporation for which the molar-average velocity is not zero (see Sec. IV,D,1).

   [11] When Eq. (39a) is written for components A and B and the two equations are added together, one gets (since $\mathbf{j}_A + \mathbf{j}_B = 0$ and $k_A + k_B = 0$):

$$(\partial \rho/\partial t) + (\nabla \cdot \rho \mathbf{v}) = 0 \tag{39c}$$

Hence for systems with constant ρ, $(\nabla \cdot \mathbf{v}) = 0$; the latter is used in going from Eq.

3. *Thermal Diffusion in Binary Systems* (G12, G11, H11)

In the preceding section binary systems are discussed for which Eq. (33) gives the sole contribution to the mass flux. In this section the discussion is extended to binary systems in which Eqs. (33) and (36) are both of importance—that is, both ordinary and thermal diffusion are considered. Then the expression for the mass flux A is

$$\mathbf{j}_A = -(c^2/\rho) M_A M_B \mathfrak{D}_{AB} \left[\frac{x_A}{RT} \left(\frac{\partial \bar{G}_A}{\partial x_A} \right)_{T,p} \nabla x_A \right] - D_A{}^T \nabla \ln T$$

$$= -(c^2/\rho) M_A M_B \mathfrak{D}_{AB} \left[\frac{x_A}{RT} \left(\frac{\partial \bar{G}_A}{\partial x_A} \right)_{T,p} \nabla x_A + k_T \nabla \ln T \right] \tag{53}$$

In the second expression the *thermal diffusion ratio* k_T has been introduced; it is defined for binary mixtures to be

$$k_T = (\rho/c^2 M_A M_B)(D_A{}^T / \mathfrak{D}_{AB}) \tag{54}$$

The quantity k_T is so defined that when k_T is positive the component A moves to the cold region, and when k_T is negative A moves to the hot region. This definition of k_T is in agreement with the accepted definitions of previous workers, and in particular with the text books of Chapman and Cowling (C3) and Grew and Ibbs (G12). In the gas phase many authors prefer to write the mass flux in terms of the *thermal diffusion factor* α, which is defined as

$$\alpha = k_T / x_A x_B \tag{55}$$

This quantity is used since it is much less concentration dependent than is k_T. In the liquid phase it is usual to use a quantity

$$\sigma = k_T / T x_A x_B \tag{56}$$

which is called the *Soret coefficient*.

(39a) to Eq. (50). Similarly when Eq. (40a) is written for both components and the two equations added, one obtains (since $\Phi_A + \Phi_B = 0$)

$$(\partial c / \partial t) + (\nabla \cdot c \mathbf{w}) = (K_A + K_B) \tag{40c}$$

Hence when c is constant, $(\nabla \cdot \mathbf{w}) = (1/c)(K_A + K_B)$. This relation is used in obtaining Eq. (51) from Eq. (40c). In the absence of chemical reaction $(\nabla \cdot \mathbf{w}) = 0$. For flow and diffusion in one direction only, $(\nabla \cdot \mathbf{w}) = 0$ implies that $\mathbf{w}$ is independent of position and a function of time alone; use is made of this in Section III,D,1. In general, however, the vanishing of the divergence of $\mathbf{w}$ implies that $\mathbf{w}$ is independent of position only if in addition $[\nabla \times \mathbf{w}] = 0$ (H11, p. 518).

For an ideal gas mixture or for a dilute solution the mass fluxes for A and B further simplify to

$$\mathbf{j}_A = -(c^2/\rho)M_A M_B \mathfrak{D}_{AB}[\nabla x_A + k_T \nabla \ln T] \qquad (57)$$
$$\mathbf{j}_B = -(c^2/\rho)M_A M_B \mathfrak{D}_{AB}[\nabla x_B - k_T \nabla \ln T] \qquad (58)$$

Equation (57) states that if there is a negative concentration gradient in the $+x$ direction, there will be a mass flux in the x direction; further a negative gradient in the temperature in the $+x$ direction will cause A to flow in the $+x$ direction (that is, toward the colder region). From Eqs. (57) and (58) it can be shown that the difference in the average velocities of species A and B is

$$(\mathbf{v}_B - \mathbf{v}_A) = (c^2/c_A c_B)\mathfrak{D}_{AB}[\nabla x_A + k_T \nabla \ln T] \qquad (59)$$

These last three equations form the starting point for the discussion of thermal diffusion.

D. Diffusion in More Complex Systems

In the preceding section it is seen how the theory of multicomponent systems gives the correct starting formulas for the analysis of ordinary and thermal diffusion in binary systems. Whereas these latter topics have been the subject of considerable investigation, there are a number of types of more complex diffusion problems of engineering interest for which little has been done. Several of these topics are discussed here, and an attempt is made to indicate to what extent they can be interpreted in terms of the theoretical development in the preceding sections.

1. *Diffusion in Multicomponent Systems*

The basic expressions for the mass fluxes and the equations of continuity for multicomponent mixtures are given in Sec. II,B. For a ν-component mixture of ideal gases in a system in which there is no pressure diffusion, forced diffusion, or thermal diffusion, the fluxes are given by

$$\mathbf{j}_i = (c^2/\rho) \sum_{j=1}^{\nu} M_i M_j D_{ij} \nabla x_j \qquad i = 1, 2, 3 \ldots \nu \qquad (60)$$

Because of the definition of the $\mathbf{j}_i$, only $\nu - 1$ of these mass fluxes are independent (since $\Sigma \mathbf{j}_i = 0$). It has been shown by Curtiss and Hirschfelder (C12) in their development of the kinetic theory of multicomponent gas mixtures that the foregoing relations may be put into an alternate form:[12]

$$\nabla x_i = \sum_{\substack{j=1 \\ j \neq i}}^{\nu} (c_i c_j/c^2)(1/\mathfrak{D}_{ij})(\mathbf{v}_j - \mathbf{v}_i) \qquad (61)$$

[12] This expression has been generalized by Curtiss and Hirschfelder (C12) to include thermal diffusion pressure diffusion, and forced diffusion [see reference (H11, p. 487)].

Note that in Eq. (60) it is the multicomponent D_{ij} which appear, whereas in Eq. (61) the binary diffusion coefficients $\mathfrak{D}_{ij}$ appear.[13] For a system at constant temperature and pressure Eq. (61) may also be written

$$\nabla p_i = \sum_{\substack{j=1 \\ j \neq i}}^{\nu} \alpha_{ij} c_i c_j (\mathbf{v}_j - \mathbf{v}_i) \tag{62}$$

in which the "resistance factors" α_{ij} are defined as $(RT)^2/p\mathfrak{D}_{ij}$. This result, originally developed by Stefan and Maxwell, has been used for the discussions on the multicomponent diffusion through gas films by Hougen and Watson (H13) and by Wilke (W9).

2. *Diffusion in Turbulent Systems*

It has already been pointed out that the equations of change are valid for describing turbulent flow. The diffusion of A in a nonreacting binary mixture is described by the equation of continuity:

$$(\partial \rho_A / \partial t) + (\nabla \cdot \rho_A \mathbf{v}) + (\nabla \cdot \mathbf{j}_A) = 0 \tag{63}$$

in which $\mathbf{j}_A$ is given by Eq. (42). The quantities ρ_A and $\mathbf{v}$ are rapidly varying functions of the time owing to the turbulent oscillations. Normally one does not care to know the complete time behavior of these quantities, but rather the time-smoothed behavior. It is therefore useful to write

$$\rho_A = \bar{\rho}_A + \rho_A' \tag{64}$$
$$\mathbf{v} = \bar{\mathbf{v}} + \mathbf{v}' \tag{65}$$

in which the barred quantities represent the time-smoothed variables and the primed quantities represent the turbulent fluctuations. Substitution of these quantities into Eq. (63) gives

$$\frac{\partial}{\partial t}(\bar{\rho}_A + \rho_A') + (\nabla \cdot \{\bar{\rho}_A + \rho_A'\}\{\bar{\mathbf{v}} + \mathbf{v}'\}) + (\nabla \cdot \mathbf{j}_A) = 0 \tag{66}$$

Equation (66) is averaged with respect to time, the average being taken over a time interval long with respect to the turbulent variations, but short with respect to the time changes in the average behavior. Such an averaging gives

$$\frac{\partial \bar{\rho}_A}{\partial t} + (\nabla \cdot \bar{\rho}_A \bar{\mathbf{v}}) + (\nabla \cdot \{\bar{\mathbf{j}}_A + \mathbf{j}_A^{(t)}\}) = 0 \tag{67}$$

in which the turbulent mass flux $\mathbf{j}_A^{(t)}$ is

$$\mathbf{j}_A^{(t)} = \overline{\rho_A' \mathbf{v}'} \tag{68}$$

In a two-component mixture in which ρ is approximately constant, the average molecular-mass flux is

$$\bar{\mathbf{j}}_A = -\mathfrak{D}_{AB} \nabla \bar{\rho}_A \tag{69}$$

---

[13] In order to obtain Eq. (61) from Eq. (60), it has to be assumed that D_{ij} and $\mathfrak{D}_{ij}$ are calculated to the first approximation only; this is not, however, a serious approximation [see reference (H11, p. 487)].

And the turbulent-mass flux is arbitrarily written as

$$j_A^{(t)} = -\mathfrak{D}_{AB}^{(t)} \nabla \bar{\rho}_A \tag{70}$$

which serves to define a coefficient of turbulent diffusion (or coefficient of eddy diffusion) $\mathfrak{D}_{AB}^{(t)}$. In fully developed turbulent flow $\mathfrak{D}_{AB}^{(t)}$ is ordinarily much larger than $\mathfrak{D}_{AB}$, and so one customarily describes turbulent-mass diffusion by substitution of Eq. (70) into Eq. (67), ignoring the molecular-mass flux altogether. This implies that all problems solved analytically for ordinary molecular diffusion can be taken over for the description of turbulent diffusion simply by replacing $\mathfrak{D}_{AB}$ by $\mathfrak{D}_{AB}^{(t)}$.

The equation of motion and the equation of energy balance can also be time averaged according to the procedure indicated above (S1, pp. 336 *et seq.*; G7, pp. 191 *et seq.*; pp. 646 *et seq.*). In this averaging process there arises in the equation of motion an additional component to the stress tensor $\tau^{(t)}$ which may be written formally in terms of a turbulent (eddy) coefficient of viscosity $\mu^{(t)}$; and in the equation of energy balance there appears an additional contribution to the energy flux $q^{(t)}$, which may be written formally in terms of the turbulent (eddy) coefficient of thermal conductivity $k^{(t)}$. Hence for an incompressible fluid, the x components of the fluxes may be written

$$j_{Ax}^{(t)} = \overline{\rho_A' v_x'} = -\mathfrak{D}_{AB}^{(t)} (\partial \bar{\rho}_A / \partial x) \tag{71}$$

$$\tau_{yx}^{(t)} = \overline{\rho v_y' v_x'} = -\mu^{(t)} (\partial \bar{v}_y / \partial x) \tag{72}$$

$$q_x^{(t)} = \overline{\rho \hat{C}_v T' v_x'} = -k^{(t)} (\partial \bar{T} / \partial x) \tag{73}$$

The quantities $\overline{\rho v_y' v_x'}$ are the Reynolds stresses and have been extensively studied by hot-wire anemometry techniques (S1, G7). The other turbulent fluxes have not yet received much attention.

It should be emphasized that the eddy transport coefficients are not fundamental physical properties characteristic of the material being studied. The eddy coefficients are in general, functions of position, velocity, and nature of the turbulence (its intensity and scale). The theoretical interpretation of the eddy transport coefficients has been primarily in terms of the "mixing-length" theory of turbulence, which is a crude (but moderately successful) theory somewhat similar to the mean-free-path development of the kinetic theory of gases. The mixing-length theory of eddy diffusion and the experimental studies of the subject have been reviewed by Sherwood and Pigford (S9, Chap. II); additional references may be obtained from an article by Schlinger and Sage (S2). This is a fertile field for further research and an important one from the standpoint of the industrial separation processes.

3. *Diffusion by Forced and Free Convection*

Diffusion problems in systems involving forced and free convection are good illustrations of the importance of presenting all three of the equations of change as a prelude to a general discussion of diffusion. Only a handful of idealized problems of this type have been solved analytically. Since they are, however, of considerable importance in chemical engineering it is worth while to make some general remarks about them.

Problems in forced convection are solved in two steps: first one solves the equation of motion to obtain the velocity distribution, and then one puts the expression for **v** back into the diffusion equation [usually as given in Eq. (50)]. An illustration of this type of problem is that of the absorption of a gas by a liquid film flowing down a vertical plate. First the Navier-Stokes equations are solved to obtain a parabolic velocity profile; then this parabolic profile is inserted into Eq. (50) (P3, V2). Since Eq. (50) for non-reaching systems is of exactly the same form as the Fourier-Poisson equation for the heat transfer in forced convection, a number of problems can be solved by analogy. Particularly valuable in this connection are the summaries of solutions of the Fourier-Poisson equation given by Drew (D3) and by Jakob (J1, pp. 451–482). Some steady state diffusion problems in forced convection are discussed in Sec. IV,C.

Problems of forced convection diffusion in non-Newtonian flow have to this author's knowledge not yet been attacked. The equations needed for solving such problems are given in this article. The equation of motion in terms of the stress tensor [Eq. (25)] can be used to describe non-Newtonian flow provided that a suitable form for the stress tensor is used; examples of two non-Newtonian stress tensors are given in Eqs. (28a) and (28b).

Problems in free convection are much more difficult to handle, inasmuch as the equations of change are strongly coupled. Furthermore the temperature and concentration dependence of the coefficient of viscosity in the Navier-Stokes equations is of considerable importance. The free convection can be brought about by temperature gradients or by concentration gradients or both. A good summary of the work done on the problem of determining the velocity and temperature profiles for the free convection near an infinite heated vertical wall has been given by Jakob (J1, p. 443 *et seq.*). This problem amounts to solving the x and y components of the equation of motion simultaneously with the energy-balance (temperature) equation. The free convection brought about by a temperature gradient between two walls has been discussed in connection with the operation of the Clusius-Dickel (thermal diffusion) column (F10, F11). The free-convection mass transfer problems have been interpreted largely by means of analogy with free-convection heat transfer and by means of dimensional analysis. The convection patterns resulting from mass transfer may well be disturbing factors in the accurate analysis of mass transfer experiments; the role of convection due to density gradients resulting from diffusion in wetted-wall experiments has been discussed by Boelter (B14). The role of free-convection mass transfer in dissolution and deposition at electrodes has recently been studied by Tobias, Eisenberg, and Wilke (T5, W12).

4. *Diffusion in Multiphase Systems*

The mass transfer between phases is, of course, the very basis for most of the "diffusional operations" of chemical engineering. A considerable amount of experimental and empirical work has been done in connection with interphase mass transfer because of its practical importance; an excellent and complete survey of this subject may be found in the text book of Sherwood and Pigford (S9, Chap. III), where dimensionless correlations for mass transfer coefficients in systems of various shapes are assembled.

In connection with the interphase mass transfer in liquid-liquid and liquid-gas systems, the diffusion equations (and indeed all the equations of change) are valid in both phases. Hence, in principle, diffusion problems in a two-phase system may be solved by solving the diffusion equations in each phase and then choosing the constants of interaction in such a way that the solutions "match up" at the interface. It is customary to require that the following two conditions be fulfilled at the interface, in a system in which the solute is being transferred from phase I to phase II: (1) the flux of mass leaving phase I must equal the flux of mass entering phase II; if the diffusion is in the z direction with the interface at $z = 0$, then

$$-\mathfrak{D}_\mathrm{I} \frac{\partial c_\mathrm{I}}{\partial z}\bigg|_{z=0-} = -\mathfrak{D}_\mathrm{II} \frac{\partial c_\mathrm{II}}{\partial z}\bigg|_{z=0+} \tag{74}$$

(2) the concentration in phase I must be in equilibrium with the concentration in phase II according to the distribution law:

$$c_\mathrm{I}(0-) = mc_\mathrm{II}(0+) \tag{75}$$

in which m is the distribution coefficient for the solute between the two phases.

Certainly the condition in Eq. (74) is valid since there must be no accumulation of solute at the interface. But the condition for equilibrium at the interface in Eq. (75) may not be adequate for the description of many mass transfer processes. It is not, for example, difficult to imagine that in the evaporation of a liquid, the vaporization may take place so rapidly that the concentration of vapor just above the liquid surface is considerably less than the concentration corresponding to the equilibrium vapor pressure. The problem of obtaining a quantitative theoretical description of this process has been attacked by Schrage (S4), who has suggested several molecular theories for describing gas-liquid and gas-solid systems.

One possible way to modify Eq. (75) was suggested in a paper by Scott, Tung, and Drickamer (S7), who proposed that the mass flux across the interface $N^{(\mathrm{int})}$ be written as the product of an interface mass transfer coefficient $k^{(\mathrm{int})}$ and a concentration difference:

$$N^{(\mathrm{int})} = k^{(\mathrm{int})}[c_\mathrm{I}(0-) - mc_\mathrm{II}(0+)] \tag{76}$$

in which m is the distribution coefficient. If there is no resistance due to the interface, then $k^{(\mathrm{int})} = \infty$ and the concentration difference factor must be zero; but the latter statement is just tantamount to the distribution law given in Eq. (75). If the interfacial resistance is infinite, the $k^{(\mathrm{int})} = 0$ and there is no interphase mass transfer. Scott, Tung, and Drickamer (S7) solved the differential equations for a closed-cell type of diffusion apparatus, taking into account the interfacial resistance as described by Eq. (76); subsequently Tung and Drickamer (T7) used the mathematical solution to interpret the results of their radio-tracer experiments on several liquid-liquid systems. Their experiments were ingeniously arranged so that they were always dealing with saturated solutions, equimolal counterdiffusion (so that Fick's law holds exactly), and an isothermal system (no heat of solution effects); prior investigators had not such refined experimental analyses (I3, H16, D2). Tung and Drickamer concluded from their experiments that the interfacial resistance is sufficiently large that the alteration in concentration profiles may be observed experimentally; they were not able to make any decision as to a connection between interfacial resistance and interfacial tension. Similar mathematical analysis and experiments have never been performed for the analogous phenomena in momentum and energy transfer. Somewhat later Sinfelt and Drickamer (S9a) investigated some more systems in order to find a correlation between interfacial resistance and other properties. They were unable to find a correlation between interfacial resistance and dipole moments or interfacial tension, but that high resistances seemed to be associated with liquids showing a high degree of hydrogen bonding.

Since the experiments of Tung and Drickamer, the resistance to diffusion through an interface has been further studied in gas-liquid systems by Emmert and Pigford (E2), who studied the absorption and desorption of CO_2 and O_2 in water in a wetted-wall tower and interpreted their results in terms of accommodation coefficients. They too reached the conclusion that the failure to include the resistance of the interface to diffusion may cause errors of about 30% in some calculations for the design of mass transfer equipment. Ward and Brooks (W5), on the other hand, found no energy barrier for interphase mass transfer of several organic acids across a water-toluene interface.

The interfacial diffusion model of Scott, Tung, and Drickamer is somewhat open to criticism in that it does not take into account the finite thickness of the interface. This objection led Auer and Murbach (A4) to consider a "three-region" model for the diffusion between two immiscible phases, the third region being an interface of finite thickness. These authors have solved the diffusion equations for their model for several special cases; their solutions should be of interest in future analysis of interphase mass transfer experiments.

In connection with multiphase diffusion another poorly understood topic should be mentioned—namely, the diffusion through porous media. This topic is of importance in connection with the drying of solids, the diffusion in catalyst pellets, and the recovery of petroleum. It is quite common to use Fick's laws to describe diffusion through porous media (J14). However, the mass transfer is possibly taking place partly by gaseous diffusion and partially by liquid-phase diffusion along the surface of the capillary tubes; if the pores are sufficiently small, Knudsen gas flow may prevail (W7, B1).

III. Calculation and Estimation of Diffusion Coefficients

For chemical engineering calculations it is necessary to know values of the coefficients of diffusion and thermal diffusion for a wide range of chemical systems. Experimental measurements of diffusion coefficients are difficult, and the data in the literature are not extensive. Hence one has to rely on theoretical and empirical relations. Ultimately it is hoped that expressions and tabulations will be available for calculating diffusion coefficients of gases and liquids from intermolecular forces and molecular properties. At the present time such calculations can be made for dilute gases, and indeed the theoretical calculations are about as good as the available experimental data. For dense gases and liquids much remains to be done in the way of establishing calculation methods based on sound molecular theory. Because several very complete surveys of the methods of predicting transport coefficients have recently appeared (H11, B12), we give here only a brief summary of the present status of the subject and discuss the importance of some recent work.

A. Diffusion Coefficients in Dilute Gases

The early theories for the transport coefficients were based on the concept of the mean free path. Excellent summaries of these older theories and their later modifications are to be found in standard text books on kinetic theory (J2, K2). The mean-free-path theories, while still very useful from a pedagogical standpoint, have to a large extent been supplanted by the rigorous mathematical theory of nonuniform gases, which is based on the solution of the Boltzmann equation. This theory is primarily the development of Chapman and of Enskog and is described in detail in the monograph of Chapman and Cowling (C3); a very readable survey of the classical and modern kinetic theory may be found in a small book by Cowling (C9).

The Chapman-Enskog theory was developed for dilute, monatomic gases for pure substances and for binary mixtures. The extension to multicomponent gas mixtures was performed by Curtiss and Hirschfelder (C12, H11), who in addition have shown that the Chapman-Enskog results may also be obtained by means of an alternate variational method. Recently Kihara (K3) has shown how expressions for the higher approximations to the transport coefficients may be obtained, which are considerably simpler than those previously proposed by Chapman and Cowling; these simpler formulas are particularly advantageous for calculating the coefficients of diffusion and thermal diffusion (M3, M4).

A kinetic theory for dilute polyatomic gases has been developed by Wang-Chang and Uhlenbeck (W3, U3). No calculations have been made of the diffusion coefficients on the basis of this theory, however. For most polyatomic gases the results of the Chapman-Enskog monatomic gas theory seem to be adequate.

1. *Intermolecular Forces*

In order to understand and use the results of the Chapman-Enskog theory and the numerous tabulations prepared for use in conjunction therewith, some knowledge of intermolecular forces is required. All the kinetic theory and statistical mechanical developments yield expressions for the bulk properties in terms of the force law governing the interaction between a pair of molecules in the fluid. In principle it is possible to determine the force between a pair of molecules by means of quantum mechanical calculations, since the mechanics of any system of nuclei and electrons is governed by the Schrödinger equation. In practice, however, such calculations have been made for only very simple molecule systems, and then only after numerous approximations have been made. The quantum mechanical approach has nevertheless provided enough information of semiquantitative character to establish the approximate functional form of the intermolecular forces. Summaries of the quantum mechanics of intermolecular forces have been given by Margenau (M1), London (L5), Hirschfelder (H11, Chapters 12, 13, 14), and de Boer (B15).

In the kinetic and statistical mechanical formulas for the bulk properties it is the potential energy of interaction $\varphi(r)$ which appears rather than the force of interaction $F(r)$ between the molecules in the fluid. For two spherical nonpolar molecules separated by a distance r the force of interaction is obtained by

$$F(r) = -d\varphi/dr \qquad (77)$$

It is customary to split the intermolecular potential energy function $\varphi(r)$ into two parts:

$$\varphi(r) = \varphi_r(r) + \varphi_a(r) \qquad (78)$$

in which $\varphi_r(r)$ is the repulsive contribution important at small separations, and

$\varphi_a(r)$ is the attractive contribution important at large separations. (This same notion is used in the van der Waals equation of state, in which the constant a accounts for molecular attraction and b for molecular repulsion.)

Spherical nonpolar molecules obey an interaction potential which has the characteristic shape shown in Fig. 2. At large values of the separation r it is known that the potential curve has the shape $-r^{-6}$, and at short distances the potential curve rises exponentially; the exact shape of the bottom part of the curve is not very well known. Numerous empirical equations of the form of Eq. (78) have been suggested for describing the molecular interaction given pictorially in Fig. 2. The discussion here is restricted to the two most important empirical functions. A rather complete summary of the contributions to intermolecular potential energy and empirical intermolecular potential energy functions used in applied statistical mechanics may be found in (H11, Sec. 1.3):

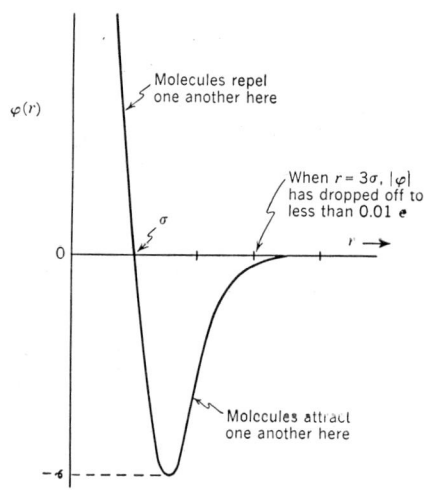

FIG. 2. Sketch showing general form of the intermolecular potential energy function.

a. *The Lennard-Jones (6-12) potential.* If the repulsive contribution is taken to be proportional to r^{-12} then one can write

$$\varphi_{AB}(r) = 4\epsilon_{AB} \left[\left(\frac{\sigma_{AB}}{r} \right)^{12} - \left(\frac{\sigma_{AB}}{r} \right)^{6} \right] \tag{79}$$

in which σ_{AB} and ϵ_{AB} have the analytical significance shown in Fig. 2. Physically σ_{AB} is a measure of the distance between centers of two molecules A and B when they are just touching; ϵ_{AB} is a measure of the strength of the attractive forces between the molecules. For the function given in Eq. (79) the ratio $(r_m)_{AB}/\sigma_{AB}$ is always equal to $2^{1/6}$.

b. *The Buckingham (6-exp) potential.* If the repulsive contribution is taken to be of an exponential form, then one can construct a function:

$$\varphi_{AB}(r) = \frac{\epsilon_{AB}}{\left(1 - \dfrac{6}{\alpha_{AB}}\right)} \left[\frac{6}{\alpha_{AB}} \exp\left\{ \alpha_{AB}\left(1 - \frac{r}{(r_m)_{AB}}\right) \right\} - \left(\frac{r}{(r_m)_{AB}} \right)^{-6} \right] \tag{80}$$

in which $(r_m)_{AB}$ and ϵ_{AB} are described in Fig. 2. The parameter α_{AB} is a measure of the steepness of the repulsive energy and varies from about 12 to 17 for most substances. The ratio $(r_m)_{AB}/\sigma_{AB}$ varies from about 1.11 to 1.14. Hence the Lennard-Jones (6-12) potential contains two parameters σ_{AB} and ϵ_{AB} which are characteristic of collisions between molecules of types A and B. The Buckingham (6-12) potential is more flexible in that it contains three parameters.

TABLE IV
Short Table of Intermolecular Potential-Energy Parameters

Substance	Lennard-Jones (6-12)[a]			Buckingham (6-exp)[b]		
	σ Å	ϵ/κ °K	r_m Å[c]	r_m Å	ϵ/κ °K	α
He	2.576	10.22	2.892	3.135	9.16	14.4
Ne	2.789	35.7	3.131	3.147	38.0	14.5
A	3.418	124.	3.837	3.866	123.2	14.0
Kr	3.610	190.	4.052	4.056	158.3	12.3
Xe	4.055	229.	4.552	4.450	231.2	13.0
H_2	2.968	33.3	3.332	3.337	37.3	14.0
N_2	3.681	91.5	4.132	4.011	101.2	17.0
CH_4	3.822[d]	137.	4.290	4.206	152.8	14.0
CO	3.590	110.	4.030	3.937	119.1	17.0

[a] The parameters given here for the Lennard-Jones (6-1) potential function are those determined from viscosity; an extensive tabulation may be found in (H11 and B12).
[b] Taken from (M3 and M4).
[c] For the Lennard-Jones (6-12) function $r_m = \sigma \sqrt[6]{2}$.
[d] Given incorrectly in (M11) as 3.882.

The transport coefficients and the second virial coefficients have been calculated for the potential energy functions given in Eqs. (79) and (80). For example, for the Lennard-Jones (6-12) potential, the coefficient of viscosity of a dilute gas may be expressed in terms of σ_{AA} and ϵ_{AA}. If one has experimental data for the viscosity of A at two temperatures, then one can solve two simultaneous equations and obtain values of σ_{AA} and ϵ_{AA} characteristic of collisions between molecules of species A. In this way a large table of values of σ_{AA} and ϵ_{AA} have been built up for the Lennard-Jones (6-12) potential both from the analysis of viscosity data and second virial coefficient data (H11, H9, H10); and a less extensive table has been prepared for the Buckingham (6-exp) potential (M5, M6). An abstract of these tables is given in Table IV. The fact that both potential functions give very nearly the same r_m and ϵ/κ for a number of substances is an indication of the progress that has been made in establishing the form of the intermolecular potential-energy function from the analysis of bulk properties. It should be noted that the parameter α for the Buckingham (6-exp) potential does not seem to follow any sort of trend for the series of noble gases; this may be an indication that the experimental data used for the calculation of the potential parameters are not sufficiently accurate.

The sets of parameters σ_{AB}, ϵ_{AB} for the Lennard-Jones (6-12) potential and $(r_m)_{AB}$, ϵ_{AB}, α_{AB} for the Buckingham (6-exp) potential, for the interactions between dissimilar molecules, can in principle be determined from experimental measurements of the properties of mixtures; the best mixture property for this purpose is binary diffusivity. Since there are not enough measurements of diffusion coefficients as a function of temperature in the literature, it has been customary to use empirical "combining laws" to obtain the interaction parameters for dissimilar molecules from the parameters for similar molecules.[14]

a. *Lennard-Jones (6-12) potential.*

$$\sigma_{AB} = \tfrac{1}{2}(\sigma_A + \sigma_B)$$
$$\epsilon_{AB} = \sqrt{\epsilon_A \epsilon_B} \tag{81}$$

b. *Buckingham (6-exp) potential.*

$$(r_m)_{AB} = \tfrac{1}{2}[(r_m)_A + (r_m)_B] \tag{82}$$
$$\epsilon_{AB} = \sqrt{\epsilon_A \epsilon_B}$$
$$\alpha_{AB} = \tfrac{1}{2}[\alpha_A + \alpha_B]$$

If the potential parameters for the pure components are not found in the tables given in (H11) and (B11), and if viscosity and second virial data are not available for their determination, then for the Lennard-Jones (6-12) potential it is possible as a last resort to estimate these parameters from the properties of the substance at its critical point c, its melting point m, or its boiling point b; these relations give ϵ/κ in °K. and σ in Ångström units (1 Å. = 10^{-8} cm.):

$$\epsilon/\kappa \doteq 0.77 T_c \qquad \sigma \doteq 0.841 \sqrt[3]{V_c} \doteq 244 \sqrt[3]{\frac{T_c}{p_c}} \tag{83}$$

$$\epsilon/\kappa \doteq 1.15 T_b \qquad \sigma \doteq 1.17 \sqrt[3]{V_b^{(\text{liq})}} \tag{84}$$
$$\epsilon/\kappa \doteq 1.92 T_m \qquad \sigma \doteq 1.22 \sqrt[3]{V_m^{(\text{sol})}} \tag{85}$$

in which the volumes are in cm.^{3} mole^{-1}, temperature is in °K., and pressure is in atm.; κ is the Boltzmann constant. There are as yet no methods available for the estimation of the intermolecular potential parameters directly from molecular structure.

2. *Classical Calculations of the Diffusion Coefficients*[15]

In order to make calculations of the transport coefficients based on the kinetic theory of Chapman and Enskog, one first assumes an empirical potential function (such as the Lennard-Jones or Buckingham functions just discussed). Then from the classical dynamics of a two-body collision (H11, §1.5) one calculates the angle of deflection χ for many values of the relative initial kinetic energy and angular momentum of the pair of colliding molecules. Next one uses the angles of deflection to calculate a set of temperature-dependent integrals $\Omega^{(l,s)\star}$ in terms of which all the transport coefficients may be expressed. These $\Omega^{(l,s)\star}$ are integrals over all possible relative kinetic energies and angular momenta; the indexes l and s indicate different ways of weighting collisions as functions of these quantities as required by the detailed kinetic theory. For any realistic

---

[14] More complex combining laws have been developed by Mason (M6) for the modified Buckingham (6-exp) potential.

[15] An expression for the coefficient of diffusion identical with that given in Chapman and Cowling (C3) has also been obtained by considering diffusion as a random-walk process (F12).

potential $\varphi(r)$ all the integrations have to be performed numerically. The integral $\Omega^{(1,1)\star}$ is used for calculating the first approximation to the coefficient of diffusion, and $\Omega^{(2,2)\star}$ gives the first approximation to the coefficients of viscosity and thermal conductivity of a pure substance. Other $\Omega^{(l,s)\star}$ are needed for higher approximations and mixture calculations.

The $\Omega^{(l,s)\star}$ have been evaluated for several simple intermolecular potential-energy functions, such as rigid spheres, point centers of repulsion, the square-well potential, and the Sutherland model (H11); here we discuss only the results for the Lennard-Jones (6-12) and the Buckingham (6-exp) functions. For the Lennard-Jones (6-12) potential the first approximation to the coefficient of ordinary diffusion for a binary dilute gas mixture of A and B is given by

$$[\mathfrak{D}_{AB}]_1 = \frac{3}{16} \frac{\sqrt{2\pi RT(M_A + M_B)/M_A M_B}}{n\pi\sigma_{AB}^2 \Omega^{(1,1)\star}(\kappa T/\epsilon_{AB})} \tag{86}$$

in which n is the number density (total number of molecules per unit volume); the perfect gas law in the form $p = n\kappa T$ may be used to rewrite Eq. (86) in terms of p and T if desired.

TABLE V

COMPARISON OF $\Omega^{(1,1)\star}$ CALCULATED BY THE LENNARD-JONES POTENTIAL AND BY THE BUCKINGHAM (6-EXP) POTENTIAL

$T^* = \dfrac{\kappa T}{\epsilon}$	$\Omega^{(1,1)\star}$ Lennard-Jones (6–12)[a]	$(r_m/\sigma)^2 \Omega^{(1,1)\star}$ Buckingham (6-exp)[b]	
		$\alpha = 12$	$\alpha = 16$
0.5	2.066	2.126	1.951
1.0	1.439	1.450	1.406
1.5	1.198	1.188	1.186
2.0	1.075	1.053	1.071
3.0	0.949	0.913	0.953
4.0	0.884	0.840	0.889
5.0	0.842	0.792	0.848
10.0	0.742	0.672	0.749
50.0	0.576	0.463	0.578
100.0	0.517	0.391	0.520

[a] These values are taken from (H11).
[b] These values are taken from (M3 and M6).

Equation (86) shows that the diffusion coefficient is inversely proportional to σ_{AB}^2 and depends upon the parameter ϵ_{AB} through the dependence of $\Omega^{(1,1)\star}$ on the reduced temperature $(\kappa T/\epsilon_{AB})$. An abbreviated table of this function is given in Table V. [An extensive table is given in (K5, H11, B12).] For rigid spherical molecules the $\Omega^{(1,1)}\star$ function is defined to be unity, and so the $\Omega^{(1,1)}\star$ for the Lennard-Jones potential may be interpreted as a measure of the deviation of the Lennard-Jones molecules from rigid-sphere behavior. From Table V it is clear that the rigid-sphere model is inadequate for describing the coefficient of diffusion over a wide temperature range. The physical interpretation of the $\Omega^{(1,1)}\star$ for the Lennard-Jones (6-12) potential is as follows. At low temperatures the molecules on the average tend to collide with one another with relatively slow speeds, with the result that they tend to spend longer periods of time in one another's attractive fields; the collisions therefore are somewhat more sluggish than for rigid spheres and the process of diffusion is slower than would be expected for rigid spheres (hence at low temperature, $\Omega^{(1,1)}\star > 1$). At high temperatures the molecules on the average tend to collide with relatively high speeds, and so it is possible for them to shoot past one another at very close distances even penetrating one another slightly. Because of the possibility of penetration, the molecules can get past one another somewhat more easily than can rigid spheres, and the process of diffusion is somewhat more rapid that one would expect for rigid spheres (hence at high temperatures, $\Omega^{(1,1)}\star < 1$).

From Eq. (86) it can be seen that measurements of $\mathfrak{D}_{AB}$ as a function of T offer a means for getting the parameters σ_{AB} and ϵ_{AB} describing the unlike interactions. For this purpose the data have to be rather good and the temperature range needs to be several hundred degrees centigrade if possible. Since there are no data of that sort, one has to calculate σ_{AB} and ϵ_{AB} from Eq. (81) by use of Eqs. (83) through (85) if necessary.

It should be pointed out that Eq. (86) is the first approximation to the coefficient of diffusion according to the Chapman-Enskog theory, and in that approximation the diffusion coefficient does not depend upon the concentration. The development of a series of higher approximations to the transport coefficients is not unique. One method of getting higher approximations to $\mathfrak{D}_{AB}$ is given by Chapman and Cowling (C3), and an alternate method has been proposed by Kihara (K3); Mason has recently made an extensive study of these two higher approximations methods and has concluded that those of Kihara are to be preferred (M4). The higher approximations to $\mathfrak{D}_{AB}$ are complex functions of the concentration and temperature, and molecular weights and give diffusion coefficients which seldom differ by more than 3% from the first approximation given in Eq. (86); hence the latter is certainly adequate for most engineering calculations. Experimental diffusion data for 26 gas pairs have been compared with the diffusion coefficients calculated by Eqs. (86) and (81) (H11, p. 579; H8; W14); the calculated values agree with the experimental values within about 5%, which is about the limit of the experimental accuracy for many of the experimental measurements. Deviations of the calculated values from the experimental have recently been studied by Wilke and Lee (W11).

Two special applications of Eq. (86) should be mentioned, namely the calculation of isotopic diffusion and the calculation of the coefficient of diffusion for polar-nonpolar mixtures. For mixtures of heavy isotopes, $M_A \doteq M_B = M$, $\sigma_A = \sigma_B = \sigma$, and $\epsilon_A = \epsilon_B = \epsilon$. Then Eq. (86) simplifies to

$$[\mathfrak{D}]_1 = \frac{3}{8} \frac{\sqrt{\pi RT/M}}{n\pi\sigma^2 \Omega^{(1,1)}\star(\kappa T/\epsilon)} \tag{87}$$

This quantity is called the *coefficient of self-diffusion*. Equation (87) also describes the interdiffusion of ortho- and para-forms.

A method for determining the interdiffusion of a polar and a nonpolar gas has been developed by Hirschfelder (H10), who showed that the intermolecular potential-energy function for polar-nonpolar interaction can be described approximately by a Lennard-Jones (6-12) potential (H11, p. 987, footnote 6). The method of calculation is discussed in detail along with an illustrative example in (H11, pp. 600–602); comparison of the calculated results with some experimental data on diffusion of water vapor into nonpolar gases indicates discrepancies of the order of 5%. This method may be regarded as quite satisfactory, in view of the fact that different experimenters disagree by about 10% in some of the water-vapor diffusion data.

The Lennard-Jones (6-12) potential has served very well as an intermolecular potential and has been widely used for statistical mechanics and kinetic-theory calculations. It suffers, however, from having only two adjustable constants, and there is no reason why it should not gradually be replaced by more flexible and more realistic functions. Recently a number of applications have been made of the Buckingham (6-exp) potential [Eq. (82)], which has three adjustable parameters. For this potential the first approximation to the coefficient of diffusion is written by Mason (M3) in the form

$$[\mathfrak{D}_{AB}]_1 = \frac{3}{16} \frac{\sqrt{2\pi R T (M_A + M_B)/M_A M_B}}{n\pi (r_m)_{AB}^2 \Omega^{(1,1)\star}(\kappa T/\epsilon_{AB}, \alpha_{AB})} \tag{88}$$

in which the parameters $(r_m)_{AB}$, ϵ_{AB}, and α_{AB} are those defined in connection with Eq. (82). Mason (M3) has calculated $\Omega^{(1,1)\star}$ and the other $\Omega^{(l,s)\star}$ as functions of the reduced temperature for $\alpha = 12, 13, 14, 15$; tables for $\alpha = 16, 17$ have also been prepared (M6) by extrapolation of the previous tables. In Table V the $\Omega^{(1,1)\star}$ for the Buckingham (6-exp) potential are compared with the $\Omega^{(1,1)\star}$ for the Lennard-Jones (6-12) potential. Mason (M4) has compared experimental data for four systems with the values calculated from Eq. (88) and finds agreement within about 5%. From existing experimental data on diffusion coefficients one cannot conclude that the Buckingham (6-exp) function is to be preferred to the Lennard-Jones (6-12) function.

The thermal diffusion ratio k_T may be calculated according to three different formulas: (a) the $[k_T]_1$ of Chapman and Cowling (C3, p. 253; H11, p. 541); (b) the $[k_T]_2$ of Chapman and Cowling worked out by Mason (M3, H11, p. 606); or (c) the formula of Kihara (K3, M3, H11, p. 609). Mason has studied these three relations extensively and has come to the following conclusions (M4). The error involved in using $[k_T]_1$ is certainly greater than the error in the experimental measurements. The most accurate expression is $[k_T]_2$ of Chapman and Cowling, which is quite complicated and tedious to use. Sample calculations indicate that Kihara's expression for k_T is more accurate than $[k_T]_1$ and usually differs from $[k_T]_2$ of Chapman and Cowling by less than the scatter in the experimental data. Hence the Kihara approximation is probably the most satisfactory for present use. None of the above-mentioned formulas are presented here since they are quite long. The formulas may be found in the references given above, where the tabulated functions needed for their application are also given.

3. *Quantum Calculations of the Diffusion Coefficients*

For the light molecules He and H_2 at low temperatures (below about 50°C.) the classical theory of transport phenomena cannot be applied because of the importance of quantum effects. The Chapman-Enskog theory has been extended to take into account quantum effects independently by Uehling and Uhlenbeck (U1, U2) and by Massey and Mohr (M7). The theory for mixtures was developed by Hellund and Uehling (H3). It is possible to distinguish between two kinds of quantum effects— "diffraction effects" and "statistics effects"; the latter are not important until one reaches temperatures below about 1°K. Recently Cohen, Offerhaus, and de Boer (C4) made calculations of the self-diffusion, binary-diffusion, and thermal-diffusion coefficients of the isotopes of helium. As yet no experimental measurements of these properties are available.

The connection between the classical and quantum formulations of the transport coefficients has been studied by applying the WKB method to the quantum formulation of the kinetic theory (B16, B17). In this way it was shown that at high temperatures the quantum formulas for the transport coefficients may be written as a power series in Planck's constant h. When the classical limit is taken (h approaches zero), then the classical formulas of Chapman and Enskog are obtained.

4. *Empirical Formulas*

Because of the excellent agreement between experimental measurements and the values calculated on the basis of the Enskog theory, empirical formulas are not needed. Sometimes, however, it is convenient to have empirical formulas available for rapid calculations or for use in analytical solutions to differential equations. Some empirical relations have been assembled by Partington (P1).

The most well-known empirical formula for diffusion coefficients is that due to Gilliland (G3, S9), which is written in terms of the molecular volumes of the diffusion species and gives $\mathfrak{D}_{AB}$ proportional to $T^{3/2}$. Since this temperature dependence has since been shown to be incorrect the Gilliland formula should not be used. Another empirical formula has been suggested by Arnold (A1). Recently Slattery (S10) has proposed the following relation on the basis of corresponding states arguments:

$$[p\mathfrak{D}_{AB}]_R = aT_r^b \tag{89}$$

in which the reduced quantities $[p\mathfrak{D}_{AB}]_R$ and T_r are defined for a binary mixture by

$$[p\mathfrak{D}_{AB}]_R = \frac{p\mathfrak{D}_{AB}[2M_AM_B/(M_A + M_B)]^{1/2}}{(p_{cA}p_{cB})^{1/3}(T_{cA}T_{cB})^{5/12}} \tag{90}$$

$$T_r = T/\sqrt{T_{cA}T_{cB}} \tag{91}$$

The constants a and b are given by

Interdiffusion of nonpolar gases and also self-diffusion: $a = 3.882 \times 10^{-4}$
$b = 1.8229$

Interdiffusion of H_2O and a nonpolar gas: $a = 5.148 \times 10^{-4}$
$b = 2.334$

These relations require only a knowledge of the molecular weights and the critical temperatures and pressures (the latter may be obtained from the summary of Lydersen (L7) or from that of Kobe and Lynn (K13). Equation (89) gives results which agree with experimental data within about 8%. To this author's knowledge no empirical relations have been proposed for k_T.

B. Diffusion Coefficients in Dense Gases

The theory which forms the basis for discussions of the transport phenomena in dense gases is Enskog's kinetic theory for a pure gas made up of rigid spheres (E3, C3, Chapter 16; H11, §9.3). To date, this theory in one of several modifications is the best theory available for calculating the temperature and density dependence of the transport coefficients. Recently Enskog's theory has been extended to a pure gas made up of nonrigid molecules by Curtiss and Snider (C10, C14). Enskog's theory has also been extended to binary gas mixtures by Thorne (C3, p. 292).

An alternate approach has been attempted for describing the transport phenomena in dense gas and liquid systems by means of the methods of nonequilibrium statistical mechanics, as developed by Kirkwood (K7, K8) and by Born and Green (B18, G10). Although considerable progress has been made in the development of a formal theory, the method does not at the present time provide a means for the practical calculation of the transport coefficients. Hence in this section we discuss only the applications based on Enskog's theory.

1. *Enskog's Rigid-Sphere Theory*

Enskog's kinetic theory for dense gases differs from his theory for dilute gases in that it takes into account the fact that the diameter of the molecules is not small with respect to the mean free path. His theory was developed for rigid spheres only; since more than two rigid spheres cannot collide simultaneously, the effects of multiple collisions are not taken into account. Although the development of the theory is quite lengthy, the results may be expressed in a simple form by giving the ratio of the transport coefficients to their dilute gas values (indicated by a superscript [0]).

$$\frac{\mu}{\mu^0} = \frac{2}{3}\pi n \sigma^3 \left[\frac{1}{y} + 0.8 + 0.761 y\right] \tag{92}$$

$$\frac{k}{k^0} = \frac{2}{3}\pi n \sigma^3 \left[\frac{1}{y} + 1.2 + 0.755 y\right] \tag{93}$$

$$\frac{n\mathfrak{D}}{(n\mathfrak{D})^0} = \frac{2}{3}\pi n \sigma^3 \left[\frac{1}{y}\right] \tag{94}[16]$$

[16] An alternate formulation of Eq. (94) is given in (H11, pp. 646 and 647).

in which n is the number density, and y is closely related to the equation of state for rigid spheres:

$$y = \frac{pV}{RT} - 1 = \left(\frac{2}{3}\pi n\sigma^3\right) + 0.6250\left(\frac{2}{3}\pi n\sigma^3\right)^2$$
$$+ 0.2869\left(\frac{2}{3}\pi n\sigma^3\right)^3 + 0.115\left(\frac{2}{3}\pi n\sigma^3\right)^4 + \cdots \quad (95)$$

Equation (95) is obtained from the virial expansion of the equation of state for rigid spheres; for higher densities the rigid-sphere equation of state obtained from the radial distribution function by Kirkwood, Maun, and Alder has to be used (K10, H11, p. 649). When Eq. (95) is substituted in Eqs. (92), (93), and (94) one then obtains the rigorous expressions for the coefficients of viscosity, thermal conductivity, and self-diffusion of a gas composed of rigid spheres.

2. Applications of Enskog's Theory to Real Gases

Enskog suggested that the foregoing results could be applied to real gases by a modification in the interpretation of the quantities y and σ. He recommended that y be obtained from experimental p-V-T data according to the relation

$$y = \frac{V}{R}\left(\frac{\partial p}{\partial T}\right)_V - 1 \quad (96)$$

and that σ then be fixed by making the experimental and calculated values of the viscosity agree at the minimum in the curve of (μ/ρ) vs. ρ. In this way Eq. (92) has been used successfully to predict the viscosity of the following substances:

	T, °C.	p, atm.	max. dev., %
CO_2	40.3	45–115	6 (E3)
N_2	50	15–960	6 (C3)
A	0–50	1–2000	10 (M9)

For the prediction of thermal conductivity Comings and Nathan (C6) and Gamson (G1) have divided Eq. (93) by Eq. (92), thereby eliminating σ and obtaining an expression for k in terms of k^0, μ/μ^0, and y; the quantity y was determined from Eq. (96) by Enskog's interpretation. This procedure led Comings and Nathan and Gamson to the preparation of generalized charts for thermal conductivity. Since their charts were prepared, a few high-density measurements of thermal conductivity have been made for several simple gases; these data were found to be in agreement with the generalized charts by about 15%. The utility of these charts for predicting the thermal conductivity of polyatomic molecules is open to question, since the Enskog theory is derived for monatomic molecules; whereas it is known that the vibrational and rotational motions of molecules have to be taken into account in calculations involving energy transfer.

For the prediction of self-diffusion, two methods have been suggested:

a. Drickamer and his collaborators have used Eq. (94), in which $(n\mathfrak{D})^0$ is calculated by the first approximation in the Chapman-Enskog scheme using the Lennard-Jones (6-12) calculations, and y is calculated according to the rigid-sphere expression [Eq. (95)] using the σ from the Lennard-Jones (6-12) potential. Jeffries and Drickamer (J3, J4) have reported quite satisfactory agreement between the self-diffusion coefficients calculated according to this method and their own self-diffusion data for CH_4; on the other hand Robb and Drickamer (R2) and Timmerhaus and Drickamer (T4) have reported considerable discrepancy between their self-diffusion data for CO_2 and this method of calculating, the disagreements being worse at higher density. The discrepancy was felt, however, to be consistent with the fact that the third virial coefficient for CO_2 deviates considerably from the Lennard-Jones (6-12) curve, possibly owing to some sort of orientation effects (B13). Becker, Vogell, and Zigan (B7), who also have measured the self-diffusion coefficient for CO_2, suggest that the pressure-dependence of the $p\mathfrak{D}$ product at moderate densities is due to dimerization. Although this method was reported to be successful for the correlation of the CH_4 data, there still remains some doubt as to its utility, in view of the fact that the same method when applied to viscosity gives very poor agreement with experimental data.

b. A second method of calculation of high-density diffusion coefficients was used by Slattery (S10), who made use of Eqs. (92), (94), and (96). Equation (94) may be written in a somewhat more convenient form:

$$\frac{p\mathfrak{D}}{(p\mathfrak{D})^0} = \frac{2}{3} \pi n \sigma^3 \frac{Z}{y} \qquad (97)$$

in which Z is the compressibility factor pV/RT. Division of Eq. (97) by Eq. (92) then gives

$$\frac{p\mathfrak{D}}{(p\mathfrak{D})^0} = \left(\frac{\mu}{\mu^0}\right)\left[\frac{Z}{1 + 0.8y + 0.761y^2}\right] \qquad (98)$$

Equation (98) contains the implicit assumption that σ for viscosity is the same as σ for diffusion. The quantity y may be expressed in terms of the reduced equation of state $Z = Z(p_r, T_r)$ thus:

$$y = Z\,\frac{1 + (\partial \ln Z/\partial \ln T_r)}{1 - (\partial \ln Z/\partial \ln p_r)} - 1 \qquad (99)$$

Hence one needs Z and its temperature and pressure derivatives, the ratio μ/μ^0, and $(p\mathcal{D})^0$ in order to calculate the diffusion coefficient by this process. When these data are not available, one may make use of the generalized Z chart of Hougen and Watson (H13), the chart of (μ/μ^0) by Comings, Mayland, and Egly (C7), and Eq. (86) or (89) for $(p\mathcal{D})^0$. The chart given by Slattery (S10) in Fig. 3 is a plot of Eq. (98) based on the best available data and correlations for (μ/μ^0), Z, and the derivatives of Z. It is not possible at the present time to assess the accuracy of this chart,

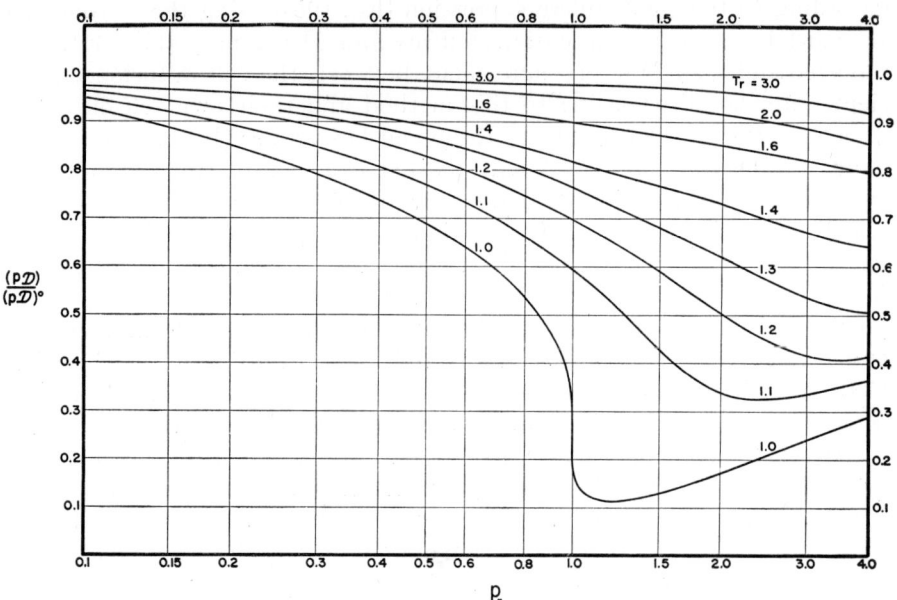

Fig. 3. Generalized chart for calculation of the coefficient of self-diffusion of gases at high densities, prepared by Slattery (S10).

but it is certainly useful for giving a good over-all picture of the behavior of the self-diffusion coefficient as a function of the reduced pressure and reduced temperature. It can also be used to estimate the diffusion coefficient in binary mixtures by use of the pseudocritical pressure and temperature (H13, p. 604), which has been used with reasonable success for the calculation of compressibilities and viscosities of multicomponent mixtures.

The field of diffusion in dense-gas systems is one in which much needs yet to be done. Enough experimental data should be taken so that a generalized chart based directly on diffusion data may be constructed; these data are also necessary for the testing of any new theory of high-density diffusion. The theory needs to be extended to include multiple collisions and molecules of various shapes and sizes. Much also needs to be done in connection with the temperature and concentration dependence of k_T at high densities. These experimental and theoretical problems are not easy, but it is important that this gap in the subject of diffusion be filled in.

C. Diffusion Coefficients in Liquids

It is previously pointed out that there are several mathematical theories for the description of transport phenomena in the liquid state, notably those of Kirkwood (K7, K8) and Born and Green (B18, G10). Since these theories are still in a state of development, most of the interpretations of diffusion phenomena have been in terms of either the hydrodynamical theories or the activated-state theories. Although none of these are rigorous approaches, they have been most useful in establishing the mechanism for diffusion. Since these rough theories do not, however, give a means for quantitative calculation of diffusion coefficients, a number of empirical formulas have been proposed. Most of these empirical formulas have used the rough theoretical relations as a starting point.

1. *Hydrodynamical Theories* (F13)

According to the Nernst-Einstein equation (N1, E1), the diffusion of a single particle or solute molecule A through a medium B may be described by the relation

$$\mathfrak{D}_{AB} = \kappa T \left(\frac{u_A}{F_A}\right) \quad (100)$$

in which (u_A/F_A) is the mobility of the particle (that is, the steady state velocity attained by a particle under the action of a unit force). A relation between force and velocity for a rigid sphere moving in "creeping flow" (that is, $Re < 1$) may be obtained from hydrodynamics. According to Basset (B6, p. 168; L1, §337) if one takes slip at the surface of the sphere into account, then

$$F_A = 6\pi\mu_B u_A R_A \left[\frac{2\mu_B + R_A\beta_{AB}}{3\mu_B + R_A\beta_{AB}}\right] \quad (101)$$

in which μ_B is the viscosity of the pure solvent, R_A is the radius of the diffusing particle, and β_{AB} is the coefficient of sliding friction. Sutherland has pointed out that there are two limiting cases of interest (S21, F13, pp. 639–641).

a. If there is no tendency for the fluid to slip at the surface of the diffusing particle ($\beta_{AB} = \infty$) then Eq. (101) becomes Stokes's law:

$$F_A = 6\pi\mu_B u_A R_A \quad (102)$$

and substitution into Eq. (100) gives

$$\frac{\mathfrak{D}_{AB}\mu_B}{\kappa T} = \frac{1}{6\pi R_A} \tag{103}$$

which is usually called the Stokes-Einstein equation. Equation (103) has been shown to be fairly good for describing the diffusion of large spherical particles or large spherical molecules, conditions under which the solvent appears to the diffusing species to be a continuum (S21).

  b. If there is no tendency for the fluid to stick at the surface of the diffusing particle ($\beta_{AB} = 0$), then Eq. (101) becomes

$$F_A = 4\pi\mu_B u_A R_A \tag{104}$$

and Eq. (100) becomes

$$\frac{\mathfrak{D}_{AB}\mu_B}{\kappa T} = \frac{1}{4\pi R_A} \tag{105}$$

If the molecules are all alike (that is, self-diffusion exists) and if they can be assumed to be arranged in a cubic lattice with all molecules just touching, then the $2R$ may be set equal to $(V/N)^{1/3}$ and

$$\frac{\mathfrak{D}\mu}{\kappa T} = \frac{1}{2\pi}\left(\frac{N}{V}\right)^{1/3} \tag{106}$$

where N is the number of molecules contained in volume V. Li and Chang (L4) have recently shown that Eq. (106) predicts the self-diffusion data for a number of liquids within about $\pm 12\%$; their comparison includes polar substances, associated substances, liquid metals, and molten sulfur.

It is thus seen that the simple hydrodynamical approach gives expressions for the diffusion coefficient for spherical molecules in dilute solution and also for the coefficient of self-diffusion. The theory further predicts that there should be a variation in $\mathfrak{D}_{AB}$ by a factor of about $2/3$ owing to the relative size of the diffusing species. The hydrodynamic theory suggests that the shape of the diffusing species may well be important, as for elongated bodies the drag coefficient varies by a factor of about 2 as the length-to-width ratio of a body goes from 1 to 10 (W2, B21, p. 76).

2. *Activated State Theories* (G6)

Another moderately successful approach to the theory of diffusion in liquids is that developed by Eyring (E4) in connection with his theory of absolute reaction rates (P6, K6). This theory attempts to explain the transport phenomena on the basis of a simple model for the liquid state and the basic molecular process occurring. It is assumed in this theory that there is some unimolecular rate process in terms of which the transport processes can be described, and it is further assumed that in this process there is some configuration that can be identified as the activated state. Then the Eyring theory of reaction rates is applied to this elementary process.

For self-diffusion this theory leads to a relation between the self-diffusion coefficient and the coefficient of viscosity:

$$\frac{\mathfrak{D}\mu}{\kappa T} = \frac{\lambda_1}{\lambda_2 \lambda_3} \tag{107}$$

in which λ_1, λ_2, and λ_3 are distances characterizing the spacing between layers of molecules in the quasicrystalline liquid lattice (G6, p. 519; H11, p. 631). When a further approximation is made that

$$\lambda_1 = \lambda_2 = \lambda_3 = (V/N)^{1/3}$$

then Eq. (107) becomes

$$\frac{\mathfrak{D}\mu}{\kappa T} = \left(\frac{N}{V}\right)^{1/3} \tag{108}$$

where N is the number of molecules contained in volume V. This relation differs by a factor of 2π from Eq. (106), the expression for the self-diffusion coefficient according to the hydrodynamical theory, but Eq. (106) has been shown to fit the experimental data better than Eq. (108). Recently Li and Chang (L4) suggested a modification in the derivation of the Eyring theory in order to account for the discrepancy.

For the diffusion of large molecules through a solvent the reaction-rate theory may be interpreted in such a way that the Stokes-Einstein relation [Eq. (103)] results (G6, pp. 519–521). The activated-state theories have also been used to study the temperature dependence and concentration dependence of the diffusion coefficient. For ideal solutions $\mathfrak{D}_{AB}\mu_B$ varies linearly with the concentration; for nonideal solutions an expression for $\mathfrak{D}_{AB}\mu_B$ in terms of the dependence of activity on concentration has been given (S16, S17, G6, p. 533). The activated-state description of diffusion does not in general give satisfactory quantitative predictions (G6, pp. 526–529; J13, pp. 474–475); it has been suggested that the basic difficulty lies in the fact that diffusion occurs by a bimolecular mechanism, whereas the theory begins by assuming that mass transfer can be described in terms of a unimolecular process (H12). Certainly much remains to be done in order to understand the diffusion process in terms of the basic molecular processes involved.

3. *Empirical Relations*

A number of useful empirical and semiempirical relations have been proposed for predicting diffusion coefficients in liquids; a rather complete summary of these has been given by Treybal (T6, pp. 102 *et seq.*). Most of these are valid only for nonionic systems at very low concentrations.

Wilke (W8) has developed a correlation for diffusion coefficients on the basis of the Stokes-Einstein equation. His results may be summarized by the approximate analytical relation (S9, p. 21) which gives the diffusion coefficient for A diffusing into B in a dilute solution:

$$\frac{\mathfrak{D}_{AB}\mu_B}{\kappa T} = \frac{N^{1/3}}{15(V_A^{1/3} - l_B)} \qquad (109)$$

in which V_A is the molar volume of the solute as calculated from a table of atomic volumes (cc./g.-mole) and l_B is a constant characteristic of the solvent (for H_2O, $l_B = 2.0$; for methyl alcohol, 2.46; for benzene, 2.84). Equation (109) is good only for dilute solutions of nondissociating solutes; it does not represent the experimental data for the diffusion of small molecules in a solvent.

Another semiempirical relation has been suggested by Othmer and Thakar (O3), which is also good only for dilute solutions of nonionic solutions; the diffusion coefficient for a dilute solution of solute A in solvent B (in cm.² sec.⁻¹) is[17]

$$\mathfrak{D}_{AB} = \frac{1.40 \times 10^{-4}}{V_A^{0.6} \mu_B^\circ \mu_w^{1.1 L_B/L_w}} \qquad (110)$$

in which V_A is the molar volume of solute A (in cc./g.-mole), μ_B° is the viscosity of solvent B at 20°C. (in cp.), μ_w is the viscosity of water at the temperature for which $\mathfrak{D}_{AB}$ is desired, and L_B and L_w are the latent heats of vaporization of solvent B and water respectively (cal./g.-mole).

Several other empirical relations for diffusion coefficients have been suggested: Olson and Walton (O1) have devised a means for estimating diffusion coefficients of organic liquids in water solution from surface-tension measurements. Hill (H5) has proposed a method based on Andrade's theory of liquids which allows for the concentration dependence of the diffusion coefficient in a binary liquid mixture. The formula of Arnold (A2, T6, p. 102) does not seem generally useful inasmuch as it contains two constants ("abnormality factors") characteristic of the solute and of the solvent.

IV. Solutions of the Diffusion Equations of Interest in Chemical Engineering

In the chemical engineering literature on the diffusional operations there are a number of relatively simple solutions to the diffusion equations which are used for the purpose of development of theoretical expressions for mass transfer processes or for interpreting diffusion experiments. It is the purpose of this section to provide a summary of these simpler solutions and to give references to the literature where the solutions of more complex or more specialized problems may be found. In each problem discussed the starting equation is taken from Sec. II,C; only two-component systems are treated here. No attempt is made here to present all the solutions of interest. Both Jost (J13, Chapter I) and Barrer (B5, Chapter I) give rather complete surveys of solutions to Fick's second law, particularly as regards the diffusion in solids. Carslaw and Jaeger's (C2) extensive summary of solutions to the heat-conduction equation is of extremely great use in solving some diffusion problems;

[17] Equation (110) is not the same as the equation given by Othmer and Thakar in their original paper; there is a typographical error in the latter, which has here been corrected.

and the summary of Drew (D3) of forced-convection heat transfer problems can be used for obtaining the solutions to forced-convection mass transfer problems.

A. Steady State Diffusion Problems in Nonflow Systems

The simplest problems are those in which the diffusion process is independent of the time. The solutions to these problems are important in the film theories of mass transfer and in steady state experiments for measuring diffusion and self-diffusion.

1. *Diffusion Through a Stagnant Fluid*

For steady state diffusion in a system in which no chemical reaction is occurring, Eq. (40b) becomes for diffusion in the z direction alone

$$dN_A/dz = 0 \quad \text{or} \quad N_A = \text{constant} \tag{111}$$

For the diffusion of A through stagnant B in a constant-pressure, constant-temperature, ideal-gas system, Eq. (49) may be written in a number of forms:

$$\begin{aligned} N_A &= -\frac{(p\mathfrak{D}_{AB}/RT)}{(1-x_A)}\frac{dx_A}{dz} \\ &= -\frac{(p\mathfrak{D}_{AB}/RT)}{(p-p_A)}\frac{dp_A}{dz} \\ &= +(p\mathfrak{D}_{AB}/RT)\frac{d\ln p_B}{dz} \end{aligned} \tag{112}$$

Since N_A is constant, this differential equation may be integrated between z_1 and z_2 to give

$$\begin{aligned} N_A &= \frac{(p\mathfrak{D}_{AB}/RT)}{(z_2-z_1)}\ln\frac{p_{B2}}{p_{B1}} \\ &= \frac{(p\mathfrak{D}_{AB}/RT)}{(z_2-z_1)}\frac{(p_{A1}-p_{A2})}{(p_B)_{l.m.}} \\ &= k_G(p_{A1}-p_{A2}) \end{aligned} \tag{113}$$

in which $(p_B)_{l.m.}$ is the logarithmic mean value of p_B defined as $(p_{B2}-p_{B1})/\ln(p_{B2}/p_{B1})$. Equation (113) is then the basis for the definition of the mass transfer coefficient k_G:

$$k_G = \frac{(p\mathfrak{D}_{AB}/RT)}{(z_2-z_1)}\frac{1}{(p_B)_{l.m.}} \tag{114}$$

for describing the mass transfer by molecular diffusion through a stagnant gas film. The diffusion coefficient $\mathfrak{D}_{AB}$ in Eq. (114) is the dilute-gas diffusion coefficient calculated according to Sec. III,A.

Equation (113) is the basis for analyzing Stefan's method of measuring diffusion coefficients by observing the rate of evaporation of a liquid in a tube. At the liquid level ($z = z_1$) the partial pressure is taken to be p_{A1} [this assumes equilibrium at the interface (see Sec. II,D,4)]; at the top of the tube ($z = z_2$) component A is continually removed so that $p_{A2} = 0$. This method has been discussed by Wilke (W9) and has been used by Lee and Wilke (L3) for obtaining diffusion coefficients in gaseous systems.

For the steady state diffusion of A through stagnant B in a very dilute liquid system at constant temperature, Eq. (49) becomes

$$N_A = -\mathfrak{D}_{AB} \frac{dc_A}{dz} \tag{115}$$

which may be integrated to give

$$N_A = \mathfrak{D}_{AB} \frac{(c_{A1} - c_{A2})}{(z_2 - z_1)} = k_L(c_{A1} - c_{A2}) \tag{116}$$

This equation is then the basis for defining the mass transfer coefficient k_L

$$k_L = \frac{\mathfrak{D}_{AB}}{(z_2 - z_1)} \tag{117}$$

in which the diffusion coefficient $\mathfrak{D}_{AB}$ is that estimated by the methods given in Sec. III,C. The use of the expressions for k_G and k_L in connection with interphase mass transfer calculations has been discussed in considerable detail by Sherwood and Pigford (S9, Chapter III).

2. *Diffusion Through a Stagnant Fluid (Spherical Coordinates)*

In problems such as the drying of droplets or diffusion through films around spherical catalyst pellets, it is more convenient to use Eqs. (40b) and (49) in spherical coordinates. Then for steady state diffusion in the radial direction alone, one has in the absence of chemical reactions

$$\frac{1}{r^2} \frac{d}{dr}(r^2 N_A) = 0 \tag{118}$$

whence it follows that

$$r^2 N_A = \text{a constant} = \frac{Q_A}{4\pi} \tag{119}$$

The factor of 4π is included so that the quantity Q_A is the total number of moles per unit time issuing forth from the source. Then the analog of Eq. (113) for spherical coordinates may be obtained by substituting the ideal-gas version of Eq. (49) into Eq. (115) and integrating:

$$Q_A = 4\pi \frac{(p\mathcal{D}_{AB}/RT)}{(r_2 - r_1)/(r_1 r_2)} \ln (p_{B2}/p_{B1})$$
$$= 4\pi \frac{(p\mathcal{D}_{AB}/RT)}{(r_2 - r_1)/(r_1 r_2)} \frac{(p_{A1} - p_{A2})}{(p_B)_{l.m.}} \quad (120)$$

in which $(p_B)_{l.m.}$ is the quantity defined in connection with Eq. (113). One may also write

$$N_A|_{r=r_1} = \left(\frac{p\mathcal{D}_{AB}}{RT}\right)\left(\frac{r_2/r_1}{r_2 - r_1}\right)\frac{(p_{A1} - p_{A2})}{(p_B)_{l.m.}}$$
$$= k_G(p_{A1} - p_{A2}) \quad (121)$$

Two special cases of this result are of interest. (a) If $(r_2 - r_1) \ll r_1$ (film thickness small compared with radius of particle), then the k_G defined by Eq. (121) reduces to that in Eq. (114). (b) If $r_2 \gg r_1$ and if only a small amount of A is diffusing (so that $(p_B)_{l.m.} = 1$), Eq. (121) becomes

$$N_A|_{r=r_1} = \frac{\mathcal{D}_{AB}}{r_1}(c_{A1} - c_{A2}) \quad (122)$$

This relation has been used for the calculation of diffusion coefficients by the rate of evaporation of liquid droplets suspended in a still gas (S9, p. 17). Clearly Eq. (122) does not describe the situation properly if the diffusion produces free convection owing to changes in density.

3. *Diffusion Through a Nonisothermal Film*

Sometimes it is necessary to calculate mass transfer rates across films under such conditions that there is a pronounced temperature gradient across the film. We consider here a simple example of the usual situation in which thermal diffusion can be neglected but the temperature dependence of the diffusion coefficient is taken into account.

Consider the steady state ideal-gas diffusion of A through a stagnant film of B surrounding a sphere of radius r_1 (as in the preceding example). The temperature is assumed to vary according to

$$(T/T_1) = (r/r_1)^n \quad (123)$$

in which T_1 is the temperature at the surface of the sphere. Furthermore, the diffusion coefficient is assumed to vary as the ($\frac{3}{2}$) power of the absolute temperature [see for a more correct power law the relation given in Eq. (89)], thus:

$$\left(\frac{\mathcal{D}_{AB}}{\mathcal{D}_{AB,1}}\right) = \left(\frac{T}{T_1}\right)^{3/2} = \left(\frac{r}{r_1}\right)^{3n/2} \quad (124)$$

in which $\mathfrak{D}_{AB,1}$ is the diffusion coefficient at the temperature at the surface of the sphere. Then according to Eqs. (119) and (45)

$$\frac{Q_A}{4\pi} = r^2 \left[-\frac{p\mathfrak{D}_{AB}/RT}{1-x_A} \frac{dx_A}{dr} \right] \tag{125}$$

or

$$Q_A = 4\pi r^2 \left[\frac{p\mathfrak{D}_{AB}}{RT} \frac{d \ln p_B}{dr} \right]$$
$$= 4\pi r^2 \left[\frac{p\mathfrak{D}_{AB,1}}{RT_1} \left(\frac{r}{r_1}\right)^{n/2} \frac{d \ln p_B}{dr} \right] \tag{126}$$

Since Q_A is constant, Eq. (126) may be integrated with respect to r to give (provided that $n \neq -2$)

$$Q_A = 4\pi r_1^{-n/2} [r_2^{-1-n/2} - r_1^{-1-n/2}] \left(\frac{p\mathfrak{D}_{AB,1}}{RT_1}\right) \ln \frac{p_{B2}}{p_{B1}} \tag{127}$$

for the rate of transfer of A through the film.

4. *Diffusion Accompanied by Chemical Reaction*

Consider the problem of the unidirectional, steady state flow of A through a stationary film B. It is assumed that the quantity of A diffusing is quite small and that A reacts with B according to a slow first-order reaction $A + B \rightarrow AB$. The amount of AB produced is so small that the system may be considered to be a two-component system. The concentrations of A at the edges of the film $z = z_1$ and $z = z_2$ are taken to be c_{A1} and c_{A2} respectively, with $c_{A1} > c_{A2}$.

For steady state diffusion with chemical reaction, Eq. (40b) becomes

$$dN_A/dz = K_A \tag{128}$$

Since A is present in small concentration only, Fick's law

$$[N_A = -\mathfrak{D}_{AB}(dc_A/dz)]$$

may be used. Since the reaction is first order, $K_A = -k'c_A$, where k' is the reaction-rate constant. If it is assumed that $\mathfrak{D}_{AB}$ is independent of concentration, then Eq. (128) becomes

$$\mathfrak{D}_{AB}(d^2c_A/dz^2) = k'c_A \tag{129}$$

The solution of this equation consistent with the boundary conditions given above is

$$c_A = \frac{c_{A1} \sinh b(z_2 - z) + c_{A2} \sinh b(z - z_1)}{\sinh b(z_2 - z_1)} \tag{130}$$

in which $b = \sqrt{k'/\mathfrak{D}_{AB}}$. This result is one of the several analytical expressions which are employed in the oversimplified models used for describing simultaneous absorption and chemical reaction. The foregoing illustration is discussed further by Sherwood and Pigford (S9, pp. 324–328). Other simple analytical solutions involving steady state diffusion with chemical reaction have been obtained by Thiele (T2) in connection with studies on catalyst activity.

5. Diffusion with Thermal Diffusion

Consider a system which is made up of two bulbs connected by a long, narrow tube; into this system is placed a mixture of A and B. If one bulb is maintained at a low temperature T_1 and the other at a high temperature T_2, there will be a tendency for A to move toward the cold region and B to the hot region (if k_T is positive). This tendency for the two species to separate from one another creates concentration gradients, which in turn cause ordinary diffusion, which opposes the process of thermal diffusion. A steady state condition is ultimately established, characterized by the fact that there is no average relative motion of A and B (that is $(\mathbf{v}_A - \mathbf{v}_B) = 0$). Hence from Eq. (59) one obtains for diffusion in the z direction only

$$\frac{dx_A}{dz} = -\frac{k_T}{T}\frac{dT}{dz} \tag{131}$$

If the dependence of k_T on composition is ignored (a valid assumption since the change in x_A is only very slight), this equation may be integrated to give

$$(x_{A2} - x_{A1}) = -\int_{T_1}^{T_2} \frac{k_T}{T} \, dT \tag{132}$$

in which k_T is kept under the integral sign since its temperature dependence cannot be ignored. Because this dependence is rather complicated, a numerical integration is customarily avoided by assuming that k_T is constant at a "mean value" $\bar{T}$ of the temperature. Then Eq. (132) becomes

$$(x_{A2} - x_{A1}) = -k_T(\bar{T}) \ln (T_2/T_1) \tag{133}$$

in which $k_T(\bar{T})$ is the thermal diffusion ratio evaluated at the temperature $\bar{T}$, for which Brown (B22) has recommended the value

$$\bar{T} = \frac{T_1 T_2}{T_2 - T_1} \ln \frac{T_2}{T_1} \tag{134}$$

based on the use of the Sutherland potential. Equation (133) has been used for the analysis of the two-bulb thermal diffusion measurements. It is also useful for estimating the order of magnitude of thermal diffusion effects in the design and interpretation of experiments.

6. *Diffusion with Pressure Diffusion*

In the previous section it is seen how a separation of two substances A and B may be effected by the establishment of a steady state in a system in which thermal and ordinary diffusion oppose one another. In this section we describe the analogous situation which is obtained when ordinary and pressure diffusion are in competition with each other. Use is made of this result in the separation of organic liquids or isotopes by centrifuging techniques.

Consider then a mixture of A and B which are being subjected to a uniform gravitational acceleration or centrifugal acceleration g in the $-z$ direction. It should be noted that there is no mass flux due to "forced diffusion" according to Eq. (35) inasmuch as $j_i(F)$ vanishes when we replace $\hat{F}$ by $-g$. The effect of the gravitational field is to produce a pressure gradient, the latter being determined by Eq. (25), which for the case under consideration becomes:

$$+ \frac{dp}{dz} = -\rho g \tag{134a}$$

Hence the mass flux in the z-direction due to the pressure gradient is given by Eq. (34) as:

$$j_A{}^{(p)} = \frac{c^2}{\rho RT} M_A M_B \mathfrak{D}_{AB} \left[x_B M_B \left(\frac{\tilde{V}_B}{M_B} - \frac{1}{\rho} \right) (-\rho g) \right] \tag{134b}$$

in which an ideal solution has been assumed so that $\tilde{V}_B = \bar{V}_B$, the molar volume of pure B. The mass flux by ordinary diffusion is given by Eq. (41), which for an ideal solution is:

$$j_A{}^{(x)} = -\frac{c^2}{\rho} M_A M_B \mathfrak{D}_{AB} \frac{dx_A}{dz} \tag{134c}$$

At steady state the sum of the fluxes given in Eqs. (134b) and (134c) must be equal to zero; this steady state solution simplifies to:

$$-(dx_A/dz) = (gx_A/RT)(M_A - \rho \bar{V}_A) \tag{134d}$$

with a similar expression for x_B. When we make the substitution

$$\rho = (x_A M_A + x_B M_B)/(x_A \tilde{V}_A + x_B \tilde{V}_B)$$

then Eq. (134d) may be integrated between the positions $z = 0$ and $z = L$ to give:

$$\left(\frac{x_{AL}}{x_{A0}}\right)^{\tilde{V}_B} \left(\frac{x_{BL}}{x_{B0}}\right)^{\tilde{V}_A} = \exp\left[(M_A\tilde{V}_B - M_B\tilde{V}_A)gL/RT\right] \quad (134e)$$

this result is the same as that obtained by thermodynamic arguments (G13, p. 359).

B. Unsteady State Diffusion Problems in Nonflow Systems

Unsteady state diffusion processes are of considerable importance in chemical engineering problems such as the rate of drying of a solid (H14), the rate of absorption or desorption from a liquid, and the rate of diffusion into or out of a catalyst pellet. Most of these problems are attacked by means of Fick's second law [Eq. (52)] even though the latter may not be strictly applicable; as mentioned previously, these problems may generally be solved simply by looking up the solution to the analogous heat-conduction problem in Carslaw and Jaeger (C2). Hence not much space is devoted to these problems here.

1. *Unbounded Equimolal Counterdiffusion*

Consider a system containing A and B with a partition at $z = 0$. Initially the concentration of A is c_A^- for $-\infty < z < 0$ and c_A^+ for $0 < z < \infty$. At time $t = 0$ the partition at $z = 0$ is removed in order to allow diffusion to take place. The differential equation to be solved is

$$\frac{\partial c_A}{\partial t} = \frac{\partial}{\partial z}\left(\mathfrak{D}_{AB}\frac{\partial c_A}{\partial z}\right) \quad (135)$$

When $\mathfrak{D}_{AB}$ is independent of the concentration, the solution is

$$\frac{c_A - c_A^-}{c_A^+ - c_A^-} = \frac{1}{2}\left[1 + \operatorname{erf}\frac{z}{\sqrt{4\mathfrak{D}_{AB}t}}\right] \quad (136)$$

in which $\operatorname{erf} u = (2/\sqrt{\pi})\int_0^u \exp(-y^2)dy$, and $\operatorname{erf}(-u) = (-\operatorname{erf} u)$. Clearly when t approaches ∞, c_A approaches $(\frac{1}{2})(c_A^+ + c_A^-)$, the average of the initial concentrations in the two half cells. Equation (136) is used in the analysis of closed-cell diffusion experiments when the time of diffusion is sufficiently short that the concentrations at the ends of the cell do not deviate from the initial values. The solution of this problem when $\mathfrak{D}_{AB}$ is a function of concentration has been studied by Stokes (S19), by Gillis and Kedem (G4), and by Fujita (F7, F8, F9).

2. Bounded Equimolal Counterdiffusion

Consider the same system as in the previous example, except that the system is bounded at $z = \pm L$. Then one has to solve Eq. (135) with the boundary and initial conditions: at $z = \pm L$, $\partial c_A/\partial z = 0$ for all t (requirement that there be no mass flux through the containing walls); and at $t = 0$, $c_A = c_A^-$ for $-L < z < 0$ and $c_A = c_A^+$ for $0 < z < +L$. This is an eigenvalue problem soluble by the method of separation of variables, and the final expression for the concentration profiles is (when $\mathfrak{D}_{AB}$ is independent of the concentration)

$$\frac{c_A - c_A^-}{c_A^+ - c_A^-} = \frac{1}{2}\left[1 + \frac{2}{\pi}\sum_{n=0}^{\infty}\frac{\sin(n+\tfrac{1}{2})\pi z/L}{(n+\tfrac{1}{2})}\exp-[(n+\tfrac{1}{2})\pi/L]^2\mathfrak{D}_{AB}t\right] \tag{137}$$

Equation (137) is used in the analysis of closed-cell diffusion experiments (B19, S20, B23, R6). The solution to obtain the concentration profiles when $\mathfrak{D}_{AB}$ is a function of concentration has been obtained by Snider and Curtiss by means of a perturbation calculation (S11).

In many problems one is concerned only with the average concentration in one of the half cells. By integration of Eq. (137) from $z = 0$ to $z = L$ one obtains

$$\frac{\bar{c}_A - c_A^-}{c_A^+ - c_A^-} = \frac{1}{2} \pm \frac{1}{\pi^2}\sum_{n=0}^{\infty}\frac{1}{(n+\tfrac{1}{2})^2}\exp-[(n+\tfrac{1}{2})\pi/L]^2\mathfrak{D}_{AB}t \tag{138}$$

Use of the $+$ sign in Eq. (138) gives the average concentration in the region $0 < z < L$ and the $-$ sign gives the average for the region $-L < z < 0$.

The results given in Eqs. (137) and (138) converge rapidly for large values of t. For small values of t it is convenient to make use of a different expansion, which may be obtained by means of Laplace transform (C2, pp. 250–252):

$$\frac{c_A - c_A^-}{c_A^+ - c_A^-} = \frac{1}{2}\left[\sum_{n=0}^{\infty}(-1)^n \operatorname{erfc}\frac{(2nL-z)}{2\sqrt{\mathfrak{D}_{AB}t}}\right.$$

$$\left.+ \sum_{n=0}^{\infty}(-1)^n \operatorname{erfc}\frac{2(n+1)L+z}{2\sqrt{\mathfrak{D}_{AB}t}}\right] \tag{138a}$$

This solution is valid for $-L < z < 0$. When $L \to \infty$, this solution simplifies to the relation given in Eq. (136). The average concentration in the half cell is given by

$$\frac{\bar{c}_A - c_A^-}{c_A^+ - c_A^-} = \sqrt{\frac{\mathfrak{D}_{AB}t}{L^2}} \left[\frac{1}{\sqrt{\pi}} + 2 \sum_{n=1}^{\infty} (-1)^n \operatorname{ierfc} \frac{n}{\sqrt{\mathfrak{D}_{AB}t}} \right] \quad (138b)$$

The expressions given in Eqs. (138a) and (138b) converge rapidly for small t; they are also useful in the analysis of closed-cell diffusion experiments.

3. *Partially Bounded Equimolal Counterdiffusion*

A diffusion tube of length $2L$ and uniform cross-sectional area S has a partition at the center of the tube ($z = 0$) which may be removed at $t = 0$ to allow diffusion to take place. At each end ($z = \pm L$) the tube is joined to a reservoir of volume V containing stirrers to maintain a uniform concentration in the reservoir. The problem is then to solve Eq. (135) with the following initial and boundary conditions: At $t = 0$, $c_A = c_A^-$ for $-L < z < 0$ and in the reservoir attached at $z = -L$, and $c_A = c_A^+$ for $0 < z < +L$ and in the reservoir attached at $z = +L$; at $z = 0$, $c_A = (\tfrac{1}{2})(c_A^+ + c_A^-)$ for all $t > 0$; and at $z = -L$,

$$\mathfrak{D}_{AB} S (\partial c_A/\partial z) = V(\partial c_A/\partial t)$$

The last boundary condition states that the flux of A out the end of the diffusion tube at $z = -L$ results in an increase in the concentration of A in the attached reservoir. This problem may be solved by the method of separation of variables to give (C2, p. 107):

$$\frac{c_A - c_A^-}{c_A^+ - c_A^-} = \frac{1}{2} \left[1 + \sum_{n=1}^{\infty} \frac{2}{\gamma_n} \left(\frac{\gamma_n^2 + N^2}{\gamma_n^2 + N^2 + N} \right) \exp(-\gamma_n^2 \mathfrak{D}_{AB} t/L^2) \sin \gamma_n z/L \right] \quad (139a)$$

in which $N = SL/V$ and the eigenvalues γ_n are the positive roots of the equation $\gamma \tan \gamma = N$. (It should be mentioned that the functions $\sin \gamma_n z/L$ do *not* form an orthogonal set in the region $-L < z < +L$.)

At time t the concentrations of A in the reservoirs are given by

$$\frac{c_A - c_A^-}{c_A^+ - c_A^-} = \frac{1}{2} \left[1 \pm \sum_{n=1}^{\infty} \frac{2N}{\gamma_n} \frac{\sqrt{\gamma_n^2 + N^2}}{\gamma_n^2 + N^2 + N} \exp(-\gamma_n^2 \mathfrak{D}_{AB} t/L^2) \right] \quad (139b)$$

the $+$ and $-$ signs corresponding to the reservoirs at $+L$ and $-L$ respectively.

4. Diffusion in a Two-Phase System

Diffusion problems in two-phase systems are of considerable importance in chemical engineering. The simplest unsteady state problem is the following. The system consists of two immiscible phases, I (in the region $-\infty < z < 0$) and II (in the region $0 < z < +\infty$), separated by a partition. The initial concentration of a solute is $c_\mathrm{I}{}^0$ in phase I and $c_\mathrm{II}{}^0$ in phase II. At time $t = 0$ the partition at $z = 0$ is removed and diffusion is allowed to take place, the diffusion coefficient of the solute in phase I being $\mathfrak{D}_\mathrm{I}$ and that in phase II being $\mathfrak{D}_\mathrm{II}$. It is assumed that the diffusion can be described by Fick's second law in both phases; and it is further assumed that there is equilibrium at the interface, which can be described by the relation $c_\mathrm{II} = mc_\mathrm{I}$, where m is the "distribution coefficient." The mathematical statement of the problem is then

Diffusion in phase I: $\quad \dfrac{\partial c_\mathrm{I}}{\partial t} = \mathfrak{D}_\mathrm{I} \dfrac{\partial^2 c_\mathrm{I}}{\partial z^2}$ (140a)

Diffusion in phase II: $\quad \dfrac{\partial c_\mathrm{II}}{\partial t} = \mathfrak{D}_\mathrm{II} \dfrac{\partial^2 c_\mathrm{II}}{\partial z^2}$ (140b)

Boundary conditions:

(a) At $z = 0$ $\qquad c_\mathrm{II} = mc_\mathrm{I}$
(b) At $z = 0$ $\qquad \mathfrak{D}_\mathrm{I}(\partial c_\mathrm{I}/\partial z) = \mathfrak{D}_\mathrm{II}(\partial c_\mathrm{II}/\partial z)$
(c) At $z = -\infty$ $\qquad \partial c_\mathrm{I}/\partial z = 0$
(d) At $z = +\infty$ $\qquad \partial c_\mathrm{II}/\partial z = 0$ (141)

Initial conditions: (a) At $t = 0$ $\quad c_\mathrm{I} = c_\mathrm{I}{}^0$
(b) At $t = 0$ $\quad c_\mathrm{II} = c_\mathrm{II}{}^0$ (142)

The boundary condition given in Eq. (141b) is the statement that all the solute leaving phase I enters phase II, that is, there are no surface reactions. The differential equations may be solved by means of Laplace transform (M2, pp. 134–136); the results are

$$\frac{c_\mathrm{I} - c_\mathrm{I}{}^0}{c_\mathrm{II}{}^0 - mc_\mathrm{I}{}^0} = \frac{1 + \operatorname{erf} z/\sqrt{4\mathfrak{D}_\mathrm{I} t}}{m + \sqrt{\mathfrak{D}_\mathrm{I}/\mathfrak{D}_\mathrm{II}}} \qquad (143)$$

$$\frac{c_\mathrm{II} - c_\mathrm{II}{}^0}{c_\mathrm{I}{}^0 - (1/m)c_\mathrm{II}{}^0} = \frac{1 - \operatorname{erf} z/\sqrt{4\mathfrak{D}_\mathrm{II} t}}{(1/m) + \sqrt{\mathfrak{D}_\mathrm{II}/\mathfrak{D}_\mathrm{I}}} \qquad (144)$$

When $m = 1$ and phase I and phase II are the same, then Eqs. (143) and (144) reduce to Eq. (136).

A number of other solutions to diffusion problems in two-phase systems have been given by Jost (J13, Chapter I) and Barrer (B5, Chapter I). More recently Scott, Tung, and Drickamer (S7) in solving the differential equations for two-phase diffusion in a bounded system, have assumed that equilibrium does not exist at the interface.

5. *Diffusion Accompanied by Rapid Chemical Reaction*

A number of unsteady state problems in diffusion have been considered, in which chemical reactions are occurring (F6, G8). Sherwood and Pigford (S9, pp. 332–337) have studied the unsteady state absorption of a substance A which diffuses into the solvent S and undergoes an infinitely fast, irreversible, second-order reaction with a solute B (that is A + B → AB). It is assumed that Fick's second law adequately describes the diffusion process and that because of the infinitely fast reaction of A and B there will be a plane parallel to the liquid surface at a distance z' from it, which separates the region containing no A from that containing no B. The distance z' is a function of t, inasmuch as the boundary between A and B retreats as B is used up in the chemical reactions.

The differential equations to be solved are

$$\frac{\partial c_A}{\partial t} = \mathfrak{D}_{AS} \frac{\partial^2 c_A}{\partial z^2} \qquad 0 < z < z'(t) \tag{145}$$

$$\frac{\partial c_B}{\partial t} = \mathfrak{D}_{BS} \frac{\partial^2 c_B}{\partial z^2} \qquad z'(t) < z < \infty \tag{146}$$

to which possible solutions are

$$c_A = a_1 + a_2 \operatorname{erf} z/\sqrt{4\mathfrak{D}_{AS}t} \tag{147}$$
$$c_B = b_1 + b_2 \operatorname{erf} z/\sqrt{4\mathfrak{D}_{BS}t} \tag{148}$$

Next one must know how the boundary z' moves with time. The analytic definition of z' is

$$c_A(z',t) = 0 \tag{149}$$

Then one forms the differential

$$dc_A = 0 = \left(\frac{\partial c_A}{\partial z'}\right)_t dz' + \left(\frac{\partial c_A}{\partial t}\right)_{z'} dt \tag{150}$$

whence

$$\frac{dz'}{dt} = -\frac{(\partial c_A/\partial t)_{z'}}{(\partial c_A/\partial z')_t} \tag{151}$$

Substituting Eq. (147) (written in terms of z') into Eq. (151) gives then an equation for z' in terms of t:

$$dz'/dt = z'/2t \tag{152}$$

which may be integrated to give

$$z' = \sqrt{4\alpha t} \tag{153}$$

in which α is a constant of integration.

The five constants a_1, a_2, b_1, b_2, and α are determined from the following five boundary and initial conditions:

(a) $c_A = c_{A0}$ at $z = 0$ for all $t > 0$
(b) $c_B = c_{B\infty}$ at $z = \infty$ for all $t > 0$
(c) $c_B = c_{B\infty}$ at $t = 0$ for all $z > 0$
(d) $c_A = c_B = 0$ at $z = z'(t)$ for all $t > 0$
(e) $\mathfrak{D}_{AS}(\partial c_A/\partial z) = -\mathfrak{D}_{BS}(\partial c_B/\partial z)$ at $z = z'(t)$ for all $t > 0$ (154)

The last boundary condition is the stoichiometric requirement that one mole of A consume one mole of B. The constants determined from these boundary conditions are given by Sherwood and Pigford (S9, p. 336) and some sample concentration profiles are shown; the results are interpreted in terms of the liquid film coefficients.

Another diffusion problem in which there is a retreating boundary because of chemical reaction has been treated by Jost (J14), who studied the chemical reaction of SO_2 with spheres of FeS. Because of the reaction the spheres became coated with a layer of Fe_3O_4, and the diffusion of SO_2 through the porous layer of Fe_3O_4 became rate controlling.

6. *Diffusion Accompanied by a Slow Chemical Reaction*

The example discussed in the preceding section illustrates the use of the concept of a regressing boundary in a system where the chemical reaction takes place rapidly with respect to the diffusion. Let us now consider an example in which the rate of reaction is slow.

Sherwood and Pigford (S9) have discussed the problem of the absorption of a solute A by a solvent S; upon solution, A may be converted into B according to the reaction $A \rightleftarrows B$ (k_f' and k_r' being the forward and reverse reaction-rate constants, and $K = k_f'/k_r'$). The concentration of A is maintained at c_{A0} at the surface of the liquid S, and it is assumed that S is semiinfinite in extent. It is further assumed that B is nonvolatile; that is, it cannot escape from solvent S. Equation (51) is then used to explain the diffusion of A and B, with $\mathfrak{D}_{AS}$ and $\mathfrak{D}_{BS}$ taken as concentration independent, and the term containing the molar average velocity **w** is neglected. Hence the mathematical statement of the problem is (for very dilute solutions of A and B)

$$\frac{\partial c_A}{\partial t} = \mathfrak{D}_{AS} \frac{\partial^2 c_A}{\partial z^2} - k_f' c_A + k_r' c_B \tag{155}$$

$$\frac{\partial c_B}{\partial t} = \mathfrak{D}_{BS} \frac{\partial^2 c_B}{\partial z^2} + k_f' c_A - k_r' c_B \tag{156}$$

These equations are to be solved with the initial and boundary conditions that

(a) $c_A = c_0/(1 + K)$ at $t = 0$
(b) $c_A = c_0/(1 + K)$ at $z = \infty$
(c) $c_A = c_{A0}$ at $z = 0$
(d) $\partial c_B/\partial z = 0$ at $z = 0$ (157)

The statement $c_A = c_0/(1 + K)$ in Eqs. (157a and b) above is tantamount to saying that $c_A + c_B = c_0$, where c_0 is the total concentration of both species of the dissolved solute. If the diffusivities $\mathfrak{D}_{AS}$ and $\mathfrak{D}_{BS}$ are assumed to be equal, then c_B can be eliminated from Eqs. (155) and (156) and a fourth-order, linear partial-differential equation is obtained. The solution of this equation consistent with the conditions in Eq. (157) is obtainable by Laplace transform techniques (S9). Sherwood and Pigford discuss the results in terms of the behavior of the liquid-film mass transfer coefficient.

More recently Perry and Pigford (P2) made a similar calculation for the absorption of a solute accompanied by a slow second-order reaction, $A + B \rightleftarrows 2C$. The problem, which involves the simultaneous solution of three coupled, second-order, partial-differential equations, was worked out by means of an electronic digital computer.

C. Steady-State Diffusion in Flow Systems

Thus far diffusion in nonflow systems has been discussed. We now turn our attention to forced-convection problems. Only steady state problems are considered here, and it is assumed that they can all be described by the differential equation [see Eq. (50)]

$$(\mathbf{v} \cdot \nabla)c_A = \mathfrak{D}_{AB}\nabla^2 c_A \tag{158}$$

The problems discussed here are basic in the description of absorption in falling films, performance of wetted-wall towers, operation of tubular reactors, and fluid blending.

1. *Diffusion into a Falling Film*

For a diffusion of a solute A into a liquid film of B moving in laminar flow, several analytical solutions have been obtained; the limits of applicability of these various solutions depend upon the assumptions which have been made.

The simplest analytical solution assumes that the contact time is very short and the following assumptions are reasonable: (*a*) the film moves with a flat velocity profile v_0 (see Fig. 4a); (*b*) the film may be taken to be infinitely thick with respect to the penetration of the absorbed material; and (*c*) the concentration at the interface $x = 0$ is c_{A0}. The analytical statement of the problem is then

$$v_0 \frac{\partial c_A}{\partial z} = \mathfrak{D}_{AB} \frac{\partial^2 c_A}{\partial x^2} \qquad (159)$$

in which the diffusion term ($\partial^2 c_A/\partial z^2$) in the direction of flow is neglected. Equation (159) is to be solved with the boundary conditions

(a)   At $x = 0$ $c_A = c_{A0}$ for all z
(b)   At $x = \infty$ $c_A = c_{A\infty}$ for all z
(c)   At $z = 0$ $c_A = c_{A\infty}$ for all x (160)

The solution to this equation is

$$\left(\frac{c_{A0} - c_{A\infty}}{c_{A0} - c_{A\infty}}\right) = 1 - \mathrm{erf}\, \frac{x}{\sqrt{4\mathfrak{D}_{AB}z/v_0}} \qquad (161)$$

This is, of course, just the same solution one obtains for the unsteady state diffusion into a slab where $t = z/v_0$; and indeed the problem considered here could just as well have been formulated in terms of the

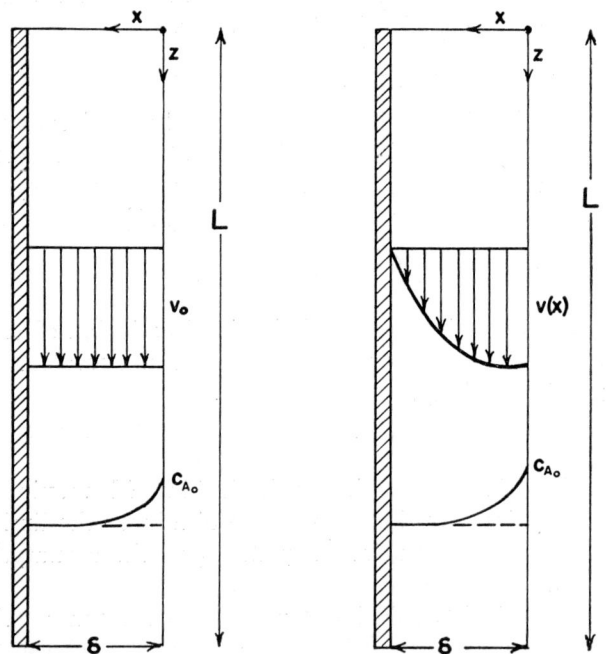

FIG. 4. Diffusion into a falling film (a) with flat velocity profile, and (b) with parabolic velocity profile.

unsteady state diffusion by choice of a coordinate system moving with the liquid surface rather than fixed as indicated in Fig. 4a. The fixed coordinate system has the advantage that it is more appropriate when the velocity profile is not constant.

Higbie (H4) has made use of the foregoing solution to obtain an expression for the liquid-phase mass transfer coefficient for short contact times for absorption in wetted-wall towers. From Eq. (161) one may calculate the total moles of A transferred per unit time per unit cross-sectional transfer area:

$$N_A = \frac{v_0}{L} \int_0^\infty (c_A - c_{A\infty})_{z=L} dx = \sqrt{\frac{4\mathfrak{D}_{AB}v_0}{\pi L}} (c_{A0} - c_{A\infty}) = k_L(c_{A0} - c_{A\infty}) \quad (162)$$

whence one obtains Higbie's liquid-film mass transfer coefficient (based on the driving force $c_{A0} - c_{A\infty}$)

$$k_L = \sqrt{\frac{4\mathfrak{D}_{AB}v_0}{\pi L}} = \sqrt{\frac{4\mathfrak{D}_{AB}}{\pi t_L}} \quad (163)$$

where t_L is the time required for the fluid film to traverse the length L.

An alternate way of obtaining (163) is by calculating the flux at the surface of the liquid film:

$$N_A(z) = -\mathfrak{D}_{AB} \left.\frac{\partial c_A}{\partial x}\right|_{x=0} = (C_{A0} - C_{A\infty})\sqrt{\frac{\mathfrak{D}_{AB}v_0}{\pi z}} \quad (164)$$

The integration with respect to z gives the average rate of transfer:

$$N_A = \frac{1}{L}\int_0^L (C_{A0} - C_{A\infty})\sqrt{\frac{\mathfrak{D}_{AB}v_0}{\pi z}}\, dz = (C_{A0} - C_{A\infty})\sqrt{\frac{4\mathfrak{D}_{AB}v_0}{\pi L}} \quad (165)$$

from which k_L in Eq. (163) may be obtained. These results are used in formulating the so-called penetration theories of mass-transfer.

Another solution to Eq. (159) for the diffusion into a falling film with a flat velocity profile is obtained by taking into account the finite thickness of the liquid film δ and its effect on the concentration profiles; that is, the boundary conditions are taken to be

(a) At $x = 0$ $c_A = c_{A0}$ for all z
(b) At $x = \delta$ $\partial c_A/\partial x = 0$ for all z
(c) At $z = 0$ $c_A = c_{A1}$ for all x (166)

For these conditions the solution to Eq. (159) is

$$\frac{c_A - c_{A1}}{c_{A0} - c_{A1}} = 1 - 2\sum_{n=0}^\infty \frac{\sin(n + \tfrac{1}{2})\pi x/\delta}{(n + \tfrac{1}{2})\pi}$$

$$\cdot \exp -\left[\left(n + \frac{1}{2}\right)\pi/\delta\right]^2 \mathfrak{D}_{AB} z/v_0 \quad (167)$$

Averaging over the distance through the film one gets for the average exit concentration:

$$\frac{\bar{c}_{A2} - c_{A1}}{c_{A0} - c_{A1}} = 1 - \frac{2}{\pi^2} \sum_{n=0}^{\infty} \frac{\exp - [(n + \tfrac{1}{2})\pi/\delta]^2 \mathcal{D}_{AB} L/v_0}{(n + \tfrac{1}{2})^2} \quad (168)$$

It is also possible to define a liquid-phase mass transfer coefficient on the basis of this expression (S9, pp. 83–84). It is customary to define k_L based on the log mean concentration difference during flow through the length of the film:

$$\begin{aligned} N_A &= k_L(c_A - c_{A0})_{l.m.} \\ &= k_L \left[\frac{\bar{c}_{A2} - c_{A1}}{\ln\left(\dfrac{c_{A0} - \bar{c}_{A2}}{c_{A0} - c_{A1}}\right)} \right] \end{aligned} \quad (169)$$

But the mass of A transferred in moles per unit time per unit area is also given by

$$N_A = \frac{\bar{c}_{A2} - c_{A1}}{L/v_0} \delta \quad (170)$$

Hence

$$k_L = \frac{v_0 \delta}{L} \ln\left(\frac{c_{A0} - c_{A1}}{c_{A0} - \bar{c}_{A2}}\right) \doteq \frac{v_0 \delta}{L} \left(\frac{\bar{c}_{A2} - c_{A1}}{c_{A0} - c_{A1}}\right) \quad (171)$$

The last expression for k_L is good for short contact times, and in order to obtain it one has to make use of the approximation $\ln(1 + \epsilon) = \epsilon$, when ϵ is small.

Clearly the assumption of a flat velocity profile is not correct. For a film in steady, laminar motion one may obtain an expression for the velocity distribution from the Navier-Stokes equations of motion [Eq. (9)]. For this case the Navier-Stokes equations simplify to

$$0 = \mu \frac{d^2 v_z}{dx^2} + \rho g \quad (172)$$

in which the coordinate system in Fig. 4b is used, and it is assumed that ρ and μ are constants. Equation (172) may be integrated with the boundary conditions $v_z = 0$ at $x = \delta$, and $dv_z/dx = 0$ at $x = 0$ to give

$$v_z(x) = \frac{\rho g \delta^2}{2\mu} \left[1 - \left(\frac{x}{\delta}\right)^2\right] = v_{max}\left[1 - \left(\frac{x}{\delta}\right)^2\right] \quad (173)$$

in which v_{max} is the maximum velocity in the film (at the surface) and $v_{avg} = (\tfrac{2}{3}) v_{max}$. It is straightforward then to show that the film thickness is given by

$$\delta = \sqrt[3]{\frac{3\mu\Gamma}{\rho^2 g}} = \sqrt[3]{\frac{3}{4}\frac{\mu^2}{\rho^2 g} Re} \quad (174)$$

where $Re = 4\Gamma/\mu$ and $\Gamma = \rho v_{avg} \delta$ is the mass rate of flow in the z direction per unit width of wetted wall in the y direction.

Pigford (P3, J8) and Vyazovov (V2) have solved Eq. (158) for the velocity profile in Eq. (173), that is, the differential equation

$$v_{max}\left(1 - \left(\frac{x}{\delta}\right)^2\right)\frac{\partial c_A}{\partial z} = \mathfrak{D}_{AB}\frac{\partial^2 c_A}{\partial x^2} \qquad (175)$$

with the boundary condition in Eq. (166), and obtained for the integrated solution [i.e., the analog of Eq. (168)]

$$\frac{\bar{c}_{A2} - c_{A1}}{c_{A0} - c_{A1}} = 1 - \sum_{n=1}^{\infty} a_n e^{-b_n \mathfrak{D}_{AB} L / V_{max} \delta^2} \qquad (176)$$

in which the constants a_n and b_n are

n	a_n	b_n
1	0.7857	5.121
2	0.1001	39.31
3	0.0360	105.6
4	0.0181	204.7

Pigford's solution may be interpreted in terms of the liquid-phase mass transfer coefficient, that is, by means of Eq. (171) in which v_0 is replaced by v_{avg}. For short contact times the substitution of Eq. (176) into the second expression for k_L in Eq. (171) gives

$$k_L = \sqrt{\frac{4\mathfrak{D}_{AB}v_{max}}{\pi L}} \qquad (177)$$

which is in agreement with Higbie's formula for k_L in Eq. (163). For very long contact times one needs to use only the first term in Eq. (176), and then the first expression for k_L in Eq. (171) gives

$$\begin{aligned} k_L &= \frac{v_{avg}\delta}{L}\ln\left(\frac{e^{5.121\mathfrak{D}_{AB}L/V_{max}\delta^2}}{0.7857}\right) \\ &= \frac{v_{avg}\delta}{L}\left[1.04 + 5.121\frac{\mathfrak{D}_{AB}L}{v_{max}\delta^2}\right] \\ &\doteq 3.41\mathfrak{D}_{AB}/\delta \end{aligned} \qquad (178)$$

These formulas are discussed and compared with experimental data by Sherwood and Pigford (S9, p. 266).

The foregoing analytical developments are quite straight forward, but they do not give the complete description of the process of absorption by falling films. Two factors not taken into account in these developments are (a) the tendency for a rippling type of motion to be the hydrodynamically stable laminar-flow pattern, even at rather small Reynolds numbers, and (b) the change in viscosity and density through the film, which affects the flow pattern. In wetted-wall-tower experiments there is also the question of the resistance offered by the interface in the absorption process. These points are discussed by Emmert and Pigford (E2).

2. Diffusion in Flow Through Circular Tubes

Next we consider a fluid flowing through a circular tube with material at the wall diffusing into the moving fluid. This situation is met with in the analysis of the mass transfer to the upward-moving gas stream in wetted-wall-tower experiments. Just as in the discussion of absorption in falling films, we consider mass transfer to a fluid moving with a constant velocity profile and also flow with a parabolic (Poiseuille) profile (see Fig. 5).

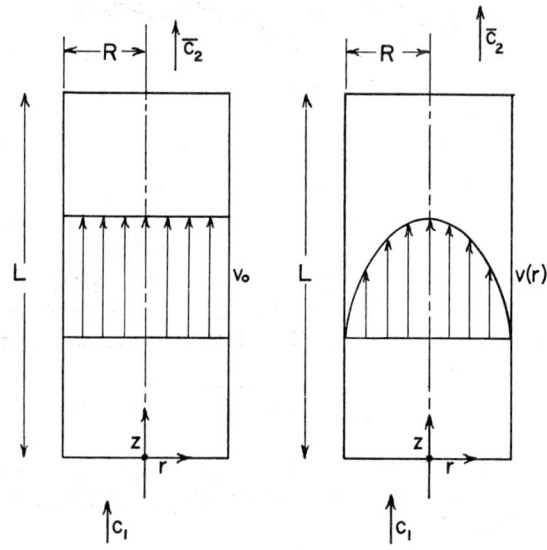

Fig. 5. Gas flow by forced convection through a wetted-wall tower, (a) with flat velocity profile, and (b) with parabolic velocity profile.

If the velocity is considered constant across the circular tube, then for the diffusion of A from the walls into the fluid B which is moving through the tube, one has to solve the differential equation

$$v_0 \frac{\partial c_A}{\partial z} = \mathfrak{D}_{AB} \frac{1}{r} \frac{\partial}{\partial r} \left(r \frac{\partial c_A}{\partial r} \right) \tag{179}$$

in which r is the cylindrical coordinate $r = \sqrt{x^2 + y^2}$. The term $\mathfrak{D}_{AB}(\partial^2 c_A/\partial z^2)$ describing the diffusion if the direction of the flow has been considered small in comparison with the convection term $v_0(\partial c_A/\partial z)$. Equation (175) is to be solved with the boundary conditions

(a)	$c_A = c_{A1}$	at $z = 0$	for all r	
(b)	$c_A = c_{AR}$	at $r = R$	for all z	(180)

This is the mass transfer analog of the Lévêque problem in heat transfer (S9, p. 81), and the solution is expressible in terms of Bessel functions (C2, p. 175):

$$\frac{c_A - c_{A1}}{c_{AR} - c_{A1}} = 1 - 2 \sum_{n=1}^{\infty} \frac{1}{\alpha_n} \frac{J_0(\alpha_n r/R)}{J_1(\alpha n)} \exp\left(-\frac{\mathfrak{D}_{AB}}{v_0 R^2} \alpha_n^2 z\right) \quad (181)$$

The average concentration at the exit is then given by the integral of (181) over r at $z = L$:

$$\frac{\bar{c}_{A2} - c_{A1}}{c_{AR} - c_{A1}} = 1 - 4 \sum_{n=1}^{\infty} \frac{1}{\alpha_n} \exp\left(-\frac{\mathfrak{D}_{AB} \alpha_n^2}{v_0 R^2} z\right) \quad (182)$$

In these expressions J_0 and J_1 are Bessel functions of the first kind and the α_n are the values of r for which $J_0(r) = 0$.

If the velocity distribution is considered to be parabolic across the circular tube (Fig. 5b) in accordance with the solution to the Navier-Stokes equations, then one has to solve the equation

$$v_{max}\left[1 - \left(\frac{r}{R}\right)^2\right]\frac{\partial c_A}{\partial z} = \mathfrak{D}_{AB} \frac{1}{r}\frac{\partial}{\partial r}\left(r \frac{\partial c_A}{\partial r}\right) \quad (183)$$

with the boundary conditions given in Eq. (180); once again the term representing mass transport by diffusion in the z direction has been omitted. Equation (183) represents the mass transfer analog of the Graetz problem in heat transfer. This problem cannot be solved in terms of any simple functions, but the first few terms in a power series have been computed (D3, J1, pp. 451–464). The integrated form of the solution analogous to Eq. (182) is then of the form

$$\frac{\bar{c}_{A2} - c_{A1}}{c_{AR} - c_{A1}} = 1 - \sum_{n=1}^{\infty} a_n \exp - b_n(\mathfrak{D}_{AB}/v_{avg} R^2)z \quad (184)$$

in which $v_{avg} = (\tfrac{1}{2}) v_{max}$ and the a_n and b_n are

n	a_n	b_n
1	0.820	3.658
2	0.0972	22.178
3	0.0135	53.05

Although at first glance the solution in Eq. (184) would seem to be more realistic than that in Eq. (182), it has been found that the solution for the flat velocity profile agrees better with experiment than does that for the parabolic. This discrepancy has been explained by Boelter (B14,

S9, p. 82) on the basis of changes in the physical properties due to diffusion, resulting in free convection.

3. *Diffusion from a Point Source into a Moving Fluid*

Another useful solution to Eq. (158) is that describing the diffusion of substance A when it is injected into solvent B which is flowing in the z direction with a constant velocity v_0. Hence one must solve the differential equation

$$v_0 \frac{\partial c_A}{\partial z} = \mathfrak{D}_{AB} \nabla^2 c_A \tag{185}$$

with boundary conditions

(a) $\quad c_A = 0 \quad$ at $r = \infty$
(b) $\quad -4\pi r^2 \mathfrak{D}_{AB}(\partial c_A/\partial r) = Q_A \quad$ as $r \to 0 \tag{186}$

in which r is the distance from the source ($r^2 = x^2 + y^2 + z^2$), z is the distance downstream from the source, and Q_A is the rate at which A enters the system. The solution satisfying this problem is (C2, p. 223; W13)

$$c_A = \frac{Q_A}{4\pi \mathfrak{D}_{AB} r} e^{-(v_0/2\mathfrak{D}_{AB})(r-z)} \tag{187}$$

This relation has been used in the analysis of some experiments on eddy diffusion, where $\mathfrak{D}_{AB}$ is then replaced by $\mathfrak{D}_{AB}{}^{(t)}$ (S9, p. 42).

The problem of diffusion from a point source has been studied under more general conditions by Klinkenberg, Krajenbrink, and Lauwerier (K12). These authors discuss the solution of the equation

$$v_0 \frac{\partial c_A}{\partial z} = \mathfrak{D}_{AB} \frac{1}{r} \frac{\partial}{\partial r} \left(r \frac{\partial c_A}{\partial r} \right) + \mathfrak{D}_{AB}' \frac{\partial^2 c_A}{\partial z^2} \tag{188}$$

with boundary conditions

(a) $\quad c_A = 0 \quad$ at $z = -\infty$
(b) $\quad \partial c_A/\partial r = 0 \quad$ at $r = R \tag{189}$

along with the conditions that there is a source of A entering the system at the origin; here r is the distance from the z axis in cylindrical coordinates and R is the radius of the tube, the axis of which is coincident with the z axis. Equation (187) can then be obtained as a special case of the solution of Eq. (188). The diffusion coefficients in the r and z directions are taken to be different, and so nonisotropic eddy diffusion can be discussed.

The diffusion from a point source in a tube has also been studied by

Bernard and Wilhelm (B8), and the diffusion from an instantaneous point source into a turbulent atmosphere has been investigated by Davies (D1).

4. Diffusion in Flow Through a Tubular Reactor

Of interest in applied kinetics is the study of chemical reactions taking place in flow systems which are hydrodynamically simple, so that the kinetics effects may be properly calculated. A simple example is the flow (with flat velocity profile v_0 in the z direction) of a fluid through a circular tube; the fluid is an inert material S containing a small quantity of substance A. The inside of the cylindrical tube is coated with a catalyst which converts A into B according to a first-order reaction, with k' as reaction-rate constant. Let it then be desired to obtain the percentage of conversion after the fluid has flowed through the reactor tube of length L and radius R.

In this problem the differential equation to be solved is the steady state equation

$$v_0 \frac{\partial c_A}{\partial z} = \mathfrak{D}_{AB} \frac{1}{r} \frac{\partial}{\partial r}\left(r \frac{\partial c_A}{\partial r}\right) \tag{190}$$

with the boundary conditions

(a) $\quad c_A = c_{A0} \quad$ at $z = 0$
(b) $\quad -\mathfrak{D}_{AB}(\partial c_A/\partial r) = k' c_A \quad$ at $r = R \tag{191}$

It is assumed that the reaction product B does not influence the diffusion of A through S; it must also be assumed that the system is isothermal. The final expression for the concentration profiles is

$$\frac{c_A}{c_{A0}} = \sum_{n=1}^{\infty} \frac{2K}{[K^2 + \beta_n^2]} \frac{J_0(\beta_n r/R)}{J_0(\beta_n)} \exp(-\beta_n^2 \mathfrak{D}_{AS} z/v_0 R^2) \tag{192}$$

and the fraction of A converted after traveling the distance $z = L$ is

$$\frac{c_{A0} - c_{AL}}{c_{A0}} = 1 - \sum_{n=1}^{\infty} \frac{4K^2/\beta_n^2}{(K^2 + \beta_n^2)} \exp(-\beta_n^2 \mathfrak{D}_{AS} L/v_0 R^2) \tag{193}$$

In these expressions $K = Rk'/\mathfrak{D}_{AS}$, a dimensionless rate constant, and the β_n are the roots of the equation $rJ_1(r)/J_0(r) = K$.

Baron, Manning, and Johnstone (B4) have discussed the experimental aspects of the tubular-reactor problem, and an analysis of the results is made by means of the solution to the differential equations for mass transfer with chemical reactions of first and second order at the wall. Some additional problems on diffusion coupled with interface reactions are given in the monograph of Barrer (B5, pp. 37–43) and several problems involving both convection and chemical reactions are discussed by Jost (J13, pp. 57–60). It should be noted that many of the problems involving surface-catalyzed reactions have analogies in the heat transfer literature. For example, the problem of the tubular reactor just discussed is exactly analogous to the unsteady state heat conduction in an infinite rod which at a given time begins to radiate heat to the surroundings; the solution to this problem is given by Carslaw and Jaeger (C2, p. 176).

D. Diffusion in More Complex Systems

Most of the problems discussed thus far have relatively simple analytical solutions. Although few systems are described exactly by these solutions, they provide a fund of relationships which find considerable value in many chemical engineering research problems. There are, of course, a number of much more complex problems involving diffusion, which when solved will be useful to the chemical engineer. In this section we indicate what progress has been made on some of these more complex problems.

1. *Unsteady State Diffusion in Evaporation*

We now consider the evaporation of a liquid A and the diffusion of this vapor through a gas B in an infinitely long tube as shown in Fig. 6. There is assumed to be a device which maintains the liquid surface of area S at a specific point in the column, which is taken to be $z = 0$. The entire system is maintained at constant temperature and pressure, and the vapors A and B are taken to be ideal gases; hence the total molar concentration c is constant through the gas column. This problem, which has been studied by Arnold (A3), is a good illustration of a situation in which Fick's laws are not applicable and serves to emphasize the danger of solving mass transfer problems entirely by analogy with heat transfer.

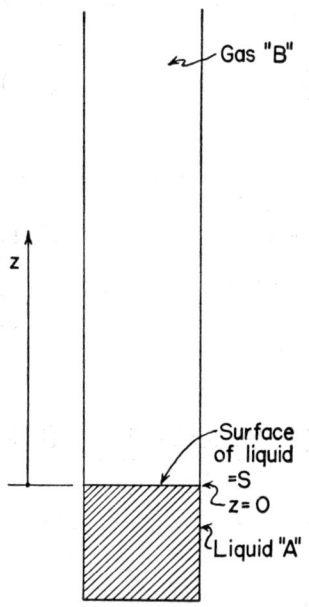

Fig. 6. Schematic diagram of the problem discussed by Arnold (A3).

The basic differential equation which describes the diffusion in this system is Eq. (51) written for diffusion in the z-direction with no chemical reaction:

$$\frac{\partial x_A}{\partial t} = \mathfrak{D}_{AB} \frac{\partial^2 x_A}{\partial z^2} - w(t) \frac{\partial x_A}{\partial z} \qquad (194)$$

in which w is the molar average velocity, which is a function of time alone, but not of position. It is clear that this differential equation does not have an analog in heat conduction. This is to be solved with the boundary and initial conditions

(a) $x_A = x_{A0}$ at $z = 0$ for all t
(b) $x_A = 0$ at $t = 0$ for all z
(c) $x_A = 0$ at $z = \infty$ for all t (195)

The concentration x_{A0} is the mole fraction of A in the vapor phase in equilibrium with the pure liquid; this equilibrium assumption is not necessarily correct, as has already been discussed in Sec. I,D,4. The problem may be transformed into simpler form by introduction of the following reduced quantities:

(a) $X = x_A/x_{A0}$ Reduced concentration
(b) $Z = z/\sqrt{4\mathfrak{D}_{AB}t}$ Reduced distance
(c) $W = w\sqrt{t/\mathfrak{D}_{AB}}$ Reduced molar-average velocity   (196)

In terms of these symbols Eq. (194) becomes

$$X'' + 2(Z - W)X' = 0 \qquad (197)$$

in which the primes represent differentiation with respect to Z. The boundary conditions are

(a) $X = 1$ at $Z = 0$
(b) $X = 0$ at $Z = \infty$ (198)

We now consider two solutions to Eq. (197).

a. Solution for $W = 0$. If it is (incorrectly) assumed that the molar-average velocity is zero, then Eq. (197) may be easily integrated to give

$$X = 1 - \operatorname{erf} Z \qquad (199)$$

From this result it may be shown that the rate of production of volume of A at time t at the surface $z = 0$ is obtained by using Fick's first law:

$$\left.\frac{\partial V_A}{\partial t}\right|_{z=0} = -S\mathfrak{D}_{AB}\left.\frac{\partial x_A}{\partial z}\right|_{z=0} = +Sx_{A0}\sqrt{\frac{\mathfrak{D}_{AB}}{\pi t}} \qquad (200)$$

in which S is the area of the surface from which evaporation is taking place. This may be integrated to give

$$V_A(t) = Sx_{A0}\sqrt{\frac{4\mathfrak{D}_{AB}t}{\pi}} \tag{201}$$

This equation (Stefan's equation) was found to give the correct dependence on $\mathfrak{D}_{AB}$ and t but not on x_{A0}.

b. *Solution for* $W = W(x_{A0})$. Arnold (A3) assumed[18] that W is not a function of time but only of the concentration of A in the gas phase at $z = 0$, that is, that $W = W(x_{A0})$. Then Eq. (197) may be integrated to give

$$X = \frac{1 - \mathrm{erf}\ (Z - W)}{1 + \mathrm{erf}\ W} \tag{202}$$

[from which Eq. (199) may be obtained as a special case].

Next the dependence of W on x_{A0} has to be determined. Since w depends on time but not on position, it can best be evaluated by getting an expression for w_0 (the molar-average velocity at $z = 0$), which is

$$w_0 = x_{A0}v_{A0} + x_{B0}v_{B0} = (1/c)(N_{A0} + N_{B0}) \tag{203}$$

At the surface $z = 0$ there is no motion of B, and so $v_{B0} = N_{B0} = 0$. The quantity N_{A0} is then calculated according to Eq. (49), and

$$w = w_0 = -\frac{\mathfrak{D}_{AB}}{(1 - x_{A0})}\frac{\partial x_A}{\partial z}\bigg|_{z=0} \tag{204}$$

or, in reduced units,

$$W = W_0 = -\frac{1}{2}\frac{x_{A0}}{1 - x_{A0}}\frac{\partial X}{\partial Z}\bigg|_{z=0} \tag{205}$$

Substituting Eq. (202) into Eq. (205) and performing the differentiations, one obtains W as an implicit function of x_{A0}; actually it is easier to solve and get x_{A0} as a function of W instead, which is

$$x_{A0} = \frac{1}{1 + [\sqrt{\pi}\ (1 + \mathrm{erf}\ W)W \exp W^2]^{-1}} \tag{206}$$

from which a numerical table of $W = W(x_{A0})$ may be prepared.

The rate of production of volume of vapor A is then

$$\frac{\partial V_A}{\partial t}\bigg|_{z=0} = Sw_0 = SW_0\sqrt{\frac{\mathfrak{D}_{AB}}{t}} = Sx_{A0}\sqrt{\frac{\mathfrak{D}_{AB}}{\pi t}}\cdot\psi_V \tag{207}$$

[18] It appears that it is not necessary to assume this: According to Eq. (205) W is a function of x_{A0} and the derivative $\partial X/\partial Z$ (evaluated at $Z = 0$). The latter, according to Eq. (197), can depend only on W; hence W is a function of x_{A0} alone.

and integrating one obtains

$$V(t) = Sx_{A0}\sqrt{4\mathfrak{D}_{AB}t/\pi} \cdot \psi_V \qquad (208)$$

in which $\psi_V = W\sqrt{\pi}/x_{A0}$ is a function of x_{A0} for which a few values are given in Table VI; the function ψ_V is then a measure of the deviation from Fickian behavior or, alternatively, the deviation from the heat-conduction analog. It can be seen that the deviations are the greater for more volatile substances.

TABLE VI
Factors Indicating Deviations from Fick's Second Law for Diffusion in an Evaporating System (A3)

x_{A0}	ψ_V	$\psi_\mathfrak{D}$
0.00	1.000	1.000
0.25	1.108	0.815
0.50	1.268	0.622
0.75	1.564	0.409
1.00	∞	0.000

Equation (208) can be used for the determination of diffusion coefficients by writing it in the form

$$\mathfrak{D}_{AB} = \frac{\pi}{t}\left(\frac{V}{2Sx_{A0}}\right)^2 \cdot \psi_\mathfrak{D} \qquad (209)$$

the function $\psi_\mathfrak{D}$ being also given in Table VI [it may be approximated by $\psi_\mathfrak{D} = (1 - x_{A0})^{2/3}$]. This method for determining diffusion coefficients from the volume of vapor produced in evaporation has been used by Fairbanks and Wilke (F1). A somewhat different experimental arrangement has been used by Schwertz and Brow (S6).

2. *Theory of the Thermal Diffusion Column*

Because the complete theoretical description of the operation of a thermal diffusion column is quite intricate, a simplified theory due to Jones and Furry (J12) is summarized here; a more extensive discussion along with a survey of column operation from the phenomenological standpoint is given by Grew and Ibbs (G11).

The following assumptions are made in the simplified treatment:

a. The diffusion-convection system is made up of ideal gases A and B contained between two parallel walls of width b and a distance $2a$ apart (see Figure 7).

b. The gas mixture is assumed to move in two countercurrent streams each occupying one half of the space between the two planes; it is further assumed that the velocity profile in each half is flat, this average velocity being obtained by taking the average over the velocity profile calculated approximately from the Navier-Stokes equations.

  c. The physical properties of the gases (that is, viscosity, thermal conductivity, and diffusion coefficients) are assumed to be independent of temperature and concentration—with the single exception that the temperature dependence of the density is accounted for in the equation of motion.

  d. The molecular weights of species A and B are assumed to be very nearly equal.

This problem is a good example of the importance of formulating a complex diffusion problem in terms of the equations of change. Hence the simplified treatment given here is discussed in terms of the simplified solutions to the three basic equations.

  a. *Approximate solution to the energy balance equation: determination of the temperature and density profiles.* For this problem the equation of energy balance [Eq. (19)] simplifies to

$$k(d^2T/dx^2) = 0 \tag{210}$$

which may be integrated to satisfy the boundary conditions indicated in Fig. 7 to give

$$T = \bar{T} + (x/2a)\Delta T \tag{211}$$

in which $\bar{T} = (\tfrac{1}{2})(T' + T'')$ and $\Delta T = T'' - T'$.

To obtain an expression for the deviation of the density ρ from the density $\bar{\rho}$ at the average temperature, ρ is expanded in a Taylor series about $\bar{T}$:

$$\begin{aligned}
\rho &= \bar{\rho} + \left(\frac{\partial \rho}{\partial T}\right)_{\bar{T}} (T - \bar{T}) + \cdots \\
&= \bar{\rho} - \frac{\bar{\rho}}{\bar{T}} (T - \bar{T}) + \cdots \\
&\doteq \bar{\rho} - \bar{\rho}\left(\frac{x}{2a}\right)\frac{\Delta T}{\bar{T}}
\end{aligned} \tag{212}$$

In the second line the ideal-gas law is used, and the third line is obtained by use of Eq. (211).

  b. *Approximate solution to the equation of motion.* The Navier-Stokes equations [Eq. (9)] for the steady state flow simplify to

$$0 = -\frac{dp}{dz} - \rho g + \mu \frac{d^2v_z}{dx^2} = -(\rho - \bar{\rho})g + \mu \frac{d^2v_z}{dx^2} \tag{213}$$

As a first approximation the pressure distribution is related to the weight of a column of fluid of density $\bar{\rho}$ (so that $-dp/dz = \bar{\rho}g$). Substitution of Eq. (212) into Eq. (213) gives after integration with respect to x

$$v_z = \frac{1}{12}\left(\frac{\bar{\rho}ga^2}{\mu}\right)\frac{x}{a}\left(1 - \frac{x^2}{a^2}\right)\frac{\Delta T}{\bar{T}} \qquad (214)$$

This distribution is sketched in Fig. 7. The average velocity over one half the slit width is then

$$\bar{v} = \frac{1}{a}\int_0^a v_z dx = \frac{1}{48}\left(\frac{\bar{\rho}ga^2}{\mu}\right)\frac{\Delta T}{\bar{T}} \qquad (215)$$

In the subsequent discussion it is assumed that the ascending and descending streams move with this velocity.

c. Approximate solutions to the equation of continuity: determination of the net transport of A down the column. Let x_A' be the mole fraction of A in the descending stream (that is, in the stream next to the cold wall with temperature T') and x_A'' the mole fraction of A in the ascending stream (next to the hot wall with temperature T''). The mass fluxes in the downward and upward stream (in g./cm.² sec.) with respect to a fixed coordinate system are then (see Table III for notation concerning mass fluxes)

$$n_A' = x_A'(n_A' + n_B') + \phi_A'$$
$$= x_A'(-\bar{v}) - \bar{\rho}\mathfrak{D}_{AB}\frac{dx_A'}{dz} \qquad (216)$$

$$n_A'' = x_A''(n_A'' + n_B'') + \phi_A''$$
$$= x_A''\bar{\rho}(+\bar{v}) - \bar{\rho}\mathfrak{D}_{AB}\frac{dx'}{dz} \qquad (217)$$

in which use has been made of the assumption that $M_A \doteq M_B$. Addition of these two gives the net transport of A in the downward direction:

$$-n_A = -(n_A' + n_A'') = \bar{\rho}\bar{v}(x_A' - x_A'') + \bar{\rho}\mathfrak{D}_{AB}\left(\frac{dx_A'}{dz} + \frac{dx_A''}{dz}\right)$$
$$\doteq \bar{\rho}\bar{v}(x_A' - x_A'') + 2\bar{\rho}\mathfrak{D}_{AB}\frac{dx_A}{dz} \qquad (218)$$

where $\bar{x}_A$ is an average of x_A' and x_A'' at any height z in the column.

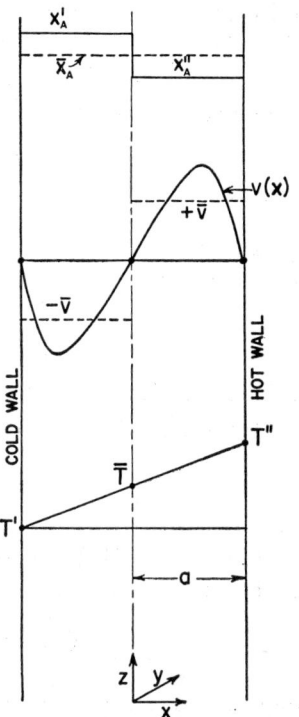

Fig. 7. Sketch of the velocity, temperature, and concentration profiles according to the simple theory (for k_T positive).

Next one needs an expression for $(x_A' - x_A'')$. The difference in concentration between the two streams results from two effects: thermal diffusion, which tends to increase the concentration difference, and convection, which tends to decrease it. Each of these effects is considered separately by obtaining an approximate integrated form of the steady state equation of continuity as applied to that particular process. If the only effect tending to produce a concentration difference were thermal diffusion, then according to Eq. (131) $dx_A/dx = -(k_T/T)(dT/dx)$; this expression may be written in difference form over the distance from $x = -(\frac{1}{2})a$ to $x = +(\frac{1}{2})a$ thus:

$$\frac{x_A' - x_A''}{a} = -\frac{\alpha \bar{x}_A \bar{x}_B}{\bar{T}} \frac{T' - T''}{2a} \tag{219}$$

where the thermal diffusion factor α [Eq. (55)] has been introduced. If the only effect were convection, then Eq. (158) applies

$$[\mathfrak{D}_{AB}(d^2 x_A/dx^2) = v(dx_A/dz)]$$

this expression may be approximated in difference form for the interval $x = -(\frac{1}{2})a$ to $x = +(\frac{1}{2})a$ thus:

$$\mathfrak{D}_{AB} \frac{x_A' - x_A''}{a^2} = \bar{v} \frac{d\bar{x}_A}{dz} \tag{220}$$

Solving Eqs. (219) and (220) for $(x_A' - x_A'')$ and adding the two effects together, one obtains:

$$(x_A' - x_A'') = \frac{\alpha}{2} \bar{x}_A \bar{x}_B \frac{\Delta T}{\bar{T}} + \frac{a^2 \bar{v}}{\mathfrak{D}_{AB}} \frac{d\bar{x}_A}{dz} \tag{221}$$

This result can be substituted into Eq. (218) to obtain the net flux of A. Finally one obtains the expression for the net transport of A down the column:

$$(-n_A ab) = H \bar{x}_A \bar{x}_B + (H_c + K_d)(d\bar{x}_A/dz) \tag{222}$$

in which

$$H = \bar{\rho} \bar{v} ab \left[\frac{1}{2} \alpha \frac{\Delta T}{\bar{T}} \right] = \frac{1}{96} \frac{\alpha \bar{\rho}^2 g a^3 b}{\mu} \left(\frac{\Delta T}{\bar{T}} \right)^2 \tag{223}$$

$$K_c = \bar{\rho} \bar{v} ab \left[\frac{a^2 \bar{v}}{\mathfrak{D}_{AB}} \right] = \frac{1}{(48)^2} \frac{\bar{\rho}^3 g^2 a^7 b}{\mu^2 \mathfrak{D}_{AB}} \left(\frac{\Delta T}{\bar{T}} \right)^2 \tag{224}$$

$$K_d = 2ab\bar{\rho} \mathfrak{D}_{AB} \tag{225}$$

where use is made of the expression in Eq. (215) for $\bar{v}$. The terms containing H, H_c, and K_d have to do with mass transfer by thermal diffusion, convection, and longitudinal ordinary diffusion respectively.

When a closed thermal diffusion column (i.e., a column with no throughput) has been running a long time a steady state is ultimately reached, for which there is no net transport through the column. Then the expression in Eq. (222) is zero and $H\bar{x}_A\bar{x}_B = -K(d\bar{x}_A/dz)$, where $K = K_c + K_d$. This equation can be integrated from the bottom of the column ($z = 0$) to the top ($z = L$) to give:

$$q_e = \frac{(\bar{x}_A/\bar{x}_B)_0}{(\bar{x}_A/\bar{x}_B)_L} = e^{(H/KL)} \qquad (226)$$

where q_e is the equilibrium separation factor.

The exact treatment yields expressions which have the same form as the expressions given above; only the numerical factors are different. The more detailed theory for the diffusion-convection problem between plane walls was developed by Furry, Jones, and Onsager (F10) and that for the column constructed from two concentric cylinders by Furry and Jones (F11). Recently more attention has been given to the rôle of the temperature dependence of the transport coefficients in column operation (B9, S15).

3. *Some Additional Solutions Discussed in the Literature*

A number of other special solutions to the diffusion equations may be found in the literature. A few of these are mentioned here:

a. The problem of obtaining *diffusion coefficients from sorption data* has been studied by Prager (P7), by Hwang (H17), and by Carman and Haul (C1). None of these investigators, however, have used Eq. (51) for their analysis.

b. The mechanism of *Liesegang ring formation* in the gaseous NH_3-HCl reaction has been studied from a diffusion standpoint by Spotz and Hirschfelder (S13, S14). This phenomenon involves unsteady state multicomponent diffusion with chemical reaction as well as phase transitions and supersaturation effects.

c. The steady state diffusion in a *chromatographic column* has been studied by Thomas (T3).

d. The theory of *flame propagation* has been formulated in terms of the equations of change by Hirschfelder and Curtiss (H6, H7); references to the computational work on flame theory may be found in ref. (H11). Recently Klein (K11) has developed some new analytical approaches to the flame problem, which are based on the successive approximations to the solution of an integral equation.

e. Sir Geoffrey Taylor (T1) has developed an analysis of *diffusion and convection in tube flow* which promises to offer a flow method for determining diffusion coefficients.

f. The *mass transfer from a flat plate* has been studied by Spalding (S12) for both forced and free convection.

g. Schwertz (S5) has discussed the separation of gases by *free double diffusion*.

h. Lauwerier (L2) solved the differential equation for the diffusion from a *source in a skew velocity field;* he solved the equation

$$\mathfrak{D}(\partial^2 c/\partial y^2) = v_0(1 + ay)(\partial c/\partial x)$$

with a source of strength Q at the origin and an absorbing wall placed at $y = -(1/a)$, i.e., at $y = -1/a, c = 0$.

i. Applications of turbulent diffusion to *mixing theory* have been made by Longwell and Weiss (L6) with possible applications to fuel injection in jet engines.

j. Considerable attention has been devoted to *diffusion into and out of spherical particles*. Rosen (R4) has studied transient kinetic behavior in a fixed bed in terms of diffusion into spherical particles. Frisch and Collins (F5) have investigated growth of aerosol particles. Geddes (G2) has pointed out an error in Eq. (96) on p. 29 of Barrer (B5).

k. The steady state diffusion into a moving sphere has been studied by Frisch (F4). The theory of heat and mass transfer to a sphere flowing at very low Reynolds numbers ($Re < 1$) has been developed by perturbation methods by Kronig and Bruijsten (K15). Kronig and Brink (K14) have discussed the theory of extraction from falling droplets.

V. Conclusions

From the standpoint of the chemical engineer and the applied physicist, the following aspects of diffusion are among those which need attention:

Theoretical	Experimental
1. Further development of the kinetic theory of dilute gases to include shape and polarity factors.	1. Further measurements of diffusion coefficients of long-chain organic molecules and highly polar substances.
2. Further development of the kinetic theory of dense gases and liquids.	2. Further measurements of self-diffusion and diffusion in dense gas regions in order to establish the corresponding states behavior in diffusion and to test further the Enskog theory.
3. Analytical studies of diffusing systems with three or more components.	3. Experimental studies of three component diffusion—particularly as regards systems encountered in simple kinetics problems.
4. Extension of analytical solutions for problems involving evaporation when Fick's second law is not applicable.	4. Experimental studies on diffusion in evaporating and condensing systems and systems with adsorption and surface reactions occurring.

Theoretical	Experimental
5. Extension of analytical solutions for interphase mass transfer, with particular attention to interfacial resistance.	5. Experimental study of more systems with interphase mass transfer, with the aim of correlating interfacial resistance with other physical properties.
6. Solution of forced convection problems for flow in non-Newtonian systems.	6. Experimental studies of diffusion in non-Newtonian flow systems.
7. Quantum theoretical analysis of the interdiffusion of hydrogen isotopes at very low temperatures; also study of ortho-para effects.	7. Experimental studies of the interdiffusion of He and H_2 and their isotopes at very low temperatures where quantum effects are of importance.
8. Further development of theory of diffusion in porous solids.	8. Systematic experiments on diffusion in porous media with aim of isolating various modes of transport.

Acknowledgments

In selecting the material for the foregoing discussion, the author has been guided by the topics discussed and the references given in the extensive course notes of diffusional operations by Professor O. A. Hougen and Dean W. R. Marshall, Jr. In writing some of the material given in Secs. IIB and IIIA, the author has drawn heavily on previous works (H11, B11, B12) written in collaboration with Professors J. O. Hirschfelder and C. F. Curtiss, to whom generous thanks are due for their permission to use this material. The author is further indebted to Professor C. F. Curtiss for a number of interesting discussions, and particularly for advice in connection with the redefinition of the diffusion coefficients introduced in Eqs. (32) to (36). The manuscript was read by Mr. David R. Longmire (University of Wisconsin) and Professor E. A. Mason (University of Maryland), both of whom made numerous corrections and suggestions. Finally the author wishes to thank the National Science Foundation for financial and during the course of this work under grant G-356.

Appendix

Note on Vector and Tensor Operations

Following are some of the important vector and tensor operations which are used in the derivations in Sec. II of this review; these relations are given for Cartesian coordinates. The quantities $\mathbf{v}$ and $\mathbf{F}$ are vectors, $\boldsymbol{\tau}$ is a second-order symmetric tensor, and T is a scalar.

$$(\mathbf{v} \cdot \mathbf{F}) = v_x F_x + v_y F_y + v_z F_z$$
$$(\boldsymbol{\tau} \cdot \mathbf{v})_x = \tau_{xx} v_x + \tau_{xy} v_y + \tau_{xz} v_z$$
$$\nabla T = \mathbf{i}(\partial T/\partial x) + \mathbf{j}(\partial T/\partial y) + \mathbf{k}(\partial T/\partial z)$$
$$(\nabla \cdot \mathbf{v}) = (\partial v_x/\partial x) + (\partial v_y/\partial y) + (\partial v_z/\partial z)$$
$$(\nabla \cdot \boldsymbol{\tau})_x = (\partial \tau_{xx}/\partial x) + (\partial \tau_{yx}/\partial y) + (\partial \tau_{zx}/\partial z)$$
$$(\mathbf{v} \cdot \nabla)T = v_x(\partial T/\partial x) + v_y(\partial T/\partial y) + v_z(\partial T/\partial z)$$
$$(\boldsymbol{\tau} : \nabla \mathbf{v}) = \tau_{xx}(\partial v_x/\partial x) + \tau_{xy}(\partial v_x/\partial y) + \tau_{xz}(\partial v_x/\partial z)$$
$$+ \tau_{yx}(\partial v_y/\partial x) + \tau_{yy}(\partial v_y/\partial y) + \tau_{yz}(\partial v_y/\partial z)$$
$$+ \tau_{zx}(\partial v_z/\partial x) + \tau_{zy}(\partial v_z/\partial y) + \tau_{zz}(\partial v_z/\partial z)$$

Additional operations may be found on pp. xxiii to xxvi of Ref. (H11). Most of these relations may be found in cylindrical, spherical, and other coordinate systems in standard reference works. Several of them do not, however, seem to be tabulated for handy reference; these operations are given here in cylindrical and spherical coordinates. Expressions for the Newtonian stress tensor in terms of the velocity gradients and the coefficient of viscosity may be found in Ref. (G7, pp. 103–105).

Cylindrical Coordinates (r, ϕ, z)

a. $(\mathbf{v} \cdot \nabla)\mathbf{v}$

r-component: $v_r \dfrac{\partial v_r}{\partial r} + \dfrac{v_\phi}{r} \dfrac{\partial v_r}{\partial \phi} - \dfrac{v_\phi^2}{r} + v_z \dfrac{\partial v_r}{\partial z}$

ϕ-component: $v_r \dfrac{\partial v_\phi}{\partial r} + \dfrac{v_\phi}{r} \dfrac{\partial v_\phi}{\partial \phi} + \dfrac{v_r v_\phi}{r} + v_z \dfrac{\partial v_\phi}{\partial z}$

z-component: $v_r \dfrac{\partial v_z}{\partial r} + \dfrac{v_\phi}{r} \dfrac{\partial v_z}{\partial \phi} + v_z \dfrac{\partial v_z}{\partial z}$

b. $(\nabla \cdot \boldsymbol{\tau})$

r-component: $\dfrac{1}{r} \dfrac{\partial}{\partial r}(r\tau_{rr}) + \dfrac{1}{r}\dfrac{\partial}{\partial \phi}\tau_{r\phi} - \dfrac{1}{r}\tau_{\phi\phi} + \dfrac{\partial \tau_{rz}}{\partial z}$

ϕ-component: $\dfrac{1}{r}\dfrac{\partial \tau_{\phi\phi}}{\partial \phi} + \dfrac{\partial \tau_{r\phi}}{\partial r} + \dfrac{2}{r}\tau_{r\phi} + \dfrac{\partial \tau_{\phi z}}{\partial z}$

z-component: $\dfrac{1}{r}\dfrac{\partial}{\partial r}(r\tau_{rz}) + \dfrac{1}{r}\dfrac{\partial \tau_{\phi z}}{\partial \phi} + \dfrac{\partial \tau_{zz}}{\partial z}$

c. $(\boldsymbol{\tau}:\nabla\mathbf{v}) = \tau_{rr}\dfrac{\partial v_r}{\partial r} + \tau_{\phi\phi}\dfrac{1}{r}\left(\dfrac{\partial v_\phi}{\partial \phi} + v_r\right) + \tau_{zz}\dfrac{\partial v_z}{\partial z}$

$+ \tau_{r\phi}\left(r\dfrac{\partial}{\partial r}\left(\dfrac{v_\phi}{r}\right) + \dfrac{1}{r}\dfrac{\partial v_r}{\partial \phi}\right)$

$+ \tau_{rz}\left(\dfrac{\partial v_z}{\partial r} + \dfrac{\partial v_r}{\partial z}\right)$

$+ \tau_{\phi z}\left(\dfrac{1}{r}\dfrac{\partial v_z}{\partial \phi} + \dfrac{\partial v_\phi}{\partial z}\right)$

Spherical Coordinates (r, θ, ϕ)

a. $(\mathbf{v} \cdot \nabla)\mathbf{v}$

r-component: $v_r \dfrac{\partial v_r}{\partial r} + \dfrac{v_\theta}{r}\dfrac{\partial v_r}{\partial \theta} + \dfrac{v_\phi}{r \sin \theta}\dfrac{\partial v_r}{\partial \phi} - \dfrac{v_\theta^2 + v_\phi^2}{r^2}$

θ-component: $v_r \dfrac{\partial v_\theta}{\partial r} + \dfrac{v_\theta}{r}\dfrac{\partial v_\theta}{\partial \theta} + \dfrac{v_\phi}{r \sin \theta}\dfrac{\partial v_\theta}{\partial \phi} + \dfrac{v_r v_\theta}{r} - \dfrac{v_\phi^2 \cot \theta}{r}$

ϕ-component: $v_r \dfrac{\partial v_\phi}{\partial r} + \dfrac{v_\theta}{r}\dfrac{\partial v_\phi}{\partial \theta} + \dfrac{v_\phi}{r \sin \theta}\dfrac{\partial v_\phi}{\partial \phi} + \dfrac{v_r v_\phi}{r} + \dfrac{v_\theta v_\phi \cot \theta}{r}$

b. $(\nabla \cdot \boldsymbol{\tau})$

r-component: $\dfrac{1}{r^2}\dfrac{\partial}{\partial r}(r^2 \tau_{rr}) + \dfrac{1}{r\sin\theta}\dfrac{\partial}{\partial\theta}(\sin\theta\,\tau_{r\theta}) + \dfrac{1}{r\sin\theta}\dfrac{\partial\tau_{r\phi}}{\partial\phi}$

$\hspace{6em} - \dfrac{\tau_{\theta\theta} + \tau_{\phi\phi}}{r}$

θ-component: $\dfrac{1}{r^2}\dfrac{\partial}{\partial r}(r^2 \tau_{r\theta}) + \dfrac{1}{r\sin\theta}\dfrac{\partial}{\partial\theta}(\sin\theta\,\tau_{\theta\theta}) + \dfrac{1}{r\sin\theta}\dfrac{\partial\tau_{\theta\phi}}{\partial\phi}$

$\hspace{6em} + \dfrac{\tau_{r\theta}}{r} - \dfrac{\cot\theta}{r}\tau_{\phi\phi}$

ϕ-component: $\dfrac{1}{r^2}\dfrac{\partial}{\partial r}(r^2 \tau_{r\phi}) + \dfrac{1}{r}\dfrac{\partial\tau_{\theta\phi}}{\partial\theta} + \dfrac{1}{r\sin\theta}\dfrac{\partial\tau_{\phi\phi}}{\partial\phi} + \dfrac{\tau_{r\phi}}{r} + \dfrac{2\cot\theta}{r}\tau_{\theta\phi}$

c. $(\boldsymbol{\tau}:\nabla\mathbf{v}) = \tau_{rr}\dfrac{\partial v_r}{\partial r} + \tau_{\theta\theta}\left(\dfrac{1}{r}\dfrac{\partial v_\theta}{\partial\theta} + \dfrac{v_r}{r}\right)$

$\hspace{3em} + \tau_{\phi\phi}\left(\dfrac{1}{r\sin\theta}\dfrac{\partial v_\phi}{\partial\phi} + \dfrac{v_r}{r} + \dfrac{v_\theta \cot\theta}{r}\right)$

$\hspace{3em} + \tau_{r\theta}\left(\dfrac{\partial v_\theta}{\partial r} + \dfrac{1}{r}\dfrac{\partial v_r}{\partial\theta} - \dfrac{v_\theta}{r}\right)$

$\hspace{3em} + \tau_{r\phi}\left(\dfrac{\partial v_\phi}{\partial r} + \dfrac{1}{r\sin\theta}\dfrac{\partial v_r}{\partial\phi} - \dfrac{v_\phi}{r}\right)$

$\hspace{3em} + \tau_{\theta\phi}\left(\dfrac{1}{r}\dfrac{\partial v_\phi}{\partial\theta} + \dfrac{1}{r\sin\theta}\dfrac{\partial v_\theta}{\partial\phi} - \dfrac{\cot\theta}{r}v_\phi\right)$

Nomenclature[19]

ROMAN LETTERS

- a_A Activity of species A (41)
- C_p Heat capacity at constant pressure (22)
- C_v Heat capacity at constant volume (21)
- c Molar density (see Table I)
- c_A Molar density of A in a two-component system (see Table I) (45)
- D_{ij} Multicomponent diffusion coefficients (33)
- D_i^T Multicomponent thermal diffusion coefficients
- $\mathfrak{D}_{AB}$ Binary diffusion coefficients (37)
- E Total energy (10)
- F External force acting on a fluid element (5)
- $F(r)$ Force of interaction between two molecules separated by a distance r (77)
- $G \equiv H - TS$ Gibbs's free energy
- g Acceleration due to gravity (134a)
- H Enthalpy (31)
- J_A Molar flux of A with respect to mass-average velocity (see Table III) (38)
- j_A Mass flux of A with respect to mass-average velocity (see Table III)
- $j^{(F)}$ Mass flux due to forced diffusion (32)

[19] Numbers in parentheses refer to the equation in which the symbol first appears.

$j^{(p)}$ Mass flux due to pressure gradient (32)

$j^{(T)}$ Mass flux due to temperature gradient (Soret effect) (32)

$j^{(x)}$ Mass flux due to concentration gradient (32)

K Kinetic energy

K_A Rate of production of species A (moles per unit volume per unit time) (40)

k Coefficient of thermal conductivity (20)

k' Reaction-rate constant (129)

k_A Rate of production of species A (mass per unit volume per unit time) (39)

k_G Gas-phase mass transfer coefficient (113)

k_L Liquid-phase mass transfer coefficient (117)

k_T Thermal diffusion ratio in a binary system (53)

L Latent heat of vaporization (110)

M Molecular weight (27)

$\mathbf{N}_A$ Molar flux of A with respect to stationary axes (see Table III) (40)

$\mathbf{n}_A$ Mass flux of A with respect to stationary axes (see Table III)

n Number density (number of molecules per unit volume) (86)

p Hydrostatic pressure (5)

Q Heat entering system (10)

Q_A Total moles of A issuing from a source during a unit time

$\mathbf{q}$ Energy flux vector with respect to the mass-average velocity (units: energy per unit area per unit time) (12)

$\mathbf{q}^{(d)}$ Energy flux accompanying diffusion (29)

$\mathbf{q}^{(T)}$ Energy flux due to temperature gradient (29)

$\mathbf{q}^{(x)}$ Energy flux due to a concentration gradient (Dufour effect) (29)

q_e Equilibrium separation factor (226)

R Gas constant (33)

Re Reynolds number (101)

r Distance between two interacting molecules (77)

r Spherical coordinate (118)

r Cylindrical coordinate (179)

$(r_m)_{AB}$ Intermolecular separation for which potential energy of interaction has a minimum (see Fig. 2)

t Time (1)

T Absolute temperature (19)

$\bar{T}$ Mean value of temperature defined by Eq. (134)

U Internal energy (11)

$\mathbf{u}_A$ Velocity of component A with respect to fixed axes (see Table II)

$\mathbf{v}$ Mass-average velocity (see Table II)

V Volume (34)

$\mathbf{v}_A$ Velocity of species A with respect to mass-average velocity (see Table II)

W Work done by system (10)

$\mathbf{w}$ Molar-average velocity (see Table II)

$\mathbf{w}_A$ Velocity of component A with respect to molar-average velocity (see Table II)

x Cartesian coordinate (1)

x_A Mole fraction of A in a two-component system (see Table I)

x_i Mole fraction of the ith component (33)

y Cartesian coordinate (1)

THEORY OF DIFFUSION

y Quantity used in Enskog's kinetic theory of dense gases (92)
z Cartesian coordinate (1)
z' Location of moving boundary (149)
Z Compressibility factor (97)

GREEK LETTERS

α Thermal diffusion factor (55)
α_n n^{th} zero of the Bessel function $J_0(r)$
α_{AB} Parameter in Buckingham potential (80)
α_{ij} "Resistance factors" (62)
β Coefficient of sliding friction (101)
β_n n^{th} root of $rJ_1/J_0 = K$ (192)
γ_n nth root of $\gamma \tan \gamma = N$ (139a)
δ Unit tensor (28)
ϵ_{AB} Parameter in Lennard-Jones potential (79) and in Buckingham potential (80)
κ Boltzmann's constant (83)
μ Coefficient of shear viscosity (7)
μ' Coefficient of bulk viscosity (7)
ν Total number of chemically distinct species in a multicomponent mixture
π 3.1416
ρ Mass density (see Table I)
ρ_A Mass density of species A in a two-component system (see Table I)
ρ_i Mass density of the ith chemical species in a multicomponent system
σ Soret coefficient (56)
σ_{AB} Parameter in Lennard-Jones potential (79)

τ Shear tensor with nine components τ_{xx}, τ_{xy}, τ_{xz}, τ_{yy}, etc. (5)
φ Intermolecular potential energy of interaction
ϕ_A Mass flux of A with respect to molar-average velocity (see Table III)
Φ_A Molar flux of A with respect to molar-average velocity (see Table III) (38)
$\Omega^{(l,s)\star}$ Complicated integrals in terms of which transport coefficients of dilute gases have been evaluated (86)
ω_A Mass fraction of A in a two-component system (see Table I)

OTHER SYMBOLS

A (subscript)   Quantity pertaining to component A in a binary mixture
c (subscript)   Quantity evaluated at the critical point
i (subscript)   Quantity pertaining to the ith chemical species in a multicomponent mixture (24)
I, II (subscripts)   Quantities associated with phases I and II in a two-phase system
0 (superscript)   Quantities evaluated for the dilute gas (92)
(t) (superscript)   Turbulent or "eddy" quantities (68)
$'$ (superscript)   Turbulent fluctuation of a physical quantity (in Sec. II,D,1)
$\wedge$ (above symbol)   Quantity per unit mass (5)
$-$ (above symbol)   Partial-molar quantity (27)

— (above quantity)   Time-smoothed quantities (in Sec. II,D,1)
— (above quantity)   Average quantity (in Sec. IV,D,2)

∇ "Nabla" or "del" operator (3)
D/Dt "Substantial derivative operator" (4)

References

A1. Arnold, J. H., *Ind. Eng. Chem.* **22**, 1091 (1930).
A2. Arnold, J. H., *J. Am. Chem. Soc.* **52**, 3937 (1930).
A3. Arnold, J. H., *Trans. Am. Inst. Chem. Engrs.* **40**, 361 (1944).
A4. Auer, P. L., and Murback, E. W., *J. Chem. Phys.* **22**, 1054 (1954).
B1. Babbitt, J. D., *Can. J. Research* **A28**, 449 (1950).
B2. Babbitt, J. D., *Can. J. Phys.* **29**, 427 (1951).
B3. Babbitt, J. D., *J. Chem. Phys.* **23**, 601 (1955).
B4. Baron, T., Manning, W. R., and Johnstone, H. F., *Chem. Engr. Progr.* **48**, 125 (1952).
B5. Barrer, R. M., "Diffusion In and Through Solids." Cambridge U. P., New York, 1941.
B6. Basset, A. B., "Hydrodynamics." Cambridge U. P., New York, 1888.
B7. Becker, E. W., Vogell, W., and Zigan, F., *Z. Naturforsch.* **8a**, 686 (1953).
B8. Bernard, R. A., and Wilhelm, R. H., *Chem. Engr. Progr.* **46**, 233 (1950).
B9. Bierlein, J. A., *J. Chem. Phys.* **23**, 10 (1955).
B10. Bird, R. B., *Soc. Plastics Engrs. J.* **11**, 35 (1955).
B11. Bird, R. B., Curtiss, C. F., and Hirschfelder, J. O., *Chemical Eng. Progr., Symposium Ser. No. 16*, **51**, 69–85 (1955).
B12. Bird, R. B., Hirschfelder, J. O., and Curtiss, C. F., *Trans. Am. Soc. Mech. Engrs.* **76**, 1011 (1954).
B13. Bird, R. B., Spotz, E. L., and Hirschfelder, J. O., *J. Chem. Phys.* **18**, 1395 (1950).
B14. Boelter, L. M. K., *Trans. Am. Inst. Chem. Engrs.* **39**, 557 (1943).
B15. de Boer, J., *Ned. Tijdschr. Natuurk.* **19**, 231 (1953).
B16. de Boer, J., and Bird, R. B., *Physica* **20**, 185 (1954).
B17. de Boer, J., and Bird, R. B., *in* "Molecular Theory of Gases and Liquids," Chapter 10. Wiley, New York, 1954.
B18. Born, M., and Green, H. S., "A General Kinetic Theory of Liquids." Cambridge U. P., New York, 1949.
B19. Boyd, C. A., Stein, N., Steingrimsson, V. B., and Rumpel, W. F., *J. Chem. Phys.* **19**, 548 (1951).
B20. Brinkman, H. C., *Appl. Sci. Research* **A2**, 120 (1951).
B21. Brown, G. G., "Unit Operations." Wiley, New York, 1950.
B22. Brown, H., *Phys. Rev.* **58**, 661 (1940).
B23. Bunde, R. E., Ph.D. Thesis, Department of Chemistry, University of Wisconsin, 1955.
C1. Carman, P. C., and Haul, R. A. W., *Proc. Roy. Soc.* **A222**, 109 (1954).
C2. Carslaw, H. S., and Jaeger, J. C., "Conduction of Heat in Solids." Oxford U. P., New York, 1947.
C3. Chapman, S., and Cowling, T. G., "Mathematical Theory of Non-Uniform Gases." Cambridge U. P., New York, 1951.
C4. Cohen, E. G. D., Offerhaus, M. J., and de Boer, J., *Physica* **20**, 501 (1954).
C5. Colburn, A. P., and Pigford, R. L., *in* "Chemical Engineers' Handbook" (J. H. Perry, ed.), Section 8, pp. 523–560. McGraw-Hill, New York, 1950.

C6. Comings, E. W., and Nathan, M. F., *Ind. Eng. Chem.* **39**, 964 (1947).
C7. Comings, E. W., Mayland, B. J., and Egly, R. S., *Illinois Engr. Expt. Sta. Bull. Ser. 354*, **42**, No. 15 (1944).
C8. Coulson, J. M., and Richardson, J. F., "Chemical Engineering." McGraw-Hill, New York, 1954.
C9. Cowling, T. G., "Molecules in Motion." Hutchinson's University Library, London, 1950.
C10. Curtiss, C. F., "The Kinetic Theory of Dense Gases." University of Wisconsin Naval Research Lab. Report, WIS-OOR-3, January 28, 1953.
C11. Curtiss, C. F., *in* "High Speed Aerodynamics and Jet Propulsion" (J. V. Charyk and M. Summerfield, eds.), Vol. 1, p. 780. Princeton U. P., Princeton, 1955.
C12. Curtiss, C. F., and Hirschfelder, J. O., *J. Chem. Phys.* **17**, 550 (1949).
C13. Curtiss, C. F., and Hirschfelder, J. O., *J. Chem. Phys.* **18**, 171 (1950).
C14. Curtiss, C. F., and Snider, R. F., "The Kinetic Theory of Moderately Dense Gases," University of Wisconsin Naval Research Lab. Report, WIS-OOR-2, May 20, 1954.
D1. Davies, D. R., *Quart. J. Mech. Appl. Math.* **7**, 462 (1954).
D2. Davies, J. T., *J. Phys. & Colloid Chem.* **54**, 185 (1950).
D3. Drew, T. B., *Trans. Am. Inst. Chem. Engrs.* **21**, 26 (1931).
E1. Einstein, A., *Ann. Physik* [4] **17**, 549 (1905); **19**, 18 (1906).
E2. Emmert, R. E., and Pigford, R. L., *Chem. Engr. Progr.* **50**, 87 (1954).
E3. Enskog, D., *Kgl. Svenska Vetenskapsakad. Handl.* **63** (1922).
E4. Eyring, H., *J. Chem. Phys.* **4**, 283 (1936).
F1. Fairbanks, D. F., and Wilke, C. R., *Ind. Eng. Chem.* **42**, 471 (1950).
F2. Feld, B. T., *in* "Experimental Nuclear Physics" (E. Segre, ed.), Vol. 2, Part VII, Wiley, New York, 1953.
F3. Frank, H. S., and Tsao, M-S., *Ann. Rev. Phys. Chem.* **5**, 56–60 (1954).
F4. Frisch, H. L., *J. Chem. Phys.* **22**, 123 (1954).
F5. Frisch, H. L., and Collins, F. C., *J. Chem. Phys.* **20**, 1797 (1952).
F6. Fujita, H., *J. Chem. Phys.* **21**, 700 (1953).
F7. Fujita, H., *Mem. Coll. Agr. Kyoto Univ.* No. **59**, 31 (1951); see ref. (F6).
F8. Fujita, H., *Bull. Japan. Soc. Sci. Fisheries* **17**, 393 (1952); see ref. (F6).
F9. Fujita, H., and Kishimoto, A., *Textile Research J.* **22**, 94 (1952); see ref. (F6).
F10. Furry, W. H., Jones, R. C., and Onsager, L., *Phys. Rev.* **55**, 1083 (1939).
F11. Furry, W. H., and Jones, R. C., *Phys. Rev.* **69**, 459 (1946).
F12. Furry, W. H., and Pitkanen, P. H., *J. Chem. Phys.* **19**, 729 (1951).
F13. Fürth, R., *in* "Handbuch der Physikalischen und Technischen Mechanik" (F. Auerbach and W. Hort, eds.), Vol. 7, p. 635. Barth, Leipzig, 1931.
F14. Fürth, R., *in* "Handbuch der Physikalischen und Technischen Mechanik" (F. Auerbach and W. Hort, eds.), Vol. 7, p. 705. Barth, Leipzig, 1931.
G1. Gamson, B. W., *Chem. Engr. Progr.* **45**, 154 (1949).
G2. Geddes, R. L., *Trans. Am. Inst. Chem. Engrs.* **42**, 79 (1946).
G3. Gilliland, E. R., *Ind. Eng. Chem.* **26**, 681 (1934).
G4. Gillis, J., and Kedem, O., *J. Polymer Sci.* **11**, 545 (1953).
G5. Glasstone, S., and Edlund, M. C., "Elements of Nuclear Reactor Theory," Chapter 5. Van Nostrand, New York, 1952.
G6. Glasstone, S., Laidler, K. J., and Eyring, H., "Theory of Rate Processes," Chapter 9. McGraw-Hill, New York, 1941.
G7. Goldstein, S., "Modern Developments in Fluid Dynamics." Oxford U. P., New York, 1938.

G8. Gomer, R., *J. Chem. Phys.* **19**, 284 (1951).
G9. Gordon, A. R., *Ann. Rev. Phys. Chem.* **1**, 59–74 (1951).
G10. Green, H. S., "Molecular Theory of Fluids." Interscience, New York, 1952.
G11. Grew, K. E., and Ibbs, T. L., "Thermal Diffusion in Gases." Cambridge U. P., New York, 1952.
G12. de Groot, S. R., "Thermodynamics of Irreversible Processes," pp. 94–123. North Holland, Amsterdam, 1951.
G13. Guggenheim, E. A., "Thermodynamics." North Holland, Amsterdam, 1950.
H1. Harned, H. S., *Ann. Rev. Phys. Chem.* **2**, 37 (1951).
H2. Harned, H. S., and Owen, B. B., "The Physical Chemistry of Electrolytic Solutions," 2nd rev. ed. Reinhold, New York, 1950.
H3. Hellund, E. J., and Uehling, E. A., *Phys. Rev.* **56**, 818 (1939).
H4. Higbie, R., *Trans. Am. Inst. Chem. Engrs.* **31**, 365 (1935).
H5. Hill, N. E., *Proc. Phys. Soc.* **B424**, 209 (1955).
H6. Hirschfelder, J. O., and Curtiss, C. F., *J. Chem. Phys.* **17**, 1076 (1949).
H7. Hirschfelder, J. O., and Curtiss, C. F., *J. Phys. & Colloid Chem.* **55**, 744 (1951).
H8. Hirschfelder, J. O., Bird, R. B., and Spotz, E. L., *J. Chem. Phys.* **16**, 968 (1948).
H9. Hirschfelder, J. O., Bird, R. B., and Spotz, E. L., *Chem. Revs.* **44**, 205 (1949).
H10. Hirschfelder, J. O., Bird, R. B., and Spotz, E. L., *Trans. Am. Soc. Mech. Engrs.* **71**, 921 (1949).
H11. Hirschfelder, J. O., Curtiss, C. F., and Bird, R. B., "Molecular Theory of Gases and Liquids." Wiley, New York, 1954.
H12. Hirschfelder, J. O., Stevenson, D. P., and Eyring, H., *J. Chem. Phys.* **5**, 896 (1937).
H13. Hougen, O. A., and Watson, K. M., "Chemical Process Principles." Wiley, New York, 1947.
H14. Hougen, O. A., McCauley, H. J., and Marshall, W. R., Jr., *Trans. Am. Inst. Chem. Engrs.* **36**, 183 (1940).
H15. Hughes, D. J., "Pile Neutron Research," Addison-Wesley, Cambridge, 1953.
H16. Hutchinson, E., *J. Phys. & Colloid Chem.* **52**, 897 (1948).
H17. Hwang, J. L., *J. Chem. Phys.* **20**, 1320 (1952).
I1. Irving, J. H., and Kirkwood, J. G., *J. Chem. Phys.* **18**, 817 (1950).
I2. Irving, J. H., and Zwanzig, R. W., *J. Chem. Phys.* **19**, 1173 (1951).
I3. Irwin, M., *Proc. Soc. Exptl. Biol. Med.* **26**, 125 (1928).
J1. Jakob, M., "Heat Transfer," Vol. 1. Wiley, New York, 1949.
J2. Jeans, Sir J., "An Introduction to the Kinetic Theory of Gases." Cambridge U. P., New York, 1940.
J3. Jeffries, Q. R., and Drickamer, H. G., *J. Chem. Phys.* **21**, 1358 (1953).
J4. Jeffries, Q. R., and Drickamer, H. G., *J. Chem. Phys.* **22**, 436 (1954).
J5. Johnson, E. F., *Ind. Eng. Chem.* **45**, 902 (1953).
J6. Johnson, E. F., *Ind. Eng. Chem.* **46**, 889 (1954).
J7. Johnson, E. F., *Ind. Eng. Chem.* **47**, 599 (1955).
J8. Johnstone, H. F., and Pigford, R. L., *Trans. Am. Inst. Chem. Engrs.* **38**, 25 (1942).
J9. Jones, A. L., *Petroleum Processing* **6**, 132 (1951).
J10. Jones, A. L., and Foreman, R. W., *Ind. Eng. Chem.* **44**, 2249 (1952).
J11. Jones, A. L., and Milberger, E. C., *Ind. Eng. Chem.* **45**, 2689 (1953).
J12. Jones, R. C., and Furry, W. H., *Revs. Mod. Phys.* **18**, 151 (1946).
J13. Jost, W., "Diffusion." Academic Press, New York, 1952.
J14. Jost, W., *Chem. Eng. Sci.* **2**, 199 (1953).

K1. Karim, S. M,. and Rosenhead, L., *Revs. Mod. Phys.* **24,** 108 (1952).
K2. Kennard, E. H., "Kinetic Theory of Gases." McGraw-Hill, New York, 1938.
K3. Kihara, T., "Imperfect Gases." Asakusa Bookstore, Tokyo, 1949 (in Japanese). Translated into English by the U. S. Office of Air Research, Wright-Patterson Air Force Base.
K4. Kihara, T., *Revs. Mod. Phys.* **24,** 45 (1952).
K5. Kihara, T., and Kotani, M., *Proc. Phys. Math. Soc. Japan* **24,** 602 (1943).
K6. Kincaid, J. F., Eyring, H., and Stearn, A. E., *Chem. Revs.* **28,** 301 (1941).
K7. Kirkwood, J. G., *J. Chem. Phys.* **14,** 180 (1946); Errata **14,** 347 (1946).
K8. Kirkwood, J. G., *J. Chem. Phys.* **15,** 72 (1947); Erratum **15,** 555 (1947).
K9. Kirkwood, J. G., and Crawford, B., Jr., *J. Phys. Chem.* **56,** 1048 (1952).
K10. Kirkwood, J. G., Maun, E. K., and Alder, B. J., *J. Chem. Phys.* **18,** 1040 (1950).
K11. Klein, G., "Equations of a Simple Flame Solved by Successive Approximations to the Solution of an Integral Equation." University of Wisconsin Naval Research Lab. Reports ONR-8 (9 June 1954), ONR-13 (Sept. 30, 1954), SQUID-1 (Feb. 25, 1955), and SQUID-3 (April 18, 1955).
K12. Klinkenberg, A., Krajenbrink, H. J., and Lauwerier, H. A., *Ind. Eng. Chem.* **45,** 1202 (1953); Errata **45,** 1515 (1953).
K13. Kobe, K. A., and Lynn, R. E., Jr., *Chem. Revs.* **52,** 121 (1953).
K14. Kronig, R., and Brink, J. C., *Appl. Sci. Research* **A2,** 142 (1950).
K15. Kronig, R., and Bruijsten, J., *Appl. Sci. Research* **A2,** 439 (1951).
L1. Lamb, H., "Hydrodynamics," 4th ed. Cambridge U. P., New York, 1916.
L2. Lauwerier, H. A., *Appl. Sci. Research* **A4,** 153 (1954).
L3. Lee, C. Y., and Wilke, C. R., *Ind. Eng. Chem.* **46,** 2381 (1954).
L4. Li, J. C. M., and Chang, P., *J. Chem. Phys.* **23,** 518 (1955).
L5. London, F., *Trans. Faraday Soc.* **33,** 8 (1937).
L6. Longwell, J. P., and Weiss, M. A., *Ind. Eng. Chem.* **45,** 667 (1953).
L7. Lydersen, A. L., "Estimation of Critical Properties of Organic Compounds by the Method of Group Contributions." University of Wisconsin Engineering Experiment Station Report, 1955.
M1. Margenau, H., *Revs. Mod. Phys.* **11,** 1 (1939).
M2. Marshall, W. R., Jr., and Pigford, R. L., "The Application of Differential Equations to Chemical Engineering Problems." U. of Delaware Press, Newark, 1947.
M3. Mason, E. A., *J. Chem. Phys.* **22,** 169 (1954).
M4. Mason, E. A., *J. Chem. Phys.* **23,** 49 (1955).
M5. Mason, E. A., and Rice, W. E., *J. Chem. Phys.* **22,** 522 (1954).
M6. Mason, E. A., and Rice, W. E., *J. Chem. Phys.* **22,** 843 (1954).
M7. Massey, H. S. W., and Mohr, C. B. O., *Proc. Roy. Soc.* **A141,** 434 (1933).
M8. Matz, W., "Die Thermodynamik des Wärme- und Stoffaustausches in der Verfahrenstechnik," Steinkopff, Frankfurt-am-Main, 1949.
M9. Michels, A. M. J. F., Botzen, A., and Schuurman, W., *Physica* **20,** 1141 (1954).
M10. Montroll, E. W., and Green, M. S., *Ann. Rev. Phys. Chem.* **5,** 449 (1954).
N1. Nernst, W., *Z. physik. Chem.* **2,** 613 (1888).
O1. Olson, R. L., and Walton, J. S., *Ind. Eng. Chem.* **43,** 703 (1955).
O2. Onsager, L., and Fuoss, R. M., *J. Phys. Chem.* **36,** 2689 (1932).
O3. Othmer, D. F., and Thakar, M. S., *Ind. Eng. Chem.* **45,** 589 (1953); Erratum **47,** 1604 (1955).
P1. Partington, J. R., "An Advanced Treatise on Physical Chemistry," Vols. 1 and 2. Longmans, Green, New York, 1949.

P2. Perry, R. H., and Pigford, R. L., *Ind. Eng. Chem.* **45,** 1247 (1953).
P3. Pigford, R. L., Ph. D. Thesis, Department of Chemistry and Chemical Engineering, University of Illinois, 1941.
P4. Pigford, R. L., *Ind. Eng. Chem.* **45,** 957 (1953).
P5. Pigford, R. L., *Ind. Eng. Chem.* **46,** 937 (1954).
P6. Powell, R. E., Roseveare, W. E., and Eyring, H., *Ind. Eng. Chem.* **33,** 430 (1941).
P7. Prager, S., *J. Chem. Phys.* **19,** 537 (1951).
P8. Prigogine, I., "Étude Thermodynamique des Phénomènes Irreversibles." Desoer, Liège, 1947.
P9. Prigogine, I., "Thermodynamics of Irreversible Processes." Thomas, Springfield, Ill., 1955.
R1. Reiner, M., "Deformation and Flow." Lewis, London, 1949.
R2. Robb, W. L., and Drickamer, H. G., *J. Chem. Phys.* **19,** 1504 (1951).
R3. Robinson, R. A., and Stokes, R. H., "Electrolyte Solutions," Chapters 10 and 11. Academic Press, New York, 1955.
R4. Rosen, J. B., *J. Chem. Phys.* **20,** 387 (1952).
R5. Rosenhead, L., and colleagues, *Proc. Roy. Soc.* **A226,** 1–69 (1954).
R6. Rumpel, W. F., Ph. D. Thesis, Department of Chemistry, University of Wisconsin, 1955.
S1. Schlichting, H., "Grenzschichttheorie." Braun, Karlsruhe, 1951.
S2. Schlinger, W. G., and Sage, B. H., *Ind. Eng. Chem.* **45,** 657 (1953).
S3. Schottky, W. F., *Z. Elektrochem.* **58,** 442 (1954).
S4. Schrage, R. W., "A Theoretical Study of Interphase Mass-Transfer." Columbia U. P., New York, 1953.
S5. Schwertz, F. A., *Ind. Eng. Chem.* **45,** 1592 (1953).
S6. Schwertz, F. A., and Brow, J. E., *J. Chem. Phys.* **19,** 640 (1951).
S7. Scott, E. J., Tung, L. H., and Drickamer, H. G., *J. Chem. Phys.* **19,** 1075 (1951).
S8. Seith, W., "Diffusion in Metallen." Springer, Berlin, 1939.
S9. Sherwood, T. K., and Pigford, R. L., "Absorption and Extraction." McGraw-Hill, New York, 1952.
S9a. Sinfelt, J. H., and Drickamer, H. G., *J. Chem. Phys.* **23,** 1095 (1955).
S10. Slattery, J. C., "Diffusion Coefficients and the Principle of Corresponding States." M. S. Thesis, University of Wisconsin, Department of Chemical Engineering, 1955.
S11. Snider, R. F., and Curtiss, C. F., "The Effects of Concentration Dependence on Diffusion Coefficients." U. of Wisconsin Report WIS-OOR-9, Feb. 17, 1954.
S12. Spalding, D. B., *Proc. Roy. Soc.* **A221,** 78 (1954).
S13. Spotz, E. L., Ph. D. Thesis, Department of Chemistry, University of Wisconsin, 1950.
S14. Spotz, E. L., and Hirschfelder, J. O., *J. Chem. Phys.* **19,** 1215 (1951).
S15. Srivastava, B. N., and Srivastava, R. C., *Physica* **20,** 237 (1954).
S16. Stearn, A. E., and Eyring, H., *J. Phys. Chem.* **44,** 955 (1940).
S17. Stearn, A. E., Irish, E. M., and Eyring, H., *J. Phys. Chem.* **44,** 981 (1940).
S18. Stewart, D. W., *Ann. Rev. Phys. Chem.* **2,** 67 (1951).
S19. Stokes, R. H., *Trans. Faraday Soc.* **48,** 887 (1952).
S20. Strehlow, R. A., *J. Chem. Phys.* **21,** 2101 (1953).
S21. Sutherland, W., *Phil. Mag.* **9,** 781 (1905).
T1. Taylor, Sir G., *Proc. Roy. Soc.* **A219,** 186 (1953).
T2. Thiele, E. W., *Ind. Eng. Chem.* **31,** 916 (1939).

T3. Thomas, H. C., *J. Chem. Phys.* **19,** 1213 (1951).
T4. Timmerhaus, K. D., and Drickamer, H. G., *J. Chem. Phys.* **20,** 981 (1952).
T5. Tobias, C. W., Eisenberg, M., and Wilke, C. R., *J. Electrochem. Soc.* **99,** 359C-365C (1952).
T6. Treybal, R. E., "Liquid Extraction." McGraw-Hill, New York, 1951.
T7. Tung, L. H., and Drickamer, H. G., *J. Chem. Phys.* **20,** 6 (1952).
U1. Uehling, E. A., *Phys. Rev.* **46,** 917 (1934).
U2. Uehling, E. A., and Uhlenbeck, G. E., *Phys. Rev.* **43,** 552 (1933).
U3. Uhlenbeck, G. E., *Ned. Tijdschr. Natuurk.*, **21,** 329 (1955).
V1. Vogelpohl, G., *Forsch. Gehiete Ingenieurw.* **B16,** Forschungsheft No. 425 (1949).
V2. Vyazovov, V. V., *J. Tech. Phys. U.S.S.R.* 10, 1519 (1940); see ref. (89).
W1. Walker, W. H., Lewis, W. K., McAdams, W. H., and Gilliland, E. R., "Principles of Chemical Engineering." McGraw-Hill, New York, 1937.
W2. Wall, F. T., and Hiller, L. A., Jr., *Ann. Rev. Phys. Chem.* **5,** 267 (1951).
W3. Waddel, H., *J. Franklin Inst.* **217,** 459 (1934).
W4. Wang Chang, C. S., and Uhlenbeck, G. E., "Transport Properties of Polyatomic Gases." U. of Michigan, C. M. 681, Project NOrd 7924, 1951.
W5. Ward, A. F. H., and Brooks, L. H., *Trans. Faraday Soc.* **48,** 1124 (1952).
W6. Weltmann, R-N., and Kuhns, P. W., *J. Colloid Sci.* **7,** 218 (1952).
W7. Wheeler, A., *in* "Catalysis" (R. H. Emmett, ed.), Vol. 2, pp. 126–131. Reinhold, New York, 1955.
W8. Wilke, C. R., *Chem. Eng. Progr.* **45,** 218 (1949).
W9. Wilke, C. R., *Chem. Eng. Progr.* **46,** 95 (1950).
W10. Wilke, C. R., *Ind. Eng. Chem.* **47,** 658 (1955).
W11. Wilke, C. R., and Lee, C. Y., *Ind. Eng. Chem.* **47,** 1253 (1955).
W12. Wilke, C. R., Tobias, C. W., and Eisenberg, M., *Chem. Engr. Progr.* **49,** 663 (1953).
W13. Wilson, H. A., *Proc. Cambridge Phil. Soc.* **12,** 406 (1904); see ref. (S9).
W14. Winter, E. R. S., *Trans. Faraday Soc.* **46,** 81 (1950).
W15. Weisskopf, V. F., *in* "Introduction to Pile Theory" (C. Goodman, ed.), Vol. 1, Addison-Wesley, Cambridge, 1952.
Y1. Yong, T. F., and Jones, A. C., *Ann. Rev. Phys. Chem.* **3,** 275 (1951).

Turbulence in Thermal and Material Transport

J. B. Opfell and B. H. Sage

California Institute of Technology, Pasadena, California

I. Introduction	242
II. Nature of Turbulence	242
A. Review of Classic References to Turbulence	242
B. Physical Manifestations of Homogeneous Turbulence	243
C. Statistical Nature of Turbulence	243
D. Fluctuating Velocity	245
E. Correlations	245
F. Fluctuating Velocity	247
III. Macroscopic Aspects of Steady, Uniform, Turbulent Flow	247
A. Definition of Eddy Properties	247
B. Velocity Distribution	247
1. Boundary Flow	247
2. Turbulent Core	250
C. Eddy Viscosity	251
D. Relative Viscosity	252
IV. Thermal Transport in Turbulent Flow	255
A. Introduction	255
B. Eddy Conductivity	256
C. Prandtl Number	258
D. Prediction of Thermal Transport	259
1. General Energy Equation	260
2. Boundary Conditions	262
3. Analytical Solutions	263
E. Macroscopic Thermal Transfer	266
V. Material Transport in Turbulent Flow	267
A. Introduction	267
B. Macroscopic Transport	267
1. Effect of Turbulence	268
2. Resistance at Interface	268
C. Diffusion	268
D. Laminar Flow	270
E. Turbulent Flow	270
1. Eddy Viscosity	271
2. Experimental Background	271
F. Schmidt Number	272
G. Prediction	275
1. Equations of State	275
2. Turbulent Streams	276

3. Analytical Method... 276
4. Limitations of Analyses...................................... 278
VI. Combined Thermal and Material Transport in Turbulent Flow........... 278
   A. Introduction... 278
   B. Eddy Properties... 280
VII. Summary.. 281
   Nomenclature... 283
   References.. 285

I. Introduction

Transfers of energy and material to, from, and through bodies of fluids occur with great frequency in a wide variety of engineering processes. The fact that turbulence in the fluid affects such transfers has long been known, but the complexity of the phenomenon is such as to make a quantitative approach difficult. However, the importance of these transport processes and the marked influence of turbulence upon them make a quantitative study of turbulence an important step in the development of engineering as a science.

The basic nature of the turbulent exchange process is not yet well enough known to allow accurate prediction of behavior without recourse to experiment. Correlation of the growing body of experimental knowledge in this field, however, offers the possibility of evaluating time-averaged point values of thermal and material transport for many conditions of industrial interest. It is the purpose of this discussion to present some of the more elementary considerations of the nature of turbulent flow with particular emphasis upon thermal and material transport.

II. Nature of Turbulence

A. Review of Classic References to Turbulence

Reynolds (R1, R2) was one of the earlier investigators to appreciate the random nature of turbulence. The dimensionless parameter bearing his name is widely used as a measure of the physical characteristics of steady, uniform flow. Such a measure is essentially macroscopic and does not describe the local or transient behavior at a point in the stream. In recent years much effort has been devoted to understanding the basic mechanism of momentum transport by turbulence. The early work of Prandtl (P6), Taylor (T1), Kármán (K1), and Howarth (K4) laid a basis for the statistical theory of turbulence which is apparently in reasonable agreement with experiment. More recently Onsager (O3), Corrsin (C6), and Kolmogoroff (K10) extended the statistical theory of turbulence to describe the available experimental data in terms of kinetic-energy spectra. Batchelor (B4, B5) presented an excellent monograph (B6) which discusses the more mathematical aspects of a microscopic description of homogeneous turbulence. However, the present status of the statistical theory of turbulence does not permit the prediction of behavior under many of the conditions of fluid flow encountered in chemical engineering practice.

B. Physical Manifestations of Homogeneous Turbulence

The recent advances in turbulence theory have been so rapid and the experimental investigations sufficiently extensive that it is beyond the scope of the present discussion to consider these matters in detail. However, before the influence of turbulence upon thermal and material transport is considered, it is worth while to describe briefly some of the characteristics of turbulent flow from an experimental point of view. The description will be limited to the characteristics of steady uniform flow

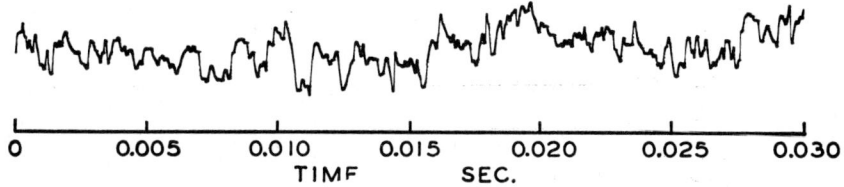

Fig. 1. Fluctuating longitudinal velocities in turbulent flow.

since little of a quantitative nature is known concerning the growth and decay of turbulence (K5). The recent experimental studies by Laufer in a cylindrical conduit (L3) have contributed much to an understanding of the variations in the microscopic properties of turbulence from point to point in nearly uniform, steady, shear flow of air.

In the interest of clarifying the discussion a number of graphical presentations of experimental data have been included. As a result of their ready availability many of these diagrams have been taken from the authors' publications.

C. Statistical Nature of Turbulence

Figure 1 presents measurements of the longitudinal fluctuations in velocity in a uniform, steady air stream as a function of time near the center of the channel, where the turbulence is essentially isotropic. An effort has been made to retain a linear scale of velocity so that the magnitude of the fluctuation may be visualized. Schubauer (S9) made these measurements by means of a hot-wire anemometer (K6) and they indicate the effectiveness of modern anemometric techniques in determining the microscopic behavior of turbulence. Figure 2 presents Laufer's experimental measurements (L3) of the longitudinal spectrum of the kinetic energy of the turbulence in shear flow as a function of wave number. These data were obtained by comparison of the energies associated with the fluctuating velocities observed by means of a hot-wire anemometer

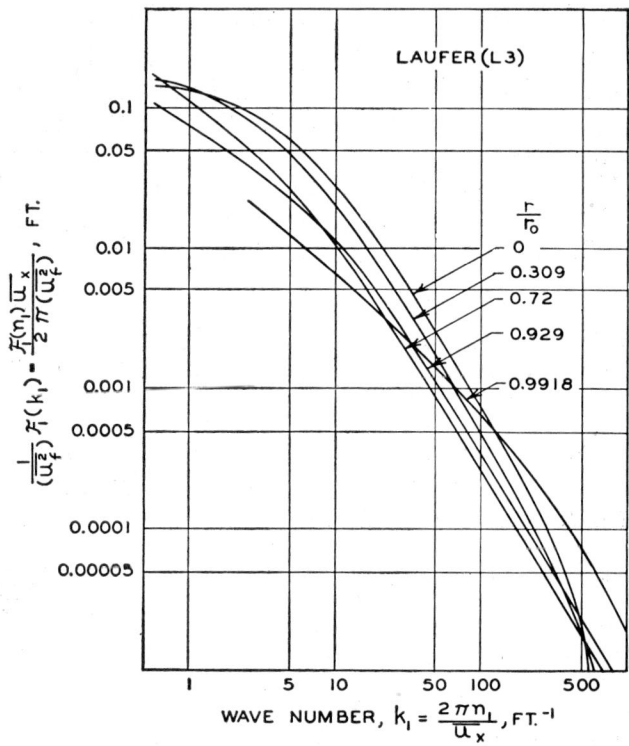

FIG. 2. Kinetic-energy spectrum of turbulence in shear flow.

with the output of a white noise generator using an electronic wave analyzer and a vacuum thermocouple. In the field of noise the work of Rice (R3) contributed to the basis for this analysis. The work of Kolmogoroff (K10) in regard to transfers of kinetic energy between the wave numbers of turbulence indicates that there should be an exchange of kinetic energy from the low wave numbers to the large wave numbers and that most of the kinetic energy in most shear flows is added to the turbulent stream at low wave numbers corresponding to the larger eddies (B6, K10).

D. Fluctuating Velocity

In the description of turbulent fluctuations it has been useful to employ the nomenclature and approach developed by Reynolds (R2). In this instance the instantaneous velocity is made up of two terms, the time-average velocity and the fluctuating velocity as indicated in the following expressions:

$$u_{x,i} = \bar{u} + u_{x,f} \tag{1}$$

$$\lim_{\theta \to \infty} \frac{1}{\theta} \int_0^\theta u_{x,f} d\theta \to 0 \tag{2}$$

Since velocity is a vector quantity, it is usually necessary to identify the component of the velocity, as was done for the rectangular Cartesian coordinate system in Eq. (1). The value of the integral as it differs from zero may be employed as a measure of the accuracy with which average characteristics (K1) of the stream may be used to describe the macroscopic aspects of turbulence. Such methods do not yield results of practical significance when applied to the solution of the Navier-Stokes equations.

E. Correlations

In the consideration of the statistical aspects of turbulence it was found to be of utility (B6, K4, R1) to establish the correlation in time (B6) of the velocity vectors as a function of the distance between two points in the turbulent stream. The correlation coefficient is defined by

$$R_y = \frac{\overline{u_{x,f} u_{y,f}}}{\overline{u_f^2}} \tag{3}$$

An estimate of the effect of separation of the points upon the correlation coefficient is given in Fig. 3 (C7). Batchelor (B6) has been able to predict many of the basic characteristics of the correlation coefficient shown in Fig. 3 for both the transverse and longitudinal fluctuating velocities. Much has been written about the characteristics of double, triple, and in a few cases higher correlations (K4, L5). It is beyond the scope of this discussion to consider these more refined measures of the statistical characteristics of turbulence. It suffices to indicate that at present a reasonable beginning has been made in the evaluation of the microscopic characteristics of turbulence but that much more experimental work must be carried out in order to supply the quantitative information required to make the extensive theoretical effort capable of quantitative application.

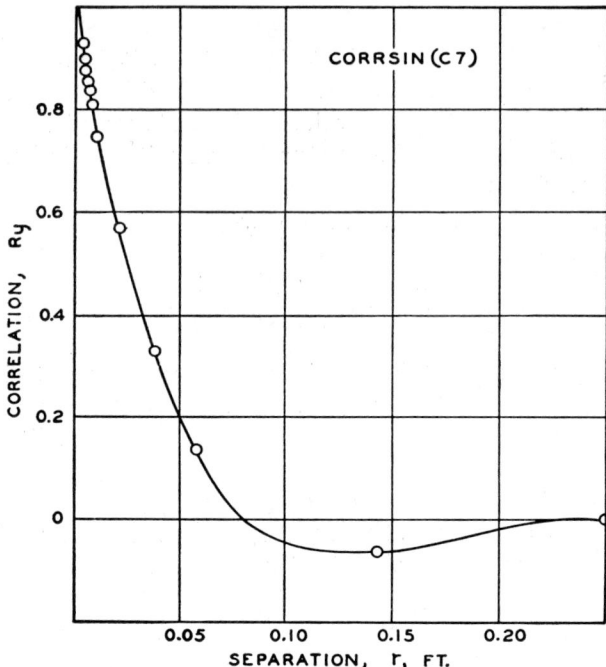

Fig. 3. Variation in correlation coefficient with distance.

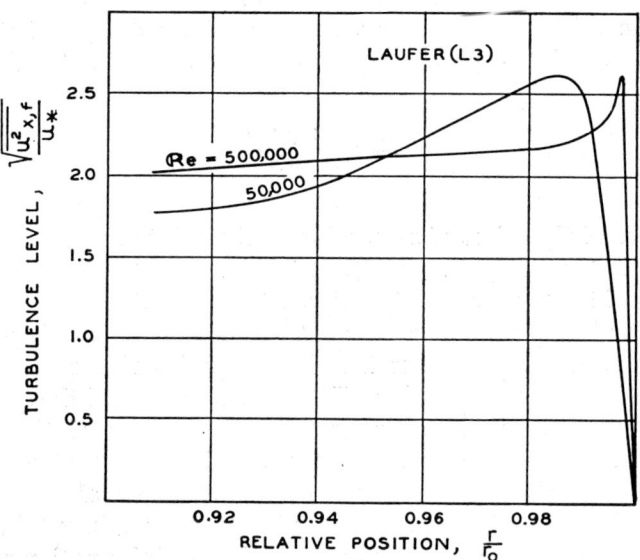

Fig. 4. Effect of position on longitudinal fluctuations in velocity.

F. Fluctuating Velocity

Figure 4 depicts the variation of the root-mean-square longitudinal fluctuation as a function of position in a flowing stream. These data were taken from the recent experimental investigation by Laufer (L3) and illustrate the complexity of behavior encountered in a steady, uniformly flowing turbulent stream. It is to be expected that fluctuations of temperature and composition are encountered in turbulent streams involving thermal or material transport.

III. Macroscopic Aspects of Steady, Uniform, Turbulent Flow

A. Definition of Eddy Properties

Because the quantitative analysis of transport processes in terms of the microscopic description of turbulence is difficult, Kármán suggested (K2) the use of a macroscopic quantity called "eddy viscosity" to describe the momentum transport in turbulent flow. This quantity, which is dimensionally and physically analogous to kinematic viscosity in the laminar motion of a Newtonian fluid, is defined by

$$\epsilon_m = \frac{\tau g}{\sigma}\left(\frac{dy_d}{du}\right) - \nu = \epsilon_m - \nu \qquad (4)$$

Equation (4) applies to the steady flow of a fluid between parallel plates. The velocity is the time average for the flow at the point in question. A corresponding definition of eddy viscosity in a circular conduit may be evolved in which the radial deficiency is substituted for the distance from the wall in Eq. (4).

B. Velocity Distribution

According to Eq. (4) the eddy viscosity is indeterminate at the plane of symmetry since both the shear and the velocity gradient approach zero as this plane is reached. However, the eddy viscosity at the plane of symmetry can be evaluated (L10) from the derivative of velocity with respect to the square of the distance from the plane of symmetry. This derivative is different from zero at the plane of symmetry and is readily determined from experimental determination of the velocity profile.

1. *Boundary Flow*

Near the boundary of steady, uniform streams it has been customary (B2) to utilize the velocity and distance parameters defined by the following equations to correlate the velocity distribution in the boundary flow:

$$y^+ = \frac{y_d}{\nu}\sqrt{\frac{\tau_0 g}{\sigma}} = \frac{y_d u_*}{\nu} \qquad (5)$$

$$u^+ = \frac{u}{\sqrt{\frac{\tau_0 g}{\sigma}}} = \frac{u}{u_*} \qquad (6)$$

It has been found (S4) that at the higher Reynolds numbers u^+ is a single-valued function of y^+ (N1). Deissler (D2) proposed an analytical expression for the variation in y^+ with u^+ in the laminar and transition regions of the boundary flow (S4).

$$y^+ = \frac{1}{n} \frac{\frac{1}{\sqrt{2\pi}} \int_0^{nu^+} \exp\{-\tfrac{1}{2}(nu^+)^2\} d(nu^+)}{\frac{1}{\sqrt{2\pi}} \exp\{-\tfrac{1}{2}(nu^+)^2\}} \qquad (7)$$

This expression reduces to the following simple form at small values of y^+:

$$y^+ = u^+ \qquad (8)$$

Equation (8) may be derived (S4) from the assumptions of laminar flow and constant shear. If the variation in shear in the boundary flow is taken into account, Page (P3) indicated that the relationship of y^+ and u^+ assumes the form

$$u^+ = y^+\left[1 - \frac{y^+}{\sqrt{3Re}}\right] \qquad (9)$$

Figure 5 presents the relationship of the distance and velocity parameter based upon Deissler's equation (D2) and the relation (P3) for laminar flow together with an expression proposed for the transition region (D7), which is described by the following equation:

$$u^+ = \frac{1}{\sqrt{K_1}} \tanh(y^+ \sqrt{K_1}) = \frac{1}{0.0695} \tanh(0.0695 y^+) \qquad (10)$$

In addition, the behavior in the turbulent core (S5) was included in Fig. 5. For this region Deissler (D2) suggests the following expression:

$$u^+ = \frac{1}{0.36} \ln y^+ + 3.8 \qquad (11)$$

If it is desired to take into account the effect of shear for steady, uniform flow, Eq. (11) assumes the form (D2)

$$u^+ = \frac{1}{0.36}\left[\sqrt{1 - \frac{y^+}{y_0{}^+}} + \ln\left(1 - \sqrt{1 - \frac{y^+}{y_0{}^+}}\right)\right] \quad (12)$$

The marked deviation of the macroscopic experimental measurements of Laufer (L2) from those of Nikuradse (N1) and Skinner (S18), shown in a part of Fig. 5, particularly for the lower Reynolds numbers, led to

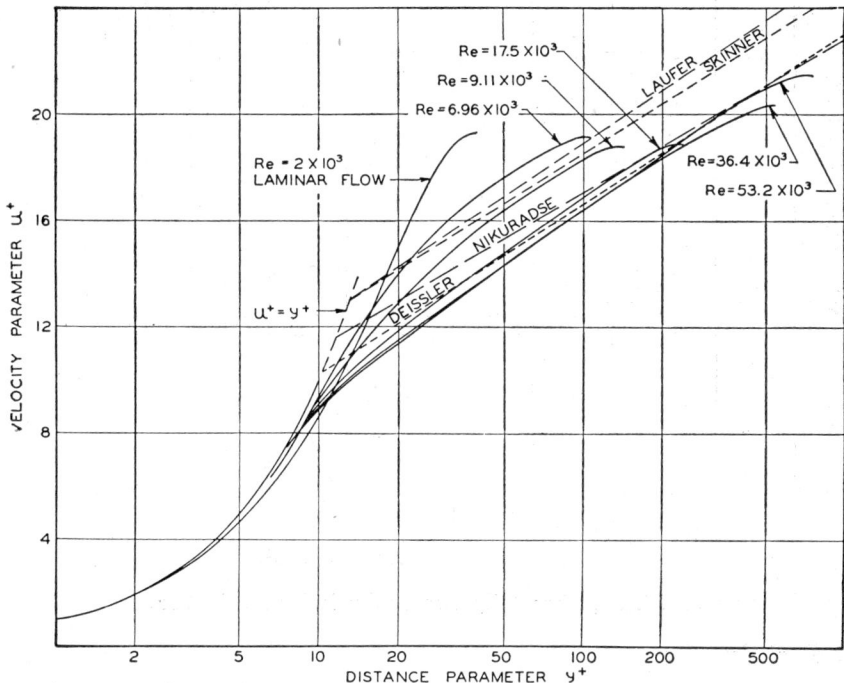

FIG. 5. Comparison of results from different investigators.

interest in further studies as to the single-valued nature of the relation of u^+ to y^+. Schlinger (S5) evaluated some earlier velocity measurements (P2) for air in turbulent, uniform flow and found a systematic variation in the relation of u^+ to y^+ at Reynolds numbers below 20,000. These results are shown in Fig. 6, which indicates that for Reynolds numbers below 35,000 a significant trend in the relationship of u^+ to y^+ with conditions of flow is experienced. The relation of u^+ to y^+ for laminar flow at a Reynolds number of 2,000 was included in Fig. 6. In laminar flow the velocity parameter u^+ is related to the distance parameter y^+ as indicated in Eq. (9). As is shown in Fig. 5 there exists a significant dispersion among the data of the several investigators. The information presented in Fig. 6 at Reynolds numbers above 35,000 is in good agreement with the correlation of Deissler (D2) for the turbulent core and the transition region as described by Eqs. (7) and (11).

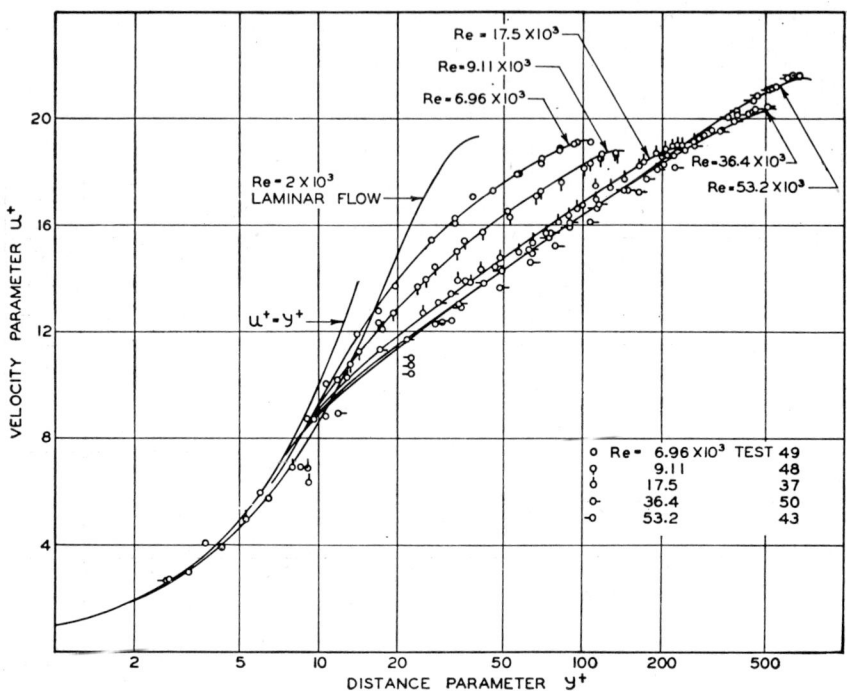

Fig. 6. Experimentally measured velocity distribution.

2. *Turbulent Core*

In the turbulent core it has been conventional to employ the velocity deficiency (B2) as a single-valued function of the relative position in the channel in order to correlate the velocity distribution outside the boundary flow. The velocity deficiency is defined by

$$\frac{u_m - u}{u_*} = \frac{u_m - u}{\sqrt{\frac{\tau_0 g}{\sigma}}} \tag{13}$$

It was recently found (S5) that there may exist a small influence of Reynolds number upon the relationship of the velocity deficiency and position for steady, uniform flows at Reynolds numbers below 35,000. Such behavior is indicated in Fig. 7. The data at the higher Reynolds numbers are in agreement with the measurements of Nikuradse (N1) and Skinner (S18). As a matter of interest the behavior in laminar flow at a Reynolds number of 2,000 was included.

C. Eddy Viscosity

Eddy viscosity was defined for steady, uniform flow by Eq. (4). The original concepts of eddy properties were presented by Reynolds although

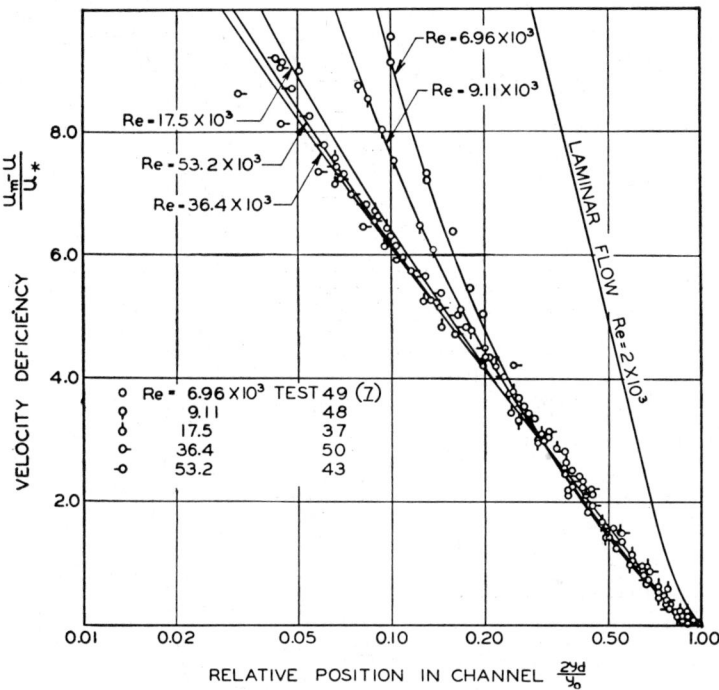

Fig. 7. Experimental values of velocity deficiency.

Kármán (K3) and Prandtl (P7) indicated analogies between momentum and thermal transport and placed the matter before the engineering profession. Using Kármán's original analysis (K1) Deissler (D2) indicated that on the basis of the similarity hypothesis the eddy viscosity may be evaluated for steady, uniform flow in the following way:

$$\epsilon_m = k^2 \frac{\left(\dfrac{du}{dy_d}\right)^3}{\left(\dfrac{d^2u}{dy_d^2}\right)^2} \tag{14}$$

Equation (14) may be deduced from dimensional analysis if it is assumed that only the first and second derivatives of velocity are important. From this expression it follows that the eddy viscosity depends only on position and local velocity for steady, uniform flow. It is convenient, however, to consider it a function of position and Reynolds number.

$$\epsilon_m = \phi_1(u, y_d) = \phi_2(Re, y_d) \tag{15}$$

D. Relative Viscosity

From a combination of Eqs. (4), (10), and (11) it follows, according to Schlinger (S7), that the relative viscosity ϵ_m/ν may be evaluated from

$$\begin{aligned}\frac{\epsilon_m}{\nu} &= \frac{\epsilon_m + \nu}{\nu} = \frac{l}{l_0} \cosh^2\left[0.0695 \frac{l_0 - l}{\nu}\right]\sqrt{\frac{\tau_0 g}{\sigma}} \\ &= \frac{l}{l_0} \cosh^2\left[0.0695 \frac{l_0 - l}{\nu}\right] U \sqrt{\frac{f}{2}}\end{aligned} \tag{16}$$

Equation (16) describes the relationship of the total viscosity ε_m and the kinematic viscosity ν throughout the transition region and yields a continuous value of the eddy viscosity with respect to position.

In the central core for Reynolds numbers above 35,000 an expression analogous to Eq. (11) but based on different experimental information may be written (S7) as

$$u^+ = 5.5 + \frac{1}{0.4} \ln y^+ \tag{17}$$

Equation (17) yields values similar to the expression suggested by Deissler (D2) and shown as Eq. (11). Schlinger (S7) indicates that if the velocity distribution is described by Eq. (17), the relative viscosity is given by

$$\frac{\epsilon_m}{\nu} = \frac{0.4}{\nu}\sqrt{\frac{\tau_0 g}{\sigma}}\left(\frac{l}{l_0}\right)(l_0 - l) \tag{18}$$

As indicated earlier, the logarithmic velocity distribution, Eq. (7), yields a zero value for the relative viscosity in the center of the channel for all values of Reynolds number. Such behavior is unreasonable in view of available information concerning eddy viscosity and eddy conductivity (P1, P3) at the plane of symmetry. For this reason Eq. (18) should not be used at values of l/l_0 less than 0.3. Figure 8 shows the relative viscosity as a function of position in a circular conduit for several different Reynolds numbers. A comparison of total viscosity as established from Eqs. (16) and (18) and as measured directly (C3) is presented in Fig. 9. Reasonable agreement is obtained between the predicted and directly measured total viscosity.

A tabulation of relative viscosity based upon Eqs. (16) and (18) is available (C5) for steady, uniform flow in pipes and between infinite parallel plates. As is seen in Fig. 9, the predicted values of the total viscosity do not yield continuous values of the first derivative of the eddy viscosity with respect to position, such as is found to be the case for the experimental data. As a first approximation, the predictions of Eqs. (16) and (18) yield a reasonable description of the variation of the transport

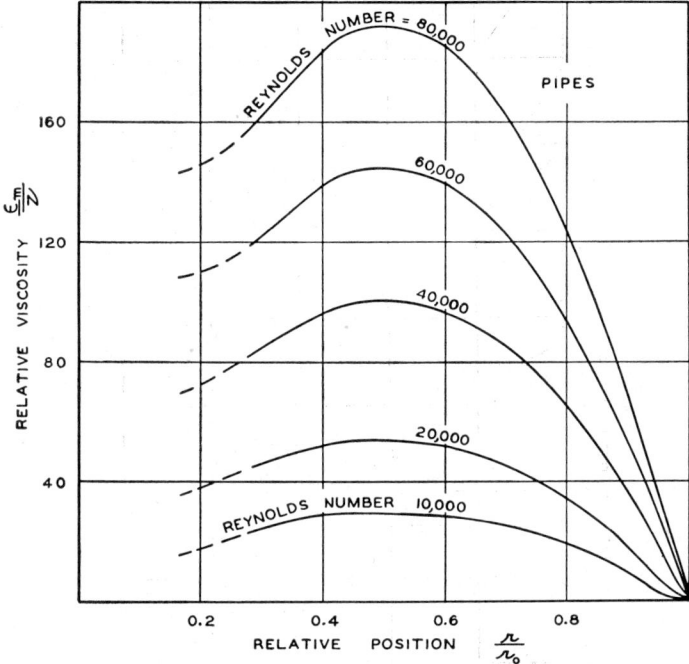

FIG. 8. Relative viscosity as a function of position in flow channel.

of momentum in turbulent flow. The decay of turbulence (K5) and its generation (M6) in nonuniform flow and in shear flow remain among the more important practical problems to be solved in the prediction of turbulent transport phenomena under the conditions encountered in chemical engineering practice. As yet it is difficult to predict the effect of nonuniform or nonsteady flow on the momentum transport.

The foregoing discussion relates to the transport of momentum under steady, uniform conditions in turbulently flowing streams. Such matters are of direct interest to the chemical engineer but usually only as they influence the power requirements for the movement of fluids. A deeper interest exists in prediction of the thermal transport and temperature distributions in turbulently flowing fluids. In the study of the behavior of such systems the characteristics in situations where the convective thermal transport is nonuniform (P4) are often of particular interest. The greater part of the early interest of chemical engineers in convective thermal transport was concerned primarily with local pseudouniform conditions of a macroscopic nature. Such methods of analysis are of

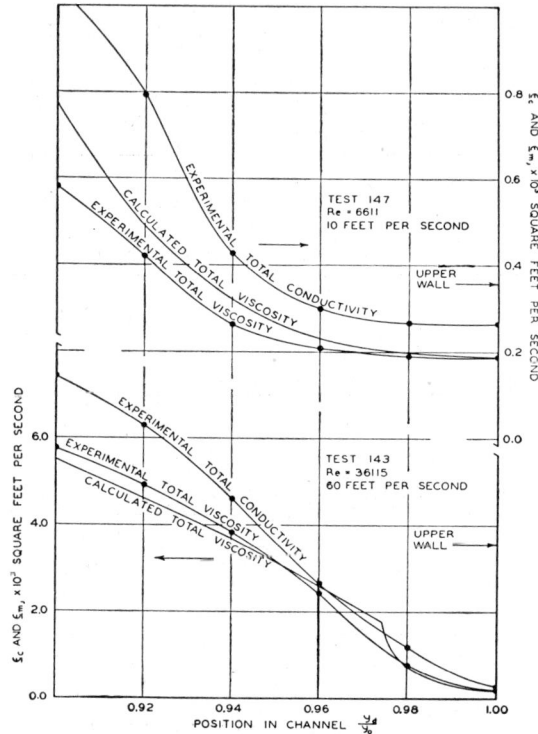

FIG. 9. Comparison of predicted and experimentally measured total viscosity.

direct usefulness and were discussed extensively in the work of McAdams (M1).

With the increasing interest in the microscopic behavior of fluids during processing operations, methods for predicting the temperature distribution in turbulently flowing fluids become of importance. Thermal transport in a turbulent stream is in large measure the result of the transport of internal energy arising from the turbulent motion of the fluid, and thus the transfer is not strictly heat (G1) in the thermodynamic sense. As has been seen, the momentum transport in a turbulent stream is markedly dependent on the Reynolds number and position under steady, uniform conditions of flow. Since the turbulent local motion of the fluid exerts a strong influence on both momentum and thermal transfer, it is to be expected that the eddy conductivity (K2) like the eddy viscosity will depend on position and Reynolds number.

IV. Thermal Transport in Turbulent Flow

A. Introduction

The present approach to the prediction of thermal transport in turbulent flow neglects the effect of thermal flux and temperature distribution upon the relationship of thermal to momentum transport. The influence of the temperature variation upon the important molecular properties of the fluid in both momentum and thermal transport may be taken into account without difficulty if such refinement is necessary.

In a description of thermal transport in steady, uniform flow it is necessary to consider the conservation of internal and kinetic energy (K7) under conditions involving gross motion of the fluid. For laminar flow

$$\frac{\partial E}{\partial \theta} + u_x \frac{\partial E}{\partial x} = -V \frac{\partial \mathring{Q}_x}{\partial x} - V \frac{\partial \mathring{Q}_y}{\partial y} + \nu \left(\frac{\partial u_x}{\partial y}\right)^2 \qquad (19)$$

Equation (19) applies to the conditions at a point in two-dimensional laminar flow. Under steady, uniform conditions it reduces to

$$\frac{\partial \mathring{Q}_y}{\partial y} = -\frac{\partial}{\partial y}\left(k \frac{\partial T}{\partial y}\right) = \eta \left(\frac{\partial u_x}{\partial y}\right)^2 \qquad (20)$$

Equation (20) assumes the following form in the case of steady, uniform flow in a circular conduit:

$$\frac{1}{r}\frac{\partial r \mathring{Q}_r}{\partial r} = -\frac{1}{r}\frac{\partial}{\partial r}\left(rk \frac{\partial T}{\partial r}\right) = \eta \left(\frac{\partial u_x}{\partial r}\right)^2 \qquad (21)$$

It is apparent that Eqs. (19) to (21) inclusive are based upon the premise that local equilibrium (K7) prevails. This premise will be assumed throughout the remaining part of this discussion. It should be emphasized that Eqs. (20) and (21) include changes in the kinetic energy of the fluid entering or leaving the volume element (S12). The effect of changes in kinetic energy is frequently assumed to be negligible. Radiant transport of energy is neglected since such a simplification appears acceptable in many situations. Extension of such energy balances to include radiant transport and changes in internal energy as a result of chemical reaction may be readily incorporated.

B. Eddy Conductivity

Reynolds (R1) suggested that the natures of turbulent momentum and thermal transport were similar. Kármán (K2) extended this analysis and defined eddy conductivity in the following way for steady, uniform, two-dimensional flow:

$$\epsilon_c = \varepsilon_c - \kappa = \frac{\mathring{Q}_y}{C_p \sigma} \frac{dy}{dt} - \kappa \tag{22}$$

The analogy between Eqs. (4) and (22) is obvious and a similar analogy can be made with regard to material transport. Measurements of the eddy

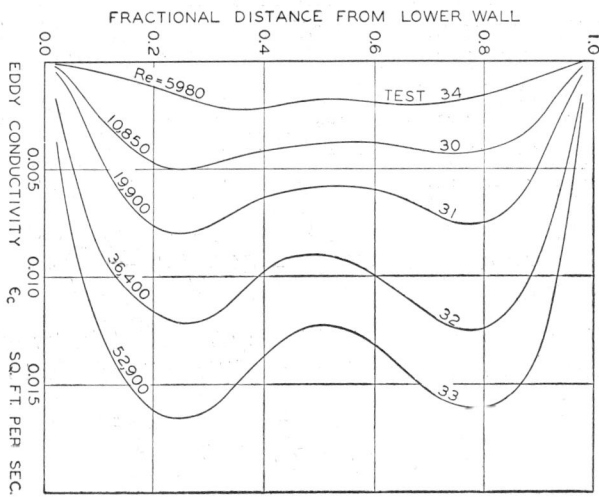

Fig. 10. Influence of Reynolds number and position upon eddy conductivity.

conductivity for air at atmospheric pressure were made (C3, P1, P3) under such conditions that both the momentum and thermal transport were substantially steady and uniform. The eddy conductivity is shown in Fig. 10 as a function of position. It increases with an increase in Reynolds number and decreases rapidly as the wall of the channel is approached. Figure 11 depicts similar information with Reynolds number as the independent variable. This diagram indicates a nearly linear increase in the eddy conductivity with an increase in Reynolds number. Near the wall it is convenient to use relative conductivity, which is the ratio of total conductivity, defined in Eq. (22), to thermometric conductivity. The product of this ratio and a linear function of position is portrayed as a single-valued function of the distance parameter y^+ in Fig. 12. The experimental data shown in this figure cover Reynolds numbers from 6,000 to 60,000, corresponding to a nominal gross velocity

TURBULENCE IN THERMAL AND MATERIAL TRANSPORT 257

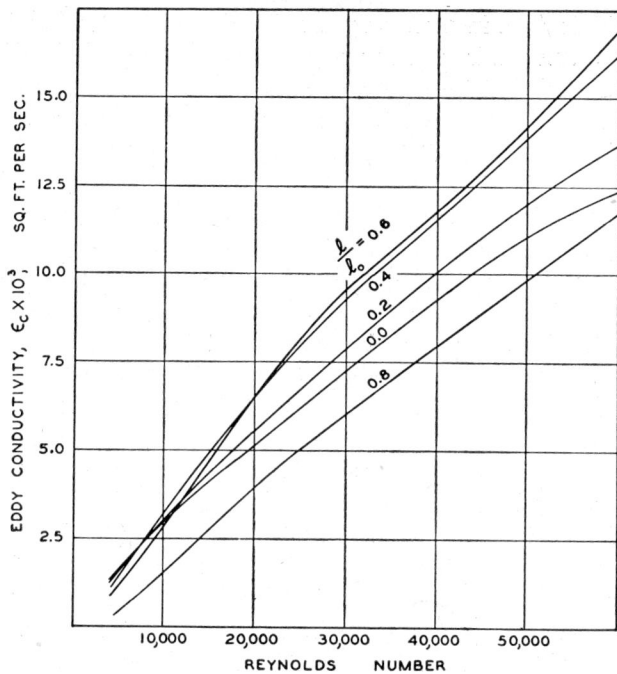

Fig. 11. Effect of Reynolds number upon eddy conductivity.

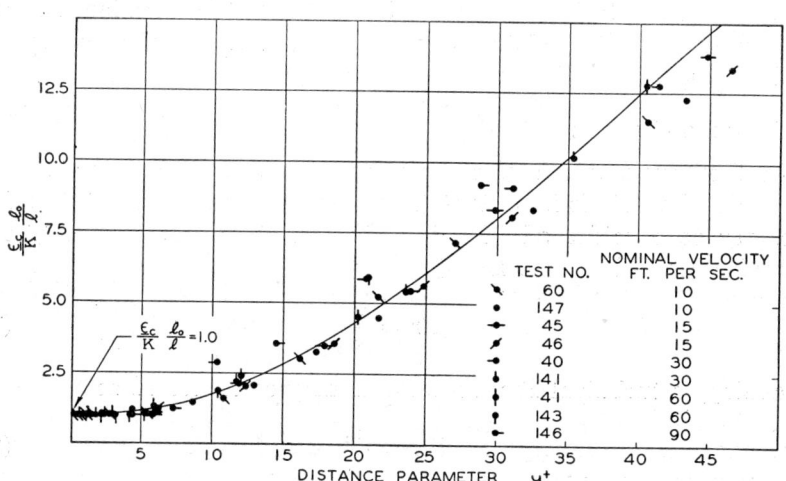

Fig. 12. Effect of distance parameter upon average relative conductivity.

from 10 to 90 ft./sec. in a rectangular channel 0.75 in. in width for air at atmospheric pressure. In the course of these experimental measurements (C3) it was found that the experimental data near the lower wall are less reproducible than those near the upper wall. For example, the standard deviation of the experimental data for the lower wall from a smooth curve, such as is shown in Fig. 12, was nearly three times that for the upper wall. Until the effect of temperature gradients upon turbulence is better understood, it does not appear worth while to attempt to account for the small differences in behavior observed between the two walls of the conduit.

The work of Laufer (L3) indicates that eddy viscosity is not isotropic in shear flow. For this reason it is unlikely that eddy conductivity is isotropic in such flows. Therefore, uncertainties in the application of eddy conductivities must arise when it is assumed that this transport coefficient is isotropic. Until additional experimental information is available, it appears reasonable to consider eddy conductivity as isotropic except in circumstances when the vectorial nature (J4, R2) of the eddy viscosity may be estimated. Such an approximation appears acceptable since the measurements available described the conductivity normal to the axis of flow, which is the direction in which most detail is required in the prediction of temperature distribution in turbulently flowing streams. Throughout the remainder of this discussion all eddy properties will be treated as isotropic. Such a simplification is open to uncertainty, and further experimentation will be required in order to determine the error introduced by neglect of the vectorial characteristics of these macroscopic transport quantities.

C. Prandtl Number

In the consideration of thermal transport the molecular Prandtl number of the fluid is important (M1). However, in turbulent flow it is necessary to consider additional Prandtl numbers. The conventional molecular Prandtl number is defined by

$$Pr = \frac{\nu}{\kappa} \tag{23}$$

By analogy a turbulent Prandtl number may be expressed as

$$Pr_\epsilon = \frac{\epsilon_m}{\epsilon_c} \tag{24}$$

From Eqs. (23) and (24), a total Prandtl number becomes

$$\mathbf{Pr} = \frac{\epsilon_m + \nu}{\epsilon_c + \kappa} \tag{25}$$

The Reynolds analogy is equivalent to setting the turbulent Prandtl number as defined in Eq. (24) equal to unity. Figure 13 shows the effect of Reynolds number upon a space average value of the turbulent Prandtl number (C3, P3, S7). Information presented in Fig. 13 is open to uncertainty since it is based upon measurements for air and represents only the space average value of this ratio throughout the turbulent portion of the stream. The turbulent Prandtl number is undoubtedly a function of position as well as of the Reynolds number for a given stream (P1).

Jenkins (J5) made some preliminary estimates of the effect of molecular Prandtl number upon eddy Prandtl number for intermediate Reynolds

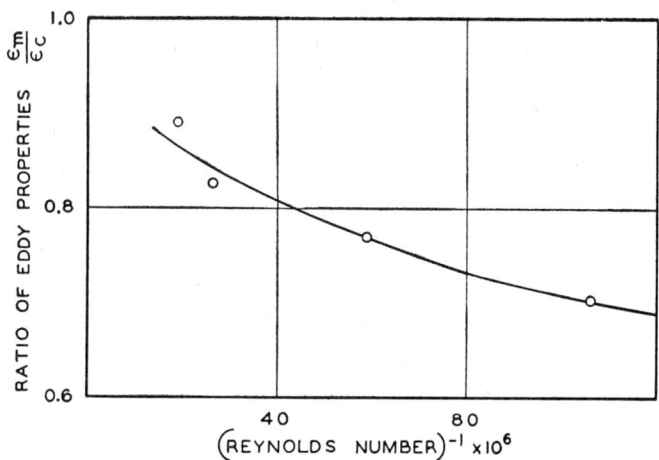

FIG. 13. Effect of Reynolds number upon ratio of eddy properties.

numbers below 100,000. Using these speculative calculations (J5) and the limited experimental information of Fig. 13, one may correlate the eddy Prandtl number as a function of the molecular Prandtl number and the ratio of the eddy and kinematic viscosities. Such estimates are shown in Fig. 14. It should be emphasized that the data of Fig. 14 are speculative and require extensive experimental confirmation. As a matter of interest the behavior predicted upon the assumption of the Reynolds analogy was included.

D. PREDICTION OF THERMAL TRANSPORT

In predicting convective thermal transport to turbulent streams it has usually been sufficient to determine the corresponding thermal flux at the boundary for a specified area. Such methods have been refined by many workers and ably summarized by McAdams (M1) and Jakob (J1).

For many purposes it is desirable to evaluate the local thermal flux at the boundary and at various points in the stream. A prediction of the temperature as a function of the spatial coordinates of the system is also of interest particularly in connection with conditions involving chemical reactions. It is beyond the scope of this discussion to consider in detail the recent developments in thermal transport from a macroscopic standpoint. The literature is replete with empirical correlations which permit the

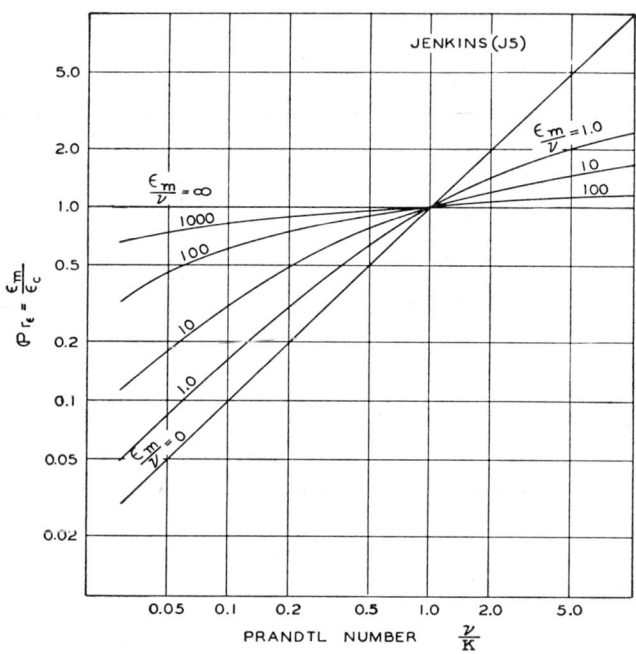

FIG. 14. Effect of molecular Prandtl number and Reynolds number upon the total Prandtl number.

prediction of the quasiuniform thermal flux at the boundary of a stream as a function of the molecular properties of the fluid and the characteristics of the flowing stream.

1. *General Energy Equation*

Pannell (P4) and Jurges (J8) laid the basis for such predictions and more recently analogue computers have been applied to this problem (J6, S6). In principle, such predictions involved the solution of the general energy equation which may be written in the following form for two-dimensional, uniform, turbulent flow:

$$\frac{\partial}{\partial x}\left(C_{p\sigma}\epsilon_c \frac{\partial t}{\partial x}\right) + \frac{\partial}{\partial y}\left(C_{p\sigma}\epsilon_c \frac{\partial t}{\partial y}\right) + \frac{\partial}{\partial z}\left(C_{p\sigma}\epsilon_c \frac{\partial t}{\partial z}\right)$$
$$- u_x C_{p\sigma}\frac{\partial t}{\partial x} - u_y C_{p\sigma}\frac{\partial t}{\partial y} - u_z C_{p\sigma}\frac{\partial t}{\partial z} = C_{p\sigma}\frac{\partial t}{\partial \theta} \quad (26)$$

In the case of steady flow the term on the right side of the equality sign is zero. Equation (26) is limited to situations in which changes in elevation, kinetic energy, and pressure may be neglected. In addition, no regard is taken of the gain in internal energy resulting from the dissipation of fine-grain turbulence (K10). In the case of the steady, uniform flow of an incompressible fluid between parallel plates with a nonuniform temperature field, Eq. (26) assumes the following simple form:

$$\frac{\partial}{\partial y}(\epsilon_c + \kappa)\frac{\partial t}{\partial y} = u_x \frac{\partial t}{\partial x} \quad (27)$$

Equation (27) may be rewritten for steady, uniform flow in a circular conduit as

$$\frac{1}{r}\frac{\partial}{\partial r}[r(\epsilon_c + \kappa)]\frac{\partial t}{\partial r} = u_x \frac{\partial t}{\partial x} \quad (28)$$

Equation (28) assumes axial symmetry but otherwise is the equivalent of Eq. (27). In the case of chemical reactions Eq. (27) may be rewritten as

$$\frac{\partial}{\partial y}(\epsilon_c + \kappa)\frac{\partial t}{\partial y} - u_x \frac{\partial t}{\partial x} = \mathring{m}_k \frac{\Delta H}{C_p} \quad (29)$$

At present it is difficult to predict the rate constant (K7) for many chemical reactions and even the proper potential to employ is open to discussion. In Eq. (29) the quantity $\mathring{m}_k$ is the rate of formation of component k per unit weight of phase and ΔH is the enthalpy change associated with the formation of a unit weight of component k from the chemical reaction. It should be emphasized that in Eq. (29) many extraneous effects associated with the change in specific weight during reaction under isobaric, isothermal conditions have been neglected. However, the solution of Eq. (29) when applied to a particular boundary condition will permit an estimation of the temperature distribution in a reacting turbulent stream. Sufficient information concerning the kinetics of the reaction must be available to permit the rate of formation of a specified component for each significant reaction to be estimated.

An example of the application of Eq. (27) to the prediction (S7) of the temperature distribution in a nonreacting turbulent flow is shown in Fig. 15. Reasonable agreement between the predicted and experimentally evaluated temperatures was found. Similar accuracy of prediction was found for a variety of situations. In the present instance the average deviation between the predicted and measured temperature was 0.12° F.

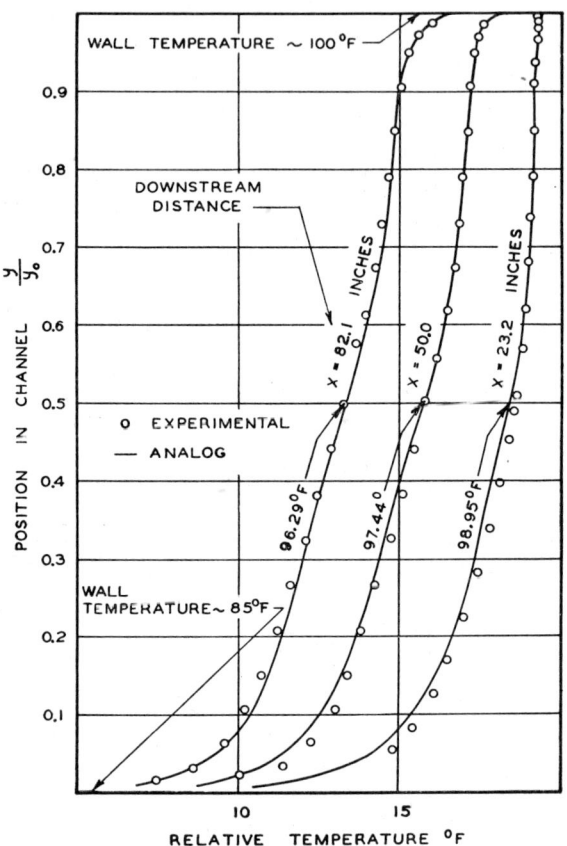

FIG. 15. Temperature distribution in a two-dimensional air stream.

The physical arrangement associated with this prediction of air temperature is shown in Fig. 16. Solutions of Eq. (27) such as were shown in Fig. 16 were obtained by the use of an electrical analogue (S7).

2. *Boundary Conditions*

In the solution of these expressions it is important to consider the boundary conditions to be employed. In the foregoing example the effect of the temperature distribution upon the steady state velocity distribution was neglected. Under these circumstances the velocity distribution was considered uniform and the eddy conductivity was treated as a function only of the Reynolds number of the flow and the position in the cross section of the channel. In the prediction of thermal transport under conditions where the transport of momentum is nonuniform, as is the case for flow around drops of liquid or across cylinders, the solution of Eq. (26) is much more difficult. The eddy viscosity and conductivity are functions of position in the wake with respect to all the pertinent coordinate axes. At the present state of knowledge of thermal transport in turbulent flow it does not appear worth while to attempt the prediction of the distribution of temperature or thermal flux as a function of the spatial coordinates

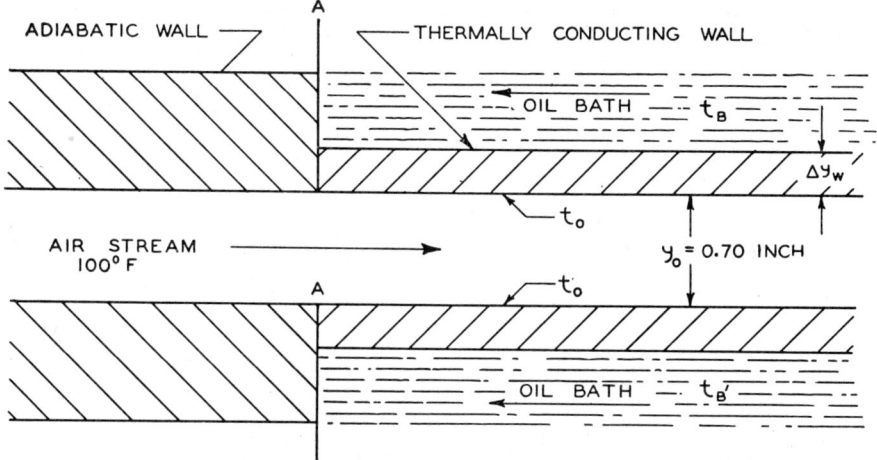

FIG. 16. Schematic diagram of two-dimensional air streams.

of the system for unsteady or nonuniform flow. The primary difficulty in making such predictions arises from the determination of the rate of generation and decay of turbulence in nonuniform or unsteady flow.

3. *Analytical Solutions*

Deissler (D3) recently extended the analysis of thermal and material transfer associated with turbulent flow in tubes to include the behavior of fluids with high molecular Prandtl and Schmidt numbers. If the variation in molecular properties of the fluid with position are neglected, the following expression for the temperature distribution was suggested (D3):

$$t^+ = \int_0^{y^+} \frac{dy^+}{\frac{1}{Pr} + n^2 u^+ y^+ (1 - \exp\{-n^2 u^+ y^+\})} \qquad (30)$$

In Eq. (30) the relationship of u^+ and y^+ may be established from

$$u^+ = \int_0^{y^+} \frac{dy^+}{1 + n^2 u^+ y^+ (1 - \exp\{-n^2 u^+ y^+\})} \qquad (31)$$

Equation (31) is similar to Eq. (7) except that it takes into account the effect of kinematic viscosity on the eddy properties near the wall. An iterative solution is required for the solution of Eqs. (30) and (31). Throughout Deissler's recent analysis (D3) the Reynolds analogy has been assumed, and throughout the turbulent core Deissler shows that

$$u^+ - u_1^+ = t^+ - t_1^+ \qquad (32)$$

In Eq. (32) the subscript 1 refers to the location of the transition between the boundary flow and the turbulent stream. It appears that these analytical considerations by Deissler represent improvements over his earlier approach (D2). For the boundary flow close to the wall the foregoing expressions are based on the assumption that the eddy properties may be evaluated from

$$\epsilon_m = \epsilon_c = \eta^2 u y \left(1 - \exp\left\{-\frac{n^2 u y}{\nu}\right\}\right) \qquad (33)$$

In dimensionless form Eq. (33) becomes

$$\frac{\epsilon_m}{\nu} = n^2 u^+ y^+ (1 - \exp\{-n^2 u^+ y^+\}) \qquad (34)$$

Figure 17 shows generalized relationships of the temperature parameter t^+ to the position parameter y^+ for the various values of the molecular Prandtl number.

In the case of macroscopic transport the foregoing analysis yields the following expressions in generalized variables for the Nusselt number:

$$Nu = \frac{2r_0 h}{k_0} = \frac{2r_0^+ Pr}{t_b^+} \qquad (35)$$

Expressing the quantities included in Eq. (35) in terms of the Reynolds number, Deissler obtained the information presented in Fig. 18. Using an approximation for t_b^+ (D3), one may express the Stanton number as follows:

$$St = \frac{2n}{\pi} \frac{\sqrt{f}}{Pr^{3/4}} = \frac{Nu}{RePr} \qquad (36)$$

Deissler (D3) reported good agreement between the predictions of Eq. (36) and experiment as applied to both material and thermal transport

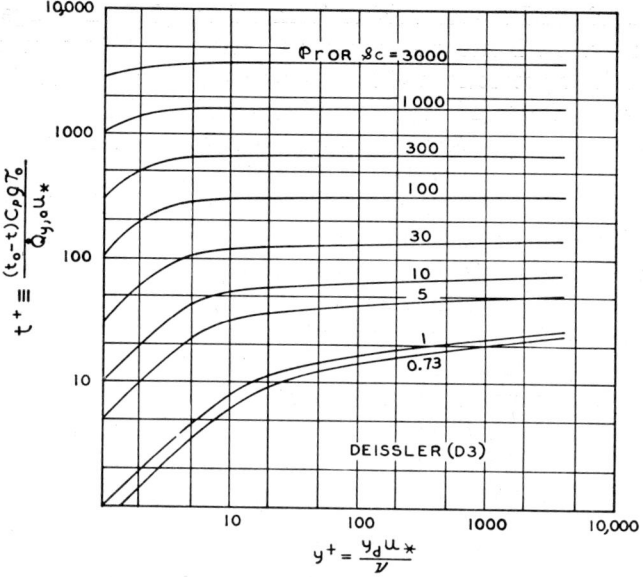

Fig. 17. Generalized temperature distribution in uniform, steady, turbulent flow.

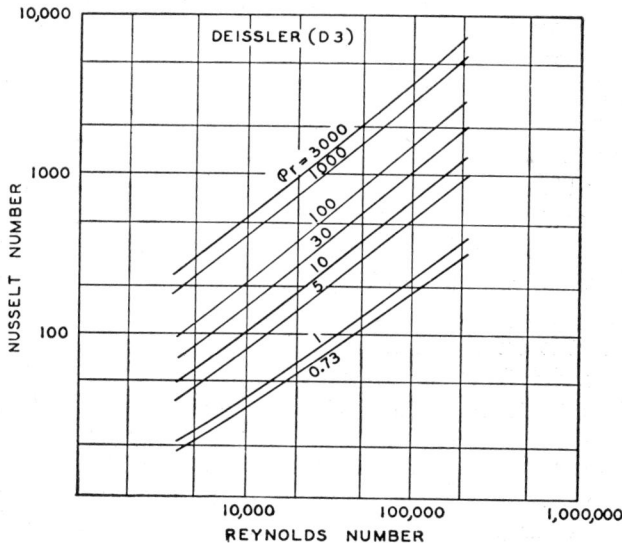

Fig. 18. Predicted influence of Reynolds number upon Nusselt number.

over ranges of the Prandtl number from 0.5 to 3,000 and large values of Reynolds number. These methods of analysis are in fair agreement with the much earlier predictions of Murphree (M8) but deviate significantly from the work of Kármán (K3) and Summerfield (S23). The agreement with experiment for macroscopic thermal transfer over a wide range of molecular Prandtl numbers lends credence to Deissler's analysis, which is based upon the Reynolds analogy, dimensional considerations, and the similarity hypothesis. Further analyses of the type carried out by Deissler, which take into account the effect of the Reynolds number and the molecular Prandtl number upon the total Prandtl number, should yield a rather satisfactory means of generalizing macroscopic thermal transport and predicting temperature distributions in situations where the turbulent flow may be treated as steady and uniform. Until further quantitative information is available concerning the growth and decay of turbulence, the application of such methods to conditions of unsteady or nonuniform turbulent flow appears subject to limitation.

E. Macroscopic Thermal Transfer

During recent years experimental work continued actively upon the macroscopic aspects of thermal transfer. Much work has been done with fluidized beds. Jakob (D5, J2) made some progress in an attempt to correlate the thermal transport to fluidized beds with transfer to plane surfaces. This contribution supplements work by Bartholomew (B3) and Wamsley (W1) upon fluidized beds and by Schuler (S10) upon transport in fixed-bed reactors. The influence of thermal convection upon laminar boundary layers and their transition to turbulent boundary layers was considered by Merk and Prins (M5). Monaghan (M7) made available a useful approach to the estimation of thermal transport associated with the supersonic flow of a compressible fluid. Monaghan's approximation of Crocco's more general solution (C9) of the momentum and thermal transport in laminar compressible boundary flow permits a rather satisfactory evaluation of the transport from supersonic compressible flow without the need for a detailed iterative solution of the boundary transport for each specific situation. None of these references bears directly on the problem of turbulence in thermal transport and for that reason they have not been treated in detail.

Fallis (F1) considered thermal transport in transitional and turbulent boundary flows and supplied a reasonable analysis of this difficult problem which is in agreement with the work of Eber (E1) and the theory of Eckert and Drewitz (E2). Callaghan (C1) contributed to the analogies between thermal and material transport in turbulent flow with particular emphasis upon the behavior near and in the boundary layer. The effect of surface roughness on thermal transport from elliptical cylinders at Reynolds numbers from 0.5 to 1.2×10^6 was investigated by Seban and co-workers (S11). The thermal transport to liquid metals in transition flow was investigated by Johnson and co-workers (J7). Further studies of the relationship of thermal and momentum transport in noncircular ducts were reported by Lowdermilk (L8). Pinkel (P5) reviewed the work of the NACA on thermal transfer for air flowing in tubes with large temperature differences. He confirmed that the evaluation of physical properties at the average bulk temperature of the fluid at the wall yielded satisfactory agreement with experiment in a Nusselt-type correlation for both circular and noncircular conduits.

Some additional work was done upon thermal transfer from cylinders in order to supplement the satisfactory correlations available (M1). The measurements of Cole and Roshko (C4) and of Brier and co-workers (B12) at normal pressures supplemented by the measurements of Bell (B7) for the free molecular flow of a nonuniform gas are of particular interest.

V. Material Transport in Turbulent Flow

A. INTRODUCTION

The transport of a component of a flowing system is a subject of distinct importance in many industrial processing operations. The following discussion relates to material transport in turbulent streams under conditions where no chemical reactions occur and where the influence of temperature and gravitational gradients can be neglected. Since the gross motion of the fluid is of particular importance, a complete understanding of the transport of momentum is usually necessary to permit an evaluation of the over-all material transport.

B. MACROSCOPIC TRANSPORT

Sherwood was one of the early workers to recognize the importance of turbulence (S15, S16, S17) in material transport. He summarized the progress in this field some years ago (S13) and contributed additional experimental work (L7, M2, M3). Kirkwood and Crawford (K7) set forth the relationships for transport in homogeneous phases with particular emphasis upon the interrelation of material and thermal flux. These contributions have laid a satisfactory basis for work in the field which has been well summarized from a macroscopic standpoint by Sherwood and Pigford (S14).

Drickamer contributed much to the knowledge of diffusion in gases and liquids. As an example, recent studies of the effect of pressure upon material transport in several inorganic systems have been made available

(C8, D4, K9). The influence of temperature upon material diffusion coefficients was studied by Strehlow (S22) and Krauss (K12). Some effort has been devoted to the analysis of diffusion problems assuming that the coefficient is some simple analytic function of composition (F3, S21). Molecular transport in a fluid moving at a uniform velocity was reported by Klinkenberg and co-workers (K8). The study of the evaporation of solids into a laminar air stream over a range of conditions of practical interest was reported by Butler and Plewes (B13), and Sherwood reported additional experimental information on material transfer between liquid phases (G3). Kronig considered material transport from the interior of falling drops (K13). Lynch and Wilke (L9) were concerned with the influence of the properties of the phase on material transfer in the gases.

1. *Effect of Turbulence*

The influence of turbulence was studied somewhat earlier by Bagotskaya (B1). Local coefficients of transport of water into an air jet were evaluated experimentally by Spielman and Jakob (S20). These essentially macroscopic studies supplement the investigations reported elsewhere in this discussion and contribute to the background of experimental information concerning material transport in fluid systems.

2. *Resistance at Interface*

There has been much interest in the possibility of the existence of a resistance to transport at the interface between liquid and gas phases. Schrage (S8) reviewed the theoretical aspects of this problem and indicated that such resistances are of primary importance at relatively low pressure. Drickamer and co-workers (T3, T4) considered interfacial resistances in binary and ternary systems and found that they may be of somewhat greater importance than was indicated from the theoretical studies of Schrage (S8). Recently Auer and Murbach (A1) made a number of analyses of the effect of interfacial resistance on the behavior in the adjacent phase. It has been the authors' experience with hydrocarbon systems that the transport of a heavy component into a gas phase or of a light component into a liquid phase across an interface does not involve resistance comparable with that associated with migration through a gas or liquid phase for distances several orders of magnitude greater than the mean free path. It appears possible that in dispersed heterogeneous systems where multiplicity of interfaces may be involved such effects may be of great practical importance.

C. Diffusion

Fick (F2) proposed one of the more useful relationships for describing molecular material transport. The Fick diffusion coefficient is identified by the following equation:

$$\mathring{m}_k - \sigma_k u = \sigma_k(u_k - u) = \sigma_k u_{d,k} = -D_{F,k}\frac{\partial \sigma_k}{\partial x} \tag{37}$$

Equation (37) also serves to define the diffusional velocity $u_{d,k}$ and the component velocity u_k. The velocity of the fluid is established from the properties of the phase as a whole in the following simple way:

$$u = \frac{\mathring{m}}{\sigma} \tag{38}$$

In Eq. (38) the hydrodynamic velocity is that used to evaluate the momentum of the phase. In viscous flow it is the term used in establishing the shear from a knowledge of the viscosity when effects of cross linking of fluxes (O1, O2) are neglected. The hydrodynamic velocity and the diffusional velocity are related by

$$u_k = u + u_{d,k} \tag{39}$$

A combination of Eqs. (37) and (39) results in

$$\mathring{m}_k = u_k \sigma_k = u\sigma_k - D_{F,k}\frac{\partial \sigma_k}{\partial x} \tag{40}$$

In the case of one-dimensional transport there follows from the conservation principle (K7):

$$\sum_{k=1}^{k=n} D_{F,k}\left(\frac{\partial \sigma_k}{\partial x}\right) = 0 \tag{41}$$

In the case of transport in the gas phase it is often convenient to use the Maxwell diffusion coefficient (M4), which is related to the Fick diffusion coefficient in the following way (O4) for a phase that may be treated as an ideal solution (L4):

$$D_{F,k} = \frac{\left(\dfrac{n_j}{\mathsf{n}_j}\right)\left(\dfrac{f_k^0}{P}\right)\left(\dfrac{Z^2}{P}\right)D_{M,k,j}}{1 + \mathring{\sigma}_k(V_j^0 - V_k^0)} \tag{42}$$

Under conditions of limiting small concentrations of component k and if the equation of state is taken as that of a perfect gas, Eq. (42) reduces to

$$D_{F,k} = \frac{D_{M,k,j}}{P} = D_{v,k} \tag{43}$$

It is possible to extend the consideration of molecular material transport in laminar flow but this does not seem worth while in the present discussion.

D. Laminar Flow

The conservation principle applied to unsteady, nonuniform, laminar, two-dimensional flow results in the following expression if the Fick diffusion coefficient is considered to be isotropic:

$$\frac{\partial \sigma_k}{\partial \theta} = -\sigma_k \left(\frac{\partial u_x}{\partial x} + \frac{\partial u_y}{\partial y} \right) + u_x \frac{\partial \sigma_k}{\partial x} + u_y \frac{\partial \sigma_k}{\partial y} + D_{F,k} \left(\frac{\partial^2 \sigma_k}{\partial x^2} + \frac{\partial^2 \sigma_k}{\partial y^2} \right) \quad (44)$$

Such expressions can be extended to permit the evaluation of the distribution of concentration throughout laminar flows. Variations in concentration at constant temperature often result in significant variation in viscosity as a function of position in the stream. Thus it is necessary to solve the basic expressions for viscous flow (L1) and to determine the velocity as a function of the spatial coordinates of the system. In the case of small variation in concentration throughout the system it is often convenient and satisfactory to neglect the effect of material transport upon the molecular properties of the phase. Under these circumstances the analysis of boundary layer as reviewed by Schlichting (S4) can be used to evaluate the velocity as a function of position in nonuniform boundary flows. Such analyses permit the determination of material transport from spheres, cylinders, and other objects where the local flow is nonuniform. In such situations it is not practical at the present state of knowledge to take into account the influence of variation in the level of turbulence in the main stream.

In thermal transport there was need of considering only temperature as the driving potential. However, in the case of multicomponent material transport it is necessary to evaluate the behavior of each component. Such relationships lead to rather complicated expressions when the diffusion of each of the components is considered individually. For this reason it is often appropriate to focus attention on one component and consider the characteristics of the remainder of the components in the phase as invariant. In the present discussion only the behavior of one component will be considered, but it should be realized that the effect of the properties of the phase upon the diffusion coefficient must be taken into account.

E. Turbulent Flow

In the case of turbulent flow it is possible to employ the same eddy concepts as were used in connection with the consideration of thermal transport. In the case of steady, uniform flow between parallel plates the eddy diffusivity may be defined by

$$\epsilon_{d,k} = \varepsilon_{d,k} - D_{F,k} = -\mathring{m}_k \frac{dy}{d\sigma_k} - D_{F,k} \quad (45)$$

It should be emphasized that the total diffusivity is a function of the component as well as the properties of the stream at the point in question. The influence of the nature of the other components on the eddy diffusivity is probably not great so long as the molecular Schmidt numbers for all the components are substantially the same. It has been difficult to obtain good experimental data concerning material transport under steady, uniform conditions of flow. The use of wetted-wall towers, which were ably investigated by Sherwood and co-workers (S16), affords a reasonable approximation to uniform material transport although there exists some change in fugacity of the component in transport along the length of the channel. It is possible that the use of horizontal equipment with cross-flowing wetted walls would avoid this difficulty to some extent. In addition, it is difficult to investigate in detail the behavior near the interface with the gas since the velocity does not approach zero and in some instances significant perturbations in the position of this interface are encountered. For this reason few experimental data of a sufficiently detailed nature are available for the direct evaluation of eddy diffusivity under conditions of uniform flow from Eq. (45) or its equivalent.

1. *Eddy Viscosity*

There are shown in Fig. 19 values of the eddy diffusivity calculated from the measurements by Sherwood (S16). These data show the same trends as were found in thermal transport, indicating that the values of eddy diffusivity are determined primarily from the transport of momentum for situations where the molecular Schmidt numbers of the components do not differ markedly from each other.

2. *Experimental Background*

More recently the use of interferometers to investigate changes in composition in two-dimensional flow have indicated a more effective means of studying material transport in nearly uniform turbulent flow. The work of Putnam (L6) is an excellent example of experimental techniques that may be effectively applied to the study of the microscopic aspects of material transport in turbulent flow particularly near the boundary of a stream. The recent study by Wilke (L9) of rotating electrodes appears to offer another interesting approach to the detailed understanding of material transport in electrolytes. At present the background of experimental data is sufficiently sparse that most of the prediction of the influence of turbulence upon transport must be made by analogy. For example, there is shown in Fig. 20 the variation in composition (L10) with position at several downstream positions in nearly fully developed, uniform, steady, turbulent air stream in which the diffusing component was introduced as an annular stream at the boundary. The corresponding values of total diffusivity are presented in Fig. 21. In the evaluation of the total diffusivity from the data of Fig. 20 the following

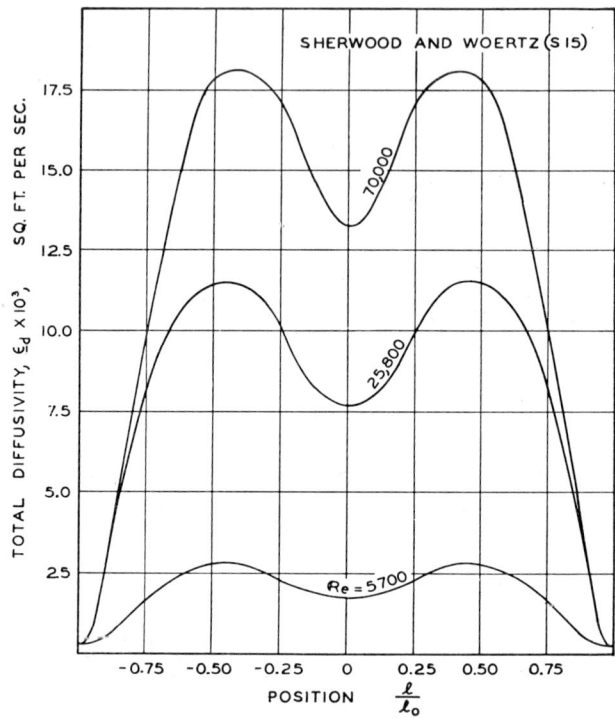

Fig. 19. Total diffusivities in a wetted-wall tower.

expression describing a component balance in cylindrical flow was employed:

$$\varepsilon_d = \frac{\frac{\partial}{\partial x}\left\{\left(\frac{r_0}{r}\right)^2 \int_0^{(r/r_0)^2} \mathbf{n}_g u_x d(r/r_0)^2\right\} - \mathbf{n}_g\left\{\left(\frac{r_0}{r}\right)^2 \int_0^{(r/r_0)^2} u_x d(r/r_0)^2\right\}}{\frac{4}{r_0^2}\sigma V_{\text{air}}{}^0 \frac{\partial \mathbf{n}_g}{\partial (r/r_0)^2}} \qquad (46)$$

F. Schmidt Number

The background of experimental data available at present does not appear to be sufficient for the effect of the several variables upon the eddy diffusivity to be determined with accuracy; however, it is worth while to consider certain relationships by analogy with thermal transport.

The molecular Schmidt number may be defined as

$$Sc_k = \frac{\nu}{D_{F,k}} \qquad (47)$$

Schmidt numbers vary widely depending on the value of the Fick diffusion coefficient, which in the case of gases is strongly dependent on the concentration of the diffusing component. It should be noted that in a mixture the Schmidt number is not a property solely of either the component or

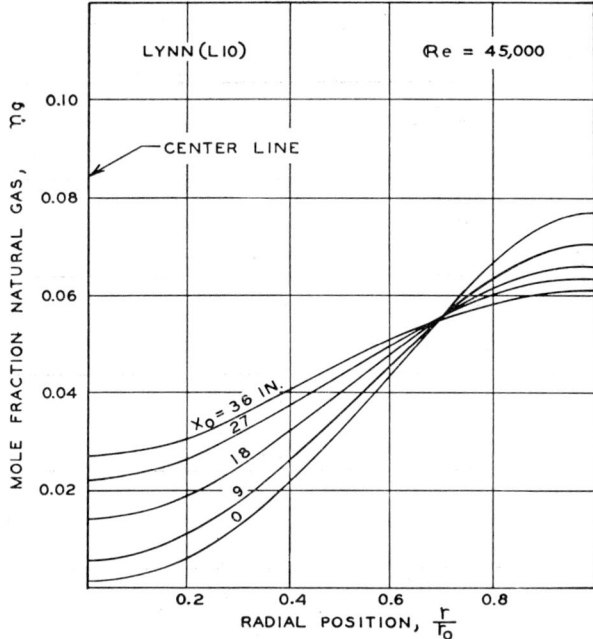

Fig. 20. Variation in composition with position in fully developed turbulent flow.

the phase but is a function of the properties of the phase and each component. The marked variation in the Fick diffusion coefficient with composition (C2) introduces somewhat larger variations in the molecular Schmidt number with position than are found for Prandtl numbers. The latter quantity is a molecular property of the phase at the point in question and is not related to the characteristics of the individual components.

By analogy with the eddy Prandtl number, the eddy Schmidt number may be defined as

$$Sc_{\epsilon,k} = \frac{\epsilon_m}{\epsilon_{d,k}} \qquad (48)$$

The eddy Schmidt number on the basis of the Reynolds analogy would be unity throughout the flowing stream. It is probable that the Reynolds analogy (R2) is no more applicable to the transport of material than it is to thermal transport. If the speculative analysis of Jenkins (J5) can be extended to material transport, the effect of molecular Schmidt number upon the eddy Schmidt number may be estimated. However, in this instance, since the molecular Schmidt number varies with each component, it is to be expected that the eddy Schmidt number will also vary

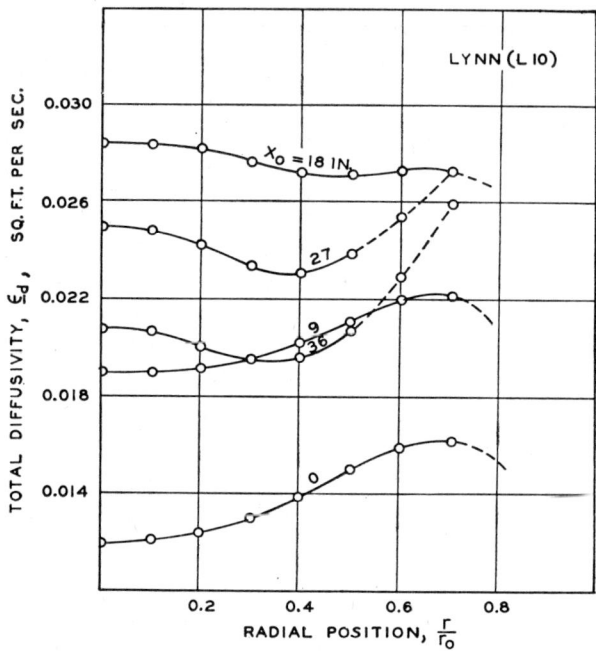

FIG. 21. Variation in eddy diffusivity with position.

from component to component. In this discussion the eddy Schmidt number will be considered to be isotropic, a consideration which may not be strictly true particularly near the outer edge of the boundary flow (L3).

The total Schmidt number is defined in a fashion analogous to the total Prandtl number in the following way:

$$\text{Sc}_k = \frac{\varepsilon_m}{\varepsilon_{d,k}} = \frac{\epsilon_m + \nu}{\epsilon_{d,k} + D_{F,k}} \qquad (49)$$

As indicated in Eq. (49), the total Schmidt number is a function of the component involved as well as of the position in the stream. It approaches the local molecular Schmidt number as the boundary flow is reached and the influence of the component on the value of the total Schmidt number becomes less important in the central portion of the turbulent flow. Throughout this discussion it will be assumed that the total Schmidt number is isotropic although there is no real justification for believing that the eddy diffusivity should be more isotropic than the eddy viscosity, which appears from the work of Laufer (L3) to be anisotropic in nature particularly near the outer edge of the transition region. Further experimental work will be required in order to determine with certainty the influence of the direction of shear upon the eddy diffusivity. In the authors' opinion, for situations where the eddy Schmidt number is near unity, the anisotropy will be in the same planes as found for the eddy viscosity.

G. Prediction

The prediction of material transport involves expressions similar to those encountered for thermal transport except that the behavior of each component must be taken into account. For example, in one-dimensional flow the flux of one component past a section is given by

$$\dot{m}_k = u\sigma_k - \varepsilon_{d,k} \frac{\partial \sigma_k}{\partial y} \tag{50}$$

It should be recognized that the boundary conditions of the problem will establish the value of the hydrodynamic velocity, u. In the case of most turbulent flows the indirect influence of molecular diffusion on the hydrodynamic velocity can be neglected. It should be emphasized that the hydrodynamic velocity is the time-average point velocity in Reynolds' sense (R2). Under unsteady, nonuniform conditions of flow between parallel plates the material balance may be expressed for turbulent flow in the following form:

$$\frac{\partial \sigma_k}{\partial \theta} = -\sigma_k \left(\frac{\partial u_x}{\partial x} + \frac{\partial u_y}{\partial y} \right) - u_x \frac{\partial \sigma_k}{\partial x} - u_y \frac{\partial \sigma_k}{\partial y} + \varepsilon_{d,k} \left(\frac{\partial^2 \sigma_k}{\partial x^2} + \frac{\partial^2 \sigma_k}{\partial y^2} \right) \tag{51}$$

Equation (51) assumes that the eddy properties are isotropic. In addition, no effect of other gradients such as temperature or gravity upon the molecular transport is taken into account. The expression was written for a single component and it is necessary to solve a set of such expressions one for each component, simultaneously if the interrelation of the material transport upon the hydrodynamic velocity is to be taken into account.

1. *Equations of State*

In the solution of equations such as Eq. (51) it is necessary to apply a detailed knowledge of the effect of composition upon the specific volume of the phase over the range of pressures, temperatures, and compositions of interest. In the case of gases the perfect gas law is often implicitly assumed. However, deviations from such a simple equation of state are significant under many of the conditions encountered in chemical engineering practice. For this reason it is often preferable to assume ideal solutions (L4) since such behavior does not impose significant complications in the solutions of the equation of change for situations where the effect of changes in pressure upon the properties of the phase can be neglected or simply treated. For more elegant solutions of the transport equations typified by Eq. (51) as applied to the gas phase it is desirable to employ an equation of state such as that proposed by Benedict (B8, B9) to describe the volumetric behavior of mixtures. It should be emphasized that a detailed knowledge of the equilibrium volumetric behavior of the phase is required to obtain a solution of Eq. (51) or its equivalent.

2. *Turbulent Streams*

It is usually convenient to solve Eq. (51) by analog methods. If detailed information concerning the variations in the total diffusivity with position is available and if the effect of material transport upon the gross motion of the phase is neglected, it is readily possible to compute the component flux and concentration as a function of position in the two-dimensional flowing stream. If the effect of material transport upon the transport of momentum is considered, it is necessary to solve the equation of change describing the component transport. The equation of state must also be included. Such complicated problems are not usually solved other than by numerical methods. The advent of large-scale automatic digital computing equipment makes possible the consideration in practical engineering work of the simultaneous solution of several partial-differential equations describing the component and momentum transport. Sufficient accuracy is often obtained even when the effect of material transport upon the distribution of momentum in the stream is neglected. Under these conditions there need be no interdependent boundary conditions for the material and momentum transport. Such simplifications are most desirable whenever they can be employed without introducing unacceptable uncertainties in the results. It is emphasized that throughout these transport calculations local equilibrium (K7) is assumed.

3. *Analytical Method*

Deissler (D3) made an interesting analysis of material transport in turbulent flow based upon the Reynolds analogy. Under these circumstances the dimensionless concentration parameter may be defined as

$$\sigma^+ = \int_0^{y^+} \frac{dy^+}{\frac{1}{Sc} + n^2 u^+ y^+ (1 - \exp\{-n^2 u^+ y^+\})} \quad (52)$$

This expression is analogous to Eq. (30) for the temperature parameter. Under such circumstances, as in the case of the temperature distribution, it follows that

$$u^+ - u_1^+ = \sigma^+ - \sigma_1^+ \quad (53)$$

A generalized concentration distribution based on Deissler's analysis is shown in Fig. 22. There is little change in the concentration parameter

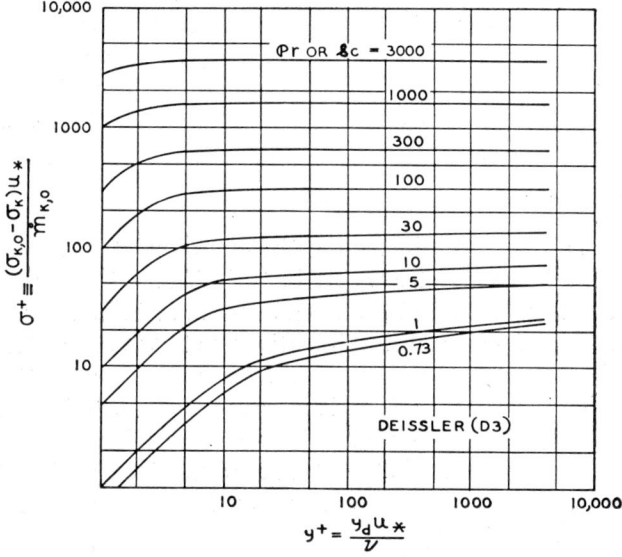

FIG. 22. Effect of position in stream on concentration parameter.

for values of the distance parameter above 20 except for situations involving a small molecular Schmidt number for the material in transport. Such behavior is to be expected since at the higher molecular Schmidt numbers all the distribution of the concentration of the component of the material in transport occurs in the laminar portion of the boundary flow. At very high Schmidt numbers Eq. (53) assumes the form

$$\sigma^+ = \int_0^{y^+} \frac{dy^+}{\frac{1}{Sc} + n^4 (y^+)^4} \quad (54)$$

The Sherwood number (S1) for turbulent transport in a circular conduit may be evaluated from

$$Sh = \frac{2r_0{}^+ \dot{m}_{k,0} Sc}{u_*(\sigma_0 - \sigma_b)} = \frac{2r_0{}^+ Sc}{\sigma_b{}^+} \tag{55}$$

Values of the concentration parameter $\sigma_b{}^+$ may be determined directly from Fig. 22 if desired. More involved expressions for material transport, similar to those established for thermal transport, have been developed by Deissler. They take into account changes in fluid properties with concentration. Since such detailed information usually is not readily available for a particular system, it does not appear worth while to discuss further ramifications of Deissler's analysis. It should be emphasized that all the foregoing is based upon the premise that the eddy diffusivity is isotropic and at all points in the shear flow is numerically equal to the eddy viscosity. As has been indicated earlier, the eddy Schmidt number undoubtedly is influenced by the local value of the molecular Schmidt number, which in turn is a function of the component in transport and the state of the phase.

4. *Limitations of Analyses*

At present analytical solutions of the equations describing the microscopic aspects of material transport in turbulent flow are not available. Nearly all the equations representing component balances are nonlinear in character even after many simplifications as to the form of the equation of state and the effect of the momentum transport upon the eddy diffusivity are made. For this reason it is not to be expected that, except by assumption of the Reynolds analogy or some simple consequence of this relationship, it will be possible to obtain analytical expressions to describe the spatial variation in concentration of a component under conditions of nonuniform material transport.

VI. Combined Thermal and Material Transport in Turbulent Flow

A. Introduction

Material transport is usually associated with thermal transport except in situations involving homogeneous phases which can be treated as ideal solutions (L4). For this reason it is necessary to consider the behavior of combined thermal and material transport in turbulent flow. The evaporation of liquids under macroscopic adiabatic conditions is a typical example of such a phenomenon. Under such circumstances the behavior in the boundary layer is similar to that found in the field of aerodynamics in a blowing boundary layer (S4). However, it is not feasible to obtain a simple analytical description of the boundary layer if variations in the physical properties as a result of the material and thermal transport are taken into account. A few such situations have been explored with numerical methods. There is shown in Fig. 23 an experimentally determined temperature distribution around a sphere from which n-heptane is evaporating into an air stream having a level of turbulence less than 0.5%. The growth of the boundary layer is clearly evident, and the vortex horns downstream of the sphere can be seen. The

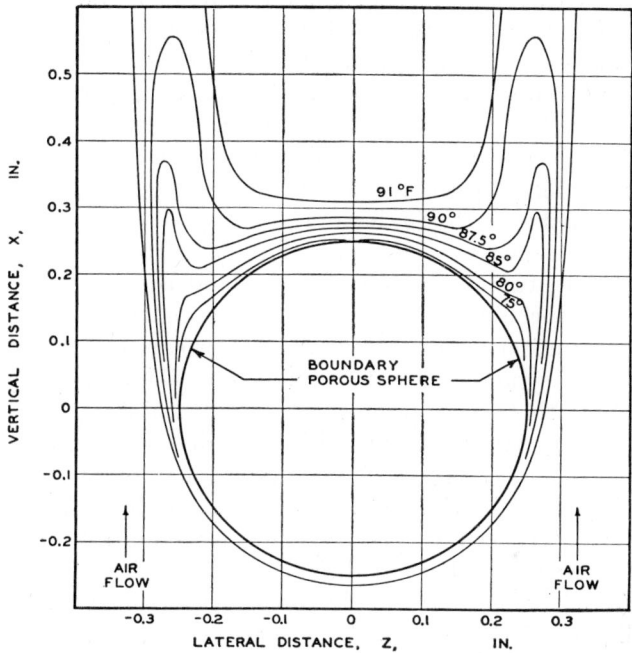

FIG. 23. Temperature distribution around a sphere with material transport.

corresponding temperature distribution is shown for a silver sphere under the same macroscopic conditions of flow. In this instance only thermal transport exists and there is no hydrodynamic velocity normal to the surface at the boundary of the sphere. A detailed comparison of the isotherms around each sphere indicates the difference in temperature distribution resulting from material transport. It does not appear that three-dimensional boundary layers have as yet been solved analytically for situations involving a local hydrodynamic velocity associated with material transport. However, conventional methods of analysis can probably be applied with numerical solutions in situations where they are worth while.

The Onsager reciprocity relationships (D1, O1, O2) become of importance in situations where there are combined material and thermal transport and under these circumstances may be expressed as

$$L_{ij} = L_{ji} \tag{56}$$

Equation (56) states that the effect of a thermal gradient on the material transport bears a reciprocal relationship to the effect of a composition gradient upon the thermal transport. Examples of L_{ij} and L_{ji} are the coefficient of thermal diffusion (S19) and the coefficient of the Dufour effect (D6). The Onsager reciprocity relationships (D1, O1, O2) are based upon certain linear approximations that have a firm physical foundation only when close to equilibrium. For this reason it is possible that under circumstances in which unusually high potential gradients are encountered the coupling between mutually related effects may be somewhat more complicated than that indicated by Eq. (56). Hirschfelder (B10, H1) discussed many aspects of these cross linkings of transport phenomena.

The large fluctuations in temperature and composition likely to be encountered in turbulence (B6) opens the possibility that the influence of these coupling effects may be even more pronounced than under the steady conditions rather close to equilibrium where Eq. (56) is strictly applicable. For this reason there exists the possibility that outside the laminar boundary layer the mutual interaction of material and thermal transfer upon the over-all transport behavior may be somewhat different from that indicated in Eq. (56). The foregoing thoughts are primarily suppositions but appear to be supported by some as yet unpublished experimental work on thermal diffusion in turbulent flow. Jeener and Thomaes (J3) have recently made some measurements on thermal diffusion in liquids. Drickamer and co-workers (G2, R4, R5, T2) studied such behavior in gases and in the critical region.

B. Eddy Properties

Little detailed experimental information is available on the value of eddy transport properties under conditions of simultaneous thermal and material transport. If it is assumed that the Reynolds analogy is applicable, it follows that the eddy diffusivity and eddy conductivity are equal and independent of cross linking. Such an assumption is probably not true since it is to be expected that a substantial part of the eddy transport is associated with molecular transport particularly as the eddies become small in accordance with Kolmogoroff's (K10) principle. For this reason it is to be expected that temperature gradients in turbulent streams will influence to some extent the material transport in the same stream. Furthermore, as a result of the fluctuating nature of the phenomenon of turbulence, the tensorial nature of these cross linkings may be changed from those predicted by the reciprocity relationships of Onsager (O2, O3).

Satterfield (S2, S3) carried out a number of interesting macroscopic studies of simultaneous thermal and material transfer. This work was done in connection with the thermal decomposition of hydrogen peroxide and yielded results indicating that for the relatively low level of turbulence experienced the thermal transport did not markedly influence the material transport. However, the results obtained deviated by 10 to 20 from the commonly accepted macroscopic methods of correlating heat and material transfer data. The final expression proposed by Satterfield (S3), neglecting the thermal diffusion effect (S19) in the boundary layer, was written as

$$j_D = 0.021 Re^{-0.2} \tag{57}$$

Equation (57) applies to material transport in tubes and yields an average deviation of 9.5% from the experimental data. An expression of similar form yielded an average deviation of 14.8% for the thermal transport. The ratio of thermal to material transport was found to be 1.09 with an average deviation of 13.7% (S3). Somewhat better agreement with predicted behavior was encountered for the studies on packed beds (S2). These data serve to illustrate the uncertainties which presently exist in the prediction of simultaneous material and thermal transfer under a variety of conditions. Satterfield's work has made a distinct contribution to understanding the macroscopic influences of combined thermal and material transport. Some of the discrepancy he noted may relate to assumptions concerning the nature of the chemical reaction associated with the decomposition of hydrogen peroxide.

VII. Summary

At present the development of more effective basic correlations of thermal and material transport in turbulent shear flow rests primarily upon an extension of the understanding of the mechanics of turbulence. Howarth and Kármán (K1, K4) and Batchelor (B6) contributed materially to the knowledge of isotropic, homogeneous turbulence, but the prediction of the behavior in shear flow still must be based on experiment (L3) even for steady, uniform flow. The absence of a basic understanding of the growth and decay of turbulence (K5) prevents a microscopic analysis of thermal and material transport under nonuniform or unsteady conditions.

Deissler (D2, D3) contributed to the understanding of thermal and material transport through laminar boundary layers by his analysis based upon the Reynolds analogy. Experimental data on air (P2, P3) indicated that the turbulent Prandtl number may increase toward a limiting value of unity as the Reynolds number is increased. It is surmised that similar behavior may be encountered with the turbulent Schmidt number. The influence of molecular Prandtl and Schmidt numbers upon the corresponding turbulent quantities is not well established but may be roughly estimated from the suppositional analysis of Jenkins (J5). No real information is available about the effects of cross linking in turbulent flow and it is probable that such effects are small. However, this is not assured, and until further experimental work is available or the concepts of Onsager are extended to unsteady conditions and to situations which differ markedly from equilibrium, the effects of simultaneous material and thernal transport must be made from available data such as those obtained by Satterfield (S2, S3). The rapid advance in automatic computing equipment of both the analog (B11) and digital (K11) types permits the prediction of temperature and composition distribution throughout laminar boundary layers and reasonable estimates in the transition regions for nearly uniform, turbulent, shear flow. The uncertainties in prediction by such methods increase under conditions of unsteady or nonuniform flow in the main body of the stream and for fluids with molecular Prandtl or Schmidt numbers which differ markedly from unity although Deissler (D3) obtained good agreement with experiment. Further experiment will be required to ensure that these predictions may be made with accuracy. It appears rather unlikely that successful analytical treatment for nonuniform flow will prove competitive with the numerical approach involving automatic computing equipment. The studies by Schlichting of blowing boundary layers (S4) are an example of the analytical approach to such problems. When properties in the boundary layer are changed as a result of material and thermal transport, the analysis becomes rather difficult. At present numerical methods of analysis permit an evaluation of the microscopic behavior of momentum, thermal, and material transport even with chemical reactions for situations where the eddy properties for each of the fluxes may be estimated.

Further advances in the field of predicting thermal and material transport in turbulent flow will rest upon a better understanding of the mechanics of the exchange and decay of kinetic energy in turbulence with particular emphasis upon unsteady or nonuniform conditions of flow. The influence of the encroachment of turbulent and laminar boundary layers into nonuniform flow is just beginning to be considered. Experimental information concerning the details of exchange of the internal energy and the concentration of components as a function of eddy size or wave number and molecular properties of the fluid is needed.

The methods of analysis involving numerical solutions appear sufficiently well advanced to permit a rapid expansion of the microscopic analysis of turbulent transport as soon as some of the basic experimental facts are obtained. The next advance of particular interest to the chemical engineer appears to be an understanding of the kinetics of chemical reactions in turbulent flow. The fluctuating temperatures and concentrations introduce perturbation in the normal approach to kinetics that may well yield interesting results in the field of combustion and perhaps in chemical processing.

Nomenclature

C_p Isobaric heat capacity, B.t.u./(lb.)(°R.)

$D_{F,k}$ Fick diffusivity for component k, sq. ft./sec.

$D_{M,k,j}$ Maxwell diffusivity, lb./sec.

$D_{v,k}$ Volumetric diffusivity, sq. ft./sec.

E Specific internal energy, B.t.u./lb.

$F_1(k_1)$ One-dimensional kinetic energy spectrum, cu. ft./sec.^{2}

$F_1(n_1)$ One-dimensional kinetic energy spectrum, sq. ft./sec.

f Fanning friction factor, dimensionless

f_k^0 Fugacity of component k in the pure state, lb./sq. ft.

g Acceleration due to gravity, ft./sec.^{2}

j_D Chilton and Colburn j factor for diffusion, dimensionless

K_1 Empirical constant, dimensionless

k Proportionality constant, ft.^{4}/sec.^{4}

k Thermal conductivity, B.t.u./(sec.)(sq. ft.)(°R./ft.)

k_1 One-dimensional wave number, reciprocal ft.

L_{ij}, L_{ji} Onsager's phenomenological coefficients

l Distance from plane of symmetry, ft.

$2l_0$ Separation of parallel plates, ft.

$\overset{\circ}{m}$ Material flux per unit area, lb./(sec.)(sq. ft.)

$\overset{\circ}{m}_k$ Material flux of component k, lb./(sec.)(sq. ft.)

$\overset{\circ}{m}_{k,o}$ Material flux at the wall, lb./(sec.)(sq. ft.)

Nu Nusselt number, dimensionless

n Empirical parameter, approximately 0.109, dimensionless

n_g Mole fraction natural gas

n_j Weight fraction of component j

n_j Mole fraction of component j

n_1 One-dimensional frequency, reciprocal sec.

P Pressure, lb./sq. ft.

Pr Prandtl number, dimensionless

$\mathbf{Pr}$ Total Prandtl number, dimensionless

Pr_ϵ Eddy Prandtl number, dimensionless

$\overset{\circ}{Q}_r$ Thermal flux in r direction, B.t.u./(sq. ft.)(sec.)

$\mathring{Q}_x$	Thermal flux in x direction, B.t.u./(sq. ft.)(sec.)	u^+	Velocity parameter, dimensionless
$\mathring{Q}_y$	Thermal flux in y direction, B.t.u./(sq. ft.)(sec.)	u_*	Friction velocity, $\sqrt{\frac{\tau_0 g}{\sigma}}$, ft./sec.
Re	Reynolds number, dimensionless	V	Specific volume, cu. ft./lb.
R_y	Coefficient of correlation, dimensionless	$V_{\mathrm{air}}{}^o$	Specific volume of pure air, cu. ft./lb.
r	Radial distance, ft.	$\mathbf{V}_j{}^o$	Molal volume of component j in pure state, cu. ft./lb. mole
r	Radial distance, ft.		
r_0	Radius of circular conduit, ft.	x	Distance in x direction, ft.
r_o	Radius of circular conduit, ft.	x_o	Distance downstream, ft.
Sc_k	Schmidt number, dimensionless	y	Distance into the stream from the plane of symmetry of the conduit, ft.
$\mathbf{Sc}_k$	Total Schmidt number, dimensionless	y_d	Distance from the wall of the conduit into the stream, ft.
$Sc_{\epsilon,k}$	Eddy Schmidt number		
Sh	Sherwood number, dimensionless	y_o	Separation of walls in parallel-plate conduit, ft.
St	Stanton number, dimensionless	$y_o{}^+$	Parameter, $\frac{u_*}{\nu}\frac{y_o}{2}$, dimensionless
T	Temperature, °R.		
t_b	Bulk temperature of stream, °F.	y^+	Distance parameter, dimensionless
t_0	Temperature of wall, °F.	Z	Compressibility factor, dimensionless
t^+	Temperature parameter, dimensionless	ϵ_c	Eddy conductivity, sq. ft./sec.
$t_b{}^+$	$C_p g \tau_o (t_o - t_b)/u_* \mathring{Q}_{y,o}$	ε_c	Total conductivity, sq. ft./sec.
U	Gross velocity, ft./sec.		
u	Hydrodynamic velocity in the x direction, ft./sec.	$\epsilon_{d,k}$	Eddy diffusivity, sq. ft./sec.
$u_{d,k}$	Diffusional velocity of component k, ft./sec.	$\varepsilon_{d,k}$	Total diffusivity, sq. ft./sec.
$\overline{u_f{}^2}$	Mean-square velocity fluctuations, sq. ft./sec.²	ϵ_m	Eddy viscosity, sq. ft./sec.
u_k	Total velocity of component k, ft./sec.	ε_m	Total viscosity, sq. ft./sec.
u_m	Maximum velocity, ft./sec.	η	Absolute viscosity, (ft.)(sec.)/sq. ft.
u_x	Time average (hydrodynamic) velocity in the x direction, ft./sec.	θ	Time, sec.
		κ	Thermometric conductivity, sq. ft./sec.
$u_{x,f}$	Fluctuating velocity in the x direction, ft./sec.	ν	Kinematic viscosity, sq. ft./sec.
$u_{x,i}$	Instantaneous component of velocity in the x direction, ft./sec.	σ	Specific weight, lb./cu. ft.

σ_b Bulk concentration of diffusing component, lb./cu. ft.

$\sigma_b{}^+$ Bulk concentration parameter, $(\sigma_o - \sigma_b)U_*/\overset{\circ}{m}_{k,0}$

σ_k Concentration of component k, lb./cu. ft.

δ_k Concentration, lb. moles/cu. ft.

σ_o Concentration of diffusing component at the wall, lb./cu. ft.

σ^+ Concentration parameter, dimensionless

τ Shear, lb./sq. ft.

τ_o Shear at the boundary of the conduit, lb./sq. ft.

$\phi_1(\ )$, $\phi_2(\ )$ Functions defined by Eq. (15), sq. ft./sec.

SUPERSCRIPT

—   Time average

REFERENCES

A1. Auer, P. L., and Murbach, P. W., *J. Chem. Phys.* **22**, 1054 (1954).
B1. Bagotskaya, I. A., *Doklady Akad. Nauk S.S.S.R.* **85**, 1057 (1952).
B2. Bakhmeteff, B. A., "The Mechanics of Turbulent Flow." Princeton U. P., Princeton, 1941.
B3. Bartholomew, R. N., and Katz, D. L., *Chem. Eng. Progr. Symposium Ser. No. 4* **48**, 3 (1952).
B4. Batchelor, G. K., *Proc. Cambridge Phil. Soc.* **43**, 533 (1947).
B5. Batchelor, G. K., *Proc. Roy. Soc.* **A195**, 513 (1949).
B6. Batchelor, G. K., "The Theory of Homogeneous Turbulence." Cambridge U. P., New York, 1953.
B7. Bell, S., Report No. HE-150-115. University of California Institute of Engineering Research, Berkeley, 1953.
B8. Benedict, M., Webb, G. B., and Rubin, L. C., *J. Chem. Phys.* **8**, 334 (1940).
B9. Benedict, M., Webb, G. B., and Rubin, L. C., *J. Chem. Phys.* **10**, 747 (1942).
B10. Bird, R. B., Curtiss, C. F., and Hirschfelder, J. O., Report ONR-7. University of Wisconsin, Madison, 1954.
B11. Booth, A. D., and Booth, H. V., "Automatic Digital Calculators." Academic Press, New York, 1953.
B12. Brier, J. C., Churchill, S. W., Engibous, D. L., and Thatcher, C. M., Report No. 52-SIb (Contract NOrd 12109), University of Michigan Engineering Research Institute, Ann Arbor, 1952.
B13. Butler, R. M., and Plewes, A. C., *Chem. Eng. Progr. Symposium Ser., No. 10*, **50**, 121 (1954).
C1. Callaghan, E. E., *Natl. Advisory Comm. Aeronaut. Tech. Notes* **3045** (1953).
C2. Carmichael, L. T., Reamer, H. H., and Sage, B. H., "Diffusion Coefficients in Hydrocarbon Systems. n-Heptane in the Gas Phase of the Methane-n-Heptane System" (submitted to *Ind. Eng. Chem.* for publication).
C3. Cavers, S. D., Hsu, N. T., Schlinger, W. G., and Sage, B. H., *Ind. Eng. Chem.* **45**, 2139 (1953).
C4. Cole, J., and Roshko, A., Preprints of papers from 1954 Heat Transfer and Fluid Mechanics Conf., pp. 13–23. Institute, University of California, Berkeley, June 30, July 1 and 2, 1954.
C5. Connell, W. R., Schlinger, W. G., and Sage, B. H., *Am. Doc. Inst. Doc.* **No. 3657** (1952).
C6. Corrsin, S., *J. Aeronaut. Sci.* **18**, 417 (1951).
C7. Corrsin, S., *Natl. Advisory Comm. Aeronaut. Tech. Notes* **2124** (1950).
C8. Cova, D. R., and Drickamer, H. G., *J. Chem. Phys.* **21**, 1364 (1953).
C9. Crocco, L., "Monografie Scientifiche di Aeronautica," No. 3. Dec., 1946 (trans-

lated from the Italian by I. Hodes and J. Castelfranco as Report CR-1038), North American Aviation, Inc., Aerophysics Lab., 1948.
D1. De Groot, S. R., "Thermodynamics of Irreversible Processes," Interscience, New York, 1952.
D2. Deissler, R. G., *Natl. Advisory Comm. Aeronaut. Tech. Notes* **2138** (1950).
D3. Deissler, R. G., *Natl. Advisory Comm. Aeronaut. Tech. Notes* **3145** (1954).
D4. Doane, E. P., and Drickamer, H. G., *J. Chem. Phys.* **21**, 1359 (1953).
D5. Dow, M. M., and Jakob, M., *Chem. Eng. Progr.* **47**, 637 (1951).
D6. Dufour, L., *Ann. phys.* [5] **28**, 490 (1873).
D7. Dunn, L. G., Powell, W. B., and Seifert, H. S., "Heat Transfer Studies Relating to Power Plant Development," *Roy. Aeronaut. Soc. 3rd Anglo-American Aeronaut. Conf.*, 1951.
E1. Eber, G., *J. Aeronaut. Sci.* **19**, 1 (1952).
E2. Eckert, E., and Drewitz, O., *Natl. Advisory Comm. Aeronaut. Tech. Mem.* **1045** (1943).
F1. Fallis, W., *J. Aeronaut. Sci.* **20**, 646 (1953).
F2. Fick, A., *Ann. phys. u. chem.* [2] **94**, 59 (1855).
F3. Fujita, H., *J. Colloid Sci.* **9**, 269 (1954).
G1. Gibbs, J. W., "Collected Works," Vol. 1. Green, New York, 1925.
G2. Giller, E. B., Duffield, R. B., and Drickamer, H. G., *J. Chem. Phys.* **18**, 1027 (1950).
G3. Gordon, K. F., and Sherwood, T. K., *Chem. Eng. Progr. Symposium Ser. No. 10* **50**, 15 (1954).
H1. Hirschfelder, J. O., Curtiss, C. F., and Bird, R. B., "Molecular Theory of Gases and Liquids." Wiley, New York, 1954.
J1. Jakob, M., "Heat Transfer," Vol. 1. Wiley, New York, 1949.
J2. Jakob, M., *J. Appl. Phys.* **23**, 1056 (1952).
J3. Jeener, J., and Thomaes, G., *J. Chem. Phys.* **22**, 566 (1954).
J4. Jeffreys, H., "Cartesian Tensors." Cambridge U. P., New York, 1931.
J5. Jenkins, R., thesis, California Institute of Technology, Pasadena, 1950.
J6. Jenkins, R., Brough, H. W., and Sage, B. H., *Ind. Eng. Chem.* **43**, 2483 (1951).
J7. Johnson, H. A., Hartnell, J. P., and Clabaugh, M. J., *Trans. Am. Soc. Mech. Engrs.* **76**, 513 (1954).
J8. Jurges, W., *Gesundh. Ing.* **1**, Suppl. **19**, 1 (1924).
K1. Kármán, von, T., *J. Aeronaut. Sci.* **1**, 1 (1934).
K2. Kármán, von, T., *Mech. Eng.* **57**, 407 (1935).
K3. Kármán, von, T., *Trans. Am. Soc. Mech. Engrs.* **61**, 705 (1939).
K4. Kármán, von, T., and Howarth, L., *Proc. Roy. Soc.* **A164**, 192 (1938).
K5. Kármán, von, T., and Lin, C. C., *Advances in Appl. Mech.* **2**, 1 (1951).
K6. King, L. V., *Phil. Trans. Roy. Soc.* **A214**, 373 (1914).
K7. Kirkwood, J. G., and Crawford, B., Jr., *J. Phys. Chem.* **56**, 1048 (1952).
K8. Klinkenberg, A., Krajenbrink, H. J., and Lauwerier, H. A., *Ind. Eng. Chem.* **45**, 1202 (1953).
K9. Koeller, R. C., and Drickamer, H. G., *J. Chem. Phys.* **21**, 575 (1953).
K10. Kolmogoroff, A. N., *Compt. rend. acad. sci. U.R.S.S.* **30**, 301 (1941).
K11. Korn, G. A., and Korn, T. M., "Electronic Analog Computers." McGraw-Hill, New York, 1952.
K12. Krauss, C. J., and Spinks, J. W., *Can. J. Chem.* **32**, 71 (1954).
K13. Kronig, R., Van Der Veen, B., and Ijzerman, P., *Appl. Sci. Research* **A3**, 103 (1953).

L1. Lamb, H., "Hydrodynamics." Cambridge U. P., New York, 1932.
L2. Laufer, J., *Natl. Advisory Comm. Aeronaut. Tech. Notes* **2123** (1950).
L3. Laufer, J., *Natl. Advisory Comm. Aeronaut. Tech. Notes* **2954** (1953).
L4. Lewis, G. N., *J. Am. Chem. Soc.* **30**, 668 (1908).
L5. Liepmann, H., *J. Appl. Math. Phys.* **3**, 321 (1952).
L6. Lin, C. S., Moulton, R. W., and Putnam, G. L., *Ind. Eng. Chem.* **45**, 636 (1953).
L7. Linton, W. H., Jr., and Sherwood, T. K., *Chem. Eng. Progr.* **46**, 258 (1950).
L8. Lowdermilk, W., Weiland, W., Jr., and Livingood, J. B., *Natl. Advisory Comm. Aeronaut. Research Mem. E53J07* (1954).
L9. Lynch, E. J., and Wilke, C. R., UCRL-2057 (Contract No. W-7405-eng-48), University of California Radiation Laboratory, Berkeley, 1953.
L10. Lynn, S., thesis, California Institute of Technology, Pasadena, 1953.
M1. McAdams, W. H., "Heat Transmission." McGraw-Hill, New York, 1954.
M2. Maisel, D. S., and Sherwood, T. K., *Chem. Eng. Progr.* **46**, 131 (1950).
M3. Maisel, D. S., and Sherwood, T. K., *Chem. Eng. Progr.* **46**, 172 (1950).
M4. Maxwell, J. C., "Scientific Papers," Vol. 2. Cambridge U. P., New York, 1890.
M5. Merk, H. J., and Prins, J. A., *Appl. Sci. Research* **A4**, 207 (1954).
M6. Mitchner, M., *Tunnel Lab. Harvard Univ. Intern. Rept.* **No. 3** (1952).
M7. Monaghan, R. K., and Cooke, J. R., *Royal Aircraft Establishment Tech. Note* **No. 2171** (1952).
M8. Murphree, E. V., *Ind. Eng. Chem.* **24**, 726 (1932).
N1. Nikuradse, J., *Forsch. Gebiete Ingenieurw. Forschungsheft* **356**, 1 (1932).
O1. Onsager, L., *Phys. Rev.* **37**, 405 (1931).
O2. Onsager, L., *Phys. Rev.* **38**, 2265 (1931).
O3. Onsager, L., *Phys. Rev.* **68**, 286 (1945).
O4. Opfell, J. B., and Sage, B. H., "Some Relations in Material Transport," accepted by *Ind. Eng. Chem.*
P1. Page, F., Jr., Corcoran, W. H., Schlinger, W. G., and Sage, B. H., *Ind. Eng. Chem.* **44**, 419 (1954).
P2. Page, F., Jr., Schlinger, W. G., Breaux, D. K., and Sage, B. H., *Am. Doc. Inst. Doc.* **No. 3294** (1951).
P3. Page, F., Jr., Schlinger, W. G., Breaux, D. K., and Sage, B. H., *Ind. Eng. Chem.* **44**, 424 (1952).
P4. Pannell, J. R., *Tech. Rept. Natl. Advisory Comm. Aeronaut. Mem.* **243**, 22–30 (1916).
P5. Pinkel, B., *Trans. Am. Soc. Mech. Engrs.* **76**, 305 (1954).
P6. Prandtl, L., in "Aerodynamic Theory" (W. F. Durand, ed.), Vol. 3, Div. G. Springer, Berlin, 1943.
P7. Prandtl, L., "Fuehrer durch die Stroemungslehre." University, Goettingen, 1944.
R1. Reynolds, O., *Phil. Trans. Roy. Soc.* **174**, 935 (1883).
R2. Reynolds, O., *Collected Papers, Phil. Trans. Roy. Soc.* **2**, 53 (1895).
R3. Rice, S. O., *Bell System Tech. J.* **23**, 282 (1944); **24**, 46 (1945).
R4. Robb, W. L., and Drickamer, H. G., *J. Chem. Phys.* **19**, 818 (1951).
R5. Rutherford, W. M., and Drickamer, H. G., *J. Chem. Phys.* **22**, 1157 (1954).
S1. Sage, B. H., *Chem. Eng. Progr.* **49**, 4 (1953).
S2. Satterfield, C. N., and Resnick, H., *Chem. Eng. Progr.* **50**, 504 (1954).
S3. Satterfield, C. N., Resnick, H., and Wentworth, R. L., *Chem. Eng. Progr.* **50**, 460 (1954).
S4. Schlichting, H., "Grenzschicht-Theorie." Braun, Karlsruhe, 1951.
S5. Schlinger, W. G., and Sage, B. H., *Ind. Eng. Chem.* **45**, 2636 (1953).

S6. Schlinger, W. G., Berry, V. J., Mason, J. L., and Sage, B. H., *Proc. Gen. Conf. on Heat Transfer*, 1951, pp. 11–13. Institution of Mechanical Engineers, London.
S7. Schlinger, W. G., Berry, V. J., Mason, J. L., and Sage, B. H., *Ind. Eng. Chem.* **45,** 662 (1953).
S8. Schrage, R. W., "A Theoretical Study of Interphase Mass Transfer," Columbia U. P., New York, 1953.
S9. Schubauer, G. B., private communication (1954).
S10. Schuler, R. W., Stalling, V. P., and Smith, J. M., *Chem. Eng. Progr. Symposium Ser. No. 4*, **48,** 19 (1954).
S11. Seban, R. A., Levy, S., Doughty, D. L., and Drake, R., Jr., *Trans. Am. Soc. Mech. Engrs.* **76,** 519 (1954).
S12. Shapiro, A. H., "The Dynamics and Thermodynamics of Compressible Fluid Flow." Ronald Press, New York, 1953.
S13. Sherwood, T. K., *Ind. Eng. Chem.* **42,** 2077 (1950).
S14. Sherwood, T. K., and Pigford, R. L., "Adsorption and Extraction." McGraw-Hill, New York, 1952.
S15. Sherwood, T. K., and Woertz, B. B., *Ind. Eng. Chem.* **31,** 1034 (1939).
S16. Sherwood, T. K., and Woertz, B. B., *Trans. Am. Inst. Chem. Engrs.* **35,** 517 (1939).
S17. Sherwood, T. K., Seder, L. A., and Towle, W. L., *Ind. Eng. Chem.* **31,** 462 (1939).
S18. Skinner, G., thesis, California Institute of Technology, Pasadena, 1950.
S19. Soret, C., *Arch. sci. (Geneva)* **29,** 4 (1893).
S20. Spielman, M., and Jakob, M., *Trans. Am. Soc. Mech. Engrs.* **75,** 385 (1953).
S21. Stokes, R. H., *Trans. Faraday Soc.* **48,** 887 (1952).
S22. Strehlow, R. A., *J. Chem. Phys.* **21,** 2101 (1953).
S23. Summerfield, M., "Heat Transfer Symposium," p. 151. Engineering Research Institute, U. of Michigan, Ann Arbor, 1953.
T1. Taylor, G. I., *Proc. Roy. Soc.* **A151,** 429 (1935).
T2. Tung, L. H., and Drickamer, H. G., *J. Chem. Phys.* **18,** 1031 (1950).
T3. Tung, L. H., and Drickamer, H. G., *J. Chem. Phys.* **20,** 6 (1952).
T4. Tung, L. H., and Drickamer, H. G., *J. Chem. Phys.* **20,** 10 (1952).
W1. Wamsley, W. W., and Johanson, L. W., *Chem. Eng. Progr.* **50,** 347 (1954).

Mechanically Aided Liquid Extraction

ROBERT E. TREYBAL

Department of Chemical Engineering
New York University, New York, New York

I. Introduction 290
   A. Inadequacy of Conventional Equipment 290
   B. Scope of This Review 291
II. Agitated Vessels 291
   A. Introduction 291
   B. Field of Usefulness 292
   C. Characteristics of Agitated Vessels 292
      1. Mixing Effectiveness 293
      2. Design of Vessel and Agitator 294
         a. Batch Operation 294
         b. Continuous Operation 295
      3. Power 296
      4. Interfacial Area 298
      5. Mass Transfer Rates, Transfer Units, and Stage Efficiencies 299
         a. Batch Operation 299
         b. Continuous Operation 302
         c. Extraction Rates 306
      6. Emulsions 308
      7. Scale-Up 308
   D. Bench-Scale Equipment 309
      1. The Pump-Mix Extractor 309
      2. Others 310
III. Agitated, Multistage Countercurrent Columns 310
   A. Field of Usefulness 311
   B. Comparison of the Extractors 311
      1. The Oldshue-Rushton Column 312
      2. The Rotating-Disk Contactor-RDC 314
      3. The Scheibel Column 315
IV. Pulsed Columns 317
   A. Pulsing Methods 317
   B. Fields of Usefulness 318
   C. Power for Pulsing 319
   D. Internal Design of Towers 319
      1. Perforated-Plate Towers 319
         a. Flow Characteristics 319
         b. Mass Transfer Characteristics 321
      2. Packed Towers 322
   E. Pulsed Mixer Settler 322

V. Centrifugal Extractors.. 323
   A. The Podbielniak Extractor... 323
   B. Other Machines... 325
      1. The Luwesta (Centri-Westa) Extractor........................... 325
      2. The Sharples Super Centactor.................................. 326
   Acknowledgment... 327
   Nomenclature... 327
   References... 328

I. Introduction

Like all the mass transfer operations, liquid extraction is a means of separating the components of a solution, and it is accomplished by bringing the solution into contact with another insoluble phase. The unequal distribution of the components of the solution between the two phases which then results provides the separation. In the case of liquid extraction, of course, the two phases in question are both liquids, but, just as in the other mass transfer operations, intimacy of contact and large interfacial area between the phases are required for rapid diffusional transfer of substance from one phase to the other.

A. Inadequacy of Conventional Equipment

As the potentialities of liquid extraction as a separation method were developed, the need for efficient, continuously operated, multistage equipment became apparent. It was natural therefore to turn to devices which had been so successful in other similar fluid-contacting operations, such as the bubble-tray tower and the packed tower of distillation. These devices have proved to be disappointing in liquid-extraction service, however; for example, bubble-tray towers provide tray efficiencies in liquid-extraction operations of less than 5% (S7), and conventional packed towers show heights of transfer units of 10 to 20 ft. or more (T3).

The reason for the disparity in performance of such devices in the two services has been clearly outlined by Hachmuth (H1). Bubble-tray towers for distillation, for example, use as the source of energy for dispersion of the gas and for developing the desirable turbulent flow conditions both the expansion of the vapor as it experiences a pressure drop in flowing through the tray, and the liquid head available between trays. In liquid extraction only the liquid head is available. When it is considered that the difference in densities of the contacted phases in distillation may be of the order of 50 to 60 lb./cu. ft., whereas in extraction it is more likely to be of the order of 5 or less, it is easy to understand that in the latter case there is simply insufficient energy available from this source to provide for adequate dispersion and interphase movement. Interfacial area between phases remains small, turbulences developed are of a low order, and mass transfer rates are disappointingly small.

B. Scope of This Review

As a result of these considerations there has been increasing interest in devices which use additional energy sources, mechanical or otherwise, for providing the necessary contact. This review considers some of these, but is limited to only a few major types, those which have been found to be of industrial importance. The discussion begins with the simplest device, the agitated vessel, because it is important in its own right and because it is felt that an adequate understanding of this extractor will eventually assist in developing designs of more compact, multistage equipment.

It will be realized as the discussion develops that adequate data for strict comparison of the various equipment types are not available. Nevertheless, some relative measure of performance is desirable, and the ratio of throughput to volume of a theoretical stage has been chosen as a simple characterizing device. Typical values of this criterion have been quoted, under conditions which represent as nearly as possible the best available performance. The figures must be used with considerable caution, since operating conditions were not usually comparable. Furthermore it will be clear from the discussion that this criterion alone is not the only one which should be applied in deciding upon the merits of a particular apparatus.

II. Agitated Vessels

A. Introduction

Agitated vessels are perhaps the oldest of the extraction devices in industrial service. They may be operated batchwise or in continuous fashion. In the former case they are simply the large-scale counterparts of the laboratory separatory funnel, and the liquids to be contacted are both agitated and settled in the same vessel. In continuous operation the liquids are continuously pumped through the agitated vessel (the mixer) in parallel flow, the resulting dispersion is continuously settled or decanted in a separate vessel (the settler), and the combination is called a mixer settler. Each mixer settler constitutes one stage of extraction, and multistage effects may be had by arranging any desired number of mixer settlers according to any desired flow sheet.

Many types of mixing devices have been used or proposed, and reviews of these developments are available (D1, M3, T3). They may generally be classified according to whether they are agitated vessels or "line" mixers. In the case of the former, the mixer consists of a vessel or tank which is mechanically stirred; whereas in the latter case the mixing is accomplished by pumping both liquids simultaneously through some constriction or other device installed in a pipe line. The energy for mixing and dispersion then usually results from the drop in pressure incurred in flow through the device. The distinction is not always clear cut. For example, both liquids may be introduced into the suction side of a centrifugal pump, and the pump action then simultaneously provides the dispersal and the movement of the liquids to the settler. This is considered a line mixer despite the application of mechanical agitation in the action of the pump impeller. This report will consider only the simplest of agitated vessels and will not consider the line mixers or the problems of settling and decanting.

B. Field of Usefulness

In the case of continuous operation on a large scale, the mixer settler is particularly suitable for liquid systems which are difficult to disperse, those of high interfacial tension and density difference, as adequate agitation power properly applied will guarantee adequate dispersal. Stage efficiencies are usually high, 80% or better, and so results are more readily guaranteed than for other types of plants. Especially where the mixer-settler plant is built horizontally along the ground, if additional stages are required they are more readily added to this type of plant than to those extractors using some form of vertical tower.

Instrumentation such as flow, flow-ratio, and liquid-level controllers which may be required for each stage, as well as extensive pumping requirements and multiple motor drives for agitators, make this type of plant inherently expensive. Solvent inventory per stage, especially in the settlers, is likely to be larger than for continuous-contact towers. The quantity of metal or other construction material is likely to be larger, per stage, than for tower-type devices of high stage efficiency, but lower than for towers of low stage efficiency. Consequently a thorough comparison of the costs of mixer settlers and of other devices is probably the only sound basis for choice of type of plant.

In relatively small-scale operations, where frequent shutdowns may be desirable, as at night or during week ends, the mixer-settler plant offers the advantage that concentration gradients are maintained within the plant indefinitely even though operations have stopped. On start up, steady-state conditions are established immediately, product specifications are met with the first effluents, and recycling of the first products withdrawn is unnecessary. The agitated vessel is also eminently suitable for purely batch processes, which are likely to be on a small scale.

C. Characteristics of Agitated Vessels

Simple agitated vessels for fluid mixing have been the object of much engineering study, especially in the applications involving stirring of miscible liquids, mixing of solids with liquids, and heat transfer to liquids. Considering the many years they have been used in this service, however, surprisingly little is known concerning their application to liquid-extraction operations, where two insoluble liquids must be brought into contact. The questions which arise in practical cases are at least the following. What should be the design of the vessel and of the agitator? In the case of continuous operation what power input is required for a given liquid throughput and holdup to provide any desired stage efficiency? In the case of batch operations with a fixed agitator power, what will be the time required to reach a given stage efficiency? Under what conditions will unsettlable emulsions form? What success can be expected in scale-up procedures; i.e., can small-scale experiments adequately predict performance of large-scale equipment?

The discussion which follows briefly considers these problems and some of their corollaries, with perhaps the greater emphasis upon what we do not know rather than upon what we do.

1. *Mixing Effectiveness*

In discussing the first two problems listed above, what we are really interested in, of course, is determining how the variables involved influence the rate of extraction we can obtain in an agitated vessel. The rate of mass transfer of solute from the solution to be extracted into an immiscible extraction solvent is usually described by some such relation as

$$N = KA\Delta c_{avg} = KaV\Delta c_{avg} \qquad (1)$$

or related expressions, where N is the rate of extraction, $A = aV$ is the interfacial surface, and Δc_{avg} is some measure of the concentration gradient existing between the phases. The mass transfer coefficient K, or Ka, is then a characterizing constant for the conditions of the operation. It is the purpose of the agitator impeller to disperse one liquid within the other, producing small droplets of one phase which will be immersed in a continuum of the other. The rate of mass transfer will be directly proportional to the interfacial surface and consequently will depend upon the fineness of the dispersion. At the same time the action of the agitator, by producing flow currents within the liquid mass, has a profound influence upon the average concentration gradient, which also influences the rate of mass transfer. The mass transfer coefficient includes the characteristics of the diffusional resistances residing within the dispersed and within the continuous phase and possibly any additional resistance which may be present at the interface. The first two of these particularly depend upon the degree of turbulence produced within the agitated fluid.

2. Design of Vessel and Agitator

These matters have a bearing principally upon the nature of the concentration gradients maintained in the vessel and upon the power requirements for operation. Most of the available information, as explained above, has unfortunately been obtained without consideration of liquid extraction, and the data for the latter are extremely meager.

a. Batch Operation. The published work on nonextraction applications has largely been done in partly filled vessels, in the presence of an air-liquid interface, and without continuous flow of liquid through the vessel. Without actual knowledge that it is so, it seems reasonable nevertheless to suppose that such work would be more nearly applicable to batch liquid extraction, since in continuous extraction the vessels are more likely to be completely filled with liquid. Most of the work has been done in connection with measurement of the power required to turn the agitator at a given speed, with little regard for the effect on the degree of mixing that may result. For adequate mixing in the case of two insoluble liquids it seems reasonable that the agitator must produce (*a*) sufficiently strong vertical circulation of the tank contents to prevent gravity settling of the liquids into two layers and (*b*) sufficiently strong radial motion to prevent stratification by centrifugal force, whereby the heavy liquid would collect in a layer about the vertical wall of the tank.

The tank seems best arranged as a vertical cylinder, fitted with a dished bottom, with the liquid depth at least equal to the tank diameter. It has been well established in the case of single liquids, and also in the case of two-phase liquid mixtures (M2, O3), that without vertical wall baffles, swirl and vortexing result, with simultaneous development of undesirable circular motion and aeration of the liquid. Vertical wall baffles are flat, narrow strips, placed radially and vertically along the tank walls, extending for the full liquid depth. The "fully baffled state" (M1) for single liquids, where vortexing is negligible, is established with a minimum of four equally spaced baffles, of width equal to one-tenth the tank diameter. Additional baffles produce no very great additional advantages.

A large number of impeller types have been studied over the years, but interest has centered on three designs: marine-type propellers, flat- or curve-bladed turbines, and flat paddles, with the first two of greater interest than the third. Propellers produce axial flow of the liquids and are turned in such fashion as to direct flow against the bottom of the tank. Turbines provide radial flow, but in any case the presence of baffles strongly influences the flow pattern in the tank. The effectiveness of these in liquid extraction has not been well established, but it appears that there are no great differences in mixing effectiveness. The flat-bladed turbine is of simple design, has the desirable "flat power characteristics" (R6) (at constant speed in baffled tanks, it can be operated in liquids of widely differing viscosity without change of power requirement), shows desirable large effects of speed on mass transfer coefficients (R4), and can probably be recommended as suitable for most extraction work. At least insofar as liquid-liquid mixing effectiveness in unbaffled tanks is concerned, it has been reasonably well established that the impeller, if placed axially in the tank, is best located at a vertical position which is below the liquid-liquid interface for the unagitated vessel (M2).

Both large-scale motion (mass flow) and small-scale motion (turbulence) are usually required to bring about effective mixing (R5). Different ratios of mass flow to turbulence can be obtained for a given impeller type for the same power input: large ratios for large values of d/T and slow speed, small ratios for small values of d/T and high speed. The requirements peculiar to batch liquid extraction have not been established, but for other services $d/T = 0.2$ to 0.5 is usually recommended for baffled tanks.

b. *Continuous Operation.* The problems here are very different, and the continuously operated agitated vessel will have to be given quite separate consideration. There are practically no data.

The two liquids to be contacted ordinarily simultaneously enter the bottom of the agitated vessel in the desired ratio, flow in parallel through the vessel, and leave at the top. If the nature of the agitation is such that very strong vertical currents are developed in the tank, there may be two undesirable results. First, the tank contents tend to become uniformly mixed, and the solute-concentration gradients responsible for mass transfer between the immiscible liquids will be essentially the same as those in the effluents. If high stage efficiencies are desired, the effluent liquids will be substantially at equilibrium, and the concentration gradients are then small. The extraction taking place within the entire tank therefore will occur under low concentration gradients, and the holding time for the liquids to remain in the vessel must therefore be great. Second, end-to-end mixing will also invite undesirable short-circuiting, by which is meant passage of a portion of the liquid completely through the tank in a time which is shorter than the average holding time. It would seem better therefore to strive for purely radial mixing currents within the vessel, so that mixing occurs only in horizontal planes but not in the vertical direction. In this way the bulk of the tank contents are contacted under conditions of high concentration differences. Since vertical wall baffles tend to produce vertical mixing currents, it would seem better to eliminate such baffles in continuously operated agitated vessels.

Furthermore, the flat-bladed turbine impeller would seem more suitable for agitation under such conditions than the marine-type impeller.

The very meager data available confirm this. Overcashier et al. (O3) extracted butyl amine from kerosene into water in both baffled and unbaffled, continuously operated, agitated vessels, with liquid depth approximately equal to the tank diameter. It was found that the agitation power required to produce a given extraction efficiency was less with unbaffled than with baffled tanks. Of the several impeller designs investigated, the flat-bladed turbine was the least sensitive to the presence of baffles in the influence of power on stage efficiency, owing to the minimal vertical mixing provided by this design. In the absence of baffles, depth of impeller submergence and phase ratio were of insignificant consequence, but with baffles these were of considerable importance. In the absence of baffles, $d/T = 0.4$ seemed best for an impeller centered in the tank, while for a baffled tank, a d/T ratio of 0.6 was indicated. We urgently need confirmation of these findings with other liquid-liquid systems.

In view of the discussion above, it would seem perhaps desirable to use an elongated tank, whose depth is larger than its diameter, possibly with multiple turbines on a single shaft. There is also the possibility of compartmenting the tank to minimize vertical mixing, using horizontal doughnut-shaped baffles and turbines in each compartment. No liquid-extraction data for such arrangements are available. Furthermore, as Rushton has pointed out (R5, S1), an impeller can be sized and operated so as to produce any ratio of radial flow from the impeller to flow through the vessel in order to minimize short-circuiting. But requirements are not yet established.

3. *Power*

With certain limitations which will be touched upon below, we can now reasonably well estimate the power required to turn an impeller of one of the standard designs at any speed in a vessel containing a liquid. The developments here are due largely to Rushton and his co-workers, who have made extensive contributions to our understanding of the many problems involved. Most of the experimental work upon which our knowledge is based has been done with single-liquid systems agitated in open vessels, i.e., in the presence of an air-liquid interface. Under these conditions Rushton et al. (R6) have shown that the power required depends upon the type, position, and speed of rotation of the impeller, the shape of the container, the presence or absence of baffles, and the physical properties of the liquid. In order to provide generalized correlations of the data taken under a wide variety of conditions, it is necessary to establish certain conditions of similarity among the circumstances of the various experiments. Under conditions of geometric similarity,

$$\phi = \frac{N_{Po}}{N_{Fr}{}^n} = kN_{Re}{}^s \tag{2}$$

The Reynolds number N_{Re} accounts for viscous forces, and the Froude number N_{Fr} for the force of gravity when this is important. For a typical impeller-tank arrangement, curves of the sort shown in Fig. 1 result, based on Eq. (2). Briefly, at low values of N_{Re} (viscous flow, A to B in the figure) for both baffled and unbaffled tanks, no vortex is produced and the Froude number is unimportant ($n = 0$). In turbulent flow the

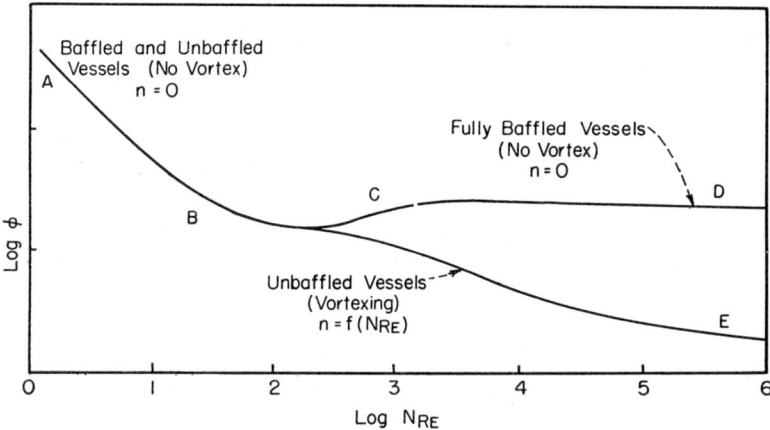

FIG. 1. Power characteristics of a mixing impeller, single-liquid batch system in an open vessel (R6).

Froude number is of importance if vortexing occurs, as in an unbaffled vessel (curve BE), but of no importance if the vessel is fully baffled (curve CD). The detailed curves for various circumstances are given elsewhere (R6), and the influence of such variables as d/T, d/Z, number of blades on the turbines, etc., can reasonably well be estimated.

In the case of batch agitation of immiscible liquid mixtures there are relatively few data. Presumably the same curves developed for single liquids could be applied provided the physical properties of the mixture could be adequately characterized. The Weber number N_{We} is presumably of importance, since it introduces the property of interfacial tension, but there has been no work establishing its influence. Miller and Mann (M2) found that use of an arithmetic average of the densities of the unmixed liquids,

$$\rho_{avg} = v_D\rho_D + v_C\rho_C \tag{3}$$

and a geometric average viscosity

$$\mu_{avg} = \mu_D{}^{v_D}\mu_C{}^{v_C} \tag{4}$$

would reasonably well permit use of single-liquid power curves for correlation of two-liquid data in unbaffled vessels. These have also been used by others (O2, O3). Certain liquid mixtures, notably kerosene-water, behaved anomalously, however, and this was ascribed to the formation of a dynamically stabilized emulsion whose viscosity is dependent upon the rate of shear. The present author has also noted peculiarities with this system. Vermeulen et al. (V2) suggest the use of

$$\mu_{avg} = \frac{\mu_C}{v_C}\left(1 + \frac{1.5 v_D \mu_D}{\mu_C + \mu_D}\right) \tag{5}$$

based on an extension of Taylor's analysis (T1), but this has not been evaluated. It would show that the viscosity of the mixture is always greater than that of the continuous phase alone. For practical purposes, perhaps, these matters are not of great moment. Power need not usually be estimated very precisely, especially since motors for driving impellers are available only in specific horsepower ratings with fairly large increments anyway.

For continuous operation with two liquids, where the vessel is operated full and consequently in the absence of vortex, very limited tests with flat-bladed turbines (F3) have tentatively indicated that (a) the effect of extent of baffling on power requirements may be different from that found in open vessels, but has not yet been established; (b) at all except low agitator speeds the ratio of phases contained within the vessel is the same as the ratio in the feed mixture; and (c) the power is independent of rate of flow through the vessel. But these conclusions are all based on measurements with small vessels and with only two systems. They need confirmation under a much wider variety of conditions.

4. Interfacial Area

Miller and Mann (M2) made the first attempt to measure the effectiveness of mixing by attempting to determine the uniformity of dispersion, really a rough measure of the interfacial area produced. Briefly, the "percentage mixed" was defined as $100\ (v_s/v)$, where v_s is the volume fraction in any sample from the vessel of that phase in which the sample is lean, and v is the volume fraction of the same phase in the entire batch. A mixing index was then defined as the average of the percentage mixed for samples taken from various parts of the vessel. Samples, of course, must not be taken in vertical positions straddling the position of the liquid-liquid interface under unagitated conditions. Values of the mixing index equal to 100 would then presumably represent perfect mixing. But it should be noted that even such an index does not really measure the fineness of dispersion, since it depends upon the size of the sample withdrawn. At any rate, for batch agitation of kerosene-water mixtures in unbaffled tanks, individual samples showed on occasion 100% mixed, but the highest value of the mixing index recorded was 98. For most of the impeller types tested, the mixing index rose with speed up to a maximum value in the neighborhood of 80 to 90, and with further increase in speed the mixing index either remained unchanged or receded to a lower level. Individual performance depended upon the phase ratio of the liquids. The maximum index occurred at power inputs in the range of 250 to 500 ft.-lb./(min.) (cu. ft. of liquid). Flat-bladed paddles gave the highest mixing indexes.

Vermeulen et al. (V2) measured the interfacial surface produced in two-liquid batch systems in baffled vessels fitted with a flat-bladed paddle impeller, using an optical technique. The time required to reach a steady-state degree of dispersion ranged from less than 1 min. to as much as 2 hr. or more. Average drop diameters were in the range 0.003 to 0.1 cm. Drop diameters increase with the distance from the impeller; in other words, coalescence occurred, especially for systems which reached steady state quickly. For two-liquid systems it was found that

$$\frac{d_p}{d} = r N_{We}'^{-0.6} \tag{6}$$

where the Weber number N_{We}' is calculated with $\rho' = 0.6\rho_D + 0.4\rho_C$ as the density, and d_p is the droplet diameter in the vicinity of the impeller tip at $v_D = 0.1$. The drop diameter increases with v_D in a manner which is not well defined for two-liquid systems. The value of r is in the range 0.054 to 0.17, with a mean value of about 0.084. Work of this sort is most certainly worthy of extension.

5. *Mass Transfer Rates, Transfer Units, and Stage Efficiencies*

Before these matters are discussed, it will be well to define the terms carefully as they apply both to batch and to continuous systems. This is especially important since, even in the meager literature of the subject now available, at least three different types of stage efficiencies have already been used to describe agitated vessels.

a. Batch Operation. Consider a batch operation where L_R moles of liquid solution containing initially x_1 mole fraction of extractable solute is contacted with L_E moles of insoluble solvent containing initially y_1 mole fraction of the same solute. For simplicity, assume that only the solute is transferred and that the solutions are dilute, so that L_E and L_R remain essentially constant. During agitation and extraction the raffinate concentration falls to x_2 and that in the extract rises to y_2 mole fraction of solute, so that a solute balance provides

$$N = L_R(x_1 - x_2) = L_E(y_2 - y_1) \tag{7}$$

where N is the number of moles of solute transferred. This operating-line equation, as it is called, is shown graphically in Fig. 2 as a line of

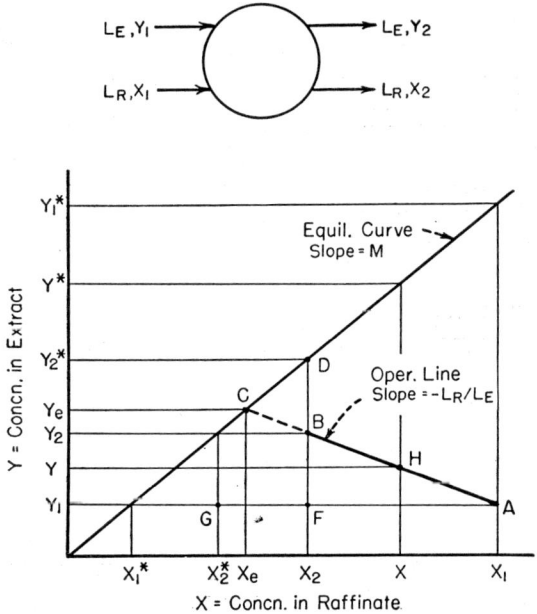

FIG. 2. Batch and continuous extraction, agitated vessels.

slope $-L_R/L_E$, passing through the points A and B whose coordinates are the initial and final concentrations of the solute in the liquids. This diagram also shows the equilibrium distribution of the solute between the liquids, which in the interest of simplicity is here taken as a straight line of slope m. If y^* represents the equilibrium concentration in the extract corresponding to any raffinate concentration x, then

$$y^* = mx \tag{8}$$

Let the solute concentration in the extract at any time be y and that in the raffinate be x, as at H in the figure. During the time $d\theta$ the number of moles of solute transferred from raffinate to extract is $dN = L_E dy$, and this rate can be described in terms of the over-all extract mass transfer coefficient K_E:

$$L_E dy = K_E A(y^* - y)d\theta = K_E a V(y^* - y)d\theta \tag{9}$$

The quantity $(y^* - y)$ is the over-all driving force expressed in terms of extract compositions, as shown in the figure. The over-all number of extract transfer units N_{tOE} for the extraction is then defined by integration of Eq. (9), assuming $K_E a$ remains constant:

$$N_{tOE} = \int_{y_1}^{y_2} \frac{dy}{y^* - y} = \frac{K_E a V \theta}{L_E} = \frac{\theta}{\theta_{tOE}} \tag{10}$$

where θ_{tOE} is the time of an over-all extract transfer unit, as defined by Hixson, Drew, and Knox (H3) and θ is the time of extraction. Where operating line and equilibrium curve are both straight, it is readily shown that

$$N_{tOE} = (y_2 - y_1)/(y^* - y)_m \tag{11}$$

where $(y^* - y)_m$ is the logarithmic average of $(y_1^* - y_1)$ and $(y_2^* - y_2)$. Similarly in terms of raffinate compositions, we have

$$N_{tOR} = \int_{x_2}^{x_1} \frac{dx}{x - x^*} = \frac{K_R a V \theta}{L_R} = \frac{\theta}{\theta_{tOR}} \tag{12}$$

Should the extraction be continued until substantial equilibrium between the phases occurs, then the material balance equation (7) shows that the concentrations in the liquids move along line AB extended until the equilibrium curve is reached at C, giving rise to the ultimate equilibrium concentrations x_e and y_e. A fractional stage efficiency E may then logically be defined (T3) as the ratio of the number of moles N of solute actually transferred in an extraction to N_e, the moles which would be transferred should equilibrium be reached:

$$E = \frac{N}{N_e} = \frac{y_2 - y_1}{y_e - y_1} = \frac{x_1 - x_2}{x_1 - x_e} = \frac{\overline{AB}}{\overline{AC}} \tag{13}$$

The Murphree stage efficiencies are also in common use. In terms of the extract the Murphree extract stage efficiency E_{ME} is defined as the ratio of the number of moles of solute actually transferred to the number which would be transferred if the final extract were in equilibrium with the actual final raffinate. Preservation of a material balance does not permit such final equilibrium concentrations actually to develop, but in terms of the concentrations shown in Fig. 2 this becomes

$$E_{ME} = \frac{y_2 - y_1}{y_2^* - y_1} = \frac{\overline{BG}}{\overline{DG}} \tag{14}$$

Similarly the Murphree raffinate stage efficiency E_{MR} is defined as

$$E_{MR} = \frac{x_1 - x_2}{x_1 - x_2^*} = \frac{\overline{AF}}{\overline{AG}} \qquad (15)$$

If E, E_{ME}, and E_{MR} are equal to unity, equilibrium between the effluents has been obtained, and the extraction is said to correspond to one ideal or theoretical stage. Usually both agitation and settling of the dispersion are included in the action of such stages.

b. Continuous Operation. Equation (7) for a material balance and Fig. 2 still apply, except that L_E and L_R must now be defined in terms of lb. moles/(hr.)(sq. ft.). If the height of the liquid is Z ft., the rate equations become

$$dN = L_E dy = K_E a(y^* - y)dZ; \qquad L_R dx = K_R a(x - x^*)dZ \qquad (16)$$

$$N_{tOE} = \int_{y_1}^{y_2} \frac{dy}{y^* - y} = \frac{K_E a Z}{L_E} = \frac{Z}{H_{tOE}} \qquad (17)$$

$$N_{tOR} = \int_{x_2}^{x_1} \frac{dx}{x - x^*} = \frac{K_R a Z}{L_R} = \frac{Z}{H_{tOR}} \qquad (18)$$

where H_{tOE} and H_{tOR} are the over-all heights of extract and raffinate transfer units, respectively. The stage efficiency E and the Murphree stage efficiencies E_{ME} and E_{MR} are still defined by Eqs. (13), (14), and (15), respectively.

In addition, in the case of cascades of stages operated in countercurrent flow, an over-all fractional stage efficiency E_O is defined as the ratio of the number of ideal stages n_i to which the cascade is equivalent divided by the number of real stages n_r:

$$E_O = n_i/n_r \qquad (19)$$

Refer to Fig. 3, where there is shown a cascade of three ideal stages operated in countercurrent. A solute balance may be written for any stage, such as stage 2, for example:

$$L_E(y_2 - y_3) = L_R(x_1 - x_2) \qquad (20)$$

which is shown graphically as the stage operating line FJ on the figure. The line touches the equilibrium curve at point F since the stages are stipulated to be ideal. Similarly a material balance for the entire cascade is

$$L_R(x_0 - x_3) = L_E(y_1 - y_4) \qquad (21)$$

which is shown as the cascade operating line CK on the figure. The familiar staircase construction between operating line and equilibrium curve, $CBDFGHK$, is then ordinarily used to represent the stepwise change in concentrations occurring in each stage of the cascade.

Owing to stage inefficiencies, or failure to reach equilibrium concentrations in the effluents from each stage, more real stages are required than the ideal number. Figure 4 represents the case where four real stages are required to accomplish the same change of concentration as in Fig. 3. If stage efficiencies are known, curve MN is drawn between the cascade

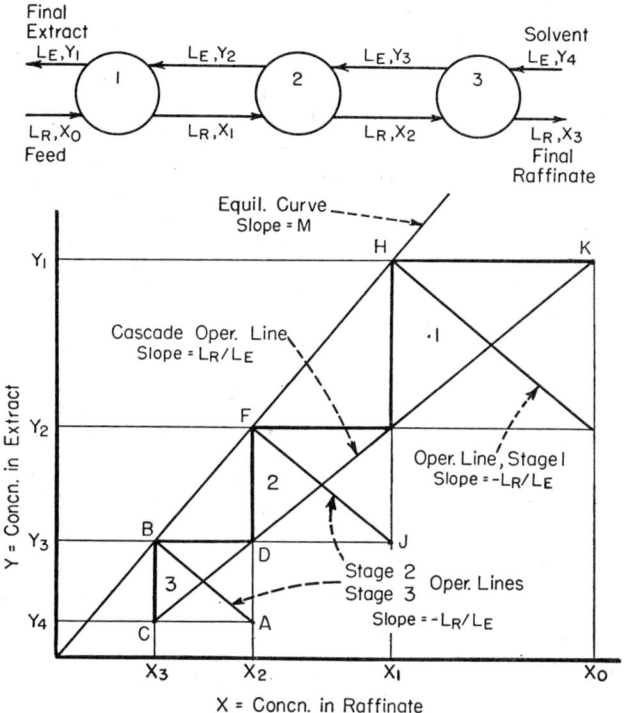

Fig. 3. Continuous, countercurrent, multistage extraction.

operating line and the equilibrium curve in accordance with the stage efficiency. Thus, for the second stage, $E = \overline{BH}/\overline{FH}$; or, in terms of Murphree efficiencies, $E_{ME} = \overline{BG}/\overline{DG}$ and $E_{MR} = \overline{BJ}/\overline{AJ}$. The position of curve MN is everywhere adjusted to provide for the efficiency appropriate to the local stage numbers and concentrations. In this particular example, the over-all stage efficiency becomes $E_O = 3/4 = 0.75$.

An additional expression of performance used for countercurrent cascades, particularly for those in the design of towers (see below), is $HETS$, the height equivalent to a theoretical stage.

The relationships among the various stage efficiencies, coefficients, and transfer units can readily be established, and since these do not appear to have been completely summarized elsewhere they are listed for convenient reference in Tables I and II. In all cases it is assumed that operating line and equilibrium lines are straight and in the case of cascades that the individual stage efficiencies are the same for each stage. When these conditions do not pertain, the interrelations are best established graphically by appropriate modification of the operating diagram.

On the basis of the familiar two-film theory, the over-all mass transfer coefficients, as well as the over-all heights and times of transfer units,

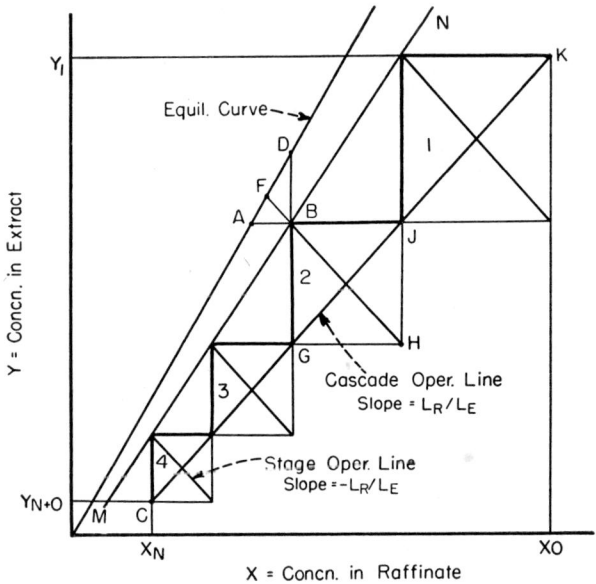

FIG. 4. Stage efficiencies in multistage cascades.

can be related to the resistances to mass transfer residing in the individual phases in a manner which is well established (S8, T4). Experimentally, a technique which has been found useful for this purpose in the study of packed and spray extraction towers involves contacting two pure liquids of limited miscibility, in the absence of a distributed solute (C4). The extent of mutual saturation in the device then provides data for determination of the individual phase resistances. This technique has not yet been applied to agitated vessels, nor, for that matter, to any of the devices described in this report, but it is most certainly one which could be applied to advantage.

For reasons of simplicity in dimensions, the height or the time of a transfer unit offers advantages over the mass transfer coefficient. For

TABLE I
Rate and Efficiency Relationships for Single Stages

$$E = 1 - \exp\left\{\frac{-K_E a Z(1 + mL_E/L_R)}{L_E}\right\} = 1 - \exp\left\{\frac{-K_E A \theta(1 + mL_E/L_R)}{L_E}\right\}$$

$$E = 1 - \exp\left\{\frac{-K_R a Z(1 + L_R/mL_E)}{L_R}\right\} = 1 - \exp\left\{\frac{-K_R A \theta(1 + L_R/mL_E)}{L_R}\right\}$$

$$E = 1 - \exp\{-N_{tOE}(1 + mL_E/L_R)\} = 1 - \exp\{-N_{tOR}(1 + L_R/mL_E)\}$$

$$Z = H_{tOR} N_{tOR} = H_{tOE} N_{tOE}$$

$$\theta = \theta_{tOR} N_{tOR} = \theta_{tOE} N_{tOE}$$

$$N_{tOR} = \frac{mL_E}{L_R}(N_{tOE}); \quad K_R a = mK_E a$$

$$E = \frac{E_{ME}(1 + mL_E/L_R)}{1 + E_{ME}(mL_E/L_R)} = \frac{E_{MR}(1 + L_R/mL_E)}{1 + E_{MR}(L_R/mL_E)}$$

$$E_{ME} = \frac{1 - \exp\{-N_{tOE}(1 + mL_E/L_R)\}}{1 + \frac{mL_E}{L_R}\exp\{-N_{tOE}(1 + mL_E/L_R)\}} = \frac{E}{1 + (1 - E)mL_E/L_R}$$

$$E_{MR} = \frac{1 - \exp\{-N_{tOR}(1 + L_R/mL_E)\}}{1 + \frac{L_R}{mL_E}\exp\{-N_{tOR}(1 + L_R/mL_E)\}} = \frac{E}{1 + (1 - E)L_R/mL_E}$$

TABLE II
Efficiency Relationships for Countercurrent Cascades of Stages

$$E_0 = \frac{\log[1 + E_{ME}(mL_E/L_R - 1)]}{\log mL_E/L_R} = \frac{\log[1 + E_{MR}(L_R/mL_E - 1)]}{\log L_R/mL_E}$$

$$E_{ME} = \frac{\left(\frac{mL_E}{L_R}\right)^{E_0} - 1}{\frac{mL_E}{L_R} - 1}$$

$$E_{MR} = \frac{\left(\frac{L_R}{mL_E}\right)^{E_0} - 1}{\frac{L_R}{mL_E} - 1}$$

$$E_0 = \frac{\log\left[\frac{1 + \frac{mL_E}{L_R} - E}{1 + \frac{mL_E}{L_R}(1 - E)}\right]}{\log \frac{mL_E}{L_R}}$$

$$E = \frac{\left(1 + \frac{mL_E}{L_R}\right)\left[1 - \left(\frac{mL_E}{L_R}\right)^{E_0}\right]}{1 + \left(\frac{mL_E}{L_R}\right)^{E_0+1}}$$

$$HETS = H_{tOR}\left(\frac{\ln \frac{mL_E}{L_R}}{1 - \frac{L_R}{mL_E}}\right) = H_{tOE}\left(\frac{\ln \frac{mL_E}{L_R}}{\frac{mL_E}{L_R} - 1}\right)$$

cases where the approach to equilibrium is very close, however, these will be awkward to use. For an ideal stage, for example, mass transfer coefficients become infinite while heights and times of transfer units become zero. In these cases stage efficiencies will probably be more convenient, and it is recommended that in the future all such experimental data be expressed in terms of E rather than over-all or Murphree efficiencies. The former is most simply related to the mass transfer rates, is just as easy to use as the Murphree expressions, and seems more logical in its definition.

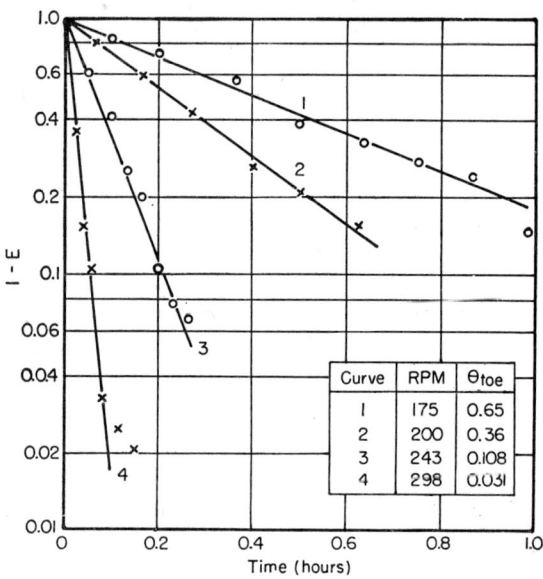

FIG. 5. Extraction of iodine between water and carbon tetrachloride; 25-cm.-diameter vessel, 1,113 ml. CCl_4, 11,130 ml. H_2O, propeller height = twice CCl_4 depth. Data of Hixson and Smith (H2).

c. Extraction Rates. Only a few investigations of this type of equipment have been published. Hixson and Smith (H2) extracted iodine from water by contact with one-tenth volume of carbon tetrachloride in three sizes of unbaffled vessels agitated with marine-type propellers. The propellers were located axially at various depths, and the experiments were conducted batchwise with an air-liquid interface. By analyzing the liquids after various times of agitation, it was possible to determine the stage efficiency E as a function of time. Assuming that, for a given set of operating conditions θ_{tOE} would remain constant, the relation (Table I) between θ and E suggests that a semilogarithmic plot such as Fig. 5 would provide a straight line for each run (H2), from the slope of which

θ_{tOE} could be computed. All the Hixson and Smith data plot well in this fashion, and the straightness of the lines indicate the utility of the time-of-a-transfer-unit concept. Hixson, Drew, and Knox (H3) showed that a characteristic "agitation number" may be defined as the product of θ_{tOE} and a velocity term for the agitated system. If then the mass transfer coefficient varies as the first power of the chosen velocity term, the agitation number would be constant for a given ratio of interfacial surface to total number of moles of extract phase. In liquid extraction, speed of agitation influences both terms of the quantity $K_E a$, and satisfactory correlation has been difficult to obtain.

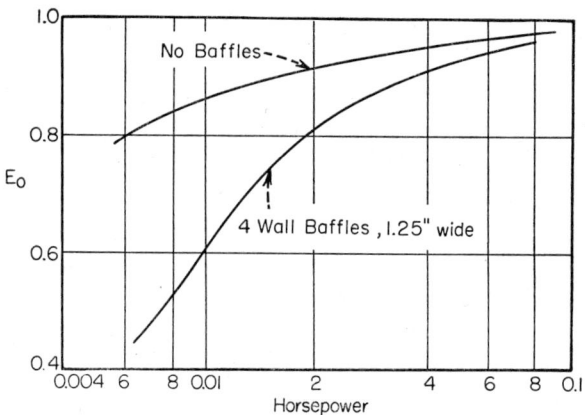

Fig. 6. Continuous extraction of n-butylamine from kerosene into water with 6-in. flat-bladed turbines. Vessel diameter = 14.75 in., 7.5 g.p.m. kerosene, 4.78 g.p.m. water, residence time = 1.08 min. Data of Overcashier et al. (O3).

Overcashier et al. (O3) extracted n-butylamine from kerosene with water in a continuously operated agitated vessel, both baffled and unbaffled, with various types of impellers placed at various depths. Some of their conclusions have already been discussed. Extraction rates were expressed as E_O, i.e., as the fraction of a countercurrent ideal stage to which the single vessel was equivalent. Without detailed equilibrium data, these cannot be converted to E. One difficulty in obtaining data from experiments of this sort, which has not been satisfactorily overcome, is the obtaining of samples of the liquids which are truly representative of the effluents: two-liquid samples must be settled prior to analysis, during which time some extraction occurs. Furthermore, it is significant that these authors found considerable fluctuation of concentrations with time in samples taken from the vessel, even to twenty times the residence time of the liquids in the vessel. It is impossible here to analyze their data in detail, but Fig. 6 is presented to indicate in part that baffles were found not to be advantageous, as discussed above. For the 6-in. flat-bladed turbines, agitator energy of roughly 1600 ft.-lb./cu. ft. of throughput were required to give over-all stage efficiencies of 0.98, and the throughput/volume of a theoretical stage for the agitator alone varied from about 42 to 54 hr.$^{-1}$ depending upon the agitator power applied. This latter figure would be considerably smaller if the settler were included in the stage volume.

Flynn and Treybal (F3) extracted benzoic acid from toluene and from kerosene into water using both 6- and 12-in. baffled vessels. These were fitted with geometrically similar flat-bladed impellers, and operation was continuous without an air-liquid interface. Considerable extraction was found to occur when no agitation was employed, the amount depending upon the flow rates. In order to correlate the data, it was necessary to express the results as the increase in stage efficiency produced by the agitation. On this basis successful correlation of the data was obtained for each system at a given phase ratio, for both vessel sizes, by use of the specific energy applied by the agitator to the liquids, expressed as foot-pounds per cubic foot of liquid. Stage efficiencies reached 1.0 at specific energies from 40 to 300 ft.-lb./cu. ft., and throughput/volume of a theoretical stage was in the range of 70 to 92 hr.$^{-1}$ for the agitator alone, depending upon the circumstances. The data cannot readily be reconciled with those of Overcashier *et al.*, and clearly much more work must be done.

6. *Emulsions*

The more finely the liquids are dispersed within one another, the more slowly will they settle, either in a separate decanter for a continuous operation or in the same vessel for a batch process. Most stable emulsions, those which settle and coalesce only very slowly if at all, are characterized by maximum particle diameters of the dispersed phase of the order of 1 to 1.5 microns. Presumably one could estimate through Eq. (6) what agitator speeds would produce such droplet sizes, but such calculations are not likely to yield completely useful results. For example, it has been observed on several occasions that the settling ability of some liquid dispersions passes through a minimum as agitator speed is increased.

7. *Scale-Up*

It seems reasonable to assume that if the same flow regime can be produced, it should be possible to predict the rates of extraction for a large vessel from data measured in a small one. Rushton (R4) has explained in detail how this may be done under conditions where dynamic and geometric similarity of the two systems is maintained. The success of such procedures for unbaffled vessels operated continuously without a liquid-gas interface but without vortexing has not yet been established.

D. BENCH-SCALE EQUIPMENT

In recent years there has been considerable interest in bench-scale mixer-settler extractors arranged in countercurrent cascades of stages for continuous operation. The advantages afforded by such equipment are several. They provide many stages in a small space, and, since they provide high stage efficiencies, the data obtainable are representative of what can be done on a large scale. They can readily be operated intermittently, with shutdowns overnight, for example, and yet provide steady-state conditions more or less immediately upon starting up. In

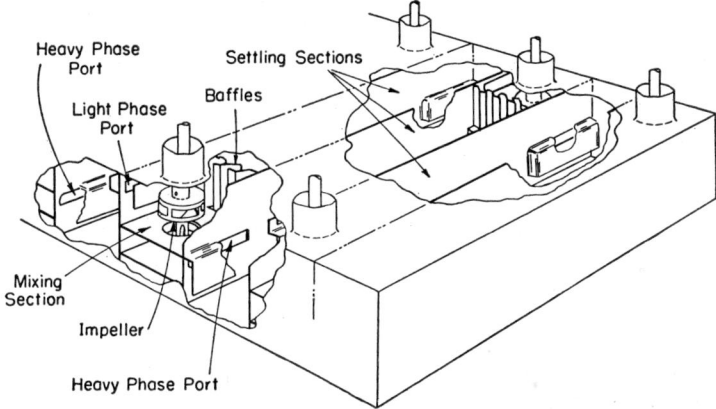

FIG. 7. Pump-mix extractor (schematic).

many process-development problems the complexity of the liquid systems involved do not permit ready interpretation in terms of orthodox equilibrium phase diagrams because of difficulty in analyzing the liquids, and so direct experimentation is necessary to determine the number of stages required to produce satisfactory products. Equipment of the sort described here readily permits this without the necessity of resorting to clumsy approximations of continuous operation with batchwise separatory funnels. Processing can actually be carried on for long periods of time with only small amounts of materials, to determine the process influence of such matters as reuse of solvent, possible accumulation of unsuspected impurities in recovered solvents, and the like.

1. *The Pump-Mix Extractor*

Figures 7 and 8 show one of the pump-mix types of extractors (C5, S2). In this case each stage consists of a mixing chamber and a settler. Adjacent stages of the cascade are alternated in right- and left-hand positions, arranged with common walls. Each stage occupies a volume 3 in. wide by 5 in. high by 11 in. long, of 2.62 liters of holdup. The impeller for each mixer is at the same time a pump, acting in the manner shown in Fig. 8. A typical cascade of 16 stages is easily placed on a small table top. Flow rates in the range 0.2 to 0.5 gal./min. total flow can be handled, depending upon the system and agitator speed, and operation is essentially automatic. Samples can readily be withdrawn from the settlers to provide intermediate concentrations of the solutes, should that be desired. In tests with acetic acid–methylisobutyl ketone–water, with an 8-stage cascade, over-all stage efficiencies E_O between 0.8 and 1.0 were obtained, depending upon the direction of extraction, throughput, and stirrer

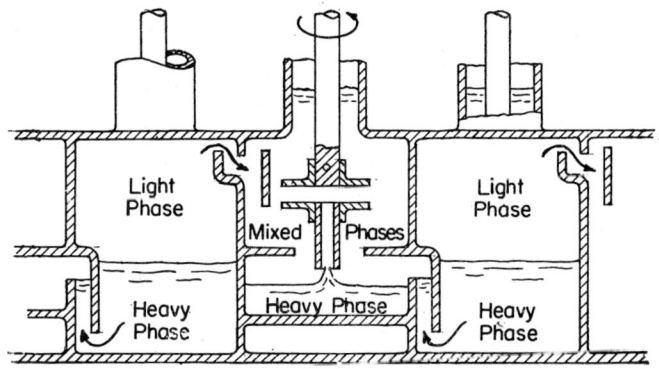

FIG. 8. Pump-mix extractor (schematic elevation).

speeds. These correspond to throughputs/volume of a theoretical stage in the range of 26 to 40 hr.$^{-1}$ for the mixer settler combined, or 94 to 145 hr.$^{-1}$ for the mixing sections alone.

2. *Others*

Alter, Codding, and Jennings (A1) have described somewhat similar equipment providing for 16 stages in a cascade 3 by 14 by 8 in., with a holdup of about 8 ml./stage, throughputs up to 4 ml./min., and over-all stage efficiencies E_O in the range of 0.7 to 0.95. Still another is described by Davis, Hicks, and Vermeulen (D1).

III. Agitated, Multistage Countercurrent Columns

The disadvantages of the countercurrent cascade of mixer settlers, including the necessity for large-volume settlers and interstage pumping of liquids, are not present in the vertical column. Here the liquids flow continuously in countercurrent fashion by virtue of their different specific gravities. With few exceptions the liquids are dispersed only once, and intermediate coalescence between stages is generally dispensed with. The problem of maintaining adequate turbulence under the condition of low rates of flow through the towers has led to the application of mechanical agitation in the towers, and the three designs in particular which are discussed below have aroused considerable interest.

A. Field of Usefulness

All these columns are limited strictly to continuous operation. Theoretically, at least, they may be built to contain the equivalent of as many stages as desired simply by making them sufficiently tall. Practical limitations of height will be dictated by cost of erection, supports, and foundations, limitations on shafting for the agitators, and perhaps even the cost of pumping the liquids to the top of the tower. Choices as between columns of this sort and mixer settlers will generally be made on the basis of cost, as discussed under agitated vessels, since there seem to be no process limitations on the use of either. The towers, in any case, do not offer the opportunity for advantageous intermittent operation which is possible with mixer settlers.

B. Comparison of the Extractors

Any evaluation of these columns must be made on the basis of the liquid-throughput rates which are possible without flooding, the mass transfer rates, however expressed, and the mechanical power required to operate them. Our knowledge of the basic mechanisms of liquid extraction is very limited, but over the past several years we have at least established that certain system variables have considerable influence on throughput and mass transfer rates. We know, for example, that the following are of importance: (1) the physical properties of the liquid systems, including at least density difference, viscosity of each phase, interfacial tension, as well as diffusivity and equilibrium distribution ratio of transferred solute, all of which for a given system change with concentration of solute; (2) the ratio of the flow rates of the insoluble phases; (3) which phase is dispersed and which continuous. The rate of mass transfer is additionally dependent upon (4) the total rate of liquid flow and (5) the direction of extraction, i.e., whether from dispersed to continuous phase or the reverse, and whether from organic to aqueous phase or the reverse. In the case of the mechanically agitated columns, both mass transfer rates and maximum permissible throughputs are also dependent upon the nature of the flow pattern developed by the agitation, in turn dependent upon the mechanical characteristics of the agitators and the internal design of the tower. These in their turn are not independent of the system variables in their influence.

This is certainly an enormous number of variables to consider, and it need hardly be emphasized that in the case of any of the devices to be described scarcely a beginning has been made on the work. It will be futile to attempt to present, in the limited space here available, a complete résumé of even the meager data which have been gathered, especially since no correlation is yet possible among them. All that can usefully be accomplished is perhaps to suggest the nature of the work that has been done and a rough indication of what the various extractors are capable of. Even this is not without danger, especially if attempts are made to compare extractors on the basis of performance with different liquid systems. Some systems seem to provide easy extraction and invariably give high mass transfer rates or stage efficiencies. Such a system is acetic acid–methylisobutyl ketone–water, one of low viscosity and low interfacial tension [only 10.4 dynes/cm. in the absence of acid and 1.44 dynes/cm. at an aqueous acid concentration of about 25% (K2)], a system which has frequently been used by experimenters in this field. Other systems are difficult, particularly those of high interfacial tension (say 30 dynes/cm., or more) and high viscosities, such as those involving kerosene and water. It has been the author's experience that frequently equipment which provides very favorable results with the first type will perform much more poorly with the second. Consequently the brief summaries given here must be used with the utmost caution.

1. *The Oldshue-Rushton Column* (O1)

The Oldshue-Rushton column is a development which stems directly from consideration of the agitated vessel as an extraction device. Figure 9 shows a schematic arrangement of this tower, and it is seen to be compartmented by a number of horizontal, doughnut-type baffles into a series of stages, each of which is agitated by a flat-bladed turbine. The turbines are all attached to a single vertical shaft driven from the top of the tower. Four vertical, radially arranged baffles extend the length of the tower, in the manner used in agitated vessels. The variables which would influence the performance of the extractor, in addition to those introduced by the liquid system as outlined above, are (1) the height of compartments, (2) the diameter of the central opening of the horizontal baffles, (3) the extent of vertical baffling employed, (4) the design and position of the impeller, and (5) the speed of the impeller.

In the single publication which describes this device (O1), tests are reported for a 6-in.-diameter tower with the system acetic acid–methylisobutyl ketone–water, ketone dispersed. Extraction rates are expressed both as height-equivalent-to-a-theoretical-stage $HETS$, and over-all stage efficiency E_O, the latter based on the definition of a real stage as one compartment. It was established that inferior performance resulted when the turbines are placed at the same level as the horizontal baffles, i.e., in the baffle opening, and the central location in each compartment, as shown in Fig. 9, seems best. As compartment height was increased from 3

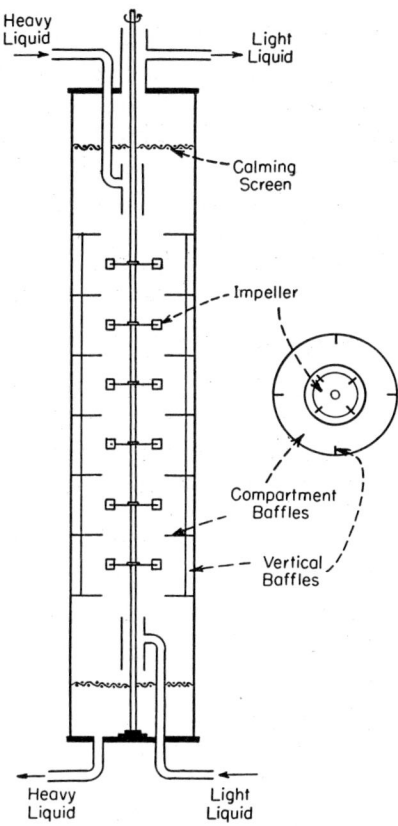

Fig. 9. The Oldshue-Rushton column (schematic).

to 6 in., E_O increased owing probably to the longer time of contact in each compartment, but the value of $HETS$ also increased. At the same time throughput capacity also increased, and so the compartment height used in a given instance would depend upon a consideration of $HETS$, throughput, and cost. In larger towers it seems quite probable that compartment heights less than one-half the tower diameter will be desirable, especially if many stages are required. Increasing the central opening of the horizontal baffles decreased the stage efficiency at a given throughput, but increased the permissible throughput, so that again a choice will be based upon the relative value of high column efficiency and high flow rates. Both 2- and 3-in.-diameter turbines were capable of producing the same E_O, the latter at a lower speed, but the power required to produce the highest E_O as a function of impeller diameter was not clearly established. Flooding occurs at high turbine speeds with low throughputs and at low speeds with high throughputs, and at any throughput E_O passes through a maximum as turbine speed increases. Total throughput for the 3-in. compartment height, 2.13-in. baffle opening, at the highest stage efficiency was 39 cu. ft./(hr.)(sq. ft.) (ketone/water = 2), with $E_O = 0.76$. This corresponds to a total throughput/volume of a theoretical stage = 118 hr.$^{-1}$. The highest value of this criterion reported is 177 hr.$^{-1}$, with 3-in. stage height, 3.25-in. baffle opening, and $E_O = 0.62$.

Towers of diameter larger than 6 in. are known to have been built and operated, but details of their performance are not publicly available.

2. *The Rotating-Disk Contactor—RDC* (R2)

Shown schematically in Fig. 10, the rotating-disk contactor is a variant of an older extractor (C2) with certain improvements in the mechanical design which have led to more satisfactory performance. The preferred design uses a compartmented tower without vertical baffles, and a rotating disk centered in each compartment, the diameter of the disks being smaller than that of the opening in the horizontal baffles. The variables, other than those introduced by the liquid system, which influence the performance are (1) diameter of compartment openings, (2) diameter of rotating disks, (3) height of compartment, and (4) speed of disk rotation.

A number of papers (R1, R3, V3) describe tests made with this device, using a wide variety of tower diameters (up to 6.5 ft.) and a large number of systems. The variables which influence performance have not all been studied systematically, but it is amply evident that systems of a very wide variety of liquid physical properties can readily be handled. The influence on performance of speed, compartment height, and baffle openings is qualitatively at least much the same as found in the case of the Oldshue-Rushton extractor, although there are of course quantitative differences and differences from system to system. Rotor speeds are generally higher for the RDC column, although this does not necessarily mean higher power requirements. With the system acetic acid–methylisobutyl ketone–water, water dispersed, in an 8-in.-diameter column, compartment height 1.6 in., baffle opening 4.9 in., disk diameter 3.1 in., the maximum throughput reported (R3) is 131 cu. ft./(hr.)(sq. ft.), with a value of *HETS* equal to 0.355 ft. and a throughput/volume of a theoretical stage = 368 hr.$^{-1}$. Changing to a compartment height of 2.4 in., baffle opening of 6.3 in., and disk diameter of 4.7 in. increased permissible total flow to 138 cu. ft./(hr.)(sq. ft.) and *HETS* to 0.57, but decreased the throughput volume of a theoretical stage to 242 hr.$^{-1}$. Note that these are not comparable to the data of Oldshue and Rushton (O1) owing to the inversion of the phase which was dispersed.

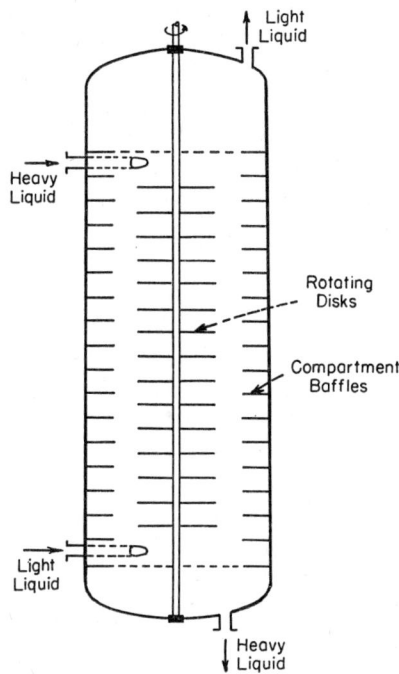

FIG. 10. The rotating-disk column, RDC (schematic).

With the system n-butylamine–kerosene–water (a difficult system) in a 16-in.-diameter tower, 10-in. disk diameter, 4-in. compartment height, 12-in. baffle opening, total flow was 91.2 cu. ft./(hr.)(sq. ft.), and *HETS* was 2 ft., corresponding to a throughput/volume of a theoretical stage = 45.6 hr.$^{-1}$.

These towers are in regular industrial use for service in the furfural extraction of lubricating oils, where they have been built to a diameter of 6.5 ft.

3. *The Scheibel Column* (S3, S4)

Figure 11 shows a schematic representation of this column. It consists of alternate agitated sections and packed sections, the former each provided with a paddle-type impeller. The packing used is a special loosely woven wire mesh of high-percentage void space. The variables which influence the performance, other than those peculiar to the liquid system used, are (1) height of mixing section, (2) height of packed section, (3) design of agitator impeller, (4) speed of impeller, and (5) nature of the packing. The last has been more or less standardized.

Columns of this design have had wide acceptance as useful laboratory devices and are available in small sizes (1-in. diameter) which are capable of producing values of $HETS$ in the neighborhood of 2 to 3 in. They are consequently frequently used for process evaluation where only small amounts of materials may be available. As a result, there have been a relatively large number of publications reporting fragmentary data taken with these small towers, and no attempt will be made to make reference to these in this review. Scheibel and Karr (S5) have reported the most completely systematic data, taken with a 12-in.-diameter tower containing mixing sections of 3-in. height, with packed sections of 9- and 13.5-in. height. With acetic acid–methylisobutyl ketone–water and a 9-in. packed section, for example, $HETS$ decreases with agitator speed and passes through a minimum; at constant agitator speed $HETS$ decreases with total liquid flow to a minimum value. Maximum total throughput reported was 73.7 cu. ft./(hr.)(sq. ft.) at a phase ratio of unity, ketone dispersed, water as extractant, and under these conditions $HETS = 1.05$ ft., corresponding to a throughput/volume of a theoretical stage = 70 hr.$^{-1}$. With water dispersed, ketone as extractant, and the same total throughput, $HETS = 0.741$ ft., corresponding to a throughput/volume of a theoretical stage = 99.5 hr.$^{-1}$. Data for other systems are also reported. The same authors (K2) have presented additional studies, where the effect of the wire-mesh packing has been eliminated by special techniques of introducing the liquids to a single mixer section.

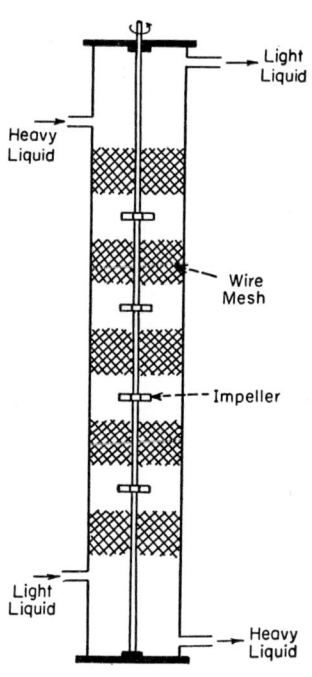

FIG. 11. The Scheibel column (schematic).

The largest reported column of this design is 18 in. in diameter and handles 500 gal./hr. of combined liquid flows (A3).

IV. Pulsed Columns

If a rapid, reciprocating motion of relatively short amplitude is applied to the liquid contents of an extraction column, the column is said to be "pulsed." The agitation thus provided has been found to give improved extraction rates. Devices of this sort were suggested as early as 1935 (V1), but it was not until the atomic energy programs in the United States and Great Britain developed a need for efficient extractors of low

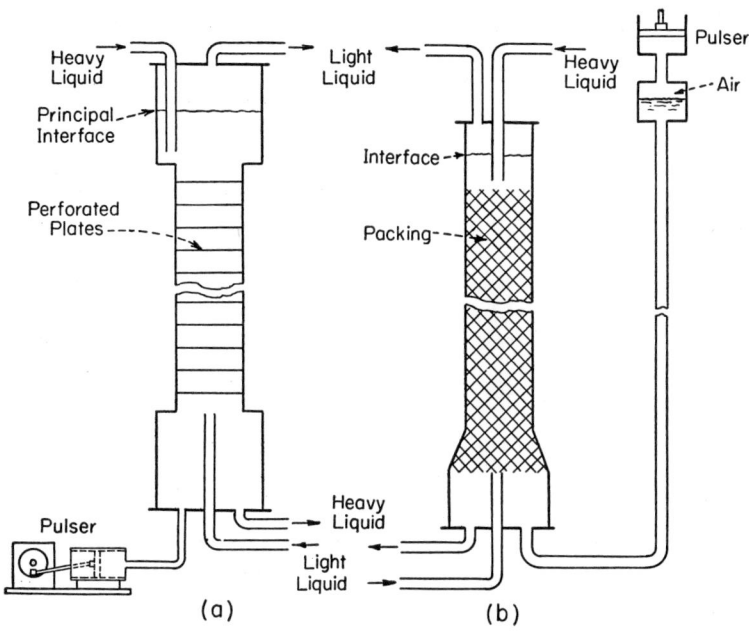

Fig. 12. Pulse columns: (a) liquid-pulsed, perforated-plate column; (b) air-pulsed, packed column (T2).

HETS that they were given serious consideration. Most of the work pertaining to them is classified, and our knowledge of these devices is consequently very fragmentary.

A. Pulsing Methods

Several methods have been used to produce the pulse action required in these columns. A reciprocating, plunger pump, from which the valves have been removed, may be connected to the extractor as at (a), Fig. 12, with a direct liquid connection between the column and the pump piston. Such an arrangement is perhaps mechanically most reliable, although there may be problems of leakage and corrosion at the piston. Alternatively, a flexible diaphragm or bellows, which is given a rapid reciprocating motion, may be in direct contact with the liquid in the tower. Such a diaphragm, made of metal or plastic, may be flexed mechanically or by an electronic transducer (T2), but will then generally suffer from short life. If hydraulically pulsed, it may be possible to extend the life of the diaphragm to 30 million cycles or more (J1). As a third possibility, an air-pulse mechanism may be arranged as at (b), Fig. 12 (T2), thereby eliminating much of the corrosion and mechanical problems at the expense of increased power for pulsing. With devices such as these, together with appropriate cam mechanisms, it has been possible to produce pulsations within the liquid of an extractor whose time-amplitude characteristics are in the form of waves of diverse shapes (sine, square, saw-tooth, etc.).

The following terms are used to describe the pulse action:

1. *Frequency* is the rate at which the reciprocating action is applied, in cycles/time.

2. *Amplitude* is the linear distance between extreme positions of the liquid in the column as a result of the pulse action.

3. *Pulsed volume* is the product of the amplitude, frequency, and cross-sectional area of the column, representing therefore the rate of movement of liquid by pulsing. It is expressed, for example, as cubic foot per hour (sometimes as cubic foot per hour per square foot).

B. Fields of Usefulness

There are no process limitations, *per se*, on the use of pulse columns, and presumably they will be of interest wherever low stage height and consequently low head room are important. In extraction of radioactive material, where these columns have had their principal utility, low *HETS* may mean easier and less costly shielding, although it must be pointed out that the total volume of the equipment may be large owing to reduced throughputs under pulsed operation. In this service, also, pulsing as a remote means of agitating the extractor is attractive since there need be no rotating or other moving mechanical parts in dangerously radioactive areas. Pulsing offers a simple means of improving the performance of existing extractors, since it may be added to existing facilities with relatively little expense. Scale-up of designs with data from small pulsed equipment is likely to be less hazardous than with other equipment, although in some cases at least it cannot be done directly (W2).

On the other hand, the cost of installation and maintenance of the pulse generator, and the power required for pulsing, must be considered. Pulsing increases the tendency toward emulsification, which may make it disadvantageous with some systems. Further, if large heads of liquids are pulsed at high frequencies, the pressure at the pulsing device on the back stroke may become less than the vapor pressure of the liquid, and cavitation will result (T2). It should be noted that this cannot happen with the air-pulse arrangement, which is self-compensating.

C. Power for Pulsing

During the pulsing action, force must be applied to the liquid in the extractor to overcome static heads, inertia, and friction. The power required then becomes the product of the force, the cross-sectional area of the tower, and the rate of linear displacement of the pulse. From these considerations, an equation has been derived (J1) for estimating the power, and this was tested on a 2-ft.-diameter tower containing 40 ft. of packing in the form of 20-mesh screens. During the test only one liquid phase, rather than two, was pulsed, and the maximum instantaneous power requirements were in the vicinity of 6 hp. It was suggested that a fly wheel on the pulse mechanism, acting as an energy reservoir, would reduce power requirements since the power is negative during one half the cycle. Alternatively, two columns might be pulsed simultaneously with the same pulse generator, with the pulse in each column out of phase by 180° (G1).

D. Internal Design of Towers

Presumably any type of extraction tower could be pulsed. Most of the applications have been made with a special perforated-plate design, although some work has been done with towers packed with the common random packings (Raschig rings, etc.) and certain regular packings (corrugated expanded metal mesh (T2) and screening (J1)).

1. *Perforated-Plate Towers*

The arrangement used is shown in Fig. 12(a). The horizontal plates are pierced with holes of small diameter (approximately $\frac{1}{8}$-in. diameter), whose total cross section may be 20 to 25% that of the column. No downspouts are provided, as for ordinary perforated-plate extractors, and both liquids must pass through the same holes. The plate spacing is usually small, 2 in. in large-diameter columns, for example. Columns of this sort may be pulsed at amplitudes of 1 in. and frequencies up to 150 cycles/min., depending upon the circumstances.

a. Flow Characteristics. The perforations in the plates are so small that the liquids will not flow in countercurrent fashion of their own accord, owing to interfacial tension. Consider, however, the situation when the column is pulsed at low frequency and amplitude. At the start of the upstroke of the pulse, the liquids are stratified between each plate, with a layer of light liquid below and one of heavy liquid above each plate. On the upstroke light liquid is forced through the perforations in each plate and is injected into the heavy liquid above the plates, dispersed into small droplets. The droplets rise through the heavy liquid and coalesce into the layer of light liquid underneath the plate above. If the pulsed volume is larger than the rate of flow of light liquid, some of the heavy liquid will ultimately be forced upward through the plate at the end of the upstroke. On the downstroke of the pulse, heavy liquid in

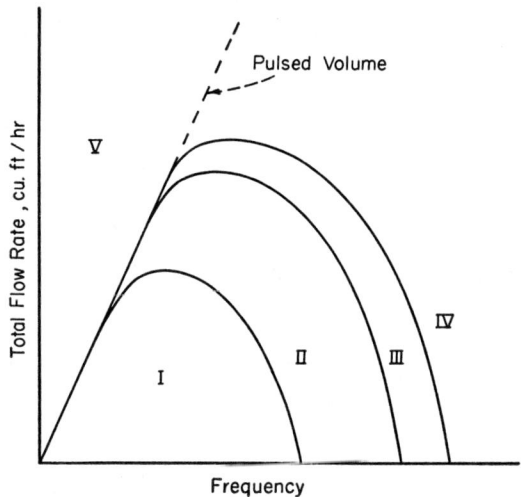

FIG. 13. Flow characteristics of pulsed perforated-plate columns (S6).

similar fashion is drawn downward through the perforations. Such operation corresponds to region I of Fig. 13 (S6). Under such conditions operation is very stable, but there is little improvement in the extraction rate over what might be accomplished in an unpulsed perforated-plate column of conventional design. As the frequency is increased (region II, Fig. 13), there is insufficient time for coalescence of the droplets, which remain more or less permanently dispersed. The drop sizes are small and the interfacial surface between the liquids is very large. The most efficient extraction occurs under these conditions. If, however, the frequency is increased unduly, coalescence of the small drops may again occur, an increased number of large drops may appear, there are irregular reversals of the dispersed and continuous phases, and operation becomes unstable (region III, Fig. 13). Further increase in frequency results in flooding owing to emulsification of the liquids (region IV, Fig. 13). Transition from regions I to IV is not abrupt, as the figure might imply, but rather continuous and gradual.

It is clear that the pulsed volume must at least equal the rate of flow of that liquid which flows at the largest volume rate, or else flooding will occur owing to insufficient pumping action of the pulse (region V, Fig. 13). Too large a pulsed volume will result in severe back mixing: for example, the heavy liquid will be forced through the perforations on the upstroke to such an extent that the concentration gradients characteristic of countercurrent flow will be destroyed, and extraction rates will suffer. The total liquid throughput which is possible is relatively independent of plate spacing provided that the spacing is large with respect to the pulse amplitude, and increases with the percentage of the column cross section devoted to perforations (S6). Furthermore, it is highly desirable that the material of the plate be preferentially wetted by that liquid which it is desired to maintain continuous, and if necessary the plates may be coated with a hydrophilic or hydrophobic substance to ensure this (S6).

While the circumstances of operation of such a device might seem to permit direct application of the results obtained on small columns to large ones, there are apparently some difficulties. In columns as large as 24 in. in diameter, for example, considerable channeling may be observed, and this may lead to as much as a 50% decrease in extraction efficiency (W2). For such columns, special redistribution plates have been developed which overcome this tendency (W2).

b. Mass Transfer Characteristics. It is possible to obtain up to a one-third reduction in height of a transfer unit in passing from the conditions of region I to those of region II, Fig. 13 (S6), owing both to increased interfacial area and increased turbulence in the continuous phase (T2). The influence of flow rate upon mass transfer rates is not great, but the data are somewhat conflicting (G1, S6, T2). At large pulse amplitudes, extraction becomes poorer owing to excessive back mixing (T2), as explained above.

In a 40-mm.-diameter column (C1) with 3-in. plate spacing and $5/64$-in. holes whose total cross section was about 6% of the column cross section, and with the system acetic acid–methylisobutyl ketone–water, ketone dispersed, the best extraction rates occurred at a total liquid flow of 35.6 cu. ft./(hr.)(sq. ft.), 29 cycles/min., and 4-mm. amplitude. The *HETS* was 10 in., corresponding to a total throughput/volume of a theoretical stage of 42.1 hr.$^{-1}$. Similarly in a 1.5-in. column (B2) with plates on a 1-in. spacing and containing $1/32$-in. holes whose area was 23% of the column cross section, pulsed at 28 cycles/min., 6-mm. amplitude, with the same system, the *HETS* was 2.59 in. at a total liquid flow of 27.2 cu. ft./(hr.)(sq. ft.), and the throughput/volume of a theoretical stage = 126 hr.$^{-1}$. This improvement is probably very largely the result of the increased number of dispersions produced by the lower plate spacing. In other columns used for more difficult extractions, over-all heights of transfer units have been reported in the range 0.6 to 1.4 ft. (S6, W2) and 0.8 to 12.8 ft. (C3).

2. Packed Towers

With conventionally packed towers the purpose of the pulse is solely to increase the degree of agitation, and the pulsed volume bears no necessary relation to the liquid throughput. It is generally the practice, however, to operate with pulsed volumes equal to several times the liquid flow rates. Owing to the decreased size of the droplets during pulsed operation, the flooding velocities of the packed towers are substantially less than their unpulsed counterparts, although no flooding correlation has yet been devised. Pulsing also causes settling of the random packings, and in some instances orientation of the packing with decreased mass-transfer rates (T2). Pulse frequencies as high as 1,000 cycles/min. have been used (F1).

The most systematic data on extraction are provided by Chantry et al. (C1), who extracted acetic acid from water by methylisobutyl ketone, ketone dispersed, in a 40-mm.-diameter tower packed with ¼-in. dumped Raschig rings. For each pulse frequency an optimum amplitude was found, and high frequencies with short amplitudes produced the greatest extraction improvement with less power. As much as 300% improvement over nonpulsed operation was possible. Under the best conditions, with 18 cu. ft./(hr.)(sq. ft.) total flow, *HETS* was 3.83 in., corresponding to a throughput/volume of a theoretical stage equal to 56.5 hr.$^{-1}$. In the case of extraction of benzoic acid between water and toluene (a difficult extraction) (F1), the height of a transfer unit was reduced from 13.4 to 0.592 ft. by pulsing (throughput/volume of a theoretical stage = 39.2 hr.$^{-1}$), although the improvement for a more readily extracted solute, acetic acid, was not so great. It is believed (W1) that the improvement in extraction resulting from pulsing is more a matter of the repeated dispersion and coalescence that occurs than of merely increased interfacial area resulting from reduced drop size.

E. Pulsed Mixer Settler

Fenske and Long (F2) have described a small, laboratory extractor of 20 countercurrent stages of unique design. The mixers are agitated by rapidly reciprocating perforated plates, reminiscent of the early Van Dijck design (V1), and produce an action similar in many ways to that of the pulsed columns. The holdup per stage is 600 ml., and 100% stage efficiency is obtainable at 0.3 gal./min. liquid flow for any system if the mixing intensity is made adequate. This corresponds to a throughput/volume of a theoretical stage of 114 hr.$^{-1}$.

The design is such that operation at high pressure (400 lb./sq. in.) is readily feasible, permitting the device to be used, for example, with such solvents as liquid ammonia.

V. Centrifugal Extractors

As explained in the Introduction, conventional countercurrent extractors are frequently ineffectual because of the small difference in specific gravity which exists in most two-liquid systems. The centrifugal extractors overcome this problem by increasing the effective specific gravity difference with the help of centrifugal force. There have been many designs proposed, but only those which have had important commercial acceptance will be mentioned here.

A. THE PODBIELNIAK EXTRACTOR

The Podbielniak extractor (P1), which is the most important of the centrifugal-extractor group, consists essentially of a horizontal cylindrical drum which is rapidly revolved about a shaft passing through its axis. The internal design of the drum, which is the part of the machine of principal interest, may take many forms (P2): in the simplest design there is a spiral plate wound about the shaft, passing from the shaft to the drum periphery; alternatively, the plate may be perforated; or there may be a series of perforated, concentric rings. The liquids to be contacted are introduced into the shaft of the machine, with the heavy liquid entering the drum at the shaft while the light liquid is led by an internal passageway to the periphery of the drum. The liquids then flow countercurrently through the internal structure of the drum, the light liquid toward the center, the heavy liquid toward the periphery, and eventually back to the shaft. In the case of the imperforate spiral, liquid contact is similar in many respects to that obtained in a wetted-wall tower, and in the case of the perforated concentric rings it is similar to that obtained in a conventional perforated-plate tower. The liquids are then discharged through the shaft. In the perforated-ring design either liquid may be made predominantly continuous or dispersed by control of the position of the principal interface. This is done by regulating the pressure on the light liquid as it flows out of the machine (B1) in a manner which is entirely analogous to the regulation employed with a conventional, upright, countercurrent extraction tower.

Machines are available in a variety of liquid capacities, from 500 ml./min. for the laboratory up to 25,000 gal./hr. for industrial applications. It is understood that still larger devices, capable of handling 1,000 gal./min., are planned (P3); the rotating drum in these extractors will be 48 in. in diameter, 52 in. long. Figure 14 shows a typical machine.

The extractor has had wide industrial acceptance. Perhaps the earliest

Courtesy, Podbielniak, Inc.
FIG. 14. Podbielniak "Duozon" centrifugal extractor.

large-scale use developed during the Second World War, where the desirable features made the machines useful in the manufacture of penicillin. Since that time they have been used in many other antibiotic extractions, for instance in the manufacture of chloromycetin (A2) and bacitracin (A4), as well as in petroleum refining and processing of petrochemicals and chemical manufacture generally. The most recent application has been to the conventional light-oil extraction of phenols from coke-oven ammonia liquor, followed by caustic treatment of the extract (K1). It is understood that plants handling up to 30,000 gal./hr. of ammonia liquor and solvent are contemplated (Z1).

Perhaps the most remarkable feature of these devices is the extremely small holdup of liquid required in the machine in comparison with the permissible throughout. For example, a machine capable of handling 137.5 gal./min. total liquid operates with a holdup of approximately 85 gal. (Z1). As a result, a number of advantages accrue: (1) an extremely short contact time in the extractor results, in turn very desirable where extraction conditions are injurious to the product, as for example in the manufacture of penicillin; (2) there is a very low inventory of expensive or flammable solvent; (3) steady state conditions are reached very quickly after startup; and (4) small over-all space is occupied by the machine, particularly with regard to head room. The high speeds at which the machines normally operate (2,000 to 5,000 rpm, depending upon the size of machine and the liquids being processed) increase the effective force of gravity acting upon the liquids up to as much as 5,000 times the normal, so that liquids which ordinarily easily emulsify are readily handled. Effluents are clear, and entrainment and settling problems are largely eliminated.

There are few systematic, publicly available data, especially for the large machines, by means of which these extractors may be evaluated. The most complete were obtained by Barson and Beyer (B1) with a laboratory model having an 18-in.-diameter rotor, and containing 17 concentric rings each fitted at 180° intervals with a small slot, 0.040 by 0.120 in. The holdup of this device is stated to be 600 ml., of which only 65 ml. are in the contacting portion of the device, the remainder being in the calming zones, inlet and outlet lines, and seals. With the system boric acid–amyl alcohol–water, the best performance was obtained with a rotor speed of 5,000 r.p.m. at 100 ml. water and 350 ml. alcohol/min. flow rates. The machine then contained the equivalent of 7.8 theoretical stages, corresponding to a throughput/volume of a theoretical stage equal to 349 hr.$^{-1}$, based on the total holdup of the machine. It is of interest to note that under pulsed operation, presumably with a different system, this model has developed 18 theoretical stages (P2).

B. OTHER MACHINES

1. *The Luwesta (Centri-Westa) Extractor*

Eisenlohr (E1) has described a three-stage centrifugal extractor based on an early design of Coutor (C6) and arranged in a vertical position, normally operated at about 3,800 r.p.m. The maximum capacity of the standard model is roughly 1,300 gal. total liquid flow/hr., depending of course upon the mixture being handled, at a total holdup of 22.8 gal. The over-all stage efficiency is said to be roughly 95%, which would correspond to a throughput/volume of a theoretical stage of 162 hr.$^{-1}$.

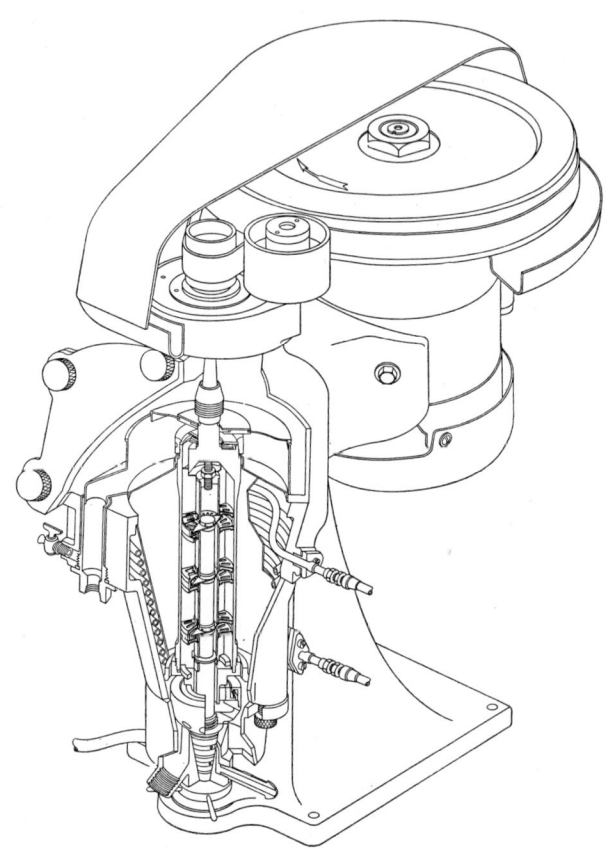

Courtesy, The Sharples Corporation.

FIG. 15. Sharples Super Centactor.

2. *The Sharples Super Centactor*

The Sharples Super Centactor, shown in Fig. 15, is also vertically arranged. It operates at high speeds, up to 25,000 r.p.m. (A5), and can effect a centrifugal force up to 15,500 times normal gravitational force. In a small machine handling 80 gal. liquid/hr., from 1 to 5 real stages can be arranged, and it is understood that larger models are contemplated.

Acknowledgment

The author wishes to thank the various individuals who provided data, advance copies of their papers, and permission to use drawings and photographs: D. K. Hart, J. H. Holmes, H. F. Johnson, R. B. Olney, W. J. Podbielniak, L. P. Sharples, R. von Berg, and G. V. Ziegenhorn.

Nomenclature

- a Interfacial surface, sq. ft./cu. ft.
- A Interfacial surface, sq. ft. (*batch*); sq. ft./sq. ft. (*continuous*)
- ΔC_{avg} Average concentration gradient
- d Impeller diameter, ft.
-     Differential operator
- d_p Drop diameter at $v_D = 0.1$, ft.
- e 2.7183
- E Stage efficiency $= N/N_e$, as a fraction, dimensionless
- E_{ME} Murphree extract stage efficiency, as a fraction, dimensionless
- E_{MR} Murphree raffinate stage efficiency, as a fraction, dimensionless
- E_O Over-all stage efficiency of a countercurrent cascade, as a fraction, dimensionless
- f A function
- g Gravitational constant, ft./sq. hr.
- H_{tOE} Height of an over-all extract transfer unit, ft.
- H_{tOR} Height of an over-all raffinate transfer unit, ft.
- $HETS$ Height equivalent to a theoretical stage, ft.
- k A constant
- K Generalized mass transfer coefficient, lb. moles/(hr.) (sq. ft.) (Δc_{avg})
- K_E Over-all extract mass transfer coefficient, lb. moles/(hr.) (sq. ft.) (mole fraction)
- K_R Over-all raffinate mass transfer coefficient, lb. moles/(hr.) (sq. ft.) (mole fraction)
- ln   Natural logarithm
- log   Common logarithm
- L_E Quantity of extract, lb. moles (*batch*); lb. moles/(hr.) (sq. ft.) (*continuous*)
- L_R Quantity of raffinate, lb. moles (*batch*); lb. moles/(hr.) (sq. ft.) (*continuous*)
- m Slope of the equilibrium-distribution curve $= dy^*/dx$
- n A constant
- n_i Number of ideal or theoretical stages, dimensionless
- n_r Number of real stages, dimensionless
- N Rate of mass transfer, moles/hr. (*batch*); lb. moles/(hr.) (sq. ft.) (*continuous*)
- N_{Fr} Froude number $= dS^2/g$, dimensionless
- N_{Po} Power number $= Pg/\rho S^3 d^5$, dimensionless
- N_{Re} Reynolds number $= d^2 S \rho/\mu$, dimensionless
- N_{tOE} Number of over-all extract transfer units, dimensionless
- N_{tOR} Number of over-all raffinate transfer units, dimensionless
- N_{We} Weber number $= d^3 S^2 \rho/\sigma$, dimensionless
- N_{We}' Special Weber number $= d^3 S^2 \rho'/\sigma$, dimensionless
- P Power, ft.-lb./hr.
- r A constant
- s A constant
- S Speed, rev./hr.
- T Tank diameter, ft.
- v Volume fraction, dimensionless
- V Volume of liquid, cu. ft.
- x Concentration of solute in raffinate, mole fraction
- y Concentration of solute in extract, mole fraction

Z Liquid depth, ft.
θ Time, hr.
θ_{tOE} Time of an over-all extract transfer unit, hr.
θ_{tOR} Time of an over-all raffinate transfer unit, hr.
μ Viscosity, lb. mass/(ft.)(hr.)
ρ Density, lb. mass/cu. ft.
ρ' $0.6\rho_D + 0.4\rho_C$, lb. mass/cu. ft.
σ Interfacial tension, lb. mass/sq. hr.
ϕ A power function [Eq. (2)]

SUBSCRIPTS

avg Average
C Continuous
D Dispersed
e At equilibrium
m Logarithmic average
1 Initial
2 Final

SUPERSCRIPT

$*$ At equilibrium

REFERENCES

A1. Alter, H. W., Codding, J. W., and Jennings, A. S., *Anal. Chem.* **26**, 1357 (1954).
A2. Anonymous, *Chem. Eng.* **56**, No. 10, 172 (1949).
A3. Anonymous, *Chem. Eng. Progr.* **47**, No. 7, 20 (1951).
A4. Anonymous, *Chem. Eng.* **60**, No. 6, 282 (1953).
A5. Anonymous, *Chem. Eng. News* **51**, 5395 (1953).
B1. Barson, N., and Beyer, G. H., *Chem. Eng. Progr.* **49**, 243 (1953).
B2. Belaga, M. W., and Bigelow, J. E., AEC Declassified Document KT-133 (1952).
C1. Chantry, A. W., von Berg, R. L., and Wiegandt, H. F., *Ind. Eng. Chem.* **47**, 1153 (1955).
C2. Coahran, J. M., U. S. Patent 1,845,128 (Feb. 16, 1932).
C3. Cohen, R. M., and Beyer, G. H., *Chem. Eng. Progr.* **49** (1953).
C4. Colburn, A. P., and Welsh, D. G., *Trans. Am. Inst. Chem. Engrs.* **38**, 179 (1942).
C5. Coplan, B. V., Davidson, J. K., and Zabroski, E. L., *Chem. Eng. Progr.* **50**, 403 (1954).
C6. Coutor, C. (to Etablissements Lambiotte frères), U. S. Patent 2,036,924 (Apr. 7, 1936).
D1. Davis, M. W., Hicks, T. E., and Vermeulen, T., *Chem. Eng. Progr.* **50**, 188 (1954).
E1. Eisenlohr, H., *Ind. Chemist* **27**, 271 (1951).
F1. Feick, G., and Anderson, H. M., *Ind. Eng. Chem.* **44**, 404 (1952).
F2. Fenske, M. R., and Long, R. B., *Chem. Eng. Progr.* **51**, 194 (1955).
F3. Flynn, A. W., and Treybal, R. E., *Am. Inst. Chem. Eng. J.* **1**, 324 (1955).
G1. Griffith, W. L., Jasny, G. R., and Tupper, H. T., AEC Declassified Document AECD-3440 (1952).
H1. Hachmuth, K. H., *Chem. Eng. Progr.* **48**, 523 (1952).
H2. Hixson, A. W., and Smith, M. I., *Ind. Eng. Chem.* **41**, 973 (1949).
H3. Hixson, A. W., Drew, T. B., and Knox, K. L., *Chem. Eng. Progr.* **50**, 592 (1954).
J1. Jealous, A. C., and Johnson, H. F., *Ind. Eng. Chem.* **47**, 1159 (1955).
K1. Kaiser, H. R., *Sewage and Ind. Wastes* **27**, No. 3, 311 (1955).
K2. Karr, A. E., and Scheibel, E. G., *Chem. Eng. Progr. Symposium Ser. No. 10*, **50**, 73 (1954).
M1. Mack, D. E., and Kroll, A. E., *Chem. Eng. Progr.* **44**, 189 (1948).
M2. Miller, S. A., and Mann, C. A., *Trans. Am. Inst. Chem. Engrs.* **40**, 709 (1944).
M3. Morello, V. S., and Poffenberger, N., *Ind. Eng. Chem.* **42**, 1021 (1950).
O1. Oldshue, J. Y., and Rushton, J. H., *Chem. Eng. Progr.* **48**, 297 (1952).
O2. Olney, R. B., and Carlson, G. J., *Chem. Eng. Progr.* **43**, 473 (1947).

O3. Overcashier, R. H., Kingsley, H. A., and Olney, R. B., *Chem. Eng. Progr.*, in press.
P1. Podbielniak, W. J., U. S. Patent 2,044,996 (June 23, 1935), and many others.
P2. Podbielniak, W. J., in discussion, ref. (B1); *Chem. Eng. Progr.* **49,** 252 (1953).
P3. Podbielniak, W. J., personal communication, May 23, 1955.
R1. Reman, G. H., *Proc. World Petroleum Congr.*, Sect. III, 121 (1951).
R2. Reman, G. H. (to Shell Development Co.), U. S. Patent 2,601,674 (June 24, 1952).
R3. Reman, G. H., and Olney, R. B., *Chem. Eng. Progr.* **51,** 141 (1955).
R4. Rushton, J. H., *Chem. Eng. Progr.* **47,** 485 (1951).
R5. Rushton, J. H., *Chem. Eng. Progr.* **50,** 587 (1954).
R6. Rushton, J. H., Costich, E. W., and Everett, H. J., *Chem. Eng. Progr.* **46,** 395, 467 (1950).
S1. Sachs, J. P., and Rushton, J. H., *Chem. Eng. Progr.* **50,** 597 (1954).
S2. Schafer, A. C., and Holmes, J. H., *Chem. Eng. Progr.*, in press.
S3. Scheibel, E. G., *Chem. Eng. Progr.* **44,** 681, 771 (1948).
S4. Scheibel, E. G. (to Hoffman-LaRoche, Inc.), U. S. Patent 2,493,265 (Jan. 3, 1950).
S5. Scheibel, E. G., and Karr, A. E., *Ind. Eng. Chem.* **42,** 1048 (1950).
S6. Sege, G., and Woodfield, F. W., *Chem. Eng. Progr.* **50,** 396 (1954); *Chem. Eng. Progr. Symposium Ser. No. 13*, **50,** 179 (1954).
S7. Sherwood, T. K., *Trans. Am. Inst. Chem. Engrs.* **31,** 670 (1935).
S8. Sherwood, T. K., and Pigford, R. L., "Absorption and Extraction," 2nd ed. McGraw-Hill, New York, 1952.
T1. Taylor, G. I., *Proc. Roy. Soc.* **A138,** 41 (1932).
T2. Thornton, J. D., *Chem. Eng. Progr. Symposium Ser. No. 13*, **50,** 39 (1954).
T3. Treybal, R. E., "Liquid Extraction." McGraw-Hill, New York, 1951.
T4. Treybal, R. E., "Mass-Transfer Operations." McGraw-Hill, New York, 1955.
V1. Van Dijck, W. J. D., U. S. Patent 2,011,186 (Aug. 13, 1935).
V2. Vermeulen, T., Williams, G. M., and Langlois, G. E., *Chem. Eng. Progr.* **51,** 85F (1955).
V3. Vermijs, H. J. A., and Kramers, H., *Chem. Eng. Sci.* **3,** 55 (1954).
W1. Wiegandt, F. H., and von Berg, R. L., *Chem. Eng.* **61,** No. 7, 183 (1954).
W2. Woodfield, F. W., and Sege, G., *Chem. Eng. Progr. Symposium Ser. No. 13*, **50,** 14 (1954).
Z1. Ziegenhorn, G. J., personal communication, June 17, 1955.

The Automatic Computer in the Control and Planning of Manufacturing Operations

ROBERT W. SCHRAGE

Esso Standard Oil Company
Linden, New Jersey

   I. Introduction.. 331
 II. Computing Equipment.. 332
     A. Analog Computers... 332
     B. Punched-Card Calculators................................... 334
     C. Stored-Program Calculators.................................. 336
     D. Cost Considerations in Computing............................ 338
III. Control of Manufacturing Operations............................. 341
     A. Operations Accounting....................................... 341
     B. Calculation of Laboratory Analyses........................... 342
     C. The Computer and Automation............................... 344
 IV. Statistical and Numerical Analysis............................... 345
     A. Statistical Analysis... 345
     B. Numerical Analysis.. 347
  V. Mathematical Models... 348
     A. Definition... 348
     B. Types of Mathematical Models................................ 349
        1. Simulation of Processes................................... 349
        2. Combinations of Models.................................. 351
        3. Blending Operations..................................... 352
        4. The Monte Carlo Method................................ 354
        5. Incorporation of Economic Factors........................ 356
 VI. Optimization Studies.. 356
     A. The General Problem.. 356
     B. Factorial Designs.. 358
     C. Sequential Designs.. 360
     D. Optimization Problems Involving Restrictions................. 363
     E. Linear Programming... 364
     References... 366

I. Introduction

Recent years have seen two important developments relating to the control and planning of manufacturing operations. One of these has been the emergence of new theoretical methods of problem analysis, most of which are predominantly mathematical in character. The second is the availability of automatic computing equipment capable of performing operations at a cost and speed which seemed unbelievable only a short time ago. The concurrence of these two developments cannot be regarded as entirely accidental. Each has been a powerful stimulus to the other.

The primary purpose of this chapter is to indicate some of the areas in which a computer can make a significant contribution in the manufacturing operations of the process industries. A second objective is to discuss some of the practical considerations involved in selecting and using a computer in this work. Consideration will be given to some of the special techniques for problem solution which are important in computing work.

A detailed discussion of specific computing applications was not felt to be within the scope of this chapter. It should not be forgotten, however, that computing is only an instrument for implementing the solution of problems. The calculations done by the computer are, in the process industries, based largely on familiar chemical engineering methods. Since almost any part of chemical engineering theory might be required in specific cases, it is neither possible nor necessary to consider this aspect of the general problem here. Several reviews (F1, R2, R3) have appeared which give specific applications of computers in such fields as distillation, heat transfer, fluid flow, and reaction kinetics. The interest of this chapter is directed more toward problems of broader scope in which the various unit operations might be represented as elemental parts.

There are, of course, many opportunities in the process industries for the use of calculation methods not normally considered a part of chemical engineering. While several somewhat unfamiliar methods are mentioned in this chapter, no general survey of all the techniques which might prove useful has been attempted. A compilation of some basic references in this area is available (H4). Problems may sometimes involve consideration of factors outside the traditional boundaries of engineering. However, in many problems the engineering aspects are so vital an element that consideration of the problem in its entirety seems as much in the domain of chemical engineering as any other field.

The portion of this chapter allotted to computing equipment can serve only as an introduction. Descriptions of specific computers have been avoided because such information would necessarily be incomplete and is of only transitory interest. Literature references on specific computers and computing methods have been provided to an extent felt appropriate. A more general bibliography on these subjects is also available (H3).

II. Computing Equipment

A. Analog Computers

Any classification of computing equipment generally starts with the distinction between analog and digital computers. This division is well founded, as the approaches to using the two types of computers can differ greatly. Most of this chapter will be concerned with digital calculators, as they have been more widely used in the subjects to be discussed here. There are, however, enough instances of analog computer applications to make it essential to be aware of the advantages and disadvantages of both types of machines.

An analog computer, as its name implies, is a device which is in some way analogous to another system. When the computer receives information, it provides measurable responses which can be correlated with the behavior of the system under consideration. The analogy may be in many forms—mechanical, electrical, electronic, etc.—but the modern commercial analog computer is almost always electronic, because of the flexibility and low cost of this type.

Analog computers are concerned with continuous variables rather than numerical data. If input information is numerical in form it must be converted to voltages, resistances, shaft rotations, etc., by setting dials or similar devices. Output is available as instrument readings, continuous plots on paper or oscilloscopes, etc. It is possible to supplement an analog computer with input or output devices for digital-to-analog or analog-to-digital conversion. These are, however, accessories which do not change the inherent nature of the computer itself.

The internal construction of an analog computer limits its accuracy in use. Consistent results, beyond this limit, cannot be expected. This is one reason why analog computers are unsuited for accounting calculations which must always give consistent results with a fixed digital accuracy. Because digital information must be converted, either manually or through auxiliary equipment, into continuous form, analog computers are not well adapted for processing large volumes of numerical data through simple arithmetic calculations. Nor is it possible to carry out efficiently long sequences of arithmetic and logical operations which are required in some problems.

The analog computer is, however, very suitable for performing some of the continuous mathematical operations, such as are required when problems involve differential equations. Its suitability for this comes about because the operation of the computer is, by design, controlled by the same mathematical equations as the system of interest. As an analog computer is low in cost compared with a digital computer, it has frequently been used in these applications unless its accuracy was inadequate. Occasionally an analog computer may be used when the fundamental equations governing a system are unknown, because by chance or trial-and-error design an analogy can be made. This situation is, however, unusual.

One of the reasons for the limited use of analog computers in industry is probably the greater flexibility of digital machines. As mentioned above, for certain types of problems digital computers are almost essential. Once available the digital computer can be adapted to deal with almost any problem that could be handled on an analog computer. Even though the analog computer might be preferable in a particular application, it often cannot be justified when a digital computer and a staff trained in its use are already available.

Anyone seriously interested in technical or scientific computing should be familiar with both types of computers. Only then can a proper decision be made as to the choice of computer for a particular problem. Soroka (S4) is an excellent source of information on analog computers in general, and Korn and Korn (K8) deal specifically with electronic analogs. Brief introductory articles on electronic analog computers have also appeared recently (H8, M5, S7).

B. Punched-Card Calculators

In digital computers input, output, and all intermediate operations involve discrete units of information. The operations of a digital computer are essentially arithmetic and clerical, with all procedures performed according to a prescribed set of rules. The means for doing the operations can be diverse, embodying mechanical, electrical, and electronic devices of many kinds. Today, however, the arithmetic and logical operations themselves are almost exclusively performed by electronic means. The input and output of digital calculators are numerical and may be in the form of punched tapes or cards, magnetic tapes, printed results, etc.

The earliest type of automatic digital calculator to become generally available, starting at about 1945, was the punched-card calculator. These machines were at first entirely electromechanical, but in their modern form are largely electronic. Their intended use was initially in accounting applications, but their utility in various technical problems was soon discovered. Many of the present users of large-scale calculators were introduced to computing by the punched-card calculator.

In its simplest form the punched-card calculator is designed to perform operations based upon data values read from continuously feeding punched cards. The calculations as each card passes through the machine are usually simple and similar. It is possible to arrange for a limited number of alternative operations to take place, depending upon information punched on the card. However, the calculation performed is generally complete for each card, and the result is immediately punched into another part of the same card from which the data were read. In certain types of problems totals may be accumulated for groups of cards. The results would eventually be punched into an answer card bearing a suitable code.

The operations of a punched-card calculator are determined by the wiring of a control panel. The panel, in effect, completes circuits between components in the machine so that desired operations are carried out. The panels may be removed from the machine and saved permanently. Thus several differently wired panels may be kept on hand for different types of problems, making it a simple matter to change the functions of the machine.

There are many problems, particularly in the field of accounting, for which the small punched-card calculator is suitable. However, for many technical or scientific calculations it is of only limited utility, primarily because of its limited speed and a basic lack of adaptability to lengthy sequential calculations. The speed limitation comes about principally because the operation of these machines is dependent upon the mechanical movement of cards. Thus, even though electronic calculations may be performed at high speed, this feature cannot be fully exploited.

The fact that these calculators are ill suited to long sequential calculations is due mainly to the limited facilities associated with their small size and relatively low cost. To some extent this is not an inherent limitation, but it does not pay to increase the size of a punched-card calculator indefinitely without making a radical change in the basic structure of the machine. Such a change will bring us in the next section to the stored-program type of calculator.

Because there was initially no alternative to making the most effective possible use of punched-card calculators, systems were designed which tended to some degree to circumvent these shortcomings. Elaborate general-purpose control panels permitted several alternative calculations. Panels were also designed for simple types of card-programmed sequential calculations (N1, S2). Each card read by the calculator would, through a suitable code, call for a particular operation to be performed, involving either data on the card or data stored internally from previous calculations. However, only very limited storage capacity for intermediate results was available.

In 1949 a machine designed especially from the viewpoint of sequential calculations became available and was soon in wide use in technical work. This was the IBM Card-Programmed Electronic Calculator. It still operated from data and instructions read from punched cards and, hence, had the same speed limitation. It was, however, augmented by additional storage capacity for intermediate results and had reading, punching, and printing facilities under one integrated control system. Although it required wired control panels, these were generally designed to provide a general-purpose system adaptable to many different problems.

C. Stored-Program Calculators

It was noted in the previous section that the operation of a punched-card calculator may be varied in two ways: first, through changes in the wiring of control panels and, second, through different codes read from punched cards. A change of control panels is a manual operation, and if this is necessary during a problem solution, the computer cannot be regarded as completely automatic. Instructions read from punched cards often permit a closer approach to automatic operation, but this idea can be made much more effective through a quite different design of computer.

The instructions which a calculator receives from punched cards are not different in form from numerical data. In fact, without knowledge of the rules of procedure for a given computer, it would be impossible to distinguish between numerical data and instructions. There is no reason, therefore, why both types of information cannot be stored within the calculator. Doing so has three important advantages:

1. The speed of instruction execution can be greatly increased, as it no longer depends upon the mechanical movement of cards.

2. The computer can follow many alternative sets of instructions, depending upon logical decisions made during calculations.

3. Instructions can be modified by the computer. It is frequently found that the same instruction (or group of instructions) can be used repetitively if changed in some simple way. Since instructions are available in storage and can be subjected to the same arithmetic operations as any other numerical data, the computer can modify its instructions as the calculation proceeds. This permits relatively short sequences of instructions to be equivalent to much longer sequences on punched-card calculators.

It is not possible to present here more than a very general description of a stored-program calculator. Figure 1 shows the organization of a typical example. The primary working storage is generally equivalent to between 1000 and 4000 ten-digit decimal numbers. It may be in the form of magnetic cores, cathode-ray tubes, sonic delay lines, magnetic drums, etc. The arithmetic and control unit executes instructions contained in the primary working storage and has random access to any data or instructions it may hold. The control unit also determines the operation of all the component parts of the complete installation.

In larger computers the primary storage is supplemented by secondary storage, which is usually in the form of magnetic tapes. Information must be brought into primary storage before it can be processed. If a magnetic drum is not used as primary working storage, it may be available as a secondary storage medium. Usually data or instructions cannot be obtained from secondary storage at random, but only as larger blocks of information. On large computers it is possible to store the equivalent of several million ten-digit decimal numbers in secondary storage.

There is also available input and output equipment such as punched-card readers and punches, printers, typewriters, etc. Magnetic tape can obviously function as an input-output device as well as a storage medium. In some computers it is possible to convert various input-output forms independently of the main computer. This permits using the fastest form of input-output (generally magnetic tape) in computing, with the slower operations of printing or punching performed separately later.

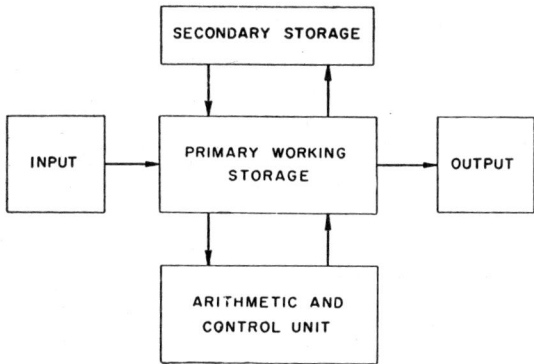

FIG. 1. Organization of a typical stored-program calculator.

Some of the larger calculators have been described specifically as scientific computers, while others have been called commercial or business calculators. The former are generally characterized by high calculating speed and relatively slower speed and versatility insofar as input-output is concerned. In the business calculators just the reverse is true. Scientific calculators are frequently designed to operate in the binary-number system, while business calculators generally operate in a coded decimal system. Most business calculators are able to process alphabetic information with greater facility. There is some likelihood that eventually the distinctions between these two types of calculators will tend to disappear, leading to a general-purpose calculator suitable for all applications (H10).

The most detailed sources of information on specific computers are, of course, the operating manuals provided by their manufacturers. These are supplemented in most cases by customer-training courses. Descriptions of some stored-program calculators have also been published in the periodical literature (B3, H1, M1, P1). The magazine

Computers and Automation has been maintaining lists of computer manufacturers and various computer models known to be in existence. References are given here to the latest of these lists at the time of this writing (A3, A5).

D. Cost Considerations in Computing

There are two aspects to the problem of costs in computing. The first of these is concerned with the lease (or equivalent purchase) cost of the computer itself, together with the costs incurred in its routine operation and maintenance. The second is the cost of the staff required for problem analysis, programming, and related activities. This second cost seems often to be underestimated by persons not familiar with computing practice.

Satisfactory estimation of the direct operating costs for a computer installation is usually not difficult. It is naturally one of the first thoughts when selection of a computer is considered. There are many factors to be taken into account, since the total cost is meaningful only when related to the computer's capabilities. Wainwright (W1) gives an extensive check list of features to be considered in any comparative evaluation of digital computers.

There is one point worthy of special mention here, because it has not received the prominence that it deserves. This is the generalization that the unit cost of computing decreases with increasing size of the computer. A convenient measure of unit computing cost is in terms of equivalent man years of hand calculations. Clippinger (C5) tabulates these costs, in dollars per man year, for several types and sizes of computers. The computers may be divided into two broad classes according to their monthly rental costs or purchase prices reduced to a roughly equivalent basis. With about $10,000 a month used as the dividing line between small- and large-scale computers, the unit costs of computing and data handling may be compared. Computing cost is more pertinent in most scientific or engineering applications, while data-handling cost is of interest chiefly in commercial applications. For large-scale computers (costing above $10,000 a month) the unit cost of either computing or data handling generally is in the range from $8 to $60 a man year. For smaller computers (costing from $2500 to $10,000 per month) the unit cost of computing is usually about $25 to $300 a man year; in data handling unit costs range from about $200 to $1000 a man year. Although these figures are only approximate, they indicate the desirability of one large computer over several smaller ones. Aside from the unit-cost advantage the larger computer is also able to handle many types of problems which would be completely impractical on smaller machines.

Naturally a large computer can be utilized to advantage only if there is a sufficient volume of work to keep it busy. Some tendency seems to be developing for the consolidation of the computing activities of scattered geographic locations in one central computer installation. This has some obvious disadvantages of inconvenience and data-transmission cost which must be taken into account. However, the development of new methods for the accurate transceiving of data over telephone and telegraph lines has aroused interest in this idea (A2). The possibility of combining technical and business computing in the same installation is another means for justifying a larger computer. Some of the newer commercial computers show great versatility in both types of work.

Today even the customer whose work volume is quite small has access to large-scale computers through service organizations. Most manufacturers of computing equipment maintain service bureaus through which their computers are available for short times on an hourly rental basis. In most cases the customer has a choice between merely using the computing facilities or also getting assistance in problem analysis and programming. A number of organizations not affiliated with computer manufacturers also offer similar services (A4, C5). The advantages and disadvantages of using computers in this manner will obviously depend upon individual circumstances. Generally speaking, these services are most helpful to smaller customers and in special spot situations.

The effective unit cost of a computer will vary inversely with its productivity in useful work. On a large computer time is a very valuable commodity, costing several dollars a minute. Productivity depends upon a number of factors: the efficiency of programs, the frequency and seriousness of program errors detected in testing, operating errors in routine production problems, machine errors, machine "down-time" for maintenance, and the efficient scheduling of successive work. Gotlieb (G2) has discussed some of these considerations.

Although an efficient program is conservative of machine time in the production running of problems, the cost of its development must be taken into account too. This brings us to the second aspect of computing cost, that is, the expense of the staff which will design the sequences of instructions, or programs, which are the basis for the machine operations. This is a part of computing cost all too often minimized in importance.

There are three phases to the development of a problem solution on a digital computer. First, the problem must be stated and the mathematical methods for its solution described. Second, the logical structure of the problem must be analyzed from the viewpoint of the calculator. Third, the program of actual machine instructions, in its prescribed coded form, must be written and tested.

It is not possible to describe completely here the varied ramifications of these three steps, all of which are closely related and to some extent overlapping. They represent the transition of the problem from the viewpoint of the scientist, engineer, or businessman to the language understood by the computer. While the relative cost of this work naturally depends upon the type of problems processed at a given installation, it usually is roughly equivalent in magnitude to the cost of the computer itself.

Beutler (B6) and Clippinger, Dimsdale, and Levin (C6, C7, C8) give examples of the procedure followed in preparing a problem for a digital computer. The latter authors also present some fairly detailed information on the staff requirements for computer installations of various sizes. Gotlieb (G3) gives some further data on the cost of programming. All of this emphasizes the importance of careful consideration of computer-program preparation. There are three approaches to the problem of programming.

Programs can be written in terms of the fundamental operational instructions of the computer. These "machine-language" programs when designed with care by skilled programmers can be more efficient than any other type, because they are tailored to fit a particular problem to a specific computer. In practice, machine-language programs are usually developed with the assistance of an assembly or compiler program. This permits the computer itself to assume a large part of the burden of coding, taking advantage of existing library programs which are integrated into the main program. The result usually shows little loss in efficiency.

The second type of programming makes use of an interpretive program. This permits the computer to act upon instructions expressed to it in a simplified form. Since the computer must interpret these instructions, translating them into its own language, efficiency is lost. However, programming is easier and can be done by persons with less skill and training. A description of one excellent interpretive coding system, developed for a large scientific calculator, has appeared (B1). The general characteristics of interpretive and assembly or compiler programs have also been described (H5).[1]

The ideal programming method is a so-called "automatic" coding system. In its ultimate form this would require only that the problem be stated in unambiguous mathematical terms. From this statement the computer develops its own efficient machine-language program. While

---

[1] To those uninitiated to the field, the special terminology in computing may be troublesome. The Association for Computing Machinery has published a glossary of commonly accepted terms (H6).

there is as yet no system which completely realizes this objective, the progress toward this goal is encouraging.

III. Control of Manufacturing Operations

A. OPERATIONS ACCOUNTING

A large part of accounting is concerned with providing control information for manufacturing operations. It is not within the scope of this chapter to discuss the many applications of computers in this work. There is, however, a part of accounting procedures which is of particular interest from a technical viewpoint. This is the subject of operations accounting, which is concerned with the current operations of process equipment. The subject is important in the planning of future operations as well as the control of current operations. It is worth while, if not indeed necessary, to know both what information is available and how it was obtained.

It is not uncommon to find that technical and accounting functions are far separated, to their mutual disadvantage. Operations accounting in the process industries depends strongly on technical interpretation of unit operating data. It is rare that data recorded directly from plant instrumentation can be used without further calculation. Many of the most significant characteristics of plant operations are derived functions of several variables. One frequently finds, however, that these functions are calculated by very approximate methods, neglecting the effects of secondary variables on the result. There may have been some justification for this when these calculations were done by hand. Not only would the cost of the calculations possibly have been prohibitive, but the additional calculations, performed by clerical personnel unfamiliar with their basis, provided opportunities for more errors. With high-speed data-processing equipment, however, the calculation can be programmed by technical personnel and, through adequate machine checking, assurance may be had that the results are correct. The incremental cost for doing the calculations properly, once the methods are set up, is usually negligible.

One of the advantages of machine processing of plant data is the quick detection of irregularities in plant operations. Calculated daily results can be compared with standard-performance figures. Significant departures from normality can be noted and the causes investigated. Appreciable savings may sometimes be realized when improper conditions can be quickly corrected.

The recording of operations accounting data in an accessible form is another matter of great importance. Too often valuable data are filed on written or printed reports which are not again consulted. The punched card has been found to be a valuable supplemental method for filing data. Not only can a large volume of data be stored in a small space, but even more important is the facility with which it can be handled. Information meeting certain criteria can be rapidly extracted from a large mass of data. The selected data can then be rapidly processed through a desired calculation routine.

Stevens and Brady (S6) have discussed the use of punched cards in recording process-control data. While the subject of record keeping itself is of little interest here, it is worth remembering that it may later be desired to process these data on a computer. The form of data arrangement should, therefore, be designed with this possibility in mind.

Magnetic tapes, used in conjunction with a stored-program digital calculator, represent another advance of great importance. Not only may data on the tape be processed much more rapidly than cards, but very complex routines can be devised for selecting and adjusting data to secure desired results. Tape has the disadvantage of not being susceptible to insertions of new data or changes in old data without being entirely rewritten. It also has a minor disadvantage in not being suitable for direct visual inspection. Because each has its special advantages, it can be expected that magnetic tapes and punched cards will continue to be complementary to one another.

B. Calculation of Laboratory Analyses

Practically concurrent with the growing utilization of automatic computers in industry has been the expanding use of various spectrometric methods for routine laboratory analyses. This coincidence is not entirely due to chance, because it seems doubtful that these analytical methods could have achieved their present wide acceptance had it not been practical for computers to do the calculations associated with them. These calculations are quite similar in the several applications, involving the solution of a system of linear simultaneous equations for each analysis.

In all cases a number of instrument readings are taken at selected points within a given spectral range. These points are chosen so that the reading will be due principally to one component of the mixture being analyzed. However, contributions of the several other components involved are large enough so that they may not be neglected. Therefore, a linear equation is set up with each term representing the unknown fractional amount of a component in the mixture multiplied by its contribution as a pure component to the instrument reading. This latter quantity is a constant which is known from calibration data on the instrument. The sum of these terms for all the components is equal to the instrument reading at that spectral value. If readings are taken at enough points, the resulting system of equations can be solved for the fractional amounts of components present.

In most routine analyses the qualitative composition of the mixture being analyzed remains the same. Generally only slight changes in the relative amounts of the components are expected. Under these conditions it will be recognized that the coefficients of the unknown fractions are always the same (at least until recalibration of the analytical instrument is necessary). Each new analysis results merely in new readings represented by the constant terms in the equations. It is then possible to eliminate a great deal of the work needed for solution by obtaining the inverse matrix of the calibration coefficients. This inverse can be multiplied by a vector consisting of readings for a particular sample, giving directly the analysis. The matrix-by-vector multiplication requires considerably fewer arithmetic operations than does the solution of entirely new sets of simultaneous equations.

Routine spectrometric analyses currently run in industry may involve mixtures of up to 20 or 30 components. The solution of systems of 20 or 30 simultaneous equations by hand calculations is so tedious an operation, and so subject to human error, that it is impractical to accomplish it on a routine basis. The necessary calculations are, however, a well-defined arithmetic procedure easily adapted to digital calculators. Analog computers can also be used in solving simultaneous equations but are subject to accuracy limitations.

Berry *et al.* (B5) describe the use of an analog computer to solve systems of up to 12 simultaneous equations in connection with both mass- and infrared-spectrometer analyses. The Gauss-Seidel iterative method of solution was employed. Morris (M8) describes another analog computer for solving up to ten simultaneous equations. This was used in infrared analyses. Opler (O1) reported on the use of only punched-card tabulating equipment to perform matrix-by-vector multiplications for infrared analyses. In this case the matrix inversion was obtained by hand calculation or by an analog computer. The use of small- and medium-scale punched-card calculators in spectrometric calculations has been described by King (K5), King and Priestly (K6, K7), and Sobcov (S3).

Dudenbostel and Priestly (D2) have reviewed the importance of the mass spectrometer in the petroleum industry, giving a brief account of the computational work involved. They also describe recent developments which have made it possible for readings from the spectrometer to be converted automatically into digital output. This output may then be fed either directly to a digital computer or through the medium of punched cards. In either case human intervention is minimized, with consequent time saving and reduction of error opportunities. It is interesting to note also that these analytical methods are now being used directly and continuously on process streams in an operating plant.

It is possible that the future may also see the use of digital calculators in qualitative spectrometric analyses. Various types of punched cards have been used as a method of recording spectral data on pure compounds. The purpose of these files is to facilitate the identification of spectral data on unknown substances. Their use in infrared analysis has been covered by Mecke and Schmid (M6), Keuntzel (K3), and Baker, Wright, and Opler (B2). The last named authors describe a file of 3150 spectra which was expected eventually to be expanded to include up to 10,000 spectra. Zemany (Z1) discussed the use of edge-notched cards in cataloging mass spectra and Matthews (M4) describes a similar application in connection with X-ray diffraction powder data. These two applications made use of only hand-sorting methods; the files of Baker *et al.* were intended to be processed by machine.

These previously described searching operations have not involved what would ordinarily be considered computing equipment. It does seem, however, that there should be a good possibility for using digital computers with magnetic-tape input in this type of work. Not only can the tape be read much more rapidly than cards, but the computer can be programmed to perform much more intricate selection tests than would be possible with simple sorting or collating equipment. In spite of the considerable attention devoted to the punched card as a medium for information handling (C1), magnetic-tape applications in this type of work have not yet appeared in the literature.

C. The Computer and Automation

During the past few years the term *automation* has come into general acceptance (not without some objections still being raised) to indicate the automatic control of processes through feedback. A computer, if it is looked upon merely as a device for producing a calculated response to given data, is always part of the control system. However, the concept of automation today goes far beyond the simple control instrumentation already quite familiar in the process industries. Existing control devices are generally applied independently at isolated control points. Today's view of automation seems to imply control in a much larger sense, from the viewpoint of a process in its entirety. While most present developments in automation are based on analog computers, it has been suggested that the future will see the use of large-scale digital calculators in process control (R1, B7).

The technology of automation is changing so rapidly that a proper perspective on its present and future is difficult. Certainly there have been some statements made which seem premature in their enthusiasm. Present large-scale digital-computing equipment is by no means so inexpensive that it can be justified by the replacement of the few men now needed on most continuous-process equipment. It is likely that these men would be required for emergency operation anyhow. The real justification for process automation must probably be found in improved operations.

A formidable obstacle, however, is the development of satisfactory mathematical criteria for automatic operation. As later sections of this chapter will indicate, the satisfactory solution of this type of problem is often difficult enough even when divorced from the urgency of direct process control. The obstacle to further development here seems to lie not so much in the technology of control-equipment design as it does in determining precisely what is expected of a computer in all the situations it can encounter.

It should be clearly recognized that the subject of computing is not synonymous with automation, as seems frequently to be suggested. The usefulness of computing, separated from direct process control, is an established fact. The relationship of computers to automation in the process industries is still so embryonic that it does not seem advisable to dwell on it in a review of this type.

IV. Statistical and Numerical Analysis

A. STATISTICAL ANALYSIS

Digital computers and accessory data-processing equipment have played an important part in making it practical to handle certain types of statistical problems. The feasibility of storing plant-operating data on punched cards or magnetic tapes was mentioned in a preceding section. One of the main benefits from so doing is the ready accessibility of this stored information for subsequent analysis. Naturally the success of this analysis will depend largely on the filing system used in originally storing the data and on the criteria used to select particular data for statistical treatment. The criteria used must, of course, be specified by the analyst, and if large volumes of data are to be processed automatically by machine, it is essential that a procedure be used which is consistent with sound statistical practice.

The main application for computers in statistical calculations has been in the field of correlation studies. The technique of multiple-regression analysis is now widely used to examine the effects of one or more independent variables on a dependent variable. This important statistical method not only permits determination of the best values for constants in a correlating equation, but also provides for examining the reliability of the resulting correlation. Multiple-regression calculations require three burdensome computational steps. The first of these is the transformation of raw data into desired forms. The second is the calculation of various sums and sums of products from the transformed data. Finally several operations of matrix algebra must be performed.

Transformations of raw data values are required whenever it is desired to investigate a regression equation which is not simply a linear function of the measured variables. Transformations to logarithms, powers, exponentials, interaction products, etc., are frequently required and, when many data are involved, these can best be done by a computer. After suitable transformations have been made, both simple sums and also sums of squares and cross products are needed from all the given data sets. For even fairly small regression problems these calculations are a very burdensome task. Not only are many arithmetic operations required, but they must be carried out with high digital accuracy to obtain arithmetically significant results in later calculations. In some problems hand calculations are almost beyond the realm of possibility because of the chances of simple arithmetic errors.

After the required sums have been obtained and normalized they become the elements of a matrix, which must be inverted. The resultant inverse matrix is the basis for the derivation of the final regression equation and testing of its significance. These last steps are accomplished in part through matrix-by-vector multiplications. Anyone who has attempted the inversion of a high-order matrix will appreciate the difficulty of performing this operation through hand calculation.

During the past 10 years a good deal of information has appeared on the use of punched-card equipment in statistical calculations. An extensive bibliography on the use of IBM punched-card equipment in statistical work has been published by the Watson Scientific Computing Laboratory of Columbia University (B4). Unfortunately many of the given references predate more recent computing equipment. The use of stored-program calculators in statistics seems not to be well documented. This should not be taken to mean that applications of importance have not been made in this area. More probably such applications have now become so routine that familiarity with them is taken for granted. Many computer installations may be expected to have available suitable programs for these operations.

With larger computers there is a justifiable temptation to make the solution of problems as automatic and self-contained as possible. In the case of statistical analysis, however, some common-sense restraints are necessary. Statistical treatment of data is most effective when it is combined with technical experience. The exercise of considerable judgment and, where possible, theoretical insight is important.

While computers are a substantial aid in statistical analysis, it is also true that statistical methods have helped in certain computer applications. In Section V the subject of mathematical models will be discussed. These are in many cases based on empirical correlations. When these have been obtained by regression methods, not only is the significance of the results better understood, but also the correlation is expressed directly in a mathematical form suitable for programming.

B. Numerical Analysis

The special branch of mathematics known as numerical analysis has assumed an added importance with the extensive use of digital computers. Since these calculators perform only the fundamental operations of arithmetic, it is necessary that all other mathematical operations be reduced to these terms. From a superficial viewpoint it might be concluded that such operations as differentiation and integration are inherently better suited to analog computers. This is not necessarily true, however, and depends upon the requirements of the particular problem at hand.

Where appropriate numerical methods of solution are available, it is usually possible to perform calculations to any desired degree of accuracy on a digital calculator, while an analog computer is always to some extent limited in this respect. On the other hand, both programming and calculating time may be excessive when numerical methods are used in some problems. If the desired accuracy is possible on an analog computer it may provide a better solution method.

Numerical analysis is important in digital-computer work from another viewpoint. Sometimes it is necessary to express complex functional relationships in a simpler form. Occasionally relationships may be given in a graphical or tabular form not directly suitable for processing on digital equipment. In these situations numerical methods for curve fitting and interpolation are techniques which will necessarily be employed.

It is not possible or necessary to present a detailed discussion of numerical analysis here. Books by Scarborough (S1), Milne (M7), Whittaker and Robinson (W2), and others are standard references on the subject. A recent work by Householder (H7) is of interest because it is written more from the viewpoint of modern computing equipment. A more extended list of references on numerical analysis has been prepared by Higgins (H3).

V. Mathematical Models

A. Definition

A model may be defined as a device which behaves in a manner similar enough to some other system so that useful knowledge about the system may be gained from a study of the model. The concept of models in the form of laboratory or pilot plant equipment is certainly very familiar in chemical engineering. The usefulness of such models in predicting the operation of present or projected plant equipment is well appreciated. However, there are circumstances in which experimentation with physical models is not the best method of study. Frequently mathematical models are more convenient to use, lower in cost, and more reliable.

A mathematical model is a concise and orderly presentation of information about a system. The information may come from diverse sources. Experimental evidence in the form of empirical correlations is often a large part of a mathematical model. Usually this is unified and extended by theoretical knowledge. Sometimes a model, to be realistic, must include important restrictions defined by managerial policy. In some cases legislation, such as tax laws, may be an important consideration.

A mathematical model is expressed as an equation or a system of equations. The variables involved correspond to measurable or calculable characteristics of a real process or operation. They are functionally related to each other so that they behave much as the variables in the real system that they simulate. Some of the variables may assume, at least over a limited range, any values at the discretion of the operations planner. Others may be determined by external factors. Still others are the dependent variables, or responses, of the system. The convenient prediction of these responses for chosen values of the controllable variables is the principal reason for the formulation of a mathematical model.

Mathematical models do not, in a literal sense, provide new information. They are rather a means for making more effective use of the information from which they were derived. Mathematical models, just as physical models, are rarely perfect in the accuracy and completeness of their correspondence to a real situation. When based largely on empirical evidence, the model must be restricted in use to the range of the original data. Often, with more effort and cost, a model can be improved, but there is always a point beyond which further refinement is unwarranted for the purpose at hand.

Since a mathematical model is a logical organization of relationships, it is adaptable to representation on an automatic computer. For all practical purposes the computer then becomes the model of the system.

This is a development of major importance. While the concept of mathematical models is not in itself new, the usefulness of the idea is greatly enlarged. Models of great complexity, which have previously been of only academic interest, are now practical tools in operations planning (H9, H10).

Mathematical models may, of course, be represented on either digital or analog computers as individual circumstances dictate. When an analog computer is used, there may be some question as to whether the mathematical model is not a somewhat artificial intermediate between the system and its analog. The analogy between the system and the model exists independently of any mathematical considerations. This point is, however, largely academic since it is usually the mathematical model which permits the analogy to be seen.

B. Types of Mathematical Models

1. *Simulation of Processes*

One of the major uses for mathematical models is in the simulation of process operations. This application is readily appreciated by chemical engineers because of the similarity of the model to a pilot plant. In spite of its importance, this subject has not received much attention in the literature. There are probably two main reasons for this. First, process models tend to be quite specific for a particular application and there is little to be said about the subject from a general viewpoint. Second, detailed accounts of specific examples would include information which most firms regard as being of competitive value.

The petroleum refining industry provides many opportunities for the use of mathematical models in process simulation. Although it is generally known in the industry that widespread interest exists in this application of computers, there have been few literature references to this work (A6, H2, U1).[2] These published remarks have been quite brief and vague. It is probably worth while to consider here, however, some of the reasons for the interest of the refining industry in mathematical models on computers, as it may serve to indicate the desirability of similar applications in other areas.

Petroleum refining involves many interrelated processes, most of which have a wide range of possible operating conditions. Many refiners are also able to exercise considerable choice in the selection of raw materials for processing. This is important, since the many available crude oils differ markedly in their refining characteristics.

[2] References will be given later to some other types of mathematical models used in the petroleum-refining industry.

It is fortunate that this flexibility exists, because the requirements for products, both in terms of relative volume and quality, are constantly changing. There are long-term trends, such as in the increasing octane level of gasoline. There are also seasonal changes, such as in the volatility specifications for gasoline or the changed product distribution between heating oil and gasoline.

These changing requirements call for a frequent reexamination of the operating conditions of individual processes and the relationships among them. For quite obvious reasons it would be impractical to do this by direct experimentation in the plant. Also, it is seldom feasible to resort to pilot plant studies for this purpose, both from a cost and time viewpoint.

Actually most of the information needed for proper decisions is already available from past research or operating experience. It is not, however, in a form which can be used effectively until a mathematical model is constructed. The design of the model and its adaptation to a computer can involve a significant amount of effort and cost. The alternatives of direct experimentation or neglecting the problem involve such large economic debits, however, that the use of computers in this work is most attractive.

Possibly the chemical industry does not have as much need for mathematical models in process simulation as does the petroleum refining industry. The operating conditions for most chemical plants do not seem subject to as broad a choice, nor do they seem to require frequent reappraisals. However, this is a matter which must be settled on the basis of individual circumstances. Sometimes the initial selection of operating conditions for a new plant is sufficiently complex to justify development of a mathematical model. Gee, Linton, Maire, and Raines describe a situation of this sort in which a mathematical model was developed for an industrial reactor (G1). Beutler describes the subsequent programming of this model on the large-scale MIT "Whirlwind" computer (B6). These two papers seem to be the most complete technical account of model development available. However, the model should not necessarily be thought typical since it relies more on theory, and less on empiricisms, than do many other process models.

The process in question involved the reaction of two materials, A and B, to produce a product C. The reaction was noncatalytic, homogeneous, and in the gas phase. It took place in a tubular reactor which could not be considered either adiabatic or isothermal. The reactor was divided into four sections, the first three of which were cooled while the fourth was adiabatic. Coking of the reactor tube introduced a time variant in the system, requiring adjustment of operating conditions and eventual shutdown for cleaning.

The entire system was described in a mathematical model consisting of four partial-differential equations. These were reduced to difference equations for integration, since it was decided that a digital computer was preferable to an analog computer in this application. A detailed flow diagram was constructed to show how the computer was to respond to all possible conditions which would arise during the calculation. From this diagram the coded set of instructions, or program, was written for the computer. The written program was finally translated to a punched paper tape used as the computer input.

The program was designed so as to start with a fixed feed rate of the reactant A at time zero. Temperature and conversion profiles along the length of the reactor tube were calculated and displayed on the oscilloscope output of the computer. If temperature exceeded a limiting value, the feed rate was automatically reduced and a new profile determined. As soon as a satisfactory profile was determined at the initial conditions, the time was increased by a predetermined increment and the calculation continued.

Two practical results were noted after experiment with the model: first, insight was gained as to methods of operating the plant reactor; second, a change in basic operating conditions, indicated by use of the model, resulted in a 25% increase in production. Using the model on a computer, rather than through hand calculations, had two advantages: first, the entire study was carried out in 42 hours of machine time compared with an estimated 20 years of man time needed for hand solution; second, the close analogy (from an external viewpoint) between the computer and the plant permitted those interested almost to get the "feel" of the plant without following the detailed arithmetic being done by the computer.

2. *Combinations of Models*

Successful construction and use of mathematical models for individual processes soon leads to more ambitious projects. It has already been mentioned, for example, that in petroleum refining most processes cannot be regarded as separate entities because they consume products of other processes and supply raw materials for still other operations. If one tries, however, to construct a mathematical model which will directly represent the operations of an entire refinery, the task can become overwhelming. The size of the model, when looked at in its entirety, is simply too large for a proper perspective.

The preferred method of attacking a large problem is to divide it into several elemental parts. For example, in the author's company mathematical models have been developed (in the form of both medium- and large-scale digital-computer programs) for all major refinery processes, such as thermal cracking, catalytic cracking, polymerization, hydroforming, distillation, gasoline blending, etc. For some time these models have been used individually, but gradually it became apparent that they could be joined together to represent complex systems of processes, culminating finally in a model of a complete refinery. Yet in the final model of the refinery each of the original processes retains its separate entity. This is advantageous. First, each individual model can be designed and kept up to date by an expert in that particular field. Second, different combinations of models can be made so as to represent different refineries, without necessarily redesigning all the individual models.[3]

An interesting application of this approach in another field has been described by Keller (K2). In the design of steam turbines rather complicated heat-balance calculations are required. While each particular installation is different, and therefore requires a different mathematical model, the components of each turbine are always similar. A large-scale computer program was developed, therefore, which would through suitable instructions combine the calculations required for each component into an over-all heat balance for the turbine.

3. *Blending Operations*

One of the important manufacturing operations of the process industries is the blending of several components so as to make a product meeting given quality specifications. Probably the most important and complex of these blending problems is encountered in the manufacture of gasoline. Later we will consider the blending problem from the viewpoint of optimization (*see* Section VI E). Here it will be considered only from the viewpoint of constructing a mathematical model.

Suppose that there are n components, each available in unlimited amounts for blending. There are associated with each of the components m quality characteristics, $q_1, \ldots, q_m$. It will be assumed that these qualities blend linearly, and the quality of the product will be represented by $Q_1, \ldots, Q_m$. Now if $v_1, \ldots, v_n$ are the fractional volumes of the components in the product, we may write

$$\sum_{j=1}^{n} v_j q_{ij} = Q_i \qquad i = 1, \ldots, m \tag{1}$$

$$\sum_{j=1}^{n} v_j = 1 \tag{2}$$

This set of linear simultaneous equations is a mathematical model of the blending operation. Given the volumes of the components, v_j, and the qualities, q_{ij}, it is possible to calculate the quality of the product, Q_i.

---

[3] In most cases the individual process models are either identical in different refineries or can be adapted to special circumstances through changes in numerical constants built into the model.

Alternatively, given the quality of the product and the qualities of the components, it is possible, provided $m + 1 = n$, to calculate the required volumes of the components.[4] If $n > m + 1$ there is no unique solution, but rather an infinity of solutions.

The case in which a unique solution is possible is not frequently encountered in practice. However, the use of an analog computer for solving such a problem in aviation-gasoline blending has been mentioned (A1). The same computer was used in calculating the quality of a motor-gasoline blend, given the volumes and qualities of the components. Other analog or digital computers of sufficient size might also be used for these calculations, which are no more than solutions of linear simultaneous equations or summations of products.

Let us consider now, however, a more realistic model. Product-quality specifications are not generally given as absolute quantities, but rather in terms of maximum or minimum values. Thus the linear equations in Eq. (1) become the linear inequalities:[5]

$$\sum_{j=1}^{n} v_j q_i \leq Q_i \qquad i = 1, \ldots, m \qquad (3)$$

These inequalities may be made into equations by introducing new unknowns, $X_1, \ldots, X_m$, which take up the "slack" in each inequality. We then have

$$\sum_{j=1}^{n} v_j q_{ij} + X_i = Q_i \qquad i = 1, \ldots, m \qquad (4)$$

The "slack" variables in these equations represent what, in the petroleum refining industry, is frequently called product-quality "giveaway." While it may seem superficially undesirable to manufacture a product exceeding specifications, this may actually be the best method of operation.

Many further complications may be introduced into the model without changing its linear form. For example, it may be required that the volumes of components used lie within certain maximum or minimum limits. There may be relationships between the specifications permitting one variable to assume different values, depending upon the level of another. These factors will introduce new equations and/or unknowns

---

[4] Provided also that the qualities of the components permit a unique and realistic solution.

[5] It is necessary, for some further discussion of this model presented in Section VI E, that the inequality signs be directed as shown and that $Q_1, \ldots, Q_m$ be greater than or equal to zero. This requirement can always be met by making linear transformations of the qualities if needed.

in the model. The more restrictions that are imposed, the smaller the range of possible operations. It is to be noted that this model has no unique solution unless values are arbitrarily assigned to some of the unknowns or there is some additional requirement for an acceptable solution.

There are, of course, some technological complications to the blending problem which we have not discussed. Some quality characteristics do not blend linearly. However, transformations of them (frequently called blending values or indexes) may sometimes be found which do blend linearly. In other cases the simple model in terms of linear equations is inadequate. Even the linear model, however, in real problems of gasoline blending may become quite large. Systems of 50 to 100 equations with 75 to 150 unknowns are not unheard of. Any calculations involving models of this size must almost necessarily use automatic computing equipment.

4. *The Monte Carlo Method*

The Monte Carlo method permits simulation, in a mathematical model, of stochastic variation in a real system. Many industrial problems involve variables which are not fixed in value, but which tend to fluctuate according to a definite pattern. For example, the demand for a given product may be fairly stable over a long time period, but vary considerably about its mean value on a day-to-day basis. Sometimes this variation is an essential element of the problem and cannot be ignored.

If the variation were completely unpredictable, there would be no hope of rational planning to take it into account. Usually, however, although it is not possible to predict that a given occurrence will certainly happen, it is possible to assign a probability for any particular occurrence. If this is done for all possible occurrences, then, in effect, a probability distribution function has been defined. Certain types of such distributions can be derived mathematically to fit special situations. The normal, Poisson, and binomial distributions are frequently encountered in practice.

When a mathematical model and the variation which affects it[6] are both simple in form, it is sometimes possible to derive analytically any desired information relating to the behavior of the system. When this is the case the Monte Carlo method may offer little or no advantage. However, it many problems it is impractical to obtain the desired results entirely by analytical methods. It is in this situation that the Monte Carlo method becomes a most valuable tool.

The Monte Carlo method subjects a mathematical model to the same

[6] The two taken collectively may properly be considered a larger mathematical model.

sort of variation as would be encountered in the real system. Studying the behavior of the model over an extended number of occurrences or "events" yields insight into the response of the system. Also, by varying the construction of the model or the nature of the distribution function, one can learn the advantages or disadvantages of changes in the system.

The synthesis of occurrences or events in the Monte Carlo method makes use of random numbers and a cumulative-distribution function. In effect the random numbers are transformed, by means of the distribution function, into a simulated sequence of events. Figure 2 shows the general procedure followed. A random number is selected and transformed

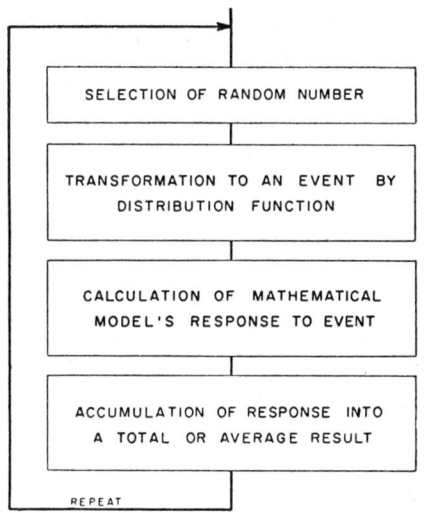

FIG. 2. The Monte Carlo method used in studying behavior of a mathematical model.

into an event. The mathematical model, through calculation, responds by giving a result. This result is accumulated into a total or average result. Then the cycle is repeated by the choosing of another random number. The process is continued long enough to give a desired degree of stability indicative of long-term effect or average behavior.

The Monte Carlo method is especially suited for use on a digital computer, particularly one of the stored-program type. The mathematical model and the distribution function, even if quite complicated, can be expressed on the computer and the necessary calculations are highly repetitive. Also, random numbers (or rather pseudorandom numbers) can be synthesized so that the computer procedure becomes fully automatic and self-contained (M9, S5).

The Monte Carlo technique has been used in some chemical engineering problems, particularly in the field of diffusion (K4). However, applications to industrial operating problems have not yet received much attention. There are probably many potential applications worth considering. One of the most apparent is in studies relating to required inventories of products or raw materials to cope with the variability of supply and demand. Another is in examining the benefits which might be achieved through closer control of process operating conditions. The author has had occasion to consider the application of the Monte Carlo method to a problem involving the design of terminal facilities at a crude shipping port. In this situation the arrival of tankers was found to follow a simple distribution law. Other applications will no doubt occur to the reader in his particular area of interest.

5. *Incorporation of Economic Factors*

There is no reason why a mathematical model should be limited to simulation of only the physical aspects of a given system. Usually the behavioral response of most interest from a management viewpoint will be an economic variable, such as cost, profit, investment, etc. In many cases the motivation for development of the model will be the optimization of one of these variables. This problem will be considered in the next section.

There are really no new considerations introduced into the model from a computational viewpoint when economic variables are included. There are, of course, certain conceptual difficulties which arise because of questions as to proper costs or realizations to be used in specific problems. These are, however, problems in economic evaluation outside the scope of this chapter.

VI. Optimization Studies

A. The General Problem

The mathematical model of a system, adapted to a computer, is a powerful aid in operations planning. The major use of the model, and indeed often the sole reason for its development, will be the improvement of some response of the system. This response will generally be some economic variable such as profit or cost. Sometimes a physical variable such as yield may be considered, but usually only because it is directly related to an economic variable. In any event the objective is to determine a set of levels for several variables subject to control, so as to achieve an optimum response.

It is assumed here that the mathematical model is sufficiently complex so that it is impractical to determine the optimum by conventional mathematical methods. This will almost always be true, or the model would not have originally been formulated for a computer. In this case it is necessary to resort to an experimental approach which is quite analogous to experimentation with physical equipment. Much of the discussion which follows will, in fact, be based on the same principles that have guided such experimental studies.

It should be appreciated at the outset that the problem of optimum determination can be quite complicated. In many cases a wide range of choice is possible for the controllable variables. As more and more variables are taken into consideration, the problem becomes more difficult at an increasing rate. Furthermore, in many cases the optimization problem is continually recurring; that is, a study may be successfully concluded and an optimum set of values for the controllable variables determined, but at a later time, owing to changes in uncontrollable variables (supply and quality of raw materials, demand and specifications for products, tax laws, etc.), the calculated optimum is no longer valid and must be determined again.

Optimization studies then, both because of their inherent difficulty and the fact that they are recurrent, will involve repeated use of the mathematical model. It is for this reason that its adaptation to a computer is important. In spite of the great reduction in calculating cost due to the automatic computer, however, the total cost of a study will usually not be so small that efficient design of the investigation is not important. The problem is quite similar to that encountered in the statistical subject of design of experiments, and it seems apparent that the two fields should have several common aspects.

There is, however, one obvious difference between a mathematical model and a physical model (or the real system itself). The response of the former to the same set of conditions is always identical. In physical experiments, where results are measured rather than calculated, there are inevitably random errors which may be appreciable. As already pointed out, mathematical models are usually to some extent imperfect; in other words, they do contain systematic errors. The important point is that these imperfections are always reproduced in the same way, even though their ultimate source may have been random errors in data on which the model was based. This point has been stressed because it is important to recognize that only partial use of methods from statistical treatments of design of experiments is involved in what follows. The use of these methods here is only for the purpose of studying the geometry of response with respect to the controllable variables. No consideration of probability or of error enters into the discussion.

There are two different approaches to determining the optimum response of a mathematical model. The first is a complete survey of the area of interest, the survey being planned in advance of any individual case calculations. The second is a sequential investigation which is continually modified as information becomes available from completed cases. The first type of experimental design has an obvious importance in a field like agricultural experimentation, where results cannot be expected except over long time periods and where it is practical (if not essential) to run parallel experiments. Investigations of model behavior on a computer can usually be made quickly, however, and, assuming use of one computer, only one case can be calculated at a time.

Both nonsequential and sequential types of designs will be considered here for several reasons. First, sequential designs consist of elements which are themselves nonsequential. Second, use of sequential designs requires a clear-cut objective which must be defined in quantitative terms. Third, sequential designs, while efficient in terms of calculations done, are subject to certain dangers because of their limited exploration of the area of possible interest.

The problem of optimization will first be considered from the viewpoint that complete freedom of choice may be exercised in assuming values for the selected controllable variables. Then the more realistic case where only a limited choice exists will be treated. This will lead finally to a consideration of linear programming for solution of certain classes of restricted optimization problems.

B. Factorial Designs

The most complete investigation of model response would establish a network of experimental cases throughout the area of interest. The more complete the coverage, obviously the more cases that will be required. For a system involving many variables the number of cases increases rapidly; for example, in a system of six controllable variables an investigation at only two levels of each variable would require $2^6 = 64$ cases. Examining three levels of each of the variables would require $3^6 = 729$ cases.

An examination of all variables at all possible combinations of the selected levels is known as a complete factorial experimental design. It is not necessary that all variables be studied at the same number of levels. The essential feature is that results are provided for all possible combinations of the chosen levels for the several variables.

For systems involving many variables and levels it is apparent that a complete factorial design leads to an excessive number of cases. However, it may be quite practical to make reasonable assumptions about the behavior of the model and, in so doing, to reduce the number of cases needed. For example, if the effect of one variable is found to be identical at two extreme levels of a second variable, it seems reasonable to assume that its behavior would be the same at any intermediate level. Of course, there is no way of being certain that this is true without examining the intermediate cases. It can usually be argued, however, that on the basis of past experience the chances of finding such irregular behavior are small. The cost of obtaining the additional information is often so great as to outweigh the risks run by not obtaining it. Even if the effects of variables do change at different levels of other variables, there is still usually enough regularity so as to make the securing of complete sets of results an unwarranted extravagance.

When some of the possible data points are omitted in factorial designs they are known as fractionally or partially replicated designs. The choice of points to be omitted is of considerable importance. There is no single fractional replicate which is "best" for any given complete factorial design. Usually the experimenter will have some idea as to the expected effects. When this sort of intuition is available, a particular design may be developed to fit a particular problem.

In order to interpret the results of a factorial design of cases it is almost always necessary to develop an equation which expresses the response in terms of the variables studied. Doing so may seem to be recreating the mathematical model, and in a sense it is. However, the new model will be specific for the particular study in hand, having embodied into it fixed values for those variables which were held constant. Since the new model is more specific it can also be expressed much more simply. Generally the response can be fitted, over a limited area at least, with a general second- (or sometimes third-) degree polynomial. The techniques used for fitting the equation would be similar to those used in multiple regression analysis, except that it must be remembered that we are not dealing with points subject to random errors. Any lack of fit of the approximating function to the original points (assuming, as one should, that degrees of freedom have been permitted) is due entirely to the inadequacy of the form of equation used to fit the original model.

Assuming that a polynomial has been found which adequately represents the response behavior, it is now possible to reduce the polynomial to its canonical form. This simply involves a transformation of coordinates so as to express the response in a form more readily interpreted. If a unique optimum (analogous to a mountain peak in three dimensions) is present, it will automatically be located. If (as is usual in multidimensional problems) a more complex form results, the canonical equation will permit proper interpretation of it.

It should be pointed out that the derivation of the fitted polynomial and its reduction to canonical form require only standard arithmetic manipulations. Since the volume of arithmetic to be done is considerable in large problems, this part of the study should also be performed on the computer. The details of the calculations required cannot be presented here, but the reader will find information on these points in Davies' book (D1).

C. SEQUENTIAL DESIGNS

The use of sequential methods in the statistical design of experiments has received increased attention in the last few years and seems particularly appropriate when the objective of a study is the optimization of a selected response. The general approach was first described by Box and Wilson (B9) and later amplified in other papers (B8, D1). It presents interesting possibilities in automatic computing with mathematical models, not only because it is efficient in terms of calculations, but also because at least part of its procedure can be well defined prior to any calculations. This permits designing a computer program to supplement the model so that the structure of successive cases is determined as the calculation progresses. It is desirable to make any such procedure as automatic as practical. If not, the machine calculations will have to be interrupted for hand calculations or human judgment, and the over-all procedure becomes less efficient.

One possible method of automatic optimization was mentioned by Kahn (K1) quite some time ago. This used the Monte Carlo technique to make a random design of cases over a broad area known to contain the optimum. After a limited number of cases had been calculated, the best case was selected and another random design of cases was made. For the second design, however, the dispersion of cases was reduced so that the investigation became more localized. After a sufficient number of repetitions of this process the optimum would be determined to a sufficient degree of accuracy. The method is perhaps the antithesis of a logical or orderly calculation procedure, but it can be programmed for automatic sequential calculation.

An adaptation of the Box method, however, seems to offer the advantage of improved efficiency while still being susceptible to automatic computation. Box's approach may be divided into two stages. The first, to which he has applied the name "method of steepest ascents," is primarily for the purpose of approximately locating the optimum response. The second is a more intensive investigation in the local region of the optimum. This will permit a precise determination of the optimum and also indicate the behavior of the response in its neighborhood.

The basis for the method of steepest ascents is easy to appreciate for a two-dimensional system. Suppose that Fig. 3 represents an area of interest for two controllable variables, X_1 and X_2. The response function is shown in this diagram by means of contour lines and there is a maximum in the given area. The cross in the lower left corner is a base case, which may be the starting point for the investigation. The classical method of studying each variable in turn would lead us along the tortuous path shown by the dotted lines. The method of steepest ascents first determines the slope of the response surface in the local neighborhood of the base case. It then progresses experimentally in the direction of steepest

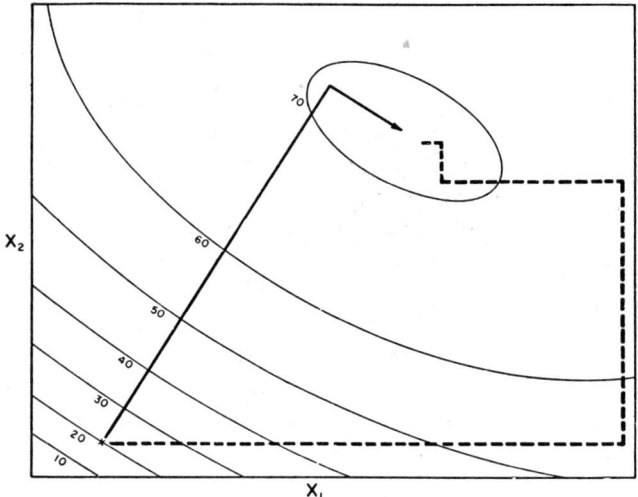

Fig. 3. The method of steepest ascents applied to a problem of maximizing a response of two variables.

ascent, both variables changing at the same time, until a point of maximum response is reached. Then the process is repeated from this point. Obviously this approach leads more directly toward the optimum. As the optimum is approached, the slopes become smaller and the improvement becomes less. At this point the method of steepest ascents is abandoned in favor of the more intensive investigation mentioned above, which is in effect a curve fitting.

It is easy to imagine how the method of steepest ascents is generalized to multidimensional studies. It is, however, difficult to portray these cases graphically. They are the most important applications for the method and it is in such large studies that the technique is most advantageous.

The mechanics of the method of steepest ascents are simple. The first step is to define the general area of interest and reduce all the controllable variables to a common basis by scaling. It is also helpful at this point, if possible, to transform the variables so that their effects on the response are not extremely curvilinear. In other words it may be advisable to consider some variables in logarithmic form, as exponential functions, etc. There will be, then, for each variable, a transformed form

$$X_i = \beta_i(x_i) \tag{5}$$

where x_i is the original variable (temperature, pressure, concentration, etc.) and β_i is a function which transforms it so that the response varies with it in an approximately linear manner at the base point. The transformation should also scale the original variable so that it becomes dimensionless and covers a convenient range (say 0 to 1) over the area of interest. Inability to obtain ideal transformations is not a matter of serious consequence; it will just make the solution a more tedious process. In many cases it will obviously be impossible to anticipate the best transformations at the start of a study.

Once the transformations have been made, a solution space has been defined. It is only within this framework that the concept of steepest ascents takes meaning. The proper direction in which to proceed can be determined by $n + 1$ cases if n is the number of controllable variables. In each of the n cases one variable is changed slightly from its value in the base case, while all other variables are held constant. This permits approximating the n partial derivatives of response with respect to each variable. The direction of steepest ascent is given by the vector which is the gradient of the response, R:

$$\nabla R = \frac{(\partial R)}{(\partial X_1)} \mathbf{i} + \frac{(\partial R)}{(\partial X_2)} \mathbf{j} + \cdots + \frac{(\partial R)}{(\partial X_n)} \mathbf{n} \tag{6}$$

where $\mathbf{i}, \mathbf{j}, \ldots, \mathbf{n}$ are unit vectors in the direction of the $X_1, X_2, \ldots, X_n$ coordinate axes.

Before proceeding further it is necessary to decide upon the size of the step to be made in the direction of steepest ascent. This step size, Δ, can be related to the several partial derivatives by a constant K:

$$\Delta^2 = K^2 \left[\frac{(\partial R)^2}{(\partial X_1)} + \frac{(\partial R)^2}{(\partial X_2)} + \cdots + \frac{(\partial R)^2}{(\partial X_n)} \right] \tag{7}$$

Once the constant K has been determined from this equation, the incremental change in each individual variable for steps in the direction of steepest ascent is

$$\delta X_i = K \frac{(\partial R)}{(\partial X_i)} \tag{8}$$

Steps are repeated until there is no further improvement in response. Preferably they should be carried a little beyond the point of maximum response. Since the point of optimum response will in general lie between two calculated points, this will permit a better interpolation to be made.

Now a new case is calculated at the best previous point and a new direction of steepest ascent is determined. The process is repeated as many times as seems advisable. The entire procedure is susceptible to automatic treatment on the computer. The only point of uncertainty is the size of the steps to be taken. It may be necessary to revise this from time to time depending upon the progress of the study. As the optimum is approached, the steps should decrease in size. Even if it is decided that for a particular problem and a particular computer completely automatic calculation is impractical, at least some fairly large combination of operations can be programmed for one computer run.

Following application of the method of steepest ascents, it will usually be advisable to investigate the neighborhood of the optimum more carefully, fitting at least a second-degree polynomial to the response surface as described in Section VI B.

D. Optimization Problems Involving Restrictions

Thus far it has been assumed possible to vary the controllable variables without any restraints. In practice this is rarely the case. Usually variables may take values only within given ranges that are determined by equipment limitations, safe operating practices, etc. These boundaries define a solution space in which the optimum response must be found. In these cases the best response will frequently be found to lie on one or more of the boundaries, rather than at some interior point. Superficially the method of steepest ascents seems equally applicable under these circumstances. When a limiting value is reached for any one of the controllable variables, it is simply held at a constant value and the calculation continued. However, consider the two-dimensional problem shown in Fig. 4, where the boundaries of the diagram constitute limiting values for the variables. Starting from the given base case, the direction of steepest ascent eventually brings us to the corner marked A. Moving in any direction from this corner would cause a falling response and it might, therefore, be concluded that this was the best possible solution. In reality, however, the best response is at the corner marked B.

The only certain way of avoiding this situation would be to make a larger exploration to cover all possibilities. Alternate optima may arise even when no restrictions are imposed on the problem. However, the difficulty is much more likely when restrictions are present. The restrictions may sometimes be expressed on derived functions of the controllable variables. For example, the temperature of a reaction may not be subject to direct control, but may depend in a complicated way on feed rate, reactant concentrations, feed temperatures, etc. If a bound is now set for reaction temperature, while an attempt is made to optimize yield, the question arises as to which variables should be adjusted so as to

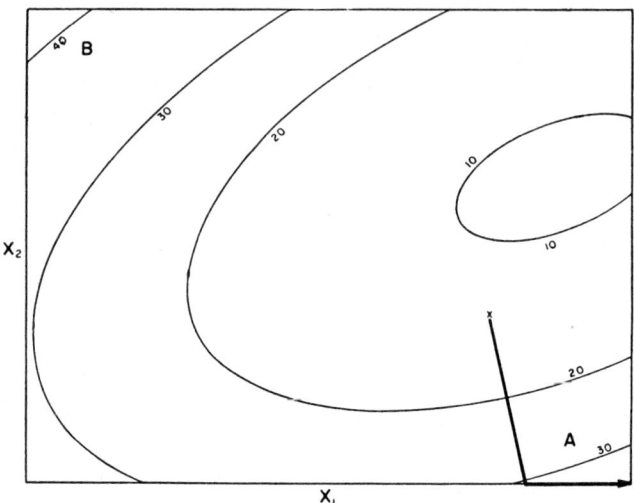

Fig. 4. Possible erroneous optimum indication in using the method of steepest ascents.

observe the restriction and still fulfill the objective. The literature does not yet contain any examples of the application of steepest ascents in situations of this sort.

E. Linear Programming

One special type of optimization problem involving restrictions or constraints has been solved quite successfully by a technique known as *linear programming*. From a mathematical viewpoint the basic form of the problem may be stated very briefly. Consider a linear response function of n variables:

$$R = \sum_{j=1}^{n} c_j X_j \qquad (9)$$

The c_j coefficients are constants in a given problem. The unknowns, X_j, are variables for which only a restricted range of values may be chosen. Necessary relationships among the variables may be expressed in the m

independent linear equations:

$$\sum_{j=1}^{n} a_{ij} X_j = b_j \qquad i = 1, 2, \ldots, m \qquad (10)$$

where the a_{ij} and b_j are constants for a particular problem, and all $b_j \geq 0$. The problem of linear programming requires finding a set of $X_j \geq 0$, such that the function R is maximized (or minimized, if that is desired). In order to do this it is, of course, necessary that $n > m$.

In Section V B.3 it will be recalled that the general problem of linear blending was expressed in a form similar to Eq. (10). It was mentioned at that point that the problem had no solution unless there was some additional requirement which permitted us to choose from the infinite number of possible solutions. If this requirement can be expressed as maximizing (or minimizing) a function in the form of Eq. (9), the problem becomes one in linear programming. It is to be noted that the requirement that all X_j be greater than or equal to zero will provide a solution in which only positive amounts of components are blended and only positive quality giveaway results. These are requirements for any acceptable solution.

The satisfactory expression of an optimization problem within the framework of linear programming is not always a simple matter. Charnes, Cooper, and Mellon have described a problem in aviation-gasoline blending (C3) and later a broader programming problem of the operations of an integrated oil company (C4). Manne (M2, M3) and Symonds (S8) have also suggested several applications of linear programming in various phases of petroleum refining. The reader is cautioned that not all the technological formulation in these examples may be acceptable under all circumstances. It is to be expected that every real application will have to be critically reviewed to be sure that in adapting it to the necessary form for linear programming vital elements of the problem have not been distorted. Frequently some ingenuity will, however, make it possible to treat problems which did not originally seem suitable for this method of attack.

Real problems are likely to be considerably more complex than the examples that have appeared in the literature. It is for this reason that the computer assumes a particular importance in this work. The method of solution for linear-programming problems is very similar, in terms of its elemental steps, to the operations required in matrix inversions. A description of the calculations required for the "Simplex" method of solution is given in Charnes, Cooper, and Henderson's introductory book on linear programming (C2). Unless the problem has special characteristics, the use of a computer in solving problems involving 20 or more equations is almost essential. Several programs have been developed for these calculations on various digital computers. One for systems of up to 100 equations has been described by Orchard-Hayes (O2). Information regarding other available programs may be had from the applications departments of some computer manufacturers.

REFERENCES

A1. Anonymous, "Refinery Blending Computer (Boeing Model 8000)," Document D-14021, Boeing Airplane, Seattle, Wash., 1953.
A2. Anonymous, *Chem. Eng. News* **33**, 222 (1955).
A3. Anonymous, *Computers and Automation* **3**, No. 9, 9, 30 (1954).
A4. Anonymous, *Computers and Automation* **4**, No. 1, 20, 22 (1955).
A5. Anonymous, *Computers and Automation* **4**, No. 2, 23 (1955).
A6. Anonymous, *Oil Gas J.* **53**, No. 32, 89 (1954).
B1. Backus, J. W., *J. Assoc. Computing Machinery* **1**, 4 (1954).
B2. Baker, A. W., Wright, N., and Opler, A., *Anal. Chem.* **25**, 1457 (1953).
B3. Bashe, C. J., Buchholz, W., and Rochester, N., *J. Assoc. Computing Machinery* **1**, 149 (1954).
B4. Beach, A. F., Hankam, E. V., and Sweeney, K. M., "Bibliography on the Use of IBM Machines in Science, Statistics, and Education," International Business Machines Corp., New York, 1954.
B5. Berry, C. E., Wilcox, D. E., Rock, S. M., and Washburn, H. W., *J. Appl. Phys.* **17**, 262 (1946).
B6. Beutler, J. A., *Chem. Eng. Progr.* **50**, 569 (1954).
B7. Bishop, J. F., *Automatic Control* **1**, No. 6, 16 (1954).
B8. Box, G. E. P., *Biometrics* **10**, 16 (1954).
B9. Box, G. E. P., and Wilson, K. B., *J. Roy. Statistical Soc.* **B13**, 1 (1951).
C1. Casey, R. S., and Perry, J. W., eds., "Punched Cards—their Applications to Science and Industry." Reinhold, New York, 1951.
C2. Charnes, A., Cooper, W. W., and Henderson, A., "An Introduction to Linear Programming." Wiley, New York, 1953.
C3. Charnes, A., Cooper, W. W., and Mellon, B., *Econometrica* **20**, 135 (1952).
C4. Charnes, A., Cooper, W. W., and Mellon, B., *Econometrica* **22**, 193 (1954).
C5. Clippinger, R. F., *Harvard Business Rev.* **33**, No. 1, 77 (1955).
C6. Clippinger, R. F., Dimsdale, B., and Levin, J. H., *J. Soc. Ind. Appl. Math.* **1**, 1 (1953).
C7. Clippinger, R. F., Dimsdale, B., and Levin, J. H., *J. Soc. Ind. Appl. Math* **1**, 91 (1953).
C8. Clippinger, R. F., Dimsdale, B., and Levin, J. H., *J. Soc. Ind. Appl. Math.* **2**, 36 (1954).
D1. Davies, O. L., "Design and Analysis of Industrial Experiments." Oliver and Boyd, London, 1954.
D2. Dudenbostel, B. F., Jr., and Priestly, W., Jr., *Chem. Eng. News* **32**, 4736 (1954).
F1. Fuchs, O., *Chem. Ing. Tech.* **25**, 377 (1953).
G1. Gee, R. E., Linton, W. H., Jr., Maire, R. E., and Raines, J. W., *Chem. Eng. Progr.* **50**, 497 (1954).
G2. Gotlieb, C. C., *J. Assoc. Computing Machinery* **1**, 124 (1954).

G3. Gotlieb, C. C., *Computers and Automation* **3**, No. 7, 14, 25 (1954).
H1. Hamilton, F. E., and Kubie, E. C., *J. Assoc. Computing Machinery* **1**, 13 (1954).
H2. Harp, W. M., *Petroleum Refiner* **32**, No. 4, 159 (1953).
H3. Higgins, T. J., *Control Eng.* **2**, No. 1, 57 (1955).
H4. Higgins, T. J., *Control Eng.* **2**, No. 2, 60 (1955).
H5. Hopper, G. M., *Computers and Automation* **2**, No. 4, 1 (1953).
H6. Hopper, G. M., et al., Report to the Association for Computing Machinery, First Glossary of Programming Terminology, June, 1954.
H7. Householder, A. S., "Principles of Numerical Analysis." McGraw-Hill, New York, 1953.
H8. Hovious, R. L., Morrill, C. D., Tomlinson, N. P., *Instruments and Automation* **28**, 594 (1955).
H9. Hurd, C. C., *J. Operations Research Soc. Amer.* **2**, 205 (1954).
H10. Hurd, C. C., *Management Sci.* **1**, 103 (1955).
K1. Kahn, H., Proc. Seminar on Scientific Computation, pp. 20–27. International Business Machines Corp., New York, November, 1949.
K2. Keller, A., *Mech. Eng.* **75**, 891 (1953).
K3. Keuntzel, L. E., *Anal. Chem.* **23**, 1413 (1951).
K4. King, G. W., *Ind. Eng. Chem.* **43**, 2475 (1951).
K5. King, G. W., Proc. Industrial Computation Seminar, pp. 32–35. International Business Machines Corp., New York, September, 1950.
K6. King, W. H., Jr., and Priestly, W., Jr., *Anal. Chem.* **23**, 1418 (1951).
K7. King, W. H., Jr., and Priestly, W., Jr., *IBM Tech. Newsletter* **No. 3**, 5 (1951).
K8. Korn, G. A., and Korn, T. M., "Electronic Analog Computers." McGraw-Hill, New York, 1952.
M1. Macdonald, N., and Berkeley, E. C., *Computers and Automation* **2**, No. 2, 23 (1953).
M2. Manne, A. S., "Concave Programming for Gasoline Blends," Report No. P-383. Rand Corp., Santa Monica, Calif., 1953.
M3. Manne, A. S., "Petroleum Refinery Operations Scheduling," Report Nos. P-481, 484, 487, 489, 493, 502, 503. Rand Corp., Santa Monica, Calif., 1954.
M4. Matthews, F. W., *Anal. Chem.* **21**, 1172 (1949).
M5. McDonnell, J. A., *Instruments and Automation* **27**, 1797 (1954).
M6. Mecke, R., and Schmid, E. D., *Angew. Chem.* **65**, 253 (1953).
M7. Milne, W. E., "Numerical Calculus." Princeton U. P., Princeton, 1949.
M8. Morris, W. L., *Ind. Eng. Chem.* **43**, 2473 (1951).
M9. Moshman, J., *J. Assoc. Computing Machinery* **1**, 88 (1954).
N1. Nims, P. T., Proc. Computation Seminar, pp. 37–47. International Business Machines Corp., New York, August, 1951.
O1. Opler, A., *Anal. Chem.* **22**, 558 (1950).
O2. Orchard-Hayes, W., "The Rand Code for the Simplex Method," Report No. P-477. Rand Corp., Santa Monica, Calif., 1954.
P1. *Proc. Inst. Radio Engrs.* **41**, 1219 (1953).
R1. Ridenour, L. N., *Sci. American* **187**, No. 3, 116 (1952).
R2. Rose, A., Schilk, J. A., and Johnson, R. C., *Ind. Eng. Chem.* **45**, 933 (1953)
R3. Rose, A., Heiny, R. L., Johnson, R. C., and Schilk, J. A., *Ind. Eng. Chem.* **46**, 916 (1954).
S1. Scarborough, J. B., "Numerical Mathematical Analysis," 2nd ed. Johns Hopkins Press, Baltimore, 1950.
S2. Schrage, R. W., *IBM Tech. Newsletter* **No. 6**, 5 (1953)

S3. Sobcov, H., *Anal. Chem.* **24,** 1386 (1952).
S4. Soroka, W. W., "Analog Methods in Computation and Simulation." McGraw-Hill, New York, 1954.
S5. Spenser, G., *Computers and Automation* **4,** No. 3, 10, 23 (1955).
S6. Stevens, R. F., and Brady, J. F., *Chem. Eng. Progr.* **50,** 493 (1954).
S7. Strong, J. D., *Instruments and Automation* **28,** 602 (1955).
S8. Symonds, G. H., "Linear Programming: The Solution of Refinery Problems." Esso Standard Oil, New York, 1955.
U1. Uhl, W. C., *Petroleum Processing* **10,** 189 (1955).
W1. Wainwright, L., *Computers and Automation* **2,** No. 6, 3 (1953).
W2. Whittaker, E. T., and Robinson, G., "The Calculus of Observations. A Treatise on Numerical Mathematics," 3rd ed. Blackie & Son, Glasgow, 1940.
Z1. Zemany, P. D., *Anal. Chem.* **22,** 920 (1950).

Ionizing Radiation Applied to Chemical Processes and to Food and Drug Processing

ERNEST J. HENLEY

Department of Chemical Engineering, Columbia University, New York, New York

AND

NATHANIEL F. BARR

Chemistry Department, Brookhaven National Laboratory, Upton, New York

I. Introduction.. 370
II. Statement of Problem.. 371
   A. Fundamental Considerations.................................... 371
      1. Use of the Unique Properties............................... 372
   B. Practical Considerations.. 373
   C. Dosimetry.. 373
III. Experimental Work on Initiation of Chemical Reactions............. 374
   A. Introduction.. 374
   B. Fundamental... 375
      1. Primary Processes... 375
      2. Chemical Intermediates..................................... 377
      3. Reactions of Chemical Intermediates........................ 383
   C. Practical Processes Investigated................................ 386
      1. Introduction... 386
      2. Inorganic Processes... 386
      3. Organic Reactions... 388
      4. Polymerizations... 390
      5. Effect of Radiation on Polymers............................ 393
IV. Radiation Sterilization.. 397
   A. Mechanism of Biological Action................................. 398
      1. Physical Interpretation..................................... 398
      2. Chemical Interpretation.................................... 399
   B. Factors in Radiation Sterilization............................... 399
      1. Type of Organism... 399
      2. Culture Age... 400
      3. Preirradiation Treatment................................... 400
      4. Induced Resistance... 401
      5. Concentration... 401
      6. Environmental Conditions during Exposure................. 401
      7. Induced Radioactivity...................................... 402
      8. Sterilization Dose... 403
      9. Postirradiation Conditions.................................. 404

C. Food Irradiation.. 404
    1. Chemical Changes... 404
    2. Loss of Nutrients.. 406
    3. Flavor and Appearance...................................... 407
    4. Toxicity.. 408
    5. Storage Properties.. 409
D. Low-Dose Irradiation... 409
    1. Meats.. 410
    2. Delicatessen.. 410
    3. Fruits.. 411
    4. Beer... 411
E. Control of Infestation... 411
    1. Grains... 411
    2. Tobacco.. 412
F. Prevention of Sprouting... 412
    1. Potatoes.. 412
    2. Onions... 412
G. Sterilization of Drugs and Pharmaceuticals......................... 413
H. Containers... 414
    1. Sterilization of Containers................................... 414
    2. "In-Container" Sterilization................................. 414
I. Engineering Design... 415
    1. Analysis.. 416
    2. Correlation of the Variables.................................. 418
    3. Contaminants... 419
Acknowledgment... 419
References.. 420

I. Introduction

The availability of energy in the form of ionizing radiations has recently increased markedly. This increase is likely to continue for some years. A portion of this energy will be potentially available as a result of accelerator development and facilities for the production of radioisotopes. Another large supply of this energy will be produced as a necessary consequence of the anticipated wide-scale use of the power reactor.

The observed and contemplated deleterious effects of this energy, which appears as delayed emission from fission products, has initiated large-scale research projects directed toward minimizing the effects resulting from the degradation of this energy in matter. Studies concerned with the practicability of using this energy have been less numerous. Evaluation of the possible practical applications of this energy is now more severely limited by lack of data on practical systems than by lack of knowledge regarding its ultimate cost. Radiation chemistry is the study of chemical effects produced by the energy-degradation processes which occur when ionizing radiations interact with matter. At present the vast majority of radiation chemical studies have been fundamental in character; they have been more concerned with the elucidation of basic processes and mechanisms than with the economic potentialities of the systems studied. As a consequence of this direction of research we have, in spite of the large number of systems studied, very few which show promise of being extended to economically feasible processes. This is a most unfortunate situation for attempting to evaluate the feasibility of utilizing ionizing radiation in the chemical process industries. Information now available gives some insight into fundamental processes and mechanisms and thus serves as a guide in searching for practical systems. This information is not complete enough to obviate the need for experimental work on practical systems.

It is most apparent that energy in the form of ionizing radiations whether produced intentionally or obtained from fission products will be more expensive than degraded thermal energy. Since ionizing radiations are expensive sources of energy, any economically sound process must make use of those properties of ionizing radiations which distinguish them from cheaper forms of energy.

When the interaction of ionizing radiation with matter is considered, it is perhaps most useful to think in terms of an energy-degradation process rather than simple energy transfer. For then attention is focused on the unique availability of this energy, and the total energy absorption is simply a measure of the energy degradation which has taken place. By the availability of the energy of ionizing radiation we refer to the large entropy associated with this form of energy. It is energy available to produce a concentration of energy states which could exist in thermal equilibrium only at high temperatures. The energy of ionizing radiation is degraded to approximately the availability of ultraviolet light, a much less expensive energy source, before chemical consequences of the degradation are noted. Extensive use of photochemical processes has not yet developed, and consequently it would be very difficult to visualize extensive use of ionizing radiations if it were not for the chemically nonspecific processes by which this energy is absorbed.

II. Statement of Problem

A. Fundamental Considerations

Usually when a new chemical or process tool is developed, one expects that it will do something unique, something no other reagent was able to accomplish effectively. This is particularly true of an expensive reagent like ionizing radiation. It behooves us therefore to examine closely the unique properties of ionizing radiation.

1. Use of the Unique Properties

a. Equilibrium. To shift the equilibrium of a mixture one must effect a change in the intensive properties of the system. This requires an energy expenditure which is comparable in magnitude to the strength of a chemical bond. Since one million roentgens are equivalent to two calories, radiation is clearly too expensive a commodity to be expended in this manner.

b. Rate. The usual manner of thinking about rate is to picture a certain distribution of energy among the molecules. By assuming that this distribution adequately describes chemical reactivity we deduce that at any temperature a certain fraction possess energy sufficiently above the activation energy to enter into reaction. For instance, if one takes a low activation energy of about 12,000 cal. at 300°K., using the integrated Maxwell-Boltzmann equation:

$$\frac{M_1}{M_0} = e^{-E/RT} = e^{-\frac{12,000}{2 \times 300}} \cong 2.00 \times 10^{-9}$$

at 310°, the rate doubles, $\frac{M_1}{M_0} \cong 4 \times 10^{-9}$ where $\frac{M_1}{M_0}$ = fraction of molecules having sufficient energy, E, to react.

Suppose we irradiate the system with 1×10^6 r.; This would give us 16×10^{17} ion pair/g. of air. If one assumes that the MW is 29, the fraction of molecules affected would be $\frac{16 \times 10^{17}}{18 \times 10^{24}} \cong 1 \times 10^{-7}$.

This would be equivalent to raising the temperature only about 70°C. Here one is equating the cost of 10^6 r. worth of radiation with 70 calories of heat; this is distinctly more advantageous than trying to shift the equilibrium. Also applicable here is the Eyring concept of an energy

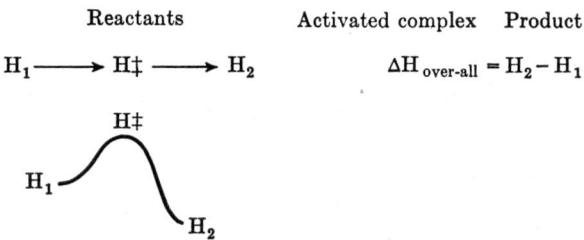

barrier. It is, however, inaccurate to state that if ΔH is negative, all one has to do is "trigger" the reaction, for this requires that the "hot," or activated, molecule is able to activate another molecule, i.e., that chain kinetics apply. In the ordinary case the ΔH of reaction is carried off as degraded, unavailable thermal energy and is not capable, except in the sense that the temperature level of the whole mass is raised, of continuing the reaction. The clear-cut implication, therefore, is that radiation would have maximum utility as an initiator of chain reactions.

B. Practical Considerations

Two independent and one dependent factor govern the applicability of radiation to any process. It is necessary to know the cost/roentgen and the quantity of product/roentgen. Knowledge of these allows one to calculate the cost/pound of the radiation processing. It must then be ascertained whether the value added to the product is greater than the cost of the treatment. Each of the independent factors will be treated in following sections.

C. Dosimetry

The energy of ionizing radiation is degraded upon interacting with matter, and the increase in temperature of an irradiated material is proportional to the amount of energy degradation which has taken place within the sample. The amount of this degradation is proportional to the amount of many of the spectrum of physical and chemical changes which occur as a result of the process of energy degradation. For this reason the energy absorbed by a sample may be used as a measure of the amount of other physical and chemical change.

Energy expenditure is however, a quantity which may be measured only with extreme difficulty. Dosimetry of ionizing radiations therefore consists of determining a more easily measured change and then relating this to energy expenditure.

For many years after the discovery of X rays and radioactivity at the turn of the century, sensitivity was of prime concern, as only very weak sources of ionizing radiation were available. Physical and chemical changes occurring upon minute amounts of energy expenditure were selected as bases of dosimetry, and convenience of measurement, precision, and relation to more fundamental processes were sacrificed. It was out of necessity that charge separation in the gas phase was selected as the basis of dosimetry. As small amounts of energy degradation produce easily measurable quantities of charge separation in air, this method of dosimetry is particularly sensitive.

Measurement of charge separation in gases, however, is not an easy operation and accuracies of a few per cent are obtainable only with extreme difficulty. Furthermore, the relation between dose measured in terms of charge separation in gases and actual energy expenditure in condensed phases is quite complicated.

Stronger sources of ionizing radiation make it feasible to sacrifice sensitivity in favor of ease of determination and precision and to permit the relation between energy degradation and quantity of change to be made easily and with greater accuracy. The outstanding example of a secondary dosimeter is the Fricke ferrous sulfate dosimeter. This is a chemical dosimeter in which ferrous ion is oxidized to ferric ion by ionizing radiations. It is convenient to use (W4), the oxidation is proportional to energy expenditure over a fairly wide dosage range, and the relation between the amount of oxidation and energy degradation has been accurately determined for radiations having low rates of energy loss (S4). The major disadvantage of the dosimeter is that the relation between amount of oxidation and energy expenditure varies with the rate of energy loss. It is however quite constant for X rays, γ rays, and high-energy electrons. This relation is independent of rate of energy loss for hydrocarbons, but the difficulty in determining the amount of chemical change in hydrocarbon systems has limited their use as dosimeters. Other useful but less well-studied secondary dosimeters are the cerous-ceric system (W3), polyvinyl chloride and cellophane films (H10, H17), and activated glasses (S5). The relation between chemical change and energy expenditure is commonly expressed in units of molecules of product produced/100 e.v. of energy absorbed. Yields will be reported either in G units or in pounds of product produced per kilowatt hour.

Unfortunately charge separation in gases as a basis of dosimetry cannot be completely disregarded since its use in radiation chemical work and especially in radiobiology continues. The roentgen unit is defined as that quantity of x or γ radiation whose associated corpuscular emission/0.001293 g. of air produces ions carrying one electrostatic unit of charge of either sign. The relation between the roentgen and units based on energy absorption are tabulated in Table I.

III. Experimental Work on Initiation of Chemical Reactions

A. INTRODUCTION

The motivation for the vast majority of radiation chemical studies has been provided by considerations other than the desire to develop economically feasible processes. The approach has been fundamental, with the result that many data exist on systems having no possible application in the chemical process industry, while data on potential systems are lacking. Though extensive work on aqueous systems has been motivated by practical applications of these data to corrosion problems and biological systems, the approach has still been a fundamental one and the systems studied have been chosen so as to yield information

TABLE I
Conversion of Units

	Conversion factors
Energy	
1 erg	$= 2.389 \times 10^{-8}$ cal.
1 rad	$= 100$ erg 2.389×10^{-6} cal.
1 rad	$= 1.2$ r. (for air)
1 electron volt	$= 1.602 \times 10^{-12}$ erg
1 joule	$= 10^7$ erg
Power	
1 curie	$= 3.7 \times 10^{10}$ disintegrations/sec.
*1 m.e.v. curie	$= 3.7 \times 10^{10}$ m.e.v./sec. $= 5.92 \times 10^4$ erg/sec.
1 erg/sec.	$= 10^{-7}$ w.
1 r./lb./hr.	$= 1.06 \times 10^{-6}$ w.
1 kw.	$= 950$ lb. material/hr. at 10^6 r.
(M.e.v.) (milliamps)	$=$ kilowatts (for generators)

* This means there is 1 Mev. of energy associated with each disintegration. For Co^{60}, for instance, 2.5 Mev. are associated with each disintegration, i.e., two gammas of 1.3 and 1.2 Mev. in cascade.

regarding fundamental processes and mechanisms. Essentially physical processes are of chief concern and the resultant chemistry has been subjugated to the role of a tool to elucidate the nature of these physical processes which take place prior to chemical changes. A brief outline of experimental work of this sort as well as the conclusions which have been reached regarding basic processes will be presented since this information is useful in an evaluation of economically feasible processes for which similar information is lacking. Emphasis, however, will not necessarily fall on systems having commercial potential, but will rather center on the more thoroughly investigated systems. Practical processes investigated will be discussed later in the light of these studies, and it will be seen that the emphasis here is on very different types of systems.

B. Fundamental

1. *Primary Processes*

Chemical change in matter interacting with ionizing radiations is preceded by a complex process of energy degradation. In this process energy states are attained which permit chemical changes to occur. First the process of energy degradation, then the nature of and evidence for the active chemical species produced, and finally the reaction of these species to produce chemical change will be reviewed.

Ionizing radiations, as their name implies, transfer energy to the electron component of matter, causing charge separation. The interaction permitting this transfer is most generally the interaction of the rapidly moving electrostatic field associated with charged particulate radiation with the essentially stationary orbital electrons of atoms. Most of the energy degradation of electromagnetic radiation also occurs in this way. For whether the electromagnetic field of the photon interacts with the orbital electrons (Compton scattering) or with the atom as a whole (photoelectric effect), the result is a degraded photon and a high-energy electron which loses its energy as mentioned above. The rate at which particulate ionizing radiations lose their energy varies with the square of their charge and inversely with their velocity. Electrons have, therefore, far greater

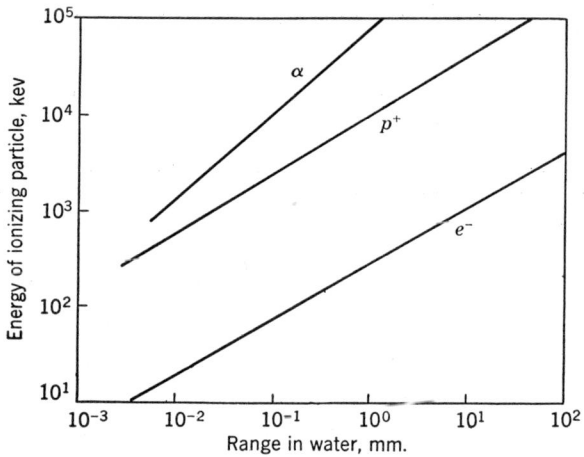

FIG. 1. Range of particulate radiation in water "Fundamentals of Radiobiology," Z. M. Bacq and P. Alexander, Academic Press, Inc., New York (1955).

ranges than protons and deuterons of equivalent energy, which in turn lose energy less rapidly than alpha particles of equal energy. The range of typical particulate radiation in water is shown in Fig. 1.

The stopping power of a material for a particular radiation is commonly expressed as the rate of energy loss (R.E.L.) or the linear energy transfer (L.E.T.) of the radiation in the material. These quantities are assumed to be proportional to the linear ion density and the specific ionization. Stopping powers range from approximately 10^6 e.v./cm. for fast electrons (1 Mev.) in water to 10^{11} e.v./cm. for fission recoils. The ranges of particles are frequently expressed in mg./cm.^{2}, which when multiplied by the density of the material yields the range.

Since electromagnetic radiation loses energy more slowly than particulate radiation of the same energy, the path of a high-energy photon in matter is characterized by a succession of widely separated interactions.

Thus energy-loss parameters are to a good approximation independent of chemical binding, i.e., molecular energy states available, and depend only on parameters involving the characteristics of the radiation and the electron density of matter. For this reason, energy degradation in matter by ionizing radiations is chemically nonspecific, and the penetration in a system is independent of the chemical nature of the system (T3). It has been recently pointed out by Platzman (P3) that in the case of low-energy electrons, energy absorption and degradation become more selective. For these reasons, ionizing radiations may be used to supply highly available energy in depth, and, while the chemical consequences of the ensuant energy degradation are not likely to be much different from those initiated by ultraviolet light, the energy may be supplied to systems in which ultraviolet light will not penetrate. It is precisely this point which indicates that application of penetrating X and gamma radiation is more likely than charged particulate radiation.

The rate of energy loss of particulate radiation of a given energy indicates the linear separation of species likely to react chemically as well as the depth to which energy degradation will occur in a system. It is apparent that heavy charged particulate radiation will produce high concentrations of active species in a thin layer of irradiated material, while the distribution will be less dense but deeper for electrons. The effective concentration of energy deposition will be independent of the rate at which this energy is supplied. This is because the volume surrounding the path of a particle passing through matter to which it transfers its energy is small compared with the distance to the track of the next particle for all but extremely high intensities. Thus in a determination of the distribution of active species it is not the rate of energy expenditure within a sample but rather the linear ion density characteristic of the radiation which is the important parameter. For example, if a given system is irradiated at the rate of x e.v./liter for a total time t, the chemical change produced by degradation of the amount of energy xt will be different for alpha rays and electrons, and this difference may be ascribed to the difference in concentration of active chemical species in the track of the ionizing particle. The effective rate of energy loss of electromagnetic radiation is the rate of energy loss of the electrons produced by the radiation. Because the rate of energy loss is independent of physical as well as chemical state, these considerations apply equally well to the solid, liquid, and gaseous state.

2. *Chemical Intermediates*

a. *Introduction.* The availability of the energy of ionizing radiations allows a broad spectrum of energy states to be formed in matter with which they interact. The principal differences between photochemistry and radiation chemistry is this large number of energy states produced by ionizing radiations as compared with the few produced by quanta of a few e.v. energy. Analysis of the energy states produced by photons of a few e.v. energy is a complicated and tedious process; description of these states for ionizing radiation chemistry is hopelessly complex. W values for gases (W equals energy in e.v. required to produce one ion pair) are approximately twice the corresponding ionization potentials. From this it is deduced that one half the energy initially manifests itself as ionization and one half as excitation energy of some kind. For monoatomic gases, it is apparent that ionized atoms will be neutralized, so that eventually all energy is degraded through excited states to thermal energies. For polyatomic gases, where fragmentation and other chemical reactions are possible, the situation is more complex, and the processes which predominate are best deduced from the reaction products. In the liquid phase, where ionization cannot be directly measured, the situation is even more complex, and the assumption commonly made that the distribution of energy between ionization and excitation is the same in liquid and gas has been the subject of much controversy. Undoubtedly for a situation of such complexity it is most profitable to reason backwards from observed chemical changes to the nature of the excited species.

b. Aqueous Systems. I. POSTULATED INTERMEDIATES. The behavior of pure water when irradiated with ionizing radiation was a subject of intensive investigation and debate in the initial stages of radiation chemistry (A4). Irreproducibility and apparently inconsistent observations plagued radiation chemists until it was realized that water containing no solute is a very poor subject of study. Because trace impurities are always present in water, the radiation chemistry of water is essentially the study of solutions containing very small concentrations of unknown solute. Water containing added solute is a far more profitable and reproducible area of study, for from the behavior of solutes in irradiated water information concerning active chemical species and primary processes may be obtained. The large number of aqueous systems which have been studied have been selected chiefly on the basis of what information they would yield concerning the intermediates and primary processes. These studies indicate that water upon irradiation with ionizing radiation behaves as if it is simultaneously dissociated into active chemical species and disproportionated to hydrogen and hydrogen peroxide. The vast majority of data concerned with the irradiation of aqueous systems with ionizing radiation may be explained by assuming the existence of the transient radical species H and OH. The mode of formation of these species is as yet an unsettled question. What is observed is that irradiated water behaves as if the following reaction were occurring:

$$X\mathrm{H_2O} \rightarrow a\mathrm{H} + b\mathrm{OH} + c\mathrm{H_2} + d\mathrm{H_2O_2}$$

where X represents the number of water molecules dissociated for each 100 e.v. absorbed by the water and the coefficients a, b, c, and d are connected with the coefficient X by an equation of material balance. Water behaves in this way quite independently of the quality of energy of the exciting radiation. The coefficients, however, vary markedly with quality and energy. In general the coefficients a and b are large, and c and d small for radiation of low rate of energy loss, while radiations that have high ionization densities are characterized by high values for coefficients c and d and low values of the coefficients a and b. Coefficients a and b describe what is commonly referred to as the radical yield; c and d, the molecular yield (A3). These coefficients as well as the coefficient X are quite well known for gamma rays, hard X rays, and high-energy electrons, for which the values, characteristic of the low ionization density of these radiations, are practically identical:

$$a = 3.6 \quad b = 2.8 \quad c = 0.39 \quad d = 0.78$$
$$X = 2c + a = 2d + b = 4.4$$

These values are representative of the extreme case of low rate of energy loss and are quite insensitive to pH and solute concentration.

The radical diffusion model for irradiated water (S6, F4, S1) assumes that energy loss along the path of an ionizing particle is nonuniform, and so regions of high-energy deposition are separated by regions of little or no energy loss. The result is the production of active chemical species in bunches or clusters along the track. These active chemical species, which are assumed to be H and OH, then diffuse from the clusters and the molecular yield results from radical-radical interactions which occur during the process of diffusion. The model thus predicts a dependence of molecular yield on concentration of reactive solute. This dependence has been investigated and found to agree closely with theory (S6).

As the rate of energy loss of the particle increases, the clusters along the track are formed closer together and radical-radical interactions between clusters occur, increasing the molecular yield. This dependence of molecular yield on linear ion density is being intensively studied by Schuler and Allen (S3).

For radiations having low rates of energy loss the molecular yield is small and quite insensitive to solute concentration. As the rate of energy loss increases, the molecular yield and its dependence on solute concentration both increase. The increased dependence on solute concentration in the case of high ionization density is due to the scavenging of radicals which would otherwise combine with radicals formed in other clusters. The radiation chemistry of aqueous systems can thus be predicted for radiations having a high proportion of radical yield if the reactions of hydrogen and hydroxyl radical radicals as well as the molecular products with the solute are known.

As the rate of energy loss increases, the marked dependence on solute concentration obscures the meaning of the coefficients a, b, c, and d. There is some indication that the coefficient X also varies with linear ion density.

II. EXPERIMENTAL EVIDENCE. Experimental evidence for the existence of H and OH in irradiated water consists chiefly of the observed behavior of dissolved solutes. From this behavior deductions are made as to the mechanism of the change as well as the initiating species. This evidence will be summarized in the discussion of the reactions of hydrogen radical and hydroxyl radical which follows. Some direct evidence for these species might however be mentioned. When acrylonitrile is polymerized in aqueous solutions, the presence of hydroxyl end groups is apparent in the infrared absorption spectrum of the polymer, and when D_2O is used as solvent the carbon deuterium stretching frequency is observed (D1). Mass spectrographic work using oxygen isotopes also provides direct evidence for these radical species.

Details of the formation of these species has been the subject of some controversy. A simple mechanism for their production is

$$H_2O \rightarrow H_2O^+ + e^-$$
$$H_2O^+ + H_2O \rightarrow H_3O^+ + OH$$
$$H_2O + e^- \rightarrow H_2O^-$$
$$H_2O^- \rightarrow H + OH^-$$

The species so formed are likely to be separated in space and are more likely to react with solute than are radicals formed from excited water molecules.

c. Nonaqueous Liquid Systems. Experimental work on nonaqueous liquids is not nearly so plentiful as work on aqueous systems. Studies of the radiolysis of pure organic liquids, paralleling the pattern of the radiation chemistry of water, were the chief concern of earlier work. G values have been obtained for total gas production and for individual volatile products as well as high-molecular-weight polymer formation from a wide variety of organic liquids. G values for gas production from aliphatic hydrocarbons are found to be approximately 2, and corresponding G values for aromatic compounds are an order of magnitude lower. Furthermore it is found that unsaturation tends to decrease the gas yield and to increase polymer production (B21).

The radiation chemistry of pure liquids is likely to be as complicated and difficult to interpret as the radiation chemistry of pure water. The few studies concerned with the radiolysis of pure liquids indicate the difficulties involved in such studies, and the complexity of the experimental results makes it most difficult to say anything in detail concerning the basic processes and mechanisms. One might mention the work of Garrison (G4) on the irradiation of acetic acid with alpha particles and deuterons where a complex mixture of reaction products ranging from hydrogen to citric acid is obtained. The work of Sheppard and Burton (S13) on fatty and naphthenic acids irradiated with alpha rays as well as the exhaustive work of Breger (B13) on a wide variety of organic liquids irradiated with alpha rays indicates that even cursory examination reveals a broad spectrum of products. It is doubtful that the irradiation of pure liquids will yield anything of fundamental importance, and the low yields obtained for any particular product preclude the possibility of practical application. These studies have however indicated a few important points. There is apparently extensive intermolecular and intramolecular energy migration within irradiated organic liquids. Studies of aromatic compounds with aliphatic side chains indicate that the aromatic ring protects the aliphatic side chain of the benzene ring. This has been termed "sponge-type" protection by Manion (M3). Fluorescence of dilute solutions of aromatic materials in hydrocarbons indicates extensive energy migration prior to degradation. Extensive energy migration is also indicated by the small amount of work which has been done with mixtures of organic liquids. The chemistry observed in a mixture tends to be that characteristic of the component with the lower ionization potential. Energy migration of this type is undoubtedly associated with the semiselective action of radiation on large molecules. While the absorption of energy from ionizing radiations is chemically nonspecific, there is some indication of preferred chemical reaction. Both decarboxylation of fatty acids (S13, S12) and rupture of bonds about —C—OH (M1) continue to be of principal importance as the length of the aliphatic chain is increased. This indicates that the chemical effects resulting from direct action on complicated molecules is not a completely random process.

It appears that the chemical change produced by a given amount of energy degradation is far less sensitive to the rate of energy loss of the radiation in hydrocarbons than in water. While pure water behaves very differently when irradiated with radiations of differing ionization densities, the chemical effects in hydrocarbons appear to be quite insensitive to the rate of energy loss. Schuler (S3) has shown that there is only a very small change in the G value for H_2 production from cyclohexane irradiated with 2 Mev. gamma-rays, 2 Mev. electrons, 20 Mev. deuterons, and 40 Mev. alpha particles.

A more fruitful approach to the irradiation of pure liquids would appear to be that being undertaken by Schuler (S2) in this country and Chapiro (C2) in France. In these studies solutes (iodine and diphenylpicrylhydrazyl) are added to organic liquids prior to irradiation and the disappearance of these solutes is taken as a measure of the number of free radicals produced by the radiation. The experimental results using the two scavengers mentioned yield values in good agreement for the number of free radicals produced in the radiolysis of organic liquids.

Radioactive iodine dissolved in organic solvents acts as a scavenger for radicals produced in organic liquids. Radicals produced by the energy-degradation process may be identified by adding carrier amounts of organic halides to an irradiated hydrocarbon containing radioactive iodine and separating the mixture by fractional distillation (G6). Results of this kind are just beginning to appear (S2) and indicate that a wide variety of radicals are produced with low G values.

d. Gases. Gases are more amenable to fundamental study than liquids. Yet the amount of radiation chemical work done on gaseous systems is more limited than the work done on liquids. This is surprising in view of the fundamental bent of radiation chemical research. An explanation of this apparent anomaly is to be found perhaps in the great complexity of the few gaseous systems studied. This complexity has apparently driven radiation chemists to liquids where the complexities are masked.

The detailed work of Essex (E5) on gas-phase radiolysis indicates clearly that four distinct primary mechanisms are involved and that the contribution of each of these primary processes is different in different systems. It also appears from this work that there is nothing involved in the radiolysis of gaseous systems which would favor the initiation of gas-phase reactions by ionizing radiations over initiation by ultraviolet light. Though the primary processes are different and more complicated in the case of radiolysis, the active chemical intermediates are the same and consequently the over-all chemistry is quite similar. Since there appears to be little possibility of practical application of ionizing radiation to vapor-phase reactions, nothing further will be said about the fundamental work concerned with the detailed processes of energy degradation in the gas phase. For reviews of this work the reader is referred to articles by Hart (H8), Allen (A1), and Burton (B22).

e. Solids. Radiation effects in solids have been the subject of extensive investigation. These studies indicate quite clearly that while energy-degradation processes in solids produce pronounced physical changes, chemical effects are at a minimum. The pronounced changes in physical properties which are observed may be ascribed to changes in the crystal lattice produced by the energy-degradation process. Very little concerning the chemical effects of energy degradation in solids is known, but it is to be expected that these will be very small owing to the restricted motion of molecules bound in a crystal lattice. Chemical effects will be especially small where no definite molecular unit exists.

A notable exception to the foregoing statements is the irradiation of solid polymers. In this case, because of the large molecular weight of the polymer, rearrangement of very few bonds produces great changes in physical properties. Massive doses of highly available energy convert organic solids into complex mixtures of unidentifiable polymers and decomposition products.

Gas evolution from ionic solids proceeds with G values of approximately 2, and all chemical effects in solids, when they occur at all, have characteristically low G values. The subject of the irradiation of solid polymers to produce cross linking will be more thoroughly discussed later.

3. Reactions of Chemical Intermediates

a. Oxidations. Reactions of the radical species H and OH in the presence of oxygen and oxidizable solutes are well illustrated by the classic reaction of radiation chemistry, the oxidation of ferrous ion in $0.8N$ sulphuric acid. The reaction scheme is

$$XH_2O \rightarrow aH + bOH + bH_2 + dH_2O_2$$
$$H_2O_2 + Fe^{++} + H^+ \rightarrow H_2O + Fe^{+++} + OH$$
$$OH + Fe^{++} + H^+ \rightarrow H_2O + Fe^{+++}$$
$$H + O_2 \rightarrow HO_2$$
$$HO_2 + Fe^{++} + H^+ \rightarrow H_2O_2 + Fe^{+++}$$

and the yield of ferric ion may be expressed in terms of the coefficients as $G_{Fe} = 3a + b + 2d$. It is to be noted that in the presence of oxygen the radical species H reacts with oxygen to form hyperoxide radical, an oxidizing species. In the absence of oxygen, hydrogen atoms are converted to oxidizing species by reaction with hydrogen ion to yield H_2^+ (R2). The arguments in favor of this mechanism, originally proposed by Weiss (W2), have recently been strengthened by the value obtained for the rate of oxidation in the absence of oxygen by Barr and King (B4) and especially by the work of Allen and Rothschild, who have studied the reaction under a wide variety of conditions and find that the experimental results can be fitted to a reaction scheme in which H and OH react as above.

Many other inorganic systems have been studied, and in general the experimental results may be explained by postulating the production of the intermediates mentioned above and supposing them to react in similar ways. Excellent reviews by Hart (H8) and Dainton (D2) contain tabulations of the various systems studied.

The important point of all these studies is to illustrate that H and OH as well as hydrogen and hydrogen peroxide may be used to explain inorganic oxidation reactions initiated by ionizing radiations and that for radiations having low rates of energy loss these are formed in the same quantity and proportion independently of the nature of dissolved solutes. These reactions are examples of indirect action wherein absorption of energy by a solvent produces active chemical species which then react with the solute. Indirect action is thus characterized by chemical reaction of species which are not primarily involved in the energy-degradation process while direct action occurs as an immediate result of this energy degradation. As has been mentioned previously, the effect of increasing linear ion density is to change the values of the coefficients a, b, c, and d in a way that favors the production of molecular products but does not affect the way the intermediate species react with solutes.

Oxidation in aquo-organic systems is a more complicated problem. If the reactions are carried to high percentage conversions the spectrum of products becomes complex owing to the complexity of the original molecule. It is only in the oxidation of simple organic molecules that the reactions of the intermediate species is not obscured by the variety of products obtained. The following reaction scheme has been suggested to explain the oxidation of aqueous solutions of formic acid containing dissolved oxygen (H7)

$$OH\cdot + HCOOH \rightarrow H_2O + HCOO\cdot$$
$$H\cdot + HCOOH \rightarrow H_2 + HCOO\cdot$$
$$HCOO\cdot + O_2 \rightarrow HO_2\cdot + CO_2$$
$$H\cdot + O_2 \rightarrow HO_2\cdot$$
$$HO_2\cdot + HO_2\cdot \rightarrow H_2O_2 + O_2$$

The oxidation of more complicated organic molecules follows much the same course initially. The oxidation of ascorbic acid (AH_2) appears to follow the following kinetic scheme:

$$H\cdot + O_2 \rightarrow HO_2\cdot$$
$$HO_2\cdot + AH_2 \rightarrow H_2O_2 + AH\cdot$$
$$HO\cdot + AH_2 \rightarrow H_2O + AH\cdot$$
$$AH\cdot + O_2 \rightarrow AHO_2\cdot$$
$$2AHO_2\cdot \rightarrow 2A + H_2O_2 + O_2$$

As the reaction proceeds, the kinetics is complicated by the reaction of the oxidation products with the radicals.

Oxidation of organic compounds generally proceeds with low G values which indicate that nonchain processes of the above-mentioned type are operative and that reactions of the type

$$AHO_2\cdot + AH_2 \to A + H_2O_2 + AH\cdot$$

which lead to chain oxidations do not occur. G values for these reactions and also for the oxidation of inorganic solutes are on the order of 10.

Studies outside the field of radiation chemistry indicate that it should be possible to initiate chain oxidations of organic compounds with ionizing radiations. Except for the studies of Mead (M8) on the chain oxidation of fatty acids by ionizing radiations, reactions in which oxygen is chain utilized have not been studied. Most organic materials in solution yield hydrogen peroxide and other oxidation products with G values which preclude the possibility of chain utilization of oxygen. For example, the work of Weiss (W2) on the irradiation of methane-oxygen mixtures in aqueous solution indicates that hydrogen peroxide is formed with $G = 1.9$ and organic peroxides with $G = 0.5$.

The widely accepted mechanism for the chain oxidation of hydrocarbons proposed by Bell (B8)

$$R\cdot + O_2 \to RO_2\cdot$$
$$RO_2\cdot + HR \to ROOH + R\cdot$$
$$ROOH \to RO\cdot + HO\cdot$$
$$RO_2\cdot + R\cdot \to ROOR$$
$$ROOR \to RO\cdot + RO\cdot$$
$$RO_2\cdot + RO_2\cdot \to O_2 + RO\cdot + RO\cdot$$
$$RO\cdot + HR \to ROH + R\cdot$$
$$R_3CO\cdot \to R_2CO + R\cdot$$

in which a chain utilization of oxygen occurs through the agency of peroxidic intermediates has had many of its individual steps verified experimentally. These reactions are commonly initiated either photochemically or chemically, but since there is strong reason to believe that the intermediate organic radicals produced by ionizing radiations will be similar to those obtained by either chemical or photochemical initiation one might expect to be able to initiate such chain reactions with ionizing radiations.

The liquid-phase oxidation of olefinic compounds initiated by azo-*bis*-isobutyronitrile has been studied by Bateman (B6, B7). A mechanism similar to that presented above is invoked to explain the data. Gas-phase oxidation of hydrocarbons involves long chains and has been the subject of considerable study.

There is indication from the work of one of the authors (H18) that chain utilization of oxygen occurs in the oxidation of ethylene when aqueous solutions of ethylene and oxygen under several atmospheres pressure are irradiated with gamma rays from Co^{60}.

b. Reduction. Reductions in aqueous solutions are less common than oxidations. Reductions may result from molecularly formed hydrogen peroxide as in the case of the reduction of ceric ion. In this case the radical yield has no net effect since the effects of H and OH cancel (A3). Reductions may also occur as a result of the reaction of hydroxyl radical with dissolved hydrogen:

$$HO\cdot + H_2 \rightarrow H_2O + H\cdot$$

Here the oxidizing species OH is converted into a hydrogen atom and the net effect is a reduction of solute by molecular hydrogen. In the case of hydrogen peroxide the over-all equation is

$$H_2O_2 + H_2 \rightarrow 2H_2O$$

and the reduction occurs by a long chain process probably involving the following reactions (H20):

$$H\cdot + H_2O_2 \rightarrow H_2O + HO\cdot$$
$$HO\cdot + H_2 \rightarrow H_2O + H\cdot$$

A similar type of chain process occurs in the case of mixtures of hydrogen peroxide and formic acid (H6).

There is some indication that hydrogen atoms react by hydrogen abstraction in air-free aquo-ethanol solutions (S21).

C. Practical Processes Investigated

1. *Introduction*

The only logical basis for labeling a reaction "practical" is that it is one which if carried out cheaply enough could have commercial applications. Since it is impossible to discuss all reactions which fall into this category an effort was made to select a few which are typical of some of the broader unit processes.

2. *Inorganic Processes*

a. Introduction. In Table II experimental yields have been recalculated to suggest the pounds of reactants which could be converted/hr. by use of a kilowatt source of radiation. The yields tabulated are somewhat arbitrary and indicate in most cases only the relative ease of propagation of the reactions. One hundred per cent utilization of the radiation was assumed, and in cases where more than one investigator reported yields for a reaction an average was used. The same procedure was followed in cases where the reaction was carried out under several conditions.

TABLE II
Tabulation of Inorganic Reactions

Reference	Reactants	Products	Conditions	Radiation	Pounds reactants converted/ hr./kw.
Oxidations					
(B23)	O_2	O_3		α, e^-	0.161
(K4)	$O_2 + N_2$	NO, O_3, N_2O, NO_2		α	0.083
(F3)	$CO + H_2O$	$CO_2, H_2, HCHO$	$pH = 1 - 7$	X ray	0.347
(S11)	$NaCN + O_2$	N_2, Na_2CO_3		γ	0.066
(S11)	$Na_2S + H_2O$	$Na_2SO_4 + 4H_2$		γ	0.167
Reductions					
(L12)	$CO_2 + H_2$	$H_2O + (H_2CO)_x$		α	0.173
(L12)	$CO + H_2$	polymer		α	0.232
Decompositions					
(L14)	HBr	$H_2 + Br_2$		α	0.651
(W9)	H_2S	$H_2 + S_2$		α	0.193
(A2)	H_2O	$H_2 + H_2O_2$		α, e^-, γ	0.039
Synthesis					
(L12)	$H_2 + N_2$	NH_3		α	0.500
(G3)	NH_3	$N_2H_4 + H_2$		α	0.057
(A5)	$CO + Cl_2$	$COCl$		e^-	10^4

b. Oxidation. Both the oxidation of air to nitrous oxide and the conversion of oxygen to ozone are strongly endothermic with very unfavorable equilibrium at low temperatures. There appear to be little hope of mitigating the severe conditions now employed in commercial operations and minimal chances of increasing the yields. The vapor-phase oxidation of sulfur dioxide has also been investigated (E3), but insignificant conversions are reported. The water-gas reaction seems to be of no real interest even if appreciable formaldehyde yields can be obtained. In basic solution formic acid is reported to be the chief product.

The radiation-induced oxidation of cyanide and sulfide have been suggested by Selke (S11) as an attractive method of waste disposal. By using radiation for destruction rather than for synthesis, side-product and purification problems are eliminated.

c. Reductions. The reactions of interest here are Fischer-Tropsch and methanol-type processes. Commercially the most significant problem here, other than separation of products, is the selection of a nonpromiscuous catalyst, and radiation does not fill the bill. The possibility of modifying commercial catalysts by use of radiation is a possibility which deserves careful study. It will be preferable to choose a catalyst which can be regenerated *in situ*. Further work in this area may be of interest, since Fischer-Tropsch reactions are exothermic and potentially very fast. Attempts by Lewis (L11) to react CO with H_2 at 680 lb./sq. in. in gamma-ray fluxes were successful.

d. Synthesis. Irradiation of H_2 and N_2 to produce ammonia has been carried out by alpha and electron radiations. No ammonia was produced when the synthesis was attempted with gamma rays (L11, S11). This, of course, indicates an ion density effect. The work of Lind (L12) has several interesting features. Since the decomposition of ammonia is favored over synthesis, it is apparent that increased yields should be attained in flow systems. Thus Lind found that he could more than double his yield by removing ammonia continuously.

e. Summary. Several general observations with regard to inorganic processes can be made. (a) Chain-type mechanisms, in cases other than chlorination or combustions, are rare indeed. One must therefore expect low yields. (b) There being only a limited number of bonds present, the possible spectrum of products is restricted. Hence "pseudo" steady states may ensue. (c) Most inorganics are cheap, heavy-tonnage chemicals. In most of these processes radiation is doing nothing except replacing a catalyst bed or mitigating severe reaction conditions. The total saving, therefore, may be limited. In a typical ammonia plant, for instance, Henley (H14) has shown that one could not afford to pay more than $50,000 for radiation, if all that is accomplished is to replace the catalyst. Savings achieved by operation at lower temperatures or pressures are slightly greater. Barrett *et al.* (B5) have shown that if a reaction which is conducted at 3,000 lb./sq. in. in ten 100-gal. reactors could be made to proceed in ten 500-gal. reactors at 60 lb./sq. in., an initial saving of $250,000 and an operating saving of $50,000 a year could be achieved.

3. *Organic Reactions*

Table III is an analogue of Table II for typical organic unit processes. Yields and conversions were calculated as in Sec. IIIC2.

a. Oxidations. The hydroxylation of aromatic hydrocarbons in aqueous solution has received considerable attention from radiation chemists who sought to prove by this means that free hydroxyl radicals were formed in water. Biphenyl, terphenyl, and more complex products are formed along with phenol. When mixed gamma and neutron sources were used, catechol formation and ring breakage were observed. The oxidation of Δ-5,6 steroids in the 5-6 positions has often been pointed to as an example of the potential specificity of radiation. This oxidation is, however, so favored that it can be carried out by vigorous shaking of the sterols in the presence of oxygen.

TABLE III
Tabulation of Organic Reactions

Reference	Reactants	Products	Radiation	Pounds reactants converted/ hr./kw.
Oxidations				
(S11)	Benzene + H_2O	Phenol	γ, X ray	1.03
(L15)	Nitrobenzene + H_2O	Hydroxylated nitrobenzene	e^-	2.23
(L16)	Benzoic acid + H_2O	Salicylic acids	X ray	0.50
(E4)	Benzene + air	Phenol	β^-, γ	12.70
(K2)	Steroids	Trans triols in 5–6 position	X ray	25.60
Decompositions				
(S13)	Palmitic acid	Pentadecane	α	0.38
(L5)	Wood	Edible matter	β^-	9.50
Condensations				
(L12)	CH_4	C_2H_6, H_2, etc.	α	0.09
(L12)	C_4H_{10}	C_8H_{16}, H_2, etc.	α	0.26
(K4)	C_2H_2	C_6H_6	α	0.21
Chlorinations				
(H4)	$C_6H_6 + Cl_2$	$C_6H_6Cl_6$	γ	1×10^5
Cross linking				
(C3)	Polyethylene	Infusible polyethylene	γ, β^-	250
(C3)	Rubber	Vulcanized	Mixed	150

   b. Chlorinations. Although high yields of products are obtainable with gamma radiation, a direct comparison with ultraviolet-induced reactions shows that these two agents are competitive. Ultraviolet light as an energy source is much cheaper than ionizing radiation. Photochemical processes, however, are concerned in the main with the initiation of exothermic reactions, and energy costs are only a small portion of photochemical processing costs.

   Ultraviolet sources are chiefly mercury-discharge lamps in various forms. These lamps are fragile and quite inefficient. The design of reactors using these lamps is hampered by these considerations, and by the limited ultraviolet transmission, of common glasses and solvents. The transmission is further decreased by coatings of opaque materials which form during the reaction. Use of penetrating ionizing radiation overcomes many of these difficulties.

The power levels of radioactive sources presently available are comparable to the output of the most intense discharge lamps available. Ultraviolet sources produce 10 to 100 w. of useful ultraviolet radiation. Kilocurie sources of 2 Mev. activity produce approximately 20 w. The total available power from such a source having a half-life of 10 years is an order of magnitude lower than the total power output available from an ultraviolet lamp operating for 1000 hours, the normal expectancy for lamps of this type.

Chlorination of aromatics by radiation has recently been reported by Harmer (H4). High yields are obtained and the product spectrum is identical with that obtained from photo-initiation. In cases where differences are claimed, the data were taken under conditions where no ultraviolet data are available, and so the situation is as yet not clear.

4. *Polymerizations*

The ability of ionizing radiation to produce free radicals suggests its use in initiation of polymerization reactions. Its use as an initiator is limited to addition-type polymerizations or to the condensation of saturated hydrocarbons. These reactions are thought to proceed via formation of radical intermediates. Condensations of low- or high-molecular-weight polymers may be induced by radiation; this leads to cross linking, the first radiation-induced reaction to be carried out industrially. Examples of some of the monomers which have been shown to be amenable to radiation-catalyzed polymerization follow; yield data, conditions, MW, and production capacity for a KW irradiator are summarized in Table IV. Inspection of Table IV makes it obvious that radiation cannot compete with conventional catalysts in standard processes. One should look for radiation applications to spring up in areas where tested recipes have failed and where cost factors do not control the choice of agents. Some of the novel features of radiation-catalyzed polymerizations have been discussed by Sun (S20) and by Ballentine (B2):

1. No foreign materials or catalyst fragments need be incorporated into the polymer. This can be important in electrical or medical applications.
2. In some systems, such as the perfluorohydrocarbons, radiation may be the only tool that can do the job.
3. Since the initiation step is independent of temperature, in some cases, molecular-weight control may be achieved by variation of temperature alone. This can be important in applications where molecular-weight control and distribution are significant.
4. Polymerization at low temperatures and low pressures can be initiated via

TABLE IV
Tabulation of Polymerization Reactions

Ref.	Monomer	Radiation	Temperature, °C.	Pressure, lb./sq. in.	Polymerization conditions	Polymer, mw.	Production capacity, lb./(hr.)(kw.)
B3	Styrene	Gamma	72	Atm.	Bulk	$1.6\text{--}5 \times 10^5$	400
		Gamma	45	Atm.	Emulsion	$1.2\text{--}2 \times 10^6$	3000
B3	Methyl Methacrylate	Gamma	72	Atm.	Bulk	380,000	4000
		Electrons	Room	Atm.	Bulk	Solid	20
		Gamma	30	Atm.	Bulk	$\sim$70,000	400
B14	Ethylene	Gamma	190	Atm.	Gas phase	—	77
L11		Gamma	20	1,900	Gas phase	20,000	135
		Gamma	200	1,200	Gas phase	Liquids	
B3	N-Vinylpyrrolidone	Gamma	50	Atm.	H$_2$O solution	$\sim$50,000	22,000
B2	Acrylamide	Gamma	30	Atm.	Solid	236,000	670
B2	Perfluoropropylene	Gamma	30	Atm.	Bulk	600	5.0
B2	Perfluorobutadiene	Gamma	30	Atm.	Bulk	"Grease"	8.3
L13	Butane	Alpha	30	Atm.	Gaseous	Solid	0.16
B9	Acrylonitrile	X	20	Atm.	H$_2$O solution	Solid	5×10^5
B14	C$_2$H$_2$	Gamma	30	Atm.	Gaseous	Powder	70

radiation catalysis. In instances where isomerization and heat stability limit the working temperatures, radiation may prove feasible.

5. Solid-state polymerization becomes a possibility.
6. Polymerizations *in situ* can be accomplished. This may be a boon in applications where intricate mold patterns are desired.

Some specific examples illustrating attempts to achieve these objectives on a laboratory scale are as follows.

a. Styrene. Ballantine (B3) and his group have studied the gamma-ray-induced polymerization of styrene, and Seitzer and Tobolsky (S10) report the effects of Sr^{-90} betas on the same system. The major findings may be summarized as follow.

i. The polymerization mechanism is free radical. Evidence for this is (1) the activation energy of 6700 cal./mole, (2) inhibition of the radiation polymerization by oxygen and benzoquinone, (3) formation of a 50-50 weight mixture resulting from the copolymerization of methacrylate with styrene.

ii. When carried out in bulk, the rate of polymerization varies inversely with radiation intensity. A first- to half-order dependence was found. The same group (B3) also reports that the molecular weight and rate vary directly with temperature. As the temperature is raised from 25° to 72° C., the rate increases from 2.2 to 10.4% and the maximum molecular weight rises from 177,000 to 500,000. If this phenomenon is substantiated and proves to be a general one, it would be a very pleasant surprise.

iii. The rate of polymerization in emulsion is roughly ten to one hundred times as great as that found in bulk at comparable conditions. For a soap-to-styrene ratio of one to nine the molecular weight was reported to be as high as 2×10^6.

iv. A comparison of Seitzer's and Ballantine's yields show Seitzer's yield to be lower by a factor of ten.

v. The Sr^{-90} work also confirms the earlier data of the French group working with acrylonitrile (P7) with regard to the comparative effectiveness of various solvents as radical transfer agents. Generally, halogen-containing compounds were the best sensitizers. Fewer halogen atoms/molecule give less sensitivity; other functional groups such as alcohols and ethers are about as effective as single halogens. Iodides are more efficient than bromides. Hydrocarbons are least efficient, aliphatics less so than aromatics. The relationship of these results to those obtained with ultraviolet is quite clear.

b. Polyethylene. Ethylene has been polymerized at both low and high pressures with gamma rays (B14, L11).

In general, the product is reported to have a higher ultimate tensile strength, but only one tenth of the elongation of the conventional product. This probably can be interpreted to mean that branching or cross linking has occurred.

c. Fluorocarbons. Perfluoropropylene, isobutylene, butadiene, and several related compounds all form viscous liquids or waxy low-molecular-weight solids when subjected to gamma doses up to 200×10^6 r. Of all the perfluoro compounds tried, only 1, 1-di-hydroperfluorobutylacrylate gave a solid product; it was rubbery, indicating that perhaps cross linking had taken place (B3). Discouraging as this original work was, indications are that more refined techniques could yield better products.

d. Acrylamide. Acrylamide is a solid at room temperature. This makes rather interesting studies of solid-state polymerizations possible. Mesrobian has conducted some experiments with this system. Apparently the radiation-produced polymer, unlike the monomer, is completely amorphous. Furthermore, a delayed reaction can be induced by irradiating the monomer at subzero temperatures. When brought to room temperatures the polymerization proceeds with explosive violence. The rate of polymerization is said to vary with the intensity to the 0.75 power.

Again, commercial interest in this compound centers on the straight-chain, water-soluble polymers and copolymers. It remains to be shown whether radiation techniques will yield a satisfactory product.

e. Methylmethacrylate. Several laboratories have worked on methylmethacrylate. Both the Brookhaven and Princeton groups (B3, S10) report comparable reaction rates. Scientists at General Electric (S7) using 800-kv. electrons at a dose rate approximately 100 times greater than that of the Brookhaven groups, published rates almost one-hundredfold lower. This is similar to the dose-rate effects reported for styrene.

In another investigation the General Electric groups found that difunctional vinyls including the acrylic type could be polymerized several orders of magnitude faster than monofunctionals like methylmethacrylate.

f. N-vinyl pyrollidone. PVP is used as a blood-plasma extender. Molecular weight and molecular-weight distribution are therefore of paramount importance. In experiments carried out at Brookhaven (B3) using gamma-ray catalysis in a water solution, it is reported that (1) it is possible to get a narrower molecular-weight distribution by use of cobalt gammas in place of the conventional ammonia catalysts; (2) increasing the radiation intensity (three fold) or temperature increases the rate (the former resulting in a lower molecular weight, the latter having no effect); (3) rate is proportional to the monomer concentration to the first power.

g. Condensations. Radiation can induce polymerization of saturated hydrocarbons provided that the condensation products are gaseous.

The simplified mechanism which accounts for this phenomenon is that the very high-energy radiation is able to abstract atoms from molecules, which then become "unsaturated" and can react further. This phenomenon was first reported in 1924 by Lind and Bardwell (L13), who bombarded simple saturated hydrocarbons up to and including butane with alpha particles. They found that hydrogen and methane in a ratio of 5 to 1 were eliminated to give higher hydrocarbons, up to those heavy enough to be liquids and solids. Low yields of condensates are obtained for these endothermic reactions.

Manion and Burton (M3) have used 1.5 Mev. electrons to polymerize benzene to a nonvolatile compound having a molecular weight of approximately 530. Inordinately high doses were necessary to accomplish this.

5. *Effect of Radiation on Polymers*

Depolymerizations and degradations always accompany atomic radiation. Aside from these statistical events, some vulnerable points in the molecule may be selectively attacked, and so nonrandom or ordered events may also be taking place. Henley (H19) has reviewed the three events which may occur: (1) an abstraction of one particular molecule or group of molecules out of the chain; (2) further polymerization, in case unsaturation was present or induced; or (3) selective decomposition of a specific side chain.

In the cases where abstraction of any group, particularly hydrogen, leads to unsaturation, the further polymerization which takes place between polymer chains is known as "cross linking." A theory which allows one to predict whether or not cross linking would occur in irradiated polymers has been advanced by Miller et al. (M10). They postulate that if a polymer has an available hydrogen (alpha hydrogen) on the carbon adjacent to the characteristic side grouping, the hydrogen will be removed and cross linking will result.

The ratio of cross links to chain scissions can be predicted. In polyethylene for instance, where there are two C—H bonds to every C—C bond, one would expect that the ratio of cross links to chain scissions would be 1 to 0.5. Experimental evidence gives a gross confirmation of this concept.

TABLE V
Cross-linked Polymers

Polymers cross linked	Polymers degraded
Polyacrylic esters	Polyvinyl methacrylate
Polystyrene	Polyvinyl chloride
Polyesters	Polyvinylidene chloride
Nylon	Polytetrafluoroethylene
Polyethylene	Polychlorotrifluoroethylene
Chlorinated polyethylene	Cellulose
Chlorosulfonated polyethylene	Polyisobutylene
Natural rubber	Furans and phenolics
GRS	Casein
Butadiene-acrylonitrile copolymer	Polyesters
Neoprene-W	Dextran
Neoprene-GN	Melamines and urea
Polydimethylsiloxanes	
Styrene-acrylonitrile copolymers	
Polyvinyl alcohol	

a. *Degradation Reaction.* Table V, due in part to Lawton (L5), is a partial list of polymers that may become degraded or cross linked during radiations. The few commercial applications for degradation reactions that have been suggested are as follows.

I. CELLULOSE. Lawton et al. (L3) have reported the effects of high-intensity electrons on wood. At sufficiently high dosages this material became digestible to rumen bacteria. Pentoses, reducing sugars, phenolic groups and solubility increase sharply at 10^7 r. Above 10^8 r. the material is again undigestible.

II. DEXTRAN. Dextran is a blood-plasma extender, and after synthesis it must be hydrolyzed to the proper molecular weight. Radiation is capable of achieving this breakdown; however, branching and other side reactions accompany the hydrolysis. Gaden (G2) has degraded dextran in solution with gamma radiation, and Price *et al.* (P8) have carried out similar experiments using dry dextran and high-velocity electrons.

III. METHYLMETHACRYLATE. Charlesby (C3) has reported that about 5×10^7 r. of pile radiation causes the entrapment of masses of bubbles in the polymer. These expand when the material is heated to produce a "foamed" plastic. Although this may make a good insulator, one cannot envision it as competing successfully with commercial foam phenolics, for instance. About 100 e.v. is required per chain scission of methyl methacrylate.

b. Cross-Linking Reactions. I. POLYETHYLENE. About 2 million roentgens transform commercial polyethylene from a soluble, low-melting polymer to a material which no longer has a sharp melting point. About 20 million roentgens link the chain together into one large molecule or gel, which is insoluble in the usual solvents. The dose required to achieve gelling is an inverse function of the initial molecular weight, a statement which holds for all cross-linking reactions.

Radiation doses up to 10^8 r. have little effect on tensile strength or other properties at room temperature. Gas permeability is also unaffected. A most striking and useful development is the increased strength above 100°C. of polymers treated with about 10^7 r. The polymer will withstand 100 to 400 lb./sq. in. at these temperatures, which is sufficient to permit containers to support their own weight. This has opened up new markets for polyethylene as a material for heat-sterilizable containers, pipe liners, motor winding material, corrosion-resistant gaskets, etc.

Since polyethylene becomes infusible after irradiation, the material should be cross linked before molding. If this is done, easy moldability, one of the prime attractions of the plastic, is lost. However, if polyethylene is irradiated after molding it can be vacuum drawn and formed in very thick sections to extremely deep draws, and with very little temperature control. One must be careful about heating cross-linked polyethylene since (1) above 230° to 240° an article will distort if it is not free from molding strains and (2) above 220° oxidative degradation reactions set in unless, of course, antioxidants have been added.

II. RUBBER. Since rubber already contains some unsaturation, it may be cross linked with somewhat lower doses of radiation than polyethylene.

In the case of rubber this cross linking is called *vulcanization*. Experiments conducted in 1946 at the Clinton Pile (D6) demonstrated a slight curing action of pile irradiation on natural rubber. Charlesby has shown that the degree of cross linking is directly proportional to radiation dose. Approximately 50×10^6 r. produces one cross link/90 isoprene units.

At very high dosages, about 100 megarep, the cross linking hinders the uncoiling and slipping of chains and the rubber becomes hard and brittle. One interesting application of radiation vulcanization is that according to Charlesby it represents a novel method of determining molecular weight and molecular-weight distributions (C3).

III. POLYSTYRENE. Here polyethylene type of effects are observed. A pronounced increase in heat stability and a decreased solubility attend radiation treatment. Owing to the inactivation effect of the benzene ring, polystyrene requires roughly fifty times the radiation dose to bring about changes comparable in magnitude to those induced in polyethylene.

IV. MISCELLANEOUS. Little detailed information regarding the cross linking of other polymers appears in the literature. Some of the ones which have been successfully treated include Carbowax, nylon, silicones, polyvinyl alcohols, etc.

c. Suggestions for Future Work. Ionizing radiation has pronounced effects on colloids (H1). Very strange behavior has been observed in the physical properties of solutions irradiated with X rays (T2, W1). Doses as small as 2 r. have an effect on the electrophoretic mobility of colloid particles (C8). In view of this, ionizing radiations may have utility in breaking emulsions and processing colloidal solutions.

Indirect effects in nonaqueous media have received very little attention. It is likely that semiselective chemical reactions will occur in organic solvents containing active solutes. Liquid ammonia and sulfur dioxide also might offer some possibilities as solvents. The direct action of densely ionizing radiation on ammonia to produce hydrazine might yield reasonable quantities of this important reagent.

Sulfochlorinations proceeding according to

$$Cl_2 \rightarrow 2Cl\cdot$$
$$RH + Cl\cdot \rightarrow HCl + R\cdot$$
$$R\cdot + SO_2 \rightarrow RSO_2\cdot$$
$$RSO_2\cdot + Cl_2 \rightarrow RSO_2Cl\cdot + Cl\cdot$$
$$2Cl\cdot \rightarrow Cl_2$$
$$RSO_2\cdot + Cl\cdot \rightarrow RSO_2Cl$$

have been suggested as suitable for initiation by ionizing radiations (B5). The sulfonochlorides are produced photochemically with quantum yields as high as 5000. The advantages of initiation with ionizing radiation have been pointed out previously and are of particular interest in this case.

The penetrating nature of gamma and X rays suggest their use in initiating polymerizations within materials. Ballentine *et al.* have attempted to improve wood quality by irradiating wood-soaked monomers.

Ionizing radiations could be used in a similar manner to affect gluing and to hasten paint drying.

An extension of polymer cross linking to ion exchange resins would make it possible to form heavily cross linked resins after functional groups had been added. Another corollary of this idea would be to use radiation to activate sites along polymer chains, thus permitting the grafting of active groups.

In materials such as catalysts where surface properties are of importance it should be possible to alter or enhance activity by use of ionizing radiation. Investigations of the possibility of using radiation to achieve graft polymerizations have recently been initiated. For this application ionizing radiation enjoys some very special advantages.

Refinery operations such as alkylation, cracking, dehydrogenation, etc. offer some interesting opportunities. Extensive programs to investigate these phenomena have recently been announced by several major petroleum companies.

IV. Radiation Sterilization

The ability of high-energy electromagnetic and particulate radiation to destroy microorganisms without appreciable temperature rise in the substrate offers an attractive means of sterilizing foods, drugs, and miscellaneous complex biological systems. To sterilize a spore-forming culture, a total energy input of approximately 2 million roentgens is required. This is equivalent to only 4 cal./g.

Since 1896, one year after Roentgen's discovery of X rays, the bactericidal effects of radiation have been under scrutiny. Patents covering both sterilization and deinfestation have been granted (W10, A7). These have long since expired. The inventors were unable to exploit their process since they could not cope with two major problems: (1) no effective or economic sources of radiation existed and (2) insufficient information regarding biological and biochemical effects of radiation was available.

Problem (1) appears solvable. The cost of sterilizing, though still high, is clearly competitive for some applications. Food sterilization probably represents the most favorable market for waste fission products. About 80 billion lb. of milk, for instance, are consumed in the United States annually (S19). Duffy (D10) has shown that about twenty 500-mw. fluid reactors would yield sufficient fission-product radiation to sterilize this entire milk supply. Here one is admittedly projecting into the future; some experts (W6) predict that for the next ten years machine-made radiation will be the most attractive source of ionizing radiation. These machines however are becoming available in increasing quantities and high quality. Biochemical and biological problems attending the use of radiation may become the only factor severely restricting the use of this agent for food sterilization. Although the sterilization doses affect chemical changes of only a few micromoles/g., in the case of foods this has serious consequences. In the extreme instance these changes can be so pernicious as actually to decrease the storage life of radiation-sterilized foods.

Literature in the food field is traditionally difficult to evaluate. The following discussion will therefore be devoted chiefly to an evaluation of the factors which influence the chemical engineering aspects of sterilization processing. A thorough and able review of the entire field has just been made by Hannan (H2). Other recent summaries are available (H5, R1, C5). These all include the sterilization of pharmaceuticals and medicinals, since the techniques parallel those used in food applications.

A. Mechanism of Biological Action

Some progress is being made toward an understanding of the fundamental processes of radiation chemistry. The ultimate lethal action of ionizing radiation remains, however, as much as a mystery as the intimate organization of the cell itself. Two somewhat overlapping theories have been advanced, each of which enjoys considerable support. Although both theories have several features in common, they differ in that one is physical and the other is chemical in nature.

1. *Physical Interpretation*

The simple one-hit-one-effect hypothesis of Dessauer (D7) still serves as the basis for most modern expositions of this type. According to this so-called "target" or "treffer" theory, the lethal effect of radiation is observed whenever an ionization is produced inside one or more "sensitive volume." The lines of evidence which support this point of view have been analyzed by Lea (L7, L8). They are

a. The survival curve (plot of surviving microorganisms vs. dose) is exponential.

b. The effect of a given dose is independent of the intensity at which it is given or the manner in which it is fractionated.

c. For the same degree of effect, the dose required with different radiations increases with increasing rate of energy loss.

The target theory is limited to situations where the action is analogous to a lethal gene mutation and may be attributed to the passage of a single ionizing particle through the target area. Quantitative observations based on the size of the "sensitive volumes" or "targets" and knowledge of interionic spacings along the track of the electrons can then be made. Confirmation of the applicability of this approach to radiation inactivation of small viruses and enzymes is not lacking, since in these cases the target is defined as the entire virus particle or enzyme molecule (P4).

Extension of this theory to larger viruses and microorganisms rapidly leads to inconsistencies. In the case of *E. coli* it must be assumed that the bug has 250 targets, each one having an average diameter of 12 mμ, i.e., of the same order of size as a gene. Serious anomalies are encountered when one considers the fact that by simply excluding oxygen from the system, the radiation sensitivity of the bacteria may be changed by a factor of 3 or 4. Is one to conclude from this that the target size is a function of the metabolic level in the microorganism? This and other considerations have led to the formulation of a theory based on the inactivation cells via a series of indirect radiation-induced chemical reactions.

2. *Chemical Interpretation*

Weiss (W2) attributes the destruction of microorganisms to the OH and H radicals formed in the water near the nuclear material (Sec. IIIB2b). These radicals inactivate cells by initiating deamination and dephosphorylation of the nucleic acids, thus weakening the nucleoprotein structure. This and other proposed chemical mechanisms have recently been discussed by Gray (G12).

While the physical and chemical viewpoints are not irreconcilable, a theory which accommodates both concepts and is consistent with the experimental data has yet to be developed.

B. Factors in Radiation Sterilization

Since lethal effects are very sensitive to experimental conditions, it is not possible to coordinate all the existing data. By selection of a few examples from experiments done by workers using carefully standardized conditions some conclusions regarding the various factors which affect sterilization can be drawn.

1. *Type of Organism*

Papers describing the destruction by radiation of approximately 150 microorganisms have been published. Since the work was carried out under varying conditions and with a wide range of objectives it is not prudent to draw quantitative comparisons between lethal curves reported by different scientists. In general, under conditions of normal aeration in buffer or saline solutions the mean lethal doses shown in Table VI may be taken as representative.

The mean lethal dose (M.L.D.) is the number of roentgens required to kill 63% of the microbiological population.

TABLE VI
Mean Lethal Doses

Organisms	Mean lethal dose, r.
Gram-negative or -positive rods and cocci	1–15,000
Yeasts	6–30,000
Spore-forming bacteria	50–400,000
Viruses	100–350,000

2. *Culture Age*

The limited data dealing with the chronological history of the culture are contradictory. For yeast cells, Lawrence (L6) reports M.L.D.'s of 15,000 and 45,000 r. for 2-day-old and 8-week-old cultures respectively. Proctor (P15) found that 48-hr. cultures of *Salmonella* offer more resistance to destruction by cathode rays than do organisms from a 24-hr. culture. Gastaldi (G5) and Dunn (D11), however, are said by Hannan (H2) to have found that old spore cells are the most sensitive to radiation.

3. *Preirradiation Treatment*

a. Desiccation. Ample experimental evidence indicates that all organisms, from frogs to viruses, enjoy greater immunity from radiation after desiccation. The magnitude of the effect, however, is variable. For *Aspergillus terreus*, in the presence of oxygen, M.L.D.'s for desiccated and suspended spores differ by a factor of 2.5 (S15).

b. Infrared and Ultraviolet. These applied prior to irradiation have some interesting effects (S22, S23, K1). Ultraviolet apparently decreases the sensitivity of cells, and infrared is reported to increase chromosone damage; but on the other hand, it decreases *E. coli's* radiation lability (K1).

c. Chemical Treatment. Certain chemicals if present in the medium during irradiation protect the organism; this phenomenon will be discussed in Sec. IVB6c. Others increase the radioresistance only if they are present during the growth period of the cell prior to exposure; for example, incubation with formate, succinate, pyruvate, serine, or ethanol has been demonstrated to protect *E. coli* B/r against the lethal action of X rays (S16). This action is most probably associated with respiratory and oxygen effects.

d. Temperature. Langendorff (L1) and Wood (W8) demonstrated that the mean lethal dose for *E. coli* and *S. cerevisiae* can be materially reduced by warming the cells to 45° and 52.5°C. prior to irradiation. Morgan and Reed (M13) found that preheating had no effect on the radiation sensitivity of bacterial spores.

4. Induced Resistance

Gaden and Henley (G1) have demonstrated that for *E. coli* gamma rays may serve as the selecting and inducing agent in the development of radioresistance. It is not yet possible to predict whether such changes in resistance will become as large a factor in radiation sterilization as they have in antibiotic therapy, but the situation bears watching.

5. Concentration

There is some experimental evidence which indicates that in dilute suspensions the inactivation is a function of concentration (B10). There is considerable doubt as to whether or not this is an artifact caused by the depletion of oxygen in the system at the higher cell concentrations (H21). (See Sec. IV6b.)

6. Environmental Conditions During Exposure

These conditions, in contradistinction to those previously discussed (Sec. IVB1 to 5), can to some measure be controlled by the processor. These then deserve special attention.

a. Temperature. Little variation in radiosensitivity with temperature has been found for either vegetative or spore-forming organisms in the range of normal ambient fluctuations (40° to 90°F.). For spore formers this immunity extends into the frozen state (L4, E2, L9).

Experiments with vegetative cells show that freezing offers the cells considerable protection from the radiation. The dose necessary to inactivate cells in the frozen state may be from two to eight times that required at room temperatures (P12, W8).

b. Oxygen Pressure. The effect of oxygen on radiosensitivity was reported as early as 1921 (P1). It is reported nearly without exception that oxygen appreciably increases the radiation sensitivity of microorganisms. Where quantitative comparisons are possible, the ratio of lethal effects in the presence of oxygen to those in its absence are usually between 2:1 and 3:1 for vegetative organisms (H23). Recent work with spore formers (M13) show that the radiation response of these organisms is less subject to changes by variation in oxygen pressure.

c. Protective Chemicals. A wide variety of additives, both organic and inorganic, protect microorganisms from the lethal effects of radiation. Among those substances which have been found effective are reducing compounds (sodium hydrosulfite), sulfhydryl compounds (cysteine, glutathione), respiratory inhibitors (sodium cyanide), alcohols, glycols, sugars, carboxylic acids, and proteins. Since nearly all these compounds either combine with oxygen directly or increase cell metabolism, thus indirectly reducing oxygen tension, it is difficult to interpret the results. The subject of protection has recently been reviewed by Patt (P1). As in the case of oxygen pressure, spore formers show considerably less variation in radiosensitivity in different substrates than do vegetative cells.

d. Hydrogen Ion Concentration. Despite the fact that with biological objects one is severely limited in the working pH range, surprisingly few published results on the variation of radiosensitivity with hydrogen ion concentration have appeared. Moriarty (M14) found that *Staphylococcus aureus* was sensitized to radiation at high and low pH. Zirkle (Z1) found that for certain cells radiosensitivity increases to a maximum and then declines as pH is decreased with carbon dioxide. Spores of *Bacillus subtilis* show little difference in radiation response with changing pH (P6).

e. Ion Density. Early workers typified by Lea (L7) in the field concluded that lethal effects diminish with increasing ion density. This observation is one of the pillars of the target theory (Sec. IVA1c). Moos has recently claimed the reverse; he found the more densely ionizing radiation to be the more potent killer (M12). Evidence to support both views exists, and all is subject to some criticism. Moos incubated his organism on irradiated agar, and Lea switched from a liquid to a solid substrate in the middle of his experiments.

In the 1 to 3 Mev. range, where the R.E.L. for electrons and gammas are of the same order of magnitude (Sec. IIIB1) one has no reason to expect any differences in biological effectiveness. Goldblith *et al.* offer some experimental proof for this (G9).

Varying the ion density by changing dose rate (Sec. IIIB1) appears to be without effect under normal conditions (L4). The Capacitron, which gives dose rates at least one-thousand-fold higher than those obtainable anywhere else, seems to produce unique results. Brasch and Huber (B12, H26, H25) report lower lethal doses, heightening of bactericidal and lowering of chemical side reactions, and an absence of protective effects at these high dose rates.

7. *Induced Radioactivity*

When the energy of the incident radiation is sufficient to penetrate the potential barrier of the nucleus, radioactivity may be induced. McElhinney *et al.* (M2), among others, have tabulated (gamma, n) and (gamma, p) reactions which may occur and give threshold energies for these reactions. Hannan (H2) has tabulated those reactions which represent potential hazards in sterilization work, listing the elements in order of their abundance in biological material.

Although several reactions have thresholds of only a few Mev., the elements involved (1) are present in only trace amounts and (2) have very short half lives. At about 10 Mev. iodine 127 undergoes a (gamma, n) reaction resulting in a radioactive product with a half-life of 13 days. Since iodine is concentrated in the thyroid this is a potential hazard.

Although there is a minimum of sixteen nuclear transformations occurring in the 10 to 15 Mev. range, these can probably be tolerated. In the 15 to 20 Mev. range the (gamma, n) threshold for both carbon and oxygen is exceeded and dangerous levels of radioactivity may result. Horsley et al. (H24) have estimated that the irradiation arising from secondary nuclear changes in human tissue may approach 0.6% of the incident dose at 24 Mev.

8. Sterilization Dose

Figures presented in various sections apply to experiments in which pure cultures have been irradiated in a controlled medium. In actual practice one is faced with the problem of producing an arbitrarily defined level of commercial sterility in an undefined medium with a mixed population. Commercial sterility can mean anywhere from 10^{-8} to 10^{-12} organism/g. depending on the product and type of contamination. Systematic data which would give the detailed shape of the survival curve in the region of commercial interests are nonexistent. The results of a number of experiments in which workers claimed to have achieved sterility are tabulated in Table VII. In nearly all cases insufficient data were obtained to allow a satisfactory statistical evaluation. Furthermore, differences and confusion in the dosimetry make the data rather approximate.

TABLE VII
Sterilization Doses

Reference	Substrate	Special conditions	Sterilization dose, r.
(B12)	Foods and drugs	Various	Ca. 1.5×10^6
(T1)	Cortisone acetate soln.	Inoculated with B. subtilis and Cl. sporogenes	2.73×10^6
(M13)	Various	C. Botulinum	2×10^6*
(D12)	Water	Various sources	2.5×10^5
(D12)	Soil	Various sources	10^6
(D12, R4)	Spices	Various sources	1.4×10^6
(P12)	Minced beef	Contaminated	1.5×10^6
(P6)	Ham, beef, and luncheon meat	Contaminated	2×10^6
(P11)	Haddock	Contaminated	1.5×10^6

* Population reduced by 10^6.

9. Postirradiation Conditions

Anderson and Langendorff (A9, L1) both have reported that *E. coli* may be reactivated if incubated at elevated temperatures after exposure (40° to 45°C.). If the cells are subjected to higher temperatures following the irradiation, the cells exhibit heat sensitization (M13, W8).

A more important phenomenon is that after low-temperature postirradiation incubation vegetative cells show a marked recovery. In some cases the effect is equivalent to reducing the dose by one third (S17, H22). A recent British symposium on radiation after effects is available (S24).

C. Food Irradiation

The interaction of radiation with bacteria having been considered, the next logical question to ask is while the radiation is decimating the flora, what is it doing to the substrate? During sterilization, in the absence of a chain reaction, radiation-induced chemical changes of a few micromoles/g. will occur. In anything but foods these changes would be of negligible consequence. Foods, however, are unique insofar as they contain a few key components which, although present in not much more than micromole/g. concentrations, regulate the flavor and nutritional value of the fare. Some of these key components could be extraordinarily radiation-labile, and much of the experimental evidence that is accumulating verifies this. Deleterious changes occurring in food during storage after treatment may be equally troublesome. Though radiation may annihilate all flora, enzymes continue to thrive, and strict attention must be paid to the consequences of postirradiation enzyme activity. A brief look at the biochemical reactions induced by radiation is enlightening.

1. *Chemical Changes*

a. Proteins. With pure amino acids oxidation of the sulfhydryl groups, deamination, and decarboxylation, in that order, will occur. The first two processes will take place with G values from 1 to 1.5, while considerably lower values are reported for decarboxylation. These processes may take place simultaneously. Lawton and Bellamy (L4) have shown that isovaleraldehyde can be formed by decarboxylation and deamination of leucine.

Deamination appears to proceed equally rapidly both in the presence and absence of oxygen. Stein and Weiss (S18) examined this phenomenon and deduced that deamination can be accompanied by both oxidation and reduction. In the absence of oxygen the H atom, rather than the HO_2 radical is the effective agent. This mechanism is satisfactory for the simple amino acids (glycine, alanine, and serine) studied by these workers. With the more complex amino acids yields of ammonia become less and side chains may be preferentially broken.

Destruction of an amino acid present in an enzyme will result in the loss of biological activity. Although the inactivation of enzymes proceeds quite readily in pure solutions, these compounds behave differently when present in complex mixtures. To an even greater extent than microorganisms, enzymes are protected by a variety of compounds. Thus either oxidizing or reducing agents markedly increase their radioresistance (F1, D4). Similar effects have been reported for a variety of "living" proteins, particularly viruses, and phages, some of which are also more readily inactivated in deoxygenated solutions (E1). Freezing also is reported to decrease drastically radiation damage (H2). The result of all this is that enzymes when present in foods are inordinately difficult to inactivate. Foodstuffs irradiated with sterilizing doses lose very little of their enzyme activity, and analogously living cells continue to metabolize although they are "dead," i.e., unable to reproduce. This lingering activity is particularly operative in postirradiation storage (Sec. IVC5).

Other phenomena of interest here are the radiation-induced polymerizations and depolymerization of protein chains. These reactions are quite analogous to those reported for other polymers (Sec. IIIB3).

b. Carbohydrates. Polymeric carbohydrates such as cellulose have been examined in some detail (S9). At doses greater than 10^7 r. depolymerization and an increase in volatile acid and reducing sugar content were noted. Wood receiving approximately 10^8 r. became digestible to rumen bacteria. At this dose the cellulose becomes soluble. G values for the depolymerization were 0.19, and for the decomposition approximately 1.0.

Phillips (P2) has studied the radiation reactions of sugars in dilute solution and reported a preferential oxidation occurring at the C 6 position, which results in conversion of the glucose to glucuronic acid. Other oxidizing reactions resulting in the formation of aldehyde offer alternative pathways.

c. Fats and Lipids. Although they are not typical of reactions which would be expected in foods, the results of the group which was trying to ascertain whether petroleum could be formed by the action of alpha rays on fatty acids are of interest (W5). The predominant reaction reported was decarboxylation. Dehydrogenation also occurred but not in the same molecule as decarboxylation. The yields were very low.

An exhaustive treatment of this topic can be found in (H2). Hannan makes the observations that three consequences are to be expected when fats are irradiated: (1) combination of free radicals to form peroxides, (2) breaking of hydrocarbon chains with recombination of the free radical fragments to form long or branched chains, and (3) decarboxylation.

At low temperatures oils autooxidize slightly more than corresponding samples irradiated at room temperature. These peroxide values continue to increase on storage; however, in the absence of oxygen the odor and flavor of the oils do not deteriorate further. Although complete oxygen removal suppressed all "irradiation" odors and flavors, it was thought possible that polymerization and cross linking were increased. The overall picture presented is a competition between the oxidative changes resulting in flavor molecules and the free radical polymerization or degradation. Removal of oxygen or increase in temperature favors polymerization-type reactions. These suggestions are in accord with what is known about polymerization reactions; they generally have a high temperature coefficient and are inhibited by oxygen. This may also be the reason that some proteins and viruses are more easily inactivated in the presence of inert gases.

Other observations of changes associated with fats are of interest. Serious bleaching and a destruction of antioxidants (tocopherols) are reported. Large amounts of antioxidants negate the foregoing effect.

Mead's work (M8, P5, M4) with fatty-acids emulsions deserves attention. X rays initiate oxidative chain reactions in these systems as measured spectrophotometrically by the formation of a conjugated double bond. Some antioxidants when added to the system prevent this reaction and are themselves destroyed.

2. *Loss of Nutrients*

Any food processing or storage usually results in the destruction of some essential nutrients—not a very critical situation, since the food can be refortified. There are, of course, practical limits, such as cost, which determine whether refortification is practicable. A close look at some of the work on the distribution of nutrients gives some idea of the order of magnitude of the problem.

a. *Pure Vitamin Preparations.* Ascorbic acid at concentrations in which it is normally found in foods is nearly completely destroyed by sterilization doses (P13). The destruction is complete; the acid is not merely oxidized to the dehydro form. Freezing apparently reduces the destruction.

The same authors (G8, G7) also found very substantial decreases in riboflavin (approx. 80%), and niacin (P9) fared little better. When mixtures were irradiated unusual events occurred. Riboflavin and ascorbic acid were each protected by niacin. Addition of cystine or cysteine apparently sensitized the niacin (P10). Since initial rates were not given, and the doses were considerably above the oxygen breakpoint (Sec. IIIA2), no mechanistic interpretation is possible. There also appears to be some doubt about the reliability of the colormetric assay used by these workers.

Fat-soluble vitamins and related carotenes suffer complete destruction at sterilization doses (G8). As has been mentioned previously, there may be some connection between the peroxide value and bleaching of carotenoid pigments in fats.

b. *Vitamins in Foods.* Vitamins in food are less sensitive to destruction than are aqueous solutions. A thorough study of the destruction of nutrients in milk has been made at Columbia University by Kung, Gaden, and King (K5). At 1×10^6 r. they report complete destruction of reduced ascorbic acid, 80% loss of vitamin A, and 45% reduction of riboflavin and carotenes. Phosphatase activity is only slightly reduced.

Solid foods have received little attention. Huber (H25) and the Massachusetts Institute of Technology group (P10) maintain that there is no destruction of water-soluble vitamins when a variety of foods is sterilized.

A complete synthetic media for the growth of *Tetrahymena pyriformis E.*, a protozoan whose nutritional requirements closely simulate those of mammals, was irradiated at Michigan (B16). When the medium failed to support the growth of *Tetrahymena*, the essential vitamins and amino acids were individually irradiated in solution. All vitamins were altered structurally by less than sterilization doses. The amino acids proved stable under the same conditions.

3. *Flavor and Appearance*

In the early days of radiation sterilization there was considerable confusion regarding the taste and appearance of irradiated foods. This situation has changed appreciably. There has been relatively uniform agreement between results reported by various laboratories for the past 2 years. Observations abstracted here will be chiefly those of the University of Michigan, the Electronized Chemicals group, and the Low Temperature Research Station (England). Some of the previously mentioned Massachusetts Institute of Technology publications also provided valuable information.

a. *Meats.* The preponderance of experiments has been conducted with beef. When treated with sterilization doses this food acquires a taste and odor characterized by food technologists and various epicures as "cheesy," "burnt," "smoked," "goaty," "cressy," "burnt-dog," and "biscuity." All agree that the change is for the worse. Various treatments mitigate, but do not completely neutralize, these effects. Among those which have been tried are (1) freezing, (2) removal of oxygen, and (3) chemical additives, mostly reducing compounds. A combination of the first two appears to solve the color problem, which is largely due to myoglobin oxidation. Not even a combination of all three completely counteracts the objectionable odor and flavor changes. Above 3.5 million r. some loss of texture may also occur.

It is quite apparent that if one must resort to such drastic and expensive measures to counteract flavor and color deterioration, any advantage that radiation originally had will be very much weakened.

Smoked meats, particularly processed pork products, show little loss of flavor and aroma after treatment. Since the storage properties of these products are usually adequate without radiation there is little point in discussing this area of endeavor. The same comment applies to cooked meats, whether they be cooked before, during, or after irradiation.

b. Fish. As in meats, considerable deleterious transformations in flavor, odor, and aesthetic quality take place. The delicate texture is also affected.

c. Dairy Foods. Radiation-induced changes in these foods are several degrees more severe than in any other animal products. Even dried milk and eggs are not immune.

Though unacceptable to humans, the radiated foods seem palatable to bacteria. Thus satisfactory cottage cheese has been made from sterilized milk.

Bakery products merit brief mention here since they are chiefly dairy in nature. As would be expected, the high-fat-content foods develop the characteristic tallowiness. Bread, on the other hand, is quite oblivious to radiation damage.

d. Vegetables and Fruit. Odor and flavor changes here are not so severe, but texture change and quality losses are inordinate. In general these products loose their characteristic flavor and crispness. Asparagus and carrots apparently sustain the ordeal somewhat better than most vegetables. Of the fruits, tomatoes and peaches may show a slightly higher order of resistance to change.

Freezing, deoxygenation, or chemical additives may reduce side effects, but the low unit costs of the products preclude serious consideration of such mitigating action.

Spices, being vegetative in origin, may be placed in this category. Their reaction however, is atypical, since they show good radiation stability, Provided the economics are favorable, this field has distinct possibilities.

4. *Toxicity*

Animal-feeding programs have been under way for approximately 5 years. Conducted chiefly at Columbia University and the University of Chicago, these tests indicate that although somewhat lower growth curves and shorter life spans may result from an irradiated diet, as far as rats are concerned, there is basically nothing wrong with irradiated foods that a vitamin supplement will not cure. The University of Michigan has recently initiated tests with chickens, and the U. S. Army Quartermaster Corps is feeding larger mammals.

Assuming that the toxic compounds are able to accumulate in the body, there are certain types of products which represent potential hazards. Three types of chemical compounds as well as the radiation itself have been tabbed for possible carcinogenicity (H2): (*a*) branching in fatty-acid carbon chains, (*b*) fatty acids with odd number of carbons, (*c*) oxidized sterols.

A discussion of some of the potential hazards has been made by officials of the Food and Drug Administration (L10). Among other things which remain to be investigated are changes occurring during postirradiation storage (Sec. IVC5).

5. *Storage Properties*

Postirradiation reactions are well known in photo and radiation chemistry. Some of the reactions which occur in polymers are of interest here. Of note also are postirradiation-induced changes which occur in proteins. These include deamination, dephosphorylation, and gelation. That radiation-treated food may either improve or deteriorate during storage is therefore a distinct possibility.

a. Meats. Apparently meats do recover somewhat from their initial shock. If they are stored in oxygen-impermeable containers, this recovery may be appreciable. However, after longer storage the meat undergoes changes which resemble a slow cooking effect. Upon cooking, after months of storage, some loss of texture (a "mousy" character) or a bitter flavor may develop.

b. Fish. Little experimental work has been done on fish, but the analogies to meat are clear. Slow cooking effects and some recovery of flavor can be expected.

c. Dairy Products. These products apparently show insufficient recovery to raise much hope. In some cases, particularly with dry products, postirradiation changes may actually reduce storage life. This may be traced to the development of oxidative changes in fats.

d. Vegetables and Fruits. No prolonged storage of irradiated vegetables and fruits has yet been reported. All indications are that the deteriorative chemical and enzymatic processes occurring in these products are not appreciably retarded. Treated peaches and cherries are said to deteriorate faster than the controls.

D. Low-Dose Irradiation

Use of radiation in relatively low doses to extend storage life is a distinctly promising possibility. For this type of service, radiation enjoys a special position. There are no regularly employed agents which are capable of accomplishing the same thing. Although ultraviolet light is frequently used in meat-storage chambers, and the use of gases and antibiotics has been proposed, the extension of storage life still depends solely on refrigeration.

By "low dose" one generally means less than 100,000 r. Considering the relative lability of the organisms (Sec. IVB1) one would expect this dose to reduce the number of vegetative bacteria by a factor of 10^3 to 10^{12}, yeast and molds by a factor of 10^{15} to $10^{1.5}$, and spore-forming bacteria by 10 or less. The process is therefore intrinsically limited to cases where pathogenic spore formers are not a problem and the material is not likely to be stored for long periods. Costwise this application presents a more favorable picture than does the use of irradiation as a sterilization agent. Even more important is the fact that low radiation doses effect almost no change in product quality in most foods. Combining radiation with a moderate heat treatment is a further possibility. Indications are that this technique may present some advantages. One moderately dangerous aspect of irradiation pasteurization has been alluded to (H2). If one kills the organisms which the consumer customarily associates with spoilage, then another less easily recognized and possibly more dangerous type of spoilage may take hold. This and the general toxicity situation remains to be carefully investigated. It may be of some benefit at this time to examine some of the typical results reported by workers who have contributed to this technology.

1. *Meats*

Experiments conducted by various groups (S8, A8, B15) indicate that after exposures ranging from 50 to 80,000 rep, the shelf life of beef, raw pork, and minced meats has been increased from 2 or 3 to 8 to 12 days. In these cases the chief result of radiation appears to be an initial delay in the reproduction of the microorganisms. An interesting sidelight of the work with pork is that pasteurization dosages are sufficient to control the Trichinella organism. This fact has been widely publicized (G11, G10). One group (B15) reports a loss of color of the meat during storage. It remains to be proved whether or not this is an experimental artifact due to the unusual packing procedures used.

2. *Delicatessen*

The packaged life of cole slaw at 40°F. is extended from 6 or 7 days to 25 days after moderate radiation treatment (B15). Another delicatessen item is smoked foods, which have an irradiated type of flavor prior to treatment (Sec. IVC2). Their flavor will, if anything, be enhanced.

Smoked fish and certain other delicatessen may therefore represent interesting potential applications.

3. *Fruits*

A few investigations (H3) have unearthed the fact that ripening of "soft" fruits can be delayed somewhat by low (10 to 50,000 r.) doses. Cabbages and similar vegetables are also reported to have a prolonged storage life after receiving 50,000 r. of gamma radiation. The commercial applications of this phenomenon must await considerable further experimentation.

4. *Beer*

Modest doses of radiation have been shown to inactivate the few residual yeast in heat-pasteurized beer. The major problem here involves flavor and odor. Breweries in common with all branches of beverage-alcohol manufacture jealously guard these features as being of prime importance (J3).

E. Control of Infestation

The lethality of ionizing radiation is proportional to the complexity of the organism. Humans, for example, cannot survive more than 800 r., but it may take 2 million r. to inactivate a microorganism. Insect life, of the type which infests field crops, show intermediate stability. Information gleaned from the best available sources (B15, B1, H9) indicate that a 1000-r. exposure will kill all eggs, 1300 r. will kill all larvae, and 3,000 r. will destroy all young pupae. The resistance of insects to radiation increases with age. For example, to kill some adult species requires 50 to 60,000 r. Adult flour beetles succumb at 10,000 to 16,000 r.

1. *Grains*

Insect infestation of grains results in an annual loss of 500 million dollars. Present methods of chemical control are relatively unsatisfactory. There can be no doubt but that radiation could do a more satisfactory job than the chemicals, since it can treat infestation both inside and outside the kernels. A complete economic and logistic evaluation of the problem has been formulated by Chamberlain (C1). The original cost estimate was low by a factor of 1.78. This mistake was corrected in a later version of the paper. He shows that isotope radiation can compete with conventional treatment methods if the irradiator can be located in the terminal warehouse and a charge of 1 cent/bu. can be assessed for the deinfestation.

The baking properties of irradiated wheat have been evaluated and found to be satisfactory (B15). The nutritive properties have not been investigated.

2. *Tobacco*

Although the losses due to poor housekeeping and infestation run into the millions, tobacco is not amenable to radiation treatment. The tobacco when stored is packaged in hogsheads which are made of slotted wood, so that although the tobacco is sterile when it goes into storage, it becomes contaminated while in the storehouse. This situation is of interest only because people unfamiliar with industrial technology have consistently suggested that radiation could be used to advantage here.

F. Prevention of Sprouting

Though one tends to associate the inhibition of sprouting with sterilization, it is a quite unrelated phenomenon. The only similarity between prevention and sterilization is that in both instances one is suspending a "life process."

1. *Potatoes*

Experimental results indicate that doses of approximately 10,000 r. are sufficient to prevent the sprouting of potatoes (S14). Indefinite storage is not possible since the radiation does not disturb normal rotting processes, and dehydration continues during storage. When stored properly the potatoes will keep for at least 18 months while losing no more than 20% of their water content. Another advantageous feature of the potato business is that the buds (eyes) are on the surface of the tuber so that treatment may be effected by electrons or gamma rays. Chamberlain (C1) has inspected the engineering feasibility of the process. His conclusions are very favorable for those who hope to see early commercialization of this process.

One potentially unfavorable observation has been made by Hannan (H2). He found some tissue damage in the irradiated tubers. It remains therefore to be shown whether the irradiated potatoes can be shipped and processed via normal channels.

2. *Onions*

White and yellow strains of sweet Spanish onions have recently been irradiated by the Brookhaven group (D5). An 8,000-r. treatment was found to be completely effective in sprout suppression although there was no indication that any of the radiation treatment affected rotting, which proceeded at a normal rate.

Unlike potatoes, where sprouts start from tissue very close to the surface, an onion shoot arises from deep within the center of the bulb. Present electron generators would therefore not be suitable for this application.

G. Sterilization of Drugs and Pharmaceuticals

1. Although food preservation has preempted the major share of public attention, sterilized drugs and pharmaceuticals have already achieved a limited market. The reasons for this are apparent when one considers the fact that almost none of the negating factors which are operative in the case of foods come into play here.

   a. Product. Instead of a complex mixture containing critical, radiation labile materials in small concentrations, one is dealing with a concentrated, reasonably well-defined single substance. Often the product is dry as well, thereby lessening the extent of destruction.

   b. Toxicity. Pharmaceutical products are not regularly consumed as is food. No concern need therefore be shown with regard to subtle chemical changes which may be the basis of accumulated poisons. For that matter many chemotherapeutic agents have distinctly toxic effects (streptomycin for example) but the value to be gained so far exceeds damage that over the short haul they are clinically useful.

   c. Costs. Pharmaceuticals are generally products of relatively high unit cost with production costs contributing a relatively smaller share of the "over-the-counter" charge than for foods. As a result a sterilization technique of demonstrated advantages with respect to product standards, public relations value, or any other criterion may find application even if it costs a trifle more. The most common technique now in use is separate sterilization of products and containers followed by aseptic filling and sealing. This is a mixed blessing since on one hand the technique is expensive, but on the other hand all large pharmaceutical houses now have extensive investments in sterile packaging which must be amortized.

2. The few coordinated, and reasonably conclusive studies made so far have been carried out in the laboratories of individual manufacturers. Much of this has been discussed publicly, but generally little has been published in this field (T1, C6, B11, M11). Examination of the foregoing sources and private discussions make several general conclusions possible.

   a. General products. General-line, nonprescription products for "mass" sale are vitamins, patent and compound medicines, household supplies. Partly because of its bearing on the food problem, the radiation stability of vitamins has received considerable attention. All results to date show that vitamins when present as concentrates do not lose much potency. This includes thiamine, riboflavin, pyridoxine, niacin, the B-complex, and polyvitamin preparations.

b. Chemotherapeutic agents. Of the antibiotics, injectionable materials, and related prescription products, some of the myriad items successfully sterilized without loss in potency include aureomycin, chloromycetin, terramycin, penicillin, streptomycin, dihydro-streptomycin, steroid hormones, and various antitoxins. Each product, however, must be tested individually since some unsatisfactory results have been reported. Included here are pitocin, penicillin in water, insulin in oil, tetanus antitoxins, blood media, and some antibody systems. It is not known whether freezing, deoxygenation, or change of media would mitigate these effects, but the indications are that they would.

c. Medical and surgical specialities. Sutures, dressings, disposable syringe kits, etc., are now in relatively common use, including aortic homografts (H27, M9) bone-bank bone, hypodermic tubing, and surgical catgut. This list will surely be extended and the markets widened.

H. Containers

The fact that the pharmaceutical industry indulges in the separate sterilization of products and containers has previously been mentioned (Sec. IVG1c). Three general applications of radiation sterilization therefore exist: radiation of (1) the product, (2) the containers, or (3) the product in the container. The first possibility has been discussed; the others deserve elaboration.

1. *Sterilization of Containers*

Conventional tin and glass containers can be satisfactorily treated by steam or hot air. Paper, cardboard, and heat-sensitive plastics require subtler techniques. Experiments with Co^{60} gamma rays indicate that 200,000 r. are sufficient to inactivate vegetative contaminants, while the customary 2×10^6 r. are required for spore formers. These doses are considerably below those that would be expected to have any effect on the physical properties of these materials (Sec. IIIC5).

2. *"In-Container" Sterilization*

Several factors must be considered here.

a. Radiation-Induced Deterioration of the Container. It has previously been indicated that sterilization doses would not affect the physical properties of plastics. This would in all likelihood include the lacquers and seaming compounds now used in tin cans. Chemical changes, particularly with vinyl halide films and certain plasticizers, are known to occur. These may result in undesirable odors or flavors being imparted to the contents.

In glasses, electron trapping leading to coloration can be expected.

This browning, which fades on standing, is undoubtedly objectionable from several standpoints. Three possible solutions are available. One can resort to (1) a noncoloring glass containing cerium oxide, for instance, (2) colored glasses, and (3) special glasses which acquire colors other than brown during the radiation.

b. Shape. To avoid wasting radiation one would undoubtedly prefer rectangular or square containers. Other than ease of fabrication, one major reason for using round cans is to mitigate stresses in heat sterilization. Rectangular cans would most certainly be a welcome sight to anyone who stacks, stores, and distributes foodstuff.

c. Penetration. If one is considering using cathode rays, limited penetration will be a factor. The penetration, being inversely proportional to density, may dictate the choice of packaging material. Aluminum, magnesium, or plastics would most certainly be preferable to tin plate if other economic considerations do not preclude their use.

I. Engineering Design

To design a radiation-process facility a designer must first know what radiation dose is required; only then can he proceed with the engineering calculations. An empirical approach to determine this dose has been suggested (P14). These scientists propose that it would be adequate to determine the first few cycles of the lethal curve (under the operating conditions) and then to extrapolate the curve to a dose sufficient to reduce the bacterial population to the desired sterility level. This method is predicated on a first-order lethal mechanism. Unfortunately many microorganisms, particularly spore formers, do not know about this and insist on expiring via a sigmoidal-type death curve. An empirical approach of this sort also requires that new experimental data be obtained any time the operating conditions are changed.

The ideal situation would be one where a file of information regarding the inactivation of pure cultures in various media is available. Armed with this information, a knowledge of the type and concentration of the flora, plus a suitable correlation, one should be able to determine the dose required to achieve any predetermined level of contamination. If the correlation were general enough, it would allow the operator to adjust the radiation exposure to any change in flora or operating variables. This approximates the techniques now used by heat processors. As a point of fact, a close analogy between radiation and thermal processing can be drawn. In both cases one needs first to establish a lethal curve for the most resistant organism. Process time in thermal canning will vary with the temperatures; in radiation processing it will be a function of the "protecting" substances. In heat transfer calculations the physical porperties of the product are of importance, and in an analogous manner they are of importance in radiation-dosage calculations.

Henley and Gaden (H12) have outlined this type of engineering

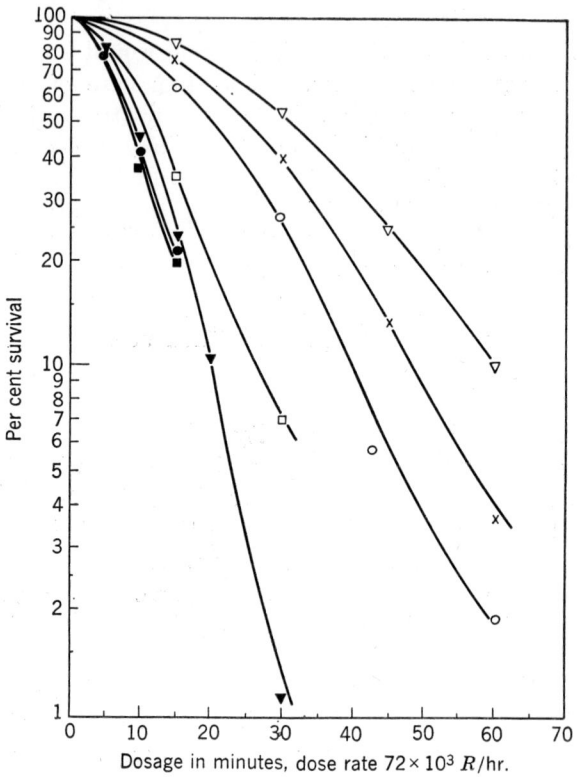

Fig. 2. Percentage of bacteria surviving vs. radiation dose, buffer irradiation

$\triangledown$-S_0 = 577 × 10^8
×-S_0 = 221 × 10^8
○-S_0 = 640 × 10^7
□-S_0 = 260 × 10^6
▼-S_0 = 141 × 10^4
●-S_0 = 131 × 10
■-S_0 = 153

approach to radiation-sterilization-process design, and the essential features of their treatment will be set forth briefly here.

1. *Analysis.* An analysis of the factors which influence the lethal action was made. These factors were as follows.

a. Concentration of Bacteria. The experimental results obtained show that there is no pronounced change in the slopes or shapes of the survival curves between 153 and 600×10^4 cells/cc. (Fig. 2). At concentrations between 100×10^4 and 600×10^7 there is a decided increase in the slope of the survival curves, but the curves retain their characteristic shapes. This can be seen by inspection of Fig. 2. The curves definitely do not fit the simple "one-hit" or "one-reaction" type of function. Rather the curves are best approximated by functions which can be derived from a consideration of the kinetics associated with the indirect-action hypothesis (Sec. IIA2). The working equation developed to correlate the experimental results is

$$\frac{S}{ds/dt} = A + \frac{S}{eI}\left[1 + \frac{1}{2 - e^{-kt}}\right] \quad (1)$$

where
- S = concentration of microorganisms
- t = time of irradiation, min.
- $A = k_2[H]/k_3 eI$
- k = probability factor in Poisson distribution formula
- eI = rate of energy absorption in $H_2O \xrightarrow{eI} OH + H$
- k_2 = rate constant for $H + OH \xrightarrow{k_2} H_2O$
- k_3 = rate constant for $OH + S \xrightarrow{k_3} K$

where K = an inactivated microorganism

b. *Protective Substances.* On the assumption that the protective action is an effect resulting from the reduction in potency of the oxidizing radiodecomposition products of water, the following modified version of Eq. (1) can be developed:

$$\frac{S}{ds/dt} = A + A'[P]^n + S/eI\left[1 + \frac{1}{2 - e^{-kt}}\right] \quad (2)$$

where
- $A' = k_7/k_3 eI$
- k_7 = the rate constant for $OH + P \xrightarrow{k_7} P'$

where
- P = concentration of protective substance
- P' = oxidized species of P
- n = constant

Values of A' were determined by experiments where the cells were irradiated in the phosphate buffer and in various nutrient-broth solutions (Fig. 3a).

2. Correlation of the Variables

Equation (2) can be integrated since S equals So at t equals to, and at high doses $\left[1 + \dfrac{1}{2 - e^{-kt}} \right] = 1.5$ and $So \gg S$.

$$t = \ln So/S[A + A'[P]^n] + 1.5 So/ei \qquad (3)$$

Equation (3) is ideally suited for the construction of a nomogram which could be used for calculating the throughput of a radiation processing plant under any operating conditions likely to be encountered. Figure 4 demonstrates what such a nomogram would look like. To use this one must first experimentally determine So. Then the values for So/eI are obtained from Fig. 4a. So/S, the reduction factor, is then chosen, and,

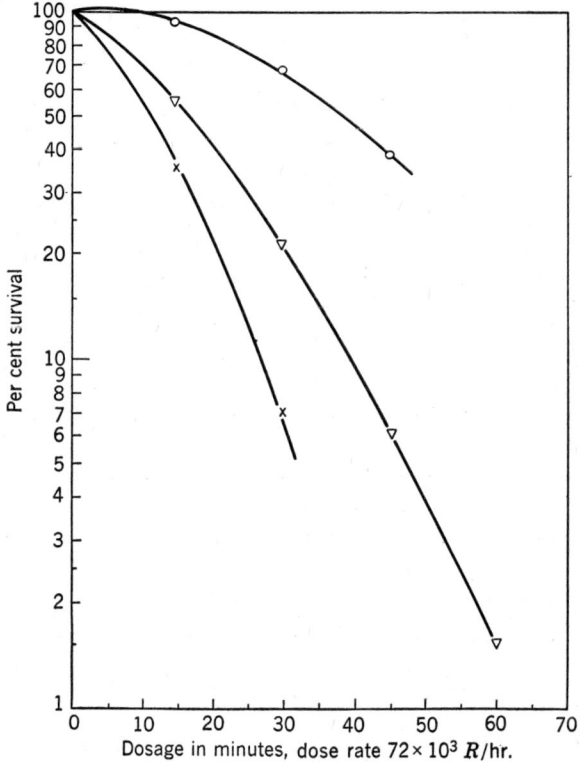

FIG. 3. Percentage of bacteria surviving vs. radiation dose, broth irradiation.

○-$S_0 = 172 \times 10^6$ — Broth dilution = $\frac{1}{5}$
▽-$S_0 = 152 \times 10^6$ — Broth dilution = $\frac{1}{50}$
×-$S_0 = 260 \times 10^6$ — In buffer

depending on the substrate, the factor $A + A'P^n$ is selected and used to obtain the dosage required (Sec. IVCb and IVCc). Figure 4 cannot be used for design purposes as it stands since it has been constructed from data obtained with *E. coli*. A more useful nomogram would be calculated with data obtained with more radioresistant organisms in a variety of media; however, the data necessary are unavailable.

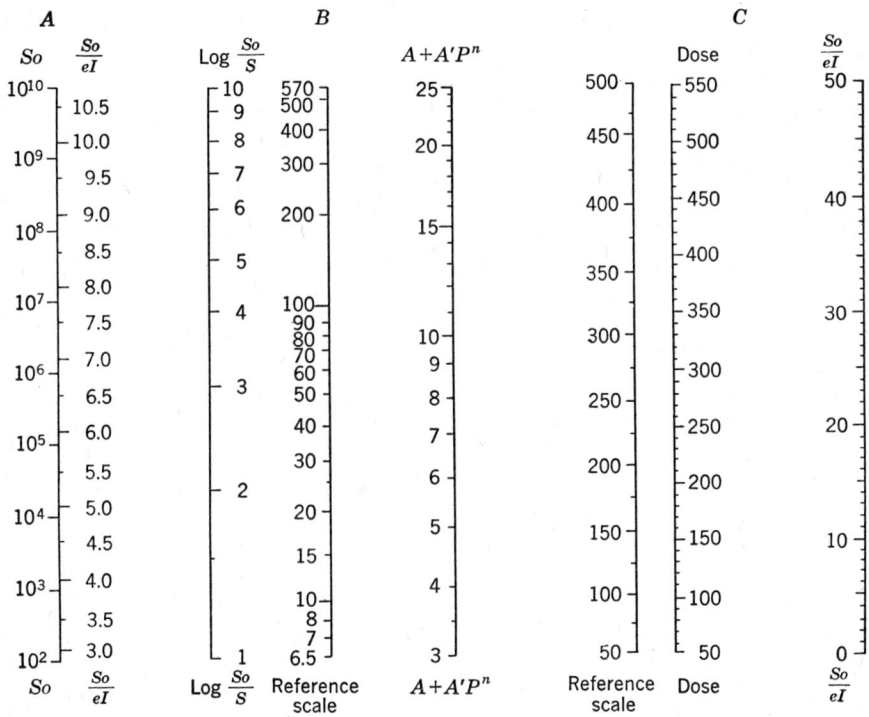

FIG. 4. Nomograph for irradiator design.

3. *Contaminants*

The final problem to consider is that owing to attenuation the radiation dosage throughout the sample, and consequently the concentration of contaminants, will not be uniform. A method for doing this is to make dosage measurements throughout the absorber by use of a technique such as has been described by Henley (H13).

Using cross plots of the depth-dose curve and S/S_o vs. dose (from the nomogram) one can calculate the concentration of contaminants.

Acknowledgment

The authors are grateful to Radiation Applications Inc., for permission to use material published in their newsletter *Radiation Applications*. Much of Section IV is due to Professor E. L. Gaden, Jr.

We would also like to acknowledge the help of Miss Lucy Palache whose work on this manuscript was indeed beyond the call of duty.

REFERENCES

A1. Allen, A. O., *Ann. Rev. Phys. Chem.* **3**, 57 (1952).
A2. Allen, A. O., *Discussions Faraday Soc.* **12**, 81 (1952).
A3. Allen, A. O., *Radiation Research* **1**, 85 (1954).
A4. Allen, A. O., Hochanadel, C. J., Ghormley, J. A., and Davis, T. W., *J. Phys. Chem.* **56**, 575 (1952).
A5. Allsopp, C. G., *Trans. Faraday Soc.* **40**, 70 (1944).
A6. Allyea, H. B., *J. Am. Chem. Soc.* **52**, 2743 (1930).
A7. U. S. Patent 924, 284, to Mr. F. Smith, Assigned to American Machine and Foundry Company (1909).
A8. American Meat Institute Foundation, Contract No. AT (11-1)-227, Summary Report (1954).
A9. Anderson, E. H., *J. Bacteriol.* **61**, 389 (1951).
B1. Baker, V. H., Tabaoda, O., and Wynant, D. E., *Agr. Engr. Pt. I* **34**, 755 (1953).
B2. Ballantine, D. S., *Modern Plastics Tech. Sect.* **33**, No. 31, 131 (1954).
B3. Ballantine, D. S., Brookhaven Natl. Lab. 294 (T-50) March (1954); Brookhaven Natl. Lab. 317 (T-53) Oct. (1954).
B4. Barr, N. F., and King, C. G., *J. Am. Chem. Soc.* **76**, 5565 (1954).
B5. Barrett, J. W., Budd, B. L., Roberts, R., and Roebuck, D. S. P., *Atomic Energy Research Establishment Circ.* **1231**, 73 (1953).
B6. Bateman, L., *Quart. Revs. (London)* **8**, 147 (1954).
B7. Bateman, L., Gee, G., Morris, A. L., and Watson, W. F., *Discussions Faraday Soc.* **10**, 250–59 (1951).
B8. Bell, E. R., Raley, J. H., Rust, F. F., Seubold, F. H., and Vaughn, W. E., *Discussions Faraday Soc.* **10**, 242–9 (1951).
B9. Bernstein, I. A., Farmer, E. C., Rothschild, W. G., and Spalding, F. F., *J. Chem. Phys.* **21**, 1303 (1953).
B10. Biagini, C., *Nature* **172**, 868 (1953).
B11. Brasch, A., and Huber, W., *Science* **105**, 112 (1947).
B12. Brasch, A., Huber, W., Friedman, V., and Taub, F. B., *Proc. Rudolf Virchow Med. Soc.* **8**, 3 (1949).
B13. Breger, I. A., American Petroleum Institute, Project 43c (1947).
B14. Bretton, R. D., Hayward, J. C., and Shair, K. A., New York Operations Office 3312, Progr. Repts. (1952–54).
B15. Brownell, L. E., *et al.*, Progr. Repts. 1–7, Contract No. AT (11-1)-162, University of Michigan (1952–55).
B16. Brownell, L. E., Progr. Rept. No. 6, 196 (1954).
B17. Brownell, L. E., Progr. Rept. No. 6, 130 (1954).
B18. Brownell, L. E., Progr. Rept. No. 6, 132 (1954).
B19. Brownell, L. E., Progr. Rept. No. 6, 135 (1954).
B20. Brownell, L. E., Progr. Rept. No. 7, 138 (1954).
B21. Burton, M., *J. Phys. & Colloid Chem.* **51**, 611 (1947).
B22. Burton, M., *Ann. Rev. Phys. Chem.* **1**, 113 (1950).
B23. Busse, W. F., and Daniels, F., *J. Am. Chem. Soc.* **59**, 3271 (1928).
C1. Chamberlain, W., Proceedings of Stanford Research Institute Atomic Forum Meeting, San Francisco, April 1955 in press.

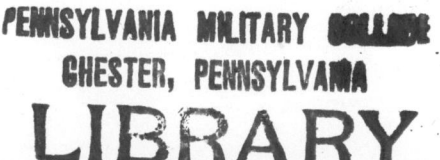

C2. Chapiro, A., *J. chim. phys.* **51**, 165 (1954).
C3. Charlesby, A., *Nucleonics* **12**, No. 6, 18 (1954).
C4. Chilton, G. H., *Chem. Eng.* **57**(4) (1950).
C5. Clifcorn, L. E., *Nucleonics* **13**, No. 1, 39 (1955).
C6. Controulis, J., Lawrence, C. A., and Brownell, L. E., *J. Am. Pharm. Assoc. Sci. Ed.* **43**, 65 (1954).
C7. Crean, L. E., Issacs, P. J., Weiss, B. J., and Fahnoe, F., *Nucleonics* **11**, No. 12, 32 (1953).
C8. Crowther, J. A., Liebman, H., and Jones, R., *Phil. Mag.* [7] **29**, 391 (1940).
D1. Dainton, F. S., *Brit. J. Radiol.* **24**, 428 (1951).
D2. Dainton, F. S., and Collinson E., *Ann. Rev. Phys. Chem.* **2**, 99 (1951).
D3. Dale, W. M., and Davis, J. V., *Nature* **163**, 64, (1949).
D4. Dale, W. M., Davis, J. V., and Meredith, W. J., *Brit. J. Cancer* **3**, 31 (1949).
D5. Dallyn, S. L., Sawyer, R. L., and Sparrow, A. H., *Nucleonics*, **13**, No. 4, 48 (1955).
D6. Davidson, W. L., and Geib, G. J., *J. Appl. Phys.* **19**, 427 (1948).
D7. Dessauer, F., *Z. Physik* **12**, 8 (1923).
D8. Dewey, D. R., Nygard, J. C., and Kelliher, M. G., *Nucleonics* **12**, No. 12, 42 (1954).
D9. Doede, C. M., and Walker, C. A., *Chem. Eng.* **62**, 159 (1955).
D10. Duffy, D., *Nucleonics* **11**, No. 10, 8 (1953).
D11. Dunn, C. G., *Bacteriol. Proc.* p. 19 (1952).
D12. Dunn, C. G., Campbell, W. L., Fram, H., and Hutchins, A., *J. Appl. Phys.* **19**, 605 (1948).
E1. Ebert, M., and Alper, T., *Nature* **173**, 987 (1954).
E2. Edwards, R. B., Peterson, L. J., and Cumings, D. G., *Food Technol.* **8**, 284 (1954).
E3. Engel, S., M. S. Thesis, Chemical Engineering, Columbia University (1953).
E4. Errera, J., and Henri, V., *J. phys. radium* **7**, 225 (1926).
E5. Essex, M., *J. Phys. Chem.* **58**, 42 (1954).
F1. Forssburg, A., *Nature* **173**, 308 (1947).
F2. Foster, F. C., Dewey, D. R., and Gale, A. J., *Nucleonics* **11**, No. 11, 14 (1953).
F3. Fricke, H., Hart, E. J., and Smith, H. P., *J. Chem. Phys.* **6**, 229 (1938).
F4. Fricke, H., *Ann. N.Y. Acad. Sci.* **59**, 567 (1955).
G1. Gaden, E. L., and Henley, E. J., *J. Bacteriol.* **65**, 727 (1953).
G2. Gaden, E. L., Progress Report, Contract AT (30-1)-1186, Columbia University, 1955 in press.
G3. Gadye, G. R., and Allibone, T. E., *Proc. Roy. Soc.* **A130**, 346 (1931).
G4. Garrison, W. M., Haymond, H. R., Morrison, D. C., Weeks, B. W., and Gile-Melchert, J., *J. Am. Chem. Soc.* **75**, 2459 (1953).
G5. Gastaldi, C., and Batt, G., *Immunology* **40**, 257 (1949).
G6. Gevantman, L. H., and Williams, R. R., Jr., *J. Phys. Chem.* **56**, 587 (1952).
G7. Goldblith, S. A., Ph.D. Thesis, Massachusetts Institute of Technology, 1949.
G8. Goldblith, S. A., and Proctor, B. E., *Nucleonics* **5**, No. 2, 50 (1949).
G9. Goldblith, S. A., Proctor, B. E., Davison, S., Kan, B., Bates, C. J., Oberle, E. M., Karel, M., and Lang, D. A., *Food Research* **18**, 659 (1953).
G10. Gomberg, H. J., Gould, S. E., Nehemias, J. V., and Brownell, L. E., *Nucleonics* **12**, No. 5, 38 (1954).
G11. Gomberg, H. J., Gould, S. E., Nehemias, J. V., and Brownell, L. E., *Food Eng.* **26**, (9), 78 (1954).

G12. Gray, L. H., *Radiation Research* **1**, 189 (1954).
H1. Haissinsky, M., *Compt. rend.* **233**, 1192 (1951).
H2. Hannan, R. S., Unpublished Report, Low Temperature Research Station, Interim Record No. 2 (1954).
H3. Hannan, R. S., *Food Sci. Abstr.* **26**, 121 (1954).
H4. Harmer, D., Ph.D. Thesis, University of Michigan, 1955.
H5. Harrington, A. C., Shaver, R. G., and Sorenson, C. W., *Nucleonics* **11**, No. 9, 34 (1953).
H6. Hart, E. J., *J. Am. Chem. Soc.* **74**, 4174 (1952).
H7. Hart, E. J., *J. Am. Chem. Soc.* **76**, 4198 (1954).
H8. Hart, E. J., *Ann. Rev. Phys. Chem.* **5**, 139 (1954).
H9. Hassett, C. C., and Jenkins, D. W., *Nucleonics* **10**, No. 12, 42 (1952).
H10. Henley, E. J., and Miller, A., *Nucleonics* **9**, No. 6, 63 (1951).
H11. Henley, E. J., *Nucleonics* **11**, No. 10, 41 (1953).
H12. Henley, E. J., and Gaden, E. L., Proc. Conference on Nuclear Engineering, University of California, Berkley, 1953.
H13. Henley, E. J., Ph.D. Thesis, Columbia University, Dept. of Chemical Engineering, 1953.
H14. Henley, E. J., Nuclear Engineering, Part III, *Chem. Eng. Progr. Symposium Ser. No. 13*, **50**, 66 (1954).
H15. Henley, E. J., *in* "Encyclopedia of Chemical Technology" (R. Kirk and D. Othmer, eds.), Vol. 12, p. 905. Interscience, New York, 1954.
H16. Henley, E. J., and Selke, W. A., *Natl. Council for Stream Improvement Tech. Bull.* **76**, 32 (1955).
H17. Henley, E. J., *Nucleonics* **12**, No. 9, 62 (1954).
H18. Henley, E. J., and Schwartz, J., *J. Am. Chem. Soc.* **77**, 3167 (1955).
H19. Henley, E. J., *Modern Plastics* **32**(7), 99 (1955).
H20. Hochanadel, C. J., *J. Am. Chem. Soc.* **56**, 587 (1952).
H21. Hollaender, A., *Federation Proc.* **9**, 61 (1950).
H22. Hollaender, A., and Stapleton, G. E., *Brit. J. Radiol.* **27**, 117 (1953).
H23. Hollaender, A., Stapleton, G. E., and Martin, F. L., Oak Ridge Natl. Lab. 832 (1951).
H24. Horsley, R. J., Johns, H. E., and Halsam, R. N. H., *Nucleonics* **11**, No. 2, 28 (1953).
H25. Huber, W., *Naturwissenschaften* **38**, 21 (1951).
H26. Huber, W., Brasch, A., and Waly, A., *Food Technol.* **7**, 109 (1953).
H27. Hui, K. L., Keefer, E. B. C., Deterling, R. A., Parshley, M. S., Humphryes, C. H., and Glenn, F., *Surgical Forum of the Am. Coll. Surgeons* **225** (1951).
J1. Jeppson, M. R., and Post, R. F., Proc. Conference on Nuclear Engineering, University of California, Los Angeles, p. C-15, April (1955).
J2. Johnson, G. R. A., and Weiss, J., *Chemistry & Industry* p. 355 (1955).
J3. John LaBatt, Ltd., Canada, News Release (1953).
K1. Kaufman, B. P., Hollaender, A., and Gay, H., *Genetics* **31**, 349 (1946).
K2. Keller, M., and Weiss, J., *J. Chem. Soc.* p. 1247 (1951).
K3. Knowlton, J. A., Mahn, G. R., and Ranftl, J. W., *Nucleonics* **11**, No. 11, 64 (1953).
K4. Kruger, F., *Physik. Z.* **13**, 1040 (1912).
K5. Kung, H. C., Gaden, E. L., and King, C. G., *Agr. Food Chem.* **1**, 142 (1953).
L1. Langendorff, H., *Z. Naturforsch* **8**, 117 (1953).
L2. Lawton, E. J., Bellamy, W. D., Hungate, R. E., Bryant, M. P., and Hall, E., *Science* **113**, 380 (1951).

L3. Lawton, E. J., Bellamy, W. D., Hungate, R. E., Bryant, M. P., and Hall, E., *Tech. Assoc. Pulp and Paper Ind.* **34**, No. 12 (1951).
L4. Lawton, E. J., and Bellamy, W. D., *Nucleonics* **12**, No. 4, 54 (1954).
L5. Lawton, E. J., Bueche, A. M., and Balwit, J. S., *Nature* **172**, 76 (1953).
L6. Lawrence, C. A., Progr. Rept. No. 2, Contract No. AT (11-1)-162, University of Michigan (1954).
L7. Lea, D. E., "Actions of Radiation on Living Cells," 2nd ed., p. 71. Cambridge U. P., New York, 1955.
L8. Lea, D. E., "Actions of Radiation on Living Cells," 2nd ed., p. 71. Cambridge U. P., New York, 1955.
L9. Lea, D. E., Haines, R. B., and Coulson, C. A., *Proc. Roy. Soc.* **B120**, 47 (1936).
L10. Lehman, A. J., and Lang, E. P., *Nucleonics* **12**, No. 1, 52 (1954).
L11. Lewis, J. G., Ph.D. Thesis, Engineering Research Institute, University of Michigan, Ann Arbor, Michigan (1954).
L12. Lind, S. C. "The Chemical Effects of Alpha Particles and Electrons," 2nd ed. Chemical Catalog Company, New York, 1928.
L13. Lind, S. C., and Bardwell, D. C., *Science* **60**, 365 (1924).
L14. Livingston, R. S., and Lind, S. C., *J. Am. Chem. Soc.* **58**, 612 (1936).
L15. Loebl, H., Stein, G., and Weiss, J., *J. Chem. Soc.* p. 1704 (1950).
L16. Loebl, H., Stein, G., and Weiss, J., *J. Chem. Soc.* p. 405 (1951).
L17. Loftness, R. L., Proc. Conference on Nuclear Engineering, University of California, Los Angeles, p. C-1 (1955).
M1. McDonell, W. R., U. S. Atomic Energy Commission Document, UCRL-1378, July (1951).
M2. McElhinney, J., Hanson, A. O., Becker, R. A., Duffield, R. B., and Diven, B. C., *Phys. Rev.* **75**, 542 (1949).
M3. Manion, J. P., and Burton, M., *J. Phys. Chem.* **56**, 560 (1952).
M4. Mead, J. F., U. S. Atomic Energy Commission Document, UCLA-160, Oct. (1951).
M5. Manowitz, B., *Nucleonics* **9**, No. 2, 10 (1951).
M6. Manowitz, B., Nucleonics **11**, No. 10, 18 (1953).
M7. Manowitz, B., *Chem. Eng. Progr. Symposium Ser. No. 12*, **50**, 203 (1954).
M8. Mead, J. F., *Science* **115**, 470 (1952).
M9. Meeker, I. A., *Science* **114**, 283 (1951).
M10. Miller, A. A., Lawton, E. J., and Balwit, J. S., *J. Polymer Sci.* **14**(77), 503 (1954).
M11. Miller, C. W., *J. Brit. Inst. Radio Engrs.* **14**(12), 637 (1954).
M12. Moos, W. G., *Nucleonics* **12**, No. 5, 46 (1954).
M13. Morgan, B. H., and Reed, J. M., *Food Research* **19**, 357 (1954).
M14. Moriarty, J. H., Progr. Rept. III, Dept. Food Technol., Mass. Inst. Technol. (1950).
P1. Patt, H. M., *Physiol. Revs.* **33**, 35 (1953).
P2. Phillips, G. O., *Nature* **173**, 1044 (1954).
P3. Platzman, R., *Radiation Research* **2**, 1 (1955).
P4. Pollard, E., Buzzell, A., Jeffreys, C., and Forro, F., *Arch. Biochem. and Biophys.* **33**, 9 (1951).
P5. Pollister, B. H., and Mead, J. F., *J. Agr. Food. Chem.* **2**, 199 (1954).
P6. Pratt, G. B., and Eklund, O. F., *Quick Frozen Foods* **16**(10), 50 (1954).
P7. Prevost-Bernas, A., Chapiro, A., Cousin, C., Landler, Y., and Magat, M., *Discussions Faraday Soc.* **12**, 98 (1952).
P8. Price, F. B., Bellamy, W. D., and Lawton, E. J., *J. Phys. Chem.* **58**, 821 (1954).
P9. Proctor, B. E., and Goldblith, S. A., *Nucleonics* **3**(2), 32 (1948).

P10. Proctor, B. E., and Goldblith, S. A., *Nucleonics* **5**(3), 56 (1949).
P11. Proctor, B. E., Goldblith, S. A., and Nickerson, J. T. R., *Refrig. Eng.* **58**, 375 (1950).
P12. Proctor, B. E., and Goldblith, S. A., *Advances in Food Research* **3**, 119 (1951).
P13. Proctor, B. E., and O'Meara, J. P., *Ind. Eng. Chem.* **43**, 718 (1951).
P14. Proctor, B. E., and Goldblith, S. A., Tech. Rept. No. 1, Contract AT (30-1)-1164, Mass. Inst. Technol. (1952).
P15. Proctor, B. E., Joslin, R. P., Nickerson, J. T. R., and Lockhart, E. E., *Food Technol.* **7**, 291 (1953).
R1. Radiation Sterilization of Foods, *Agr. Food Chem.* **2**, 1260 (1954).
R2. Rigg, T., Stein, G., and Weiss, J., *Proc. Roy. Soc.* **A211**, 375 (1952).
R3. Robinson, R. F., *Food Technol.* **8**, 191 (1954).
R4. Robinson, R. F., Overbeck, R. C., and Porter, F. E., *Coffee and Tea Inds.* **77**, No. 1, 61 (1953).
R5. Rosenblum, C., *J. Phys. & Colloid. Chem.* **52**, 474 (1948).
S1. Samuel, A. H., and Magee, J. L., *J. Chem. Phys.* **21**, 1080 (1953).
S2. Weber, E. N., Forsyth, P. F., and Schuler, R. H., *Radiation Research* **3**, 68 (1955).
S3. Schuler, R. H., and Allen, A. O., *J. Am. Chem. Soc.* **77**, 507 (1955).
S4. Schuler, R. H., and Allen, A. O., *J. Chem. Phys.* in press.
S5. Schulman, James H., NRL-Memo-266, Feb. (1954).
S6. Schwarz, H., *J. Am. Chem. Soc.* **77** in press.
S7. Schmitz, L., and Lawton, E. J., *Science* **113**, 718 (1951).
S8. Schweigert, B. E., *Food Eng.* **26**(9), 70 (1954).
S9. Seaman, J. F., Merrill, A. M., and Lawton, E. J., *Ind. Eng. Chem.* **113**, 718 (1951).
S10. Seitzer, W. H., and Tobolsky, A. V., OOR No. 1073, Tech. Repts. 3 and 4 (1954).
S11. Selke, W. A., Czikk, A., and Dempsey, J., NYO 3330, Columbia University, Department of Chemical Engineering (1955).
S12. Sheppard, C. W., and Whitehead, W. L., *Bull. Am. Assoc. Petroleum Geologists* **30**, 32 (1946).
S13. Sheppard, C. W., and Burton, V. L., *J. Am. Chem. Soc.* **68**, 1636 (1946).
S14. Sparrow, A. H., and Christensen, E., *Nucleonics* **12**, No. 8, 16 (1954).
S15. Stapleton, G. E., and Hollaender, A., AECU 1892, Biology Division, Oak Ridge Natl. Lab. (1951).
S16. Stapleton, G. E., and Hollaender, A., *J. Bacteriol.* **63**, 805 (1952).
S17. Stapleton, G. E., Billeni, D., and Hollaender, A., *J. Cellular Comp. Physiol.* **41**, 345 (1953).
S18. Stein, G., and Weiss, J., *J. Chem. Soc.* 3256 (1949).
S19. "Statistical Abstracts of the U. S.," p. 956. U. S. Govt. Printing Office, Washington, D. C., 1952.
S20. Sun, K. H., *Modern Plastics* **32**, No. 1, 141 (1954).
S21. Swallow, A. J., *Radiation Research* **1**, 570 (1954).
S22. Swanson, C. P., *Genetics* **29**, 61 (1944).
S23. Swanson, C. P., and Yost, H. T., *Proc. Natl. Acad. Sci. (U.S.)* **37**, 796 (1951).
S24. Symposium on the Chemistry of Biological After-Effects of Ultra-Violet and Ionizing Radiations, Brit. J. Radiol., 27, 36–80, 117–144 (1954).
T1. Tarpley, W., Yudis, M., Manowitz, B., Horrigan, R. V., and Weiss, J., *Ind. Eng. Chem.* **46**, 1458 (1954).
T2. Taylor, P., *Arch. Biochem.* **16**, 19 (1948).

T3. Thompson, T., *Univ. Calif. Radiation Lab*, 1910 (1952).
W1. Warburton, P., *Brit. J. Radiol.* **14,** 30 (1951).
W2. Weiss, J., *Nature* **169,** 460 (1952).
W3. Weiss, J., *Nucleonics* **10,** No. 7, 28 (1952).
W4. Weiss, J. Allen, A. O., and Schwarz, H., International Conference on the Peaceful Uses of Atomic Energy, Geneva (1955).
W5. Whitehead, W. L., Goodman, C., and Breger, I. A., *J. Chem. Phys.* **48,** 184 (1951).
W6. Wild, W., *Agr. Food Chem.* **2,** 224 (1954).
W7. Wild, W., and Wright, J., Atomic Energy Research Establishment C/R 1231, p. 3 (1953).
W8. Wood, T. H., *Arch. Biochem. and Biophys.* **52,** 157 (1954).
W9. Wourtzel, E. E., *J. phys. radium* **6,** 1 (1920).
W10. Wüst, O., "Preserving Food," *French Patent* 701,302 (1930).
Z1. Zirkle, R. E., *Am. J. Roentgenol. Radium Therapy* **35,** 230 (1936).

Author Index

Numbers in parentheses are reference numbers and are included to assist in locating references in which the authors' names are not mentioned in the text. Numbers in italics indicate the page on which the reference is listed.

A

Addoms, J. N. 15, 16, 51, 59(M3), *73*, *75*
Agoston, G. A., 83(A1), 84(A1), 86(A1), 104(A1), *150*
Alder, B. J., 192, *237*
Alexander, P., *376*
Allen, A. O., 374(S4, W4), 378(A4), 379, 382, 386(A3), 387(A2), *420*, *424*, *425*
Allibone, T. E., 387(G3), *421*
Allsopp, C. G., 387(A5), *420*
Allyea, H. B., *420*
Alper, T., 405(E1), *421*
Alter, H. W., 310, *328*
Alves, G. E., 84(A3), 86(A3), 87(A3), 88(A3), 94(A3), 96, 103(A3), 106(A3), 140, 141(A3), 142(A3), 143(A3), *150*
Ambrose, H. A., 142, 143(A4), *150*
Anderson, E. H., 404, *420*
Anderson, H. M., 322(F1), *328*
da C. Andrade, E. N., 87, *150*
Arnold, J. H., 190, 198, 220, 222, 223(A3), *234*
Aspden, R. L., *73*
Auer, P. L., 182, *234*, 268, *285*
Austin, J. M., 144(S8), *153*

B

Babbitt, H. E., 91, 94(C1), 103(C1), *151*
Babbitt, J. D., 175, 182(B1), *234*
Backus, J. W., 340(B1), *366*
Bacq, Z. M., *376*
Badger, W. L., 61, *76*, 79, *150*
Bagotskaya, I. A., 268, *285*
Baker, A. W., 344, *366*
Baker, E. M., 61, *76*
Baker, V. H., 411(B1), *420*

Bakhmeteff, B. A., 247(B2), 250(B2), *285*
Ballantine, D. S., 390, 391(B2, B3), 392(B3), 393(B3), *420*
Balwit, J. S., 389(L5), 394(L5, M10), *423*
Banchero, J. T., *73*
Bardwell, D. C., 391(L13), 393, *423*
Barker, G. E., *73*
Baron, T., 219, *234*
Barr, N. F., 383, *420*
Barrer, R. M., 159, 198, 208, 220, 228, *234*
Barrett, J. W., 388, 396(B5), *420*
Barson, N., 323(B1), 325, *328*
Bartholomew, R. N., 266, *285*
Bashe, C. J., 337(B3), *366*
Basset, A. B., 195, *234*
Batchelor, G. K., 243, 244, 245(B6), 280(B6), 281, *285*
Bateman, L., 385, *420*
Bates, C. J., 402(G9), *421*
Batt, G., 400(G5), *421*
Baumann, G. P., 106(W7), 126(W7), *153*
Beach, A. F., 346(B4), *366*
Beatty, K. O., Jr., 65(P1), *76*
Becker, E. W., 193, *234*
Becker, R., 23, *73*
Becker, R. A., 402(M2), *423*
Beecher, N., 62, *73*
Belaga, M. W., 321(B2), *328*
Belcher, H. V., 86(B2), 110(B2), *150*
Bell, E. R., 385, *420*
Bell, S., 267, *285*
Bellamy, W. D., 394(L3), 395(P8), 401(L4), 402(L4), 404, *422*, *423*
Benedict, M., 276, *285*
Berkeley, E. C., 337(M1), *367*
Bernard, R. A., 219, *234*
Bernath, L., 23, 30, 33, 49, 55(B5), 71, *73*
Bernstein, I. A., 391(B9), *420*
Berry, C. E., 343, *366*

Berry, V. J., 252(S7), 259(S7), 260(S6), 261(S7), 262(S7), *288*
Bestul, A. B., 86(B2), 110(B2), *150*
Beutler, J. A., 340, 350, *366*
Beyer, C. E., 118, *150*
Beyer, G. H., 322(C3), 323(B1), 325, *328*
Biagini, C., 401(B10), *420*
Bierlein, J. A., 227(B9), *234*
Bigelow, J. E., 321(B2), *328*
Billeni, D., 404(S17), *424*
Binder, R. C., 94(B5), 106(B5), *150*
Bird, R. B., 34(H5), 46(H6), 75, 158(B11, B12, H11), 162(B11), 164(B10), 165(H11), 166(B11), 167(H11), 168(H11), 169(B11, H11), 170(H11), 173(H11), 176(H11), 177(H11), 178(H11), 182(B12, H11), 183(H11), 184(H11), 185(B12, H9–H11), 186(B11, H11), 187(B12, H11), 188(H8, H11), 189(H11), 190(B16, B17), 191(H11), 192(H11), 193(B13), 197(H11), 227(H11), 229(B11, B12, H11), 230(H11), *234*, *235*, *236*, 280(B10, H1), *285*, *286*
Bishop, J. F., 344(B7), *366*
Black, S. A., 114, *152*
Bliss, H., 21, *75*
Blocker, H. G., 89(O4), *152*
Boelter, L. M. K., 180, 217, *234*
Boll, R. H., *73*
Bonilla, C. F., 15(C2), 20, 22, 52, 54(B7), 55(B7), 60, 65(B6), 66(B6), *73*, *74*, 121(B6), 124, 131(B6), 132, 134, *150*
Booth, A. D., 282(B11), *285*
Booth, H. V., 282(B11), *285*
Born, M., 191, 195, *234*
Bosworth, R. C. L., 14(B8), *73*
Botzen, A., *237*
Boucher, D. F., 84(A3), 86(A3), 87(A3), 88(A3), 94(A3), 96, 103(A3), 106(A3), 140, 141(A3), 142(A3), 143(A3), *150*
Box, G. E. P., 360, *366*
Boyd, C. A., 206(B19), *234*
Brady, J. F., 342, *368*
Branch, R. E., Jr., 134, 135, 136(B7), 137, *150*
Brasch, A., 402, 403(B12), 413(B11), *420*, *422*

Breaux, D. K., 248(P3), 249(P2), 252(P3), 259(P3), 282(P2, P3), *287*
Breger, I. A., 381, 405(W5), *420*, *425*
Bretton, R. D., 391(B14), 392(B14), *420*
Brier, J. C., 267, *285*
Briggs, L. J., 30, 45, 46, 49, 50, *73*, *74*
Brink, J. C., 228, *237*
Brinkman, H. C., 164(B20), *234*
Broder, J., 3, 51, 65(M10), *75*
Brodkey, R. S., *73*
Bromley, L. A., 56(B9), 61(M9), *73*, *75*
Brooks, L. H., 182, *239*
Brough, H. W., 260(J6), *286*
Broughton, G., 88(L2), *151*
Brow, J. E., 223, *238*
Brown, F., 131(C3), 134, 136(B8), 144(B8, C3), *151*
Brown, G. A., 120, *150*
Brown, G. G., 196(B21), *234*
Brown, H., 203, *234*
Brownell, L. E., 407(B17), 410(B15, G10, G11), 411(B15), 412(B15), 413(C6), *420*, *421*
Bruijsten, J., 228, *237*
Bryant, M. P., 394(L3), *422*, *423*
Buchberg, H., 64(B16), *74*
Buchholz, W., 337(B3), *366*
Buckingham, E., 90, *151*
Budd, B. L., 388(B5), 396(B5), *420*
Bueche, A. M., 389(L5), 394(L5), *423*
Bunde, R. E., 206(B23), *234*
Burridge, K. G., 131(C3), 134, 136(B8), 144(B8, C3), *151*
Burton, M., 381(B21), 382, 393, *420*, *423*
Burton, V. L., 381, 389(S13), *424*
Busher, J. E., 94(B5), 106(B5), *150*
Busse, W. F., 387(B23), *420*
Butler, R. M., 268, *285*
Buzzell, A., 399 (P4), *423*

C

Caldwell, D. H., 91, 94(C1), 103(C1), *151*
Callaghan, E. E., 266, *285*
Campbell, W. L., 403(D12), *421*
Carl, R., *75*
Carley, J. F., 118, *151*
Carlson, G. J., 298(O2), *328*

Carman, P. C., 227, *234*
Carmichael, L. T., 273(C2), *285*
Carslaw, H. S., 159, 198, 205, 206(C2), 207(C2), 217(C2), 218(C2), 220, *234*
Casey, R. S., 344(C1), *366*
Castles, J. T., 51, 55, *74*
Cavers, S. D., 252(C3), 256(C3), 258(C3), 259(C3), *285*
Cervi, A., 121(B6), 124(B6), 131(B6), 132(B6), 134(B6), *150*
Chamberlain, W., 411, 412, *420*
Chang, P., 196, 197, *237*
Chantry, A. W., 321(C1), 322(C1), *328*
Chapiro, A., 382, 392(P7), *421*, *423*
Chapman, S., 158, 167, 168(C3), 170(C3), 176, 183, 186, 188, 189, 190, 191(C3), 193, *234*
Charlesby, A., 389(C3), 395, 396, *421*
Charnes, A., 365, *366*
Chilton, G. H., *421*
Ching, G. P. K., 114(H6), *151*
Christensen, E., 412(S14), *424*
Christiansen, E. B., 83(R10), 98(R10), 106(R10), *152*, *153*
Chu, J. C., 131(C3), 134, 144(C3), *151*
Churchill, S. W., 267(B12), *285*
Cichelli, M. T., 15(C2), 20, 52, 60, *74*
Clabaugh, M. J., 267(J7), *286*
Clark, J. A., 13(R10, R11), *76*
Clifcorn, L. E., 398(C5), *421*
Clippinger, R. F., 338, 339(C5), 340, *366*
Coahran, J. M., 314(C2), *328*
Codding, J. W., 310, *328*
Cohen, E. G. D., 190, *234*
Cohen, R. M., 322(C3), *328*
Colburn, A. P., *74*, 159, *234*, 304(C4), *328*
Cole, J., 267, *285*
Collins, F. C., 228, *235*
Collinson, E., 384(D2), *421*
Colven, T. J., 121(B6), 124(B6), 131(B6), 132(B6), 134(B6), *150*
Comings, E. W., 192, 194, *235*
Connell, W. R., 253(C5), *285*
Controulis, J., 413(C6), *421*
Cooke, J. R., 266(M7), *287*
Cooper, W. W., 365, *366*
Coplan, B. V., 309(C5), *328*

Corcoran, W. H., 252(P1), 256(P1), 259(P1), *287*
Corrsin, S., 242, 245(C7), 246, *285*
Corty, C., 7, 8, 54(C4), 56, 57(C4), 58(C4), *74*
Costich, E. W., 120(R9), *152*, 295(R6), 296(R6), 297(R6), *329*
Coulson, C. A., 401(L9), *423*
Coulson, J. M., 159, *235*
Cousin, C., 392(P7), *423*
Coutor, C., 325, *328*
Cova, D. R., 268(C8), *285*
Cowling, T. G., 158, 167, 168(C3), 170(C3), 176, 183, 186, 188, 189, 191(C3), *234*, *235*
Crawford, B., Jr., 159, *237*, 255(K7), 261(K7), 267, 269(K7), 276(K7), *286*
Crean, L. E., *421*
Crocco, L., 266, *285*
Crowther, J. A., 396(C8), *421*
Cryder, D. S., 15(C6), 21, 51, 65(C6), *74*
Cumings, D. G., 401(E2), *421*
Curtiss, C. F., 34(H5), 46(H6), *75*, 158, 162(B11, C12), 165(C11, C13, H11), 166(B11, C11), 167(C11, H11), 168(H11), 169, 170(H11), 173(H11), 176(H11), 177, 178(H11), 182(B12, H11), 183, 184(H11), 185(B12, H11), 186(B11, H11), 187(B12, H11), 189(H11), 191, 192(H11), 197(H11), 206, 227, 229, 230(H11), *234*, *235*, *236*, *238*, 280(B10, H1), *285*, *286*
Czikk, A., 387(S11), 388(S11), 389(S11), *424*

D

Dainton, F. S., 380(D1), 384, *421*
Dale, W. M., 405(D4), *421*
Dalla Valle, J. M., 115, 121(O5), 122, 124, 125(O5), 128, 129(O5), 130, 145, 146(W2), *152*, *153*
Dallyn, S. L., 412(D5), *421*
Daniel, F. K., 87(D1), *151*
Daniels, F., 387(B23), *420*
Davidson, J. K., 309(C5), *328*
Davidson, W. L., 395(D6), *421*

Davies, D. R., 219, *235*
Davies, J. T., 181(D2), *235*
Davies, O. L., 360, *366*
Davis, B. I., 12(M1), *75*
Davis, J. V., 405(D4), *421*
Davis, M. W., 291(D1), 310, *328*
Davis, T. W., 378(A4), *420*
Davison, S., 402(G9), *421*
Day, R. S., *75*
de Boer, J. H., 87, *153*, 158, 183, 190, *234*
de Groot, S. R., 158, 162(G12), 166(G12), 174, 176(G12), *236*, 280(D1), *286*
Deissler, R. G., 248, 249(D2), 250, 251, 252, 263, 264, 265, 276, 277, 281, 282, *286*
Dempsey, J., 387(S11), 388(S11), 389(S11), *424*
Dergerabedian, P., 10, 69(D1), *74*
Dessauer, F., 398, *421*
Deterling, R. A., 414(H27), *422*
Dew, J. E., 8, *74*, *75*
Dewey, D. R., *421*
Dillon, R. E., 107, 118(S9), *153*
Dimsdale, B., 340, *366*
Dittman, F. W., 121(W8), 124(W8), 125, 128(W8), 129(W8), *153*
Diven, B. C., 402(M2), *423*
Doane, E. P., 268(D4), *286*
Dodge, D., 113, *151*
Doede, C. M., *421*
Döring, W., 23, *73*
Dougherty, E. L., *74*
Doughty, D. L., 267(S11), *288*
Dow, M. M., 266(D5), *286*
Drake, R., Jr., 267(S11), *288*
Drew, T. B., 7, 8, *74*, 121(H5), *151*, 180, 199, 217(D3), *235*, 301, 307, *328*
Drewitz, O., 266, *286*
Drickamer, H. G., 181, 182, 193, 208, *236*, *238*, *239*, 268(C8, D4, K9), 280, *285*, *286*, *287*, *288*
Dudenbostel, B. F., Jr., 343, *366*
Duffield, R. B., 280(G2), *286*, 402(M2), *423*
Duffy, D., 397, *421*
Dufour, L., 280, *286*
Dunn, C. G., 400, 403(D12), *421*
Dunn, L. G., *286*
Durst, R. E., 106(D3), *151*

E

Eber, G., 266, *286*
Ebert, M., 405(E1), *421*
Eckert, E., 266, *286*
Edlund, M. C., 159(G5), *235*
Edwards, R. B., 401(E2), *421*
Egly, R. S., 194, *235*
Einstein, A., 195, *235*
Eisenberg, A. A., 65(B6), 66(B6), *73*
Eisenberg, M., 180, *239*
Eisenlohr, H., 325, *328*
Eklund, O. F., 402(P6), 403(P6), *423*
Ellion, M. E., 19, 20, *74*
Ellis, A., 46, *76*
Emmert, R. E., 181, 215, *235*
Engel, S., 387(E3), *421*
Engibous, D. L., 267(B12), *285*
Enskog, D., 183, 186, 188, 190, 191, 192, 193, *235*
Errera, J., 389(E4), *421*
Essex, M., 382, *421*
Everett, H. J., 120(R9), *152*, 295(R6), 296(R6), 297(R6), *329*
Eyring, H., *74*, 173(P6, S16, S17), 196, 197(G6, H12, K6), *235*, *236*, *237*, *238*

F

Fahnoe, F., *421*
Fairbanks, D. F., 223, *235*
Fallis, W., 266, *286*
Farber, E. A., 51, 56(F1), *74*
Farmer, E. C., 391(B9), *420*
Farmer, W. S., *74*
Feick, G., 322(F1), *328*
Feld, B. T., 159(F2), *235*
Fenske, M. R., 322, *328*
Fick, A., 268, *286*
Filatov, B. S., 95, *151*
Finalborgo, A. C., 15(C6), 21, 51, 65(C6), *74*
Fisher, J. C., 23, 29, 30, 35(F3), 37, 49, *74*
Fishman, N., *73*
Flynn, A. W., 298(F3), 308, *328*
Fok, S. M., 86(F2), 89(F2), 99(F2), *151*
Foreman, R. W., 169(J10), *236*
Forro, F., 399(P4), *423*
Forssburg, A., 405(F1), *421*
Forster, H. K., 13, 16, 17, 18, 19, 20, 21, 22, 59, 61, 67, 68, 69, 70, 71, *74*

AUTHOR INDEX 431

Forsyth, P. F., 382(S2), *424*
Foster, F. C., *421*
Foust, A. S., 7, 8, 51(L3), 52(L3), 53(L3), 54(C4), 56(S2), 57(C4), 58(C4), 62, 64(S2), *74, 75, 76*
Fox, J. W., 87, *150*
Fram, H., 403(D12), *421*
Frank, H. S., 159(F3), *235*
Frenkel, J., 23, 36, *74*
Fricke, H., 379(F4), 387(F3), *421*
Friedman, V., 403(B12), *420*
Frisch, H. L., 228, *235*
Fritz, W., 14(F8), 47, 64(J3), *74, 75*
Fuchs, O., 332(F1), *366*
Fürth, R., 159, 195(F13), *235*
Fujita, H., 205, 209(F8), *235*, 268(F3), *286*
Fuoss, R. M., 115, 144(H1), *151*, 159(O2), 173, *237*
Furry, W. H., 169(F10, F11), 180(F10, F11), 186(F12), 223, 227, *235, 236*

G

Gaden, E. L., 395, 401, 407, 416, *421, 422*
Gadye, G. R., 387(G3), *421*
Gale, A. J., *421*
Gamson, B. W., 192, *235*
Garner, F. H., 115(W9), *153*
Garrison, W. M., 381, *421*
Gastaldi, C., 400, *421*
Gay, H., 400(K1), *422*
Geddes, R. L., 228, *235*
Gee, G., 385(B7), *420*
Gee, R. E., 350, *366*
Geib, G. J., 395(D6), *421*
Gevantman, L. H., 382(G6), *421*
Ghormley, J. A., 378(A4), *420*
Gibbs, J. W., 254(G1), *286*
Gilbert, C. S., 28(K2), 54(K2), 65(K2), *75*
Gile-Melchert, J., 381(G4), *421*
Giller, E. B., 280(G2), *286*
Gilliland, E. R., *74*, 174(W1), 190, *235, 239*
Gillis, J., 205, *235*
Glaser, D. A., 38, 39, *74*
Glasstone, S., 28(G2), *74*, 159(G5), 196(G6), 197(G6), *235*
Glenn, F., 414(H27), *422*

Glick, J. J., 92, *152*
Goldblith, S. A., 401(P12), 402, 403(P11, P12), 406(G7, G8, P9, P10), 407(G8, P10), 415(P14), *421, 423, 424*
Goldman, K., 11, *74*
Goldstein, S., 179(G7), 230(G7), *235*
Gomberg, H. J., 410(G10, G11), *421*
Gomer, R., 209(G8), *236*
Goodman, C., 405(W5), *425*
Gordon, A. R., 159(G9), *236*
Gordon, K. F., 268(G3), *286*
Gotlieb, C. C., 339, 340, *366, 367*
Gould, S. E., 410(G10, G11), *421*
Govier, G. W., 91, 93, 102, 103(G1), 110(G1), *151*
Gray, L. H., 399, *422*
Green, H., 87(G3), 88, 94(G3), 107(G2, G3, G4), 143(G3), 144(G3), *151*
Green, H. S., 160, 191, 195, *234, 236*
Green, M. S., 158, *237*
Greenfield, M., 64(B16), *74*
Greenfield, M. L., 16(L2), *75*
Gregory, W. B., 103(G5), *151*
Grew, K. E., 169(G11), 176, 223, *236*
Griffith, P., 13(R11), 51, 55(R12), *76*
Griffith, W. L., 319(G1), 321(G1), *328*
Guggenheim, E. A., 205(G13), *236*
Guibert, A. G., 54(M6), *75*
Gunther, F. C., 10, 20, *74*

H

Haberman, W. L., 70, *74*
Hachmuth, K. H., 290, *328*
Haines, R. B., 401(L9), *423*
Haissinsky, M., 396(H1), *422*
Hall, E., 394(L3), *422, 423*
Hall, H. T., 115, 144(H1), *151*
Halsam, R. N. H., 403(H24), *422*
Hamilton, F. E., 337(H1), *367*
Hankam, E. V., 346(B4), *366*
Hannan, R. S., 398, 400, 402, 405(H2), 409(H2), 410(H2), 411(H3), 412, *422*
Hanson, A. O., 402(M2), *423*
Harmer, D., 389(H4), 390, *422*
Harned, H. S., 159(H1), *236*
Harp, W. M., 349(H2), *367*
Harper, R. C., Jr., 84(H2), *151*
Harrington, A. C., 398(H5), *422*
Hart, E. J., 382, 384(H7), 386(H6), 387(F3), *421, 422*

Harte, W. H., 83(A1), 84(A1), 86(A1), 104(A1), *150*
Hartnell, J. P., 267(J7), *286*
Haslam, G., 107(G4), *151*
Hassett, C. C., 411(H9), *422*
Haul, R. A. W., 227, *234*
Haymond, H. R., 381(G4), *421*
Hayward, J. C., 391(B14), 392(B14), *420*
Head, V. P., 107, 114, *151*
Hedstrom, B. O. A., 91, *151*
Heiny, R. L., 332(R3), *367*
Hellund, E. J., 190, *236*
Henderson, A., 365, *366*
Henley, E. J., 374(H10, H17), 386(H18), 388, 393, 401, 416, 419, *421, 422*
Henri, V., 389(E4), *421*
Hickman, K. C. D., *75*
Hicks, T. E., 291(D1), 310, *328*
Higbie, R., 213, 215, *236*
Higgins, T. J., 332(H3, H4), 347, *367*
Hill, N. E., 198, *236*
Hiller, L. A., Jr., 159(W2), 196(W2), *239*
Hirschfelder, J. O., 34(H5), 46, 75, 158, 162(B11, C12), 165(C13), 166(B11), 167(H11), 168(H11), 169, 170(H11), 173(H11), 176(H11), 177, 178(H11), 182(B12, H11), 183, 184(H11), 185 (B12, H9–H11), 186(B11, H11), 187 (B12, H11), 188, 189(H11), 191 (H11), 192(H11), 193(B13), 197 (H11, H12), 227, 229, 230(H11), *234, 235, 236, 238*, 280, *285, 286*
Hixson, A. W., 121, *151*, 301, 306, 307, *328*
Hochanadel, C. J., 378(A4), 386(H20), *420, 422*
Hollaender, A., 400(K1, S15, S16), 401 (H21, H23), 404(H22, S17), *422, 424*
Holmes, J. H., 309(S2), *329*
Hopper, G. M., 340(H5, H6), *367*
Horrigan, R. V., 403(T1), 413(T1), *424*
Horsley, R. J., 403, *422*
Hottel, H. C., 83(A1), 84(A1), 86(A1), 104(A1), *150*
Hougen, O. A., 178, 194, 205(H14), 229, *236*
Houghton, G. L., 114(H6), *151*
Householder, A. S., 347, *367*
Hovious, R. L., 334(H8), *367*
Howarth, L., 242, 245(K4), 281, *286*

Hsu, N. T., 252(C3), 256(C3), 258(C3), 259(C3), *285*
Huber, W., 402, 403(B12), 407, 413(B11), *420, 422*
Hughes, D. J., 159(H15), *236*
Hui, K. L., 414(H27), *422*
Humphryes, C. H., 414(H27), *422*
Hungate, R. E., 394(L3), *422, 423*
Hurd, C. C., 337(H10), 349(H9, H10), *367*
Hutchins, A., 403(D12), *421*
Hutchinson, E., 181(H16), *236*
Hwang, J. L., 227, *236*

I

Ibbs, T. L., 169(G11), 176, 223, *236*
Ibrahin, A. K., 106, *151*
Ijzerman, P., 268(K13), *286*
Insinger, T. H., 21, *75*
Irish, E. M., 173(S17), 197(S17), *238*
Irving, J. H., *236*
Irwin, M., 181(I3), *236*
Issacs, P. J., *421*
Ito, S., 95, *152*

J

Jaeger, J. C., 159, 198, 205, 206(C2), 207 (C2), 217(C2), 218(C2), 220, *234*
Jakob, M., 7, 8, 14(J2), 21, 22, 47, 51, 58, 61, 64(J3), 71, 75, 180, 217(J1), *236*, 259, 266, 268, *286, 288*
Jasny, G. R., 319(G1), 321(G1), *328*
Jealous, A. C., 318(J1), 319(J1), *328*
Jeans, Sir J., 182(J2), *236*
Jeener, J., 280, *286*
Jeffrey, J. O., 56(J6), *75*
Jeffreys, C., 399(P4), *423*
Jeffreys, H., 258(J4), *286*
Jeffries, Q. R., 193, *236*
Jenkins, D. W., 411(H9), *422*
Jenkins, R., 259, 260, 274, 282, *286*
Jenness, L. C., 106(D3), *151*
Jennings, A. S., 310, *328*
Jens, W. H., *75*
Jeppson, M. R., *422*
Johanson, L. W., 266(W1), *288*
Johns, H. E., 403(H24), *422*
Johnson, E. F., 158, *236*

Johnson, G. R. A., *422*
Johnson, H. A., 267, *286*
Johnson, H. F., 318(J1), 319(J1), *328*
Johnson, R. C., 332(R2, R3), *367*
Johnstone, H. F., 215(J8), 219, *234*, *236*
Jones, A. C., 159(Y1), *239*
Jones, A. L., 169(J9, J10, J11), *236*
Jones, R., 396(C8), *421*
Jones, R. C., 169(F10, F11), 180(F10, F11), 223, 227, *235*, *236*
Joslin, R. P., 400(P15), *424*
Jost, W., 159, 182(J14), 197(J13), 198, 208, 210, 220, *236*
Jurges, W., 260, *286*

K

Kahn, H., *360*, *367*
Kaiser, H. R., 325(K1), *328*
Kan, B., 402(G9), *421*
Karel, M., 402(G9), *421*
Karim, S. M., 162, *237*
Kármán, von, T., 242, 243(K5), 245(K1, K4), 247, 251, 253(K5), 255(K2), 256, 266, 281, *286*
Karr, A. E., 312(K2), 316, *328*, *329*
Katz, D. L., 51(L3), 52(L3), 53(L3), 58, 62, *75*, *76*, 266(B3), *285*
Kaufman, B. P., 400(K1), *422*
Kazakova, E. A., 20, 51, *75*
Kedem, O., 205, *235*
Keefer, E. B. C., 414(H27), *422*
Keller, A., 352, *367*
Keller, M., 389(K2), *422*
Kelliher, M. G., *421*
Kendrick, F. B., 28(K2), 54(K2), 65(K2), *75*
Kennard, E. H., 182(K2), *237*
Kennel, W. E., *75*
Keuntzel, L. E., 344, *367*
Kihara, T., 159(K4), 168(K4), 183, 187 (K5), 188, 189, *237*
Kincaid, J. F., 196(K6), *237*
King, C. G., 383, 407, *420*, *422*
King, G. W., 343, 356(K4), *367*
King, L. V., 243(K6), *286*
King, W. H., Jr., 343, *367*
Kingsley, H. A., 294(O3), 296(O3), 298 (O3), 307, *329*

Kirkwood, J. G., 159, 191, 192, 195, *236*, *237*, 255(K7), 261(K7), 267, 269 (K7), 276(K7), *286*
Kishimoto, A., 205(F9), *235*
Klein, G., 192(K10), 227, *237*
Klemm, W. A., 83(A1), 84(A1), 86(A1), 104(A1), *150*
Klinkenberg, A., 218, *237*, 268, *286*
Knowlton, J. A., *422*
Knox, K. L., 121(H5), *151*, 301, 307, *328*
Kobe, K. A., 191, *237*
Koeller, R. C., 268(K9), *286*
Kolmogoroff, A. N., 242, 244, 261(K10), 280, *286*
Korn, G. A., 282(K11), *286*, 334, *367*
Korn, T. M., 282(K11), *286*, 334, *367*
Kotani, M., 187(K5), *237*
Krajenbrink, H. J., 218, *237*, 268(K8), *286*
Kramers, H., 314(V3), *329*
Kranich, W. L., 106(W7), 121(W8), 124 (W8), 125, 126(W7), 128(W8), 129 (W8), *153*
Krauss, C. J., 268, *286*
Kreith, F., 20, *74*, *75*
Krieger, I. M., 139, 141(K1), 145(M2), *151*
Kroll, A. E., 294(M1), *328*
Kronig, R., 228, *237*, 268, *286*
Kruger, F., 387(K4), 389(K4), *422*
Kruyt, H. R., 84(K2), *151*
Kubie, E. C., 337(H1), *367*
Kuhns, P. W., 146, *153*, 164(W6), *239*
Kung, H. C., 407, *422*
Kutateladze, S. S., 61, *75*

L

Laidler, K. J., *74*, 196(G6), 197(G6), *235*
Lamb, H., 195, *237*, 270(L1), *287*
Landler, Y., 392(P7), *423*
Lang, D. A., 402(G9), *421*
Lang, E. P., 409(L10), *423*
Langendorff, H., 400, 404, *422*
Langlois, G. E., 298(V2), 299(V2), *329*
Laufer, J., 243, 244, 246, 247, 249, 258, 274(L3), 275, 281(L3), *287*
Lauwerier, H. A., 218, 228, *237*, 268(K8), *286*

Lawrence, C. A., 400, 413(C6), *421*, *423*
Lawton, E. J., 389(L5), 392(S7), 394, 395 (P8), 401(L4), 402(L4), 404, 405(S9), *422*, *423*, *424*
Lea, D. E., 398, 401(L9), 402, *423*
Lee, C. Y., 188, 200, *237*, *239*
Lehman, A. J., 409(L10), *423*
Leidenfrost, J. G., 3(L1), *75*
LeRoy, N. R., *73*
Levin, J. H., 340, *366*
Levy, S., 267(S11), *288*
Lewis, E. E., 117, *151*
Lewis, G. N., 269(L4), 276(L4), 278(L4), *287*
Lewis, J. G., 388, 391(L11), 392(L11), *423*
Lewis, W. K., 88(L2), *151*, 174(W1), *239*
Li, J. C. M., 196, 197, *237*
Liebman, H., 396(C8), *421*
Liepmann, H., 245(L5), *287*
Lin, C. C., 243(K5), 253(K5), 281(K5), *286*
Lin, C. S., 271(L6), *287*
Lind, S. C., 387(L12, L14), 388, 389(L12), 391(L13), 393, *423*
Linke, W., 21, 58, 61, *75*
Linton, W. H., Jr., 267(L7), *287*, 350, *366*
Lipkis, R., 64(B16), *74*
Lipkis, R. P., 16(L2), *75*
Liu, C., 16(L2), *75*
Livingood, J. B., 267(L8), *287*
Livingston, R. S., 387(L14), *423*
Lockhart, E. E., 400(P15), *424*
Lockhart, R. W., 116, *151*
Loebl, H., 389(L15, L16), *423*
Loeffel, W. F., 95(W6), 103(W6), 106 (W6), *153*
Loftness, R. L., *423*
London, F., 183, *237*
Long, R. B., 322, *328*
Longwell, J. P., 228, *237*
Loomis, A. G., 142, 143(A4), *150*
Lottes, P. A., *75*
Lowdermilk, W., 267, *287*
Lower, G. W., 88(L4), 89(Z1), 143(L4), *151*, *153*
Lowery, A. J., 11(W2), *76*
Lydersen, A. L., 191, *237*
Lynch, E. J., 268, 271(L9), *287*
Lynn, R. E., Jr., 191, *237*

Lynn, S., 247(L10), 271(L10), 273, 274, *287*
Lyon, R. E., 51, 52(L3), 53(L3), *75*

M

McAdams, W. H., 51, 58, 59(M3), 65, *75*, 91(M4), 112(M4), 113(M4), 116, 126 (M4), 129(M4), 132, 134(M4), *151*, 174(W1), *239*, 254, 258(M1), 259, 267(M1), *287*
McCabe, W. L., 79, *150*
McCauley, H. J., 205(H14), *236*
Macdonald, N., 337(M1), *367*
McDonell, W. R., *423*
McDonnell, J. A., 334(M5), *367*
McElhinney, J., 402, *423*
Mack, D. E., 294(M1), *328*
McMillen, E. L., 95, 112, 113, 114, *151*
McNelly, M. J., 61, *75*
Magat, M., 392(P7), *423*
Magee, J. L., 379(S1), *424*
Magnusson, K., 119, *151*
Mahn, G. R., *422*
Maire, R. E., 350, *366*
Maisel, D. S., 267(M2, M3), *287*
Manion, J. P., 381, 393, *423*
Mann, C. A., 294(M2), 295(M2), 297, 298, *328*
Manne, A. S., 365, *367*
Manning, W. R., 219, *234*
Manowitz, B., 403(T1), 413(T1), *424*
Margenau, H., 183, *237*
Maron, S. H., 139, 141(K1), 145(M2), *151*
Marshall, W. R., Jr., 208(M2), 229, *236*, *237*
Martin, F. L., 401(H23), *422*
Martinelli, R. C., 116, *151*
Marx, J. W., 12(M1), *75*
Mason, E. A., 183(M3, M4), 185(M3–M6), 186, 187(M3, M6), 188, 189, 229, *237*
Mason, J. L., 252(S7), 259(S7), 260(S6), 261(S7), 262(S7), *288*
Massey, H. S. W., 190, *237*
Matthews, F. W., 344, *367*
Matthews, T. A., II, 142(M3), *151*
Matz, W., 159, *237*
Maun, E. K., 192, *237*
Maxwell, J. C., 269(M4), *287*

Mayland, B. J., 194, *235*
Mead, B. R., 54, 65, *75*
Mead, J. F., 385, 406, *423*
Mecke, R., 344, *367*
Meeker, I. A., 414(M9), *423*
Mellon, B., 365, *366*
Meredith, W. J., 405(D4), *421*
Merk, H. J., 266, *287*
Merrill, A. M., 405(S9), *424*
Merrill, E. W., 86(M7), 86(M6), 110 (M7), 147, *151*
Meskat, W., 118, *151*
Metzner, A. B., 83(M11), 84(M10, M11), 86(M10), 89(M9, M10, M11), 90 (O6), 96, 98, 99(M11), 100(M11), 101(M11), 102(M11), 104(M11), 105 (M11), 106(M11), 107(M11), 111 (M11), 120(O6), 133(M9), 141(O6, M11), 147(M10), *152*
Meyers, J. E., 58, *75*
Michels, A. M. J. F., *237*
Milberger, E. C., 169(J11), *236*
Mill, C. C., 88(M12), *152*
Miller, A., 374(H10), *422*
Miller, A. A., 394, *423*
Miller, A. P., Jr., 86(M13), 121(M13), 123(M13), 124, 125, 127, 128, 129 (M13), *152*
Miller, C. W., 413(M11), *423*
Miller, L. B., 51, 56(M8), *75*
Miller, P. D., Jr., 65(P1), *76*
Miller, S. A., 294(M2), 295(M2), 297, 298, *328*
Milne, W. E., 347, *367*
Minden, C. S., *75*
Mitchner, M., 253(M6), *287*
Mohr, C. B. O., 190, *237*
Monaghan, R. K., 266, *287*
Montroll, E. W., 158, *237*
Mooney, M., 95, 96, 98, 114, 144, 147, *152*
Moos, W. G., 402, *423*
Morello, V. S., 291(M3), *328*
Morgan, A. I., 61(M9), *75*
Morgan, B. H., 400, 401(M13), 403 (M13), 404(M13), *423*
Mori, Y., 115, 117(M18, M19), 147, *152*
Moriarity, J. H., 402, *423*
Morrill, C. D., 334(H8), *367*
Morris, A. L., 385(B7), *420*
Morris, W. L., 343, *367*

Morrison, D. C., 381(G4), *421*
Morton, R. K., 70, *74*
Moscicki, I., 3, 51, 65(M10), *75*
Moshman, J., 355(M9), *367*
Moulton, R. W., 271(L6), *287*
Mueller, A. C., 7, 8, *74*
Murbach, P. W., 268, *285*
Murback, E. W., 182, *234*
Murphree, E. V., 266, *287*
Mysels, K. J., 83(A1), 84(A1), 86(A1), 104(A1), *150*

N

Nakagawa, Y., 61, *75*
Nathan, M. F., 192, *235*
Nehemias, J. V., 410(G10, G11), *421*
Nernst, W., 195, *237*
Newman, M., 121(S1), 123(S1), 124, 128, 129, *153*
Nickerson, J. T. R., 400(P15), 403(P11), *424*
Nikuradse, J., 248(N1), 249, 251, *287*
Nims, P. T., 335(N1), *367*
Nissan, A. H., 115(W9), *153*
Nukiyama, S., 3, 4, 43, 51, *75*
Nygard, J. C., *421*

O

Oberle, E. M., 402(G9), *421*
Offerhaus, M. J., 190, *234*
Oldroyd, J. G., 104(O1), *152*
Oldshue, J. Y., 312, 315, *328*
Olney, R. B., 294(O3), 296(O3), 298(O2, O3), 307(O3), 314(R3), *328, 329*
Olson, R. L., 198, *237*
O'Meara, J. P., 406(P13), *424*
Onsager, L., 159(O2), 169(F10), 173, 180 (F10, F11), 227, *235, 237*, 242, 269 (O1, O2), 280, 281, *287*
Ooyama, Y., 95, *152*
Opfell, J. B., *287*
Opler, A., 343, 344, *366, 367*
Orchard-Hayes, W., 366, *367*
Orr, Clyde, Jr., 89(O4), 121(O5), 122, 124, 125, 128, 129(O5), 130, 145, *152*
Othmer, D. F., 198, *237*
Ototake, N., 115, 117(M18, M19), 147, *152*

Otto, R. E., 90(O6), 120(O6), 141(O6), *152*
Overbeck, R. C., 403(R4), *424*
Overcashier, R. H., 294(O3), 296, 298 (O3), 307, *329*
Owen, B. B., *236*

P

Page, F., Jr., 248, 249(P2), 252(P1, P3), 256(P1, P3), 259(P1, P3), 282(P2, P3), *287*
Pannell, J. R., 254(P4), 260, *287*
Parshley, M. S., 414(H27), *422*
Partington, J. R., 190, *237*
Patt, H. M., 401(P1), 402, *423*
Pawlowski, J., 139, *152*
Perkins, A., 92, *152*
Perry, C. W., 22, 54(B7), 55(B7), *73*
Perry, J. H., 87, 114(P3), *152*
Perry, J. W., 344(C1), *366*
Perry, R. H., 211, *238*
Peterson, L. J., 401(E2), *421*
Petsiavas, D. N., 120, *151*
Philippoff, W., 82, 88, 141(P4), 147, *152*
Phillips, G. O., 405, *423*
Picornell, P. M., *75*
Pigford, R. L., 84(A3), 86(A3), 87(A3), 88(A3), 94(A3), 96, 103(A3), 106 (A3), 131, 137, 140, 141(A3), 142 (A3), 143(A3), *150, 152*, 159, 179, 180, 181, 190(S9), 198(S9), 200, 201 (S9), 203, 208(M2), 209, 210, 211, 214(S9), 215, 217(S9), 218(S9), *234, 235, 236, 237, 238*, 267, *288*, 304(S8) *329*
Pike, F. R., 65, *76*
Pinkel, B., 267, *287*
Pitkanen, P. H., 186(F12), *235*
Platzman, R., 377, *423*
Plesset, M. S., 46, 67, 68, *76*
Plewes, A. C., 268, *285*
Podbielniak, W. J., 323(P1, P2), 324(P3), 325(P2), *329*
Poffenberger, N., 291(M3), *328*
Pollard, E., 399(P4), *423*
Pollister, B. H., 406(P5), *423*
Pomeroy, H. H., 83(A1), 84(A1), 86(A1), 104(A1), *150*
Porter, F. E., 403(R4), *424*

Post, R. F., *422*
Powell, R. E., 173(P6), 196(P6), *238*
Powell, W. B., *286*
Prager, S., 227, *238*
Pramuk, F. S., 63, *76*
Prandtl, L., 242, 251, *287*
Pratt, G. B., 402(P6), 403(P6), *423*
Prevost-Bernas, A., 392(P7), *423*
Price, F. B., 395, *423*
Priestley, W., Jr., 343, *366*, 367
Prigogine, I., 158, *238*
Prins, J. A., 266, *287*
Proctor, B. E., 400, 401(P12), 402(G9), 403(P11, P12), 406(G7, G8, P9, P10, P13), 407(G8, P10), 415(P14), *421, 423, 424*
Pryce-Jones, J., 87, *152*
Putnam, G. L., 271(L6), *287*

R

Rabinowitsch, B., 95, 98, *152*
Rahm, D. C., *74*
Raines, J. W., 350, *366*
Raley, J. H., 385(B8), *420*
Ranftl, J. W., *422*
Rankin, S., 90(R2), 97, 118, 141(R2), 144 (R2), *152*
Rayleigh, Lord, 16, *76*
Reamer, H. H., 273(C2), *285*
Reed, J. C., 83(M11), 84(M11), 89(M11), 98, 99(M11), 100(M11), 101(M11), 102(M11), 104(M11), 105(M11), 106 (M11), 107(M11), 109(R3), 111 (M11), 141(M11), *152*
Reed, J. M., 400, 401(M13), 403(M13), 404(M13), *423*
Reiner, M., 79, 87(R4), 88, 90, 94(R5), 97, 144(R4), 147, *152*, 167, *238*
Reiss, H., 40, *76*
Reman, G. H., 314(R1, R2, R3), *329*
Resnick, H., 281(S2, S3), 282(S2, S3), *287*
Reynolds, O., 242, 245, 256, 258(R2), 274, 275, *287*
Rice, S. O., 244, *287*
Rice, W. E., 185(M5, M6), 187(M6), 189 (M6), *237*
Rich, S. R., 147(R6), *152*
Richardson, J. F., 159, *235*
Ridenour, L. N., 344(R1), *367*

Rigg, T., 383(R2), *424*
Rinaldo, P. M., 51, 59(M3), *75*
Riseman, J., 84(H2), *151*
Riwlin, R., 94(R5), *152*
Robb, W. L., 193, *238*, 280(R4), *287*
Robbers, J. A., *73*
Roberts, R., 388(B5), 396(B5), *420*
Robinson, D. B., 62, *76*
Robinson, G., 347, *368*
Robinson, R. A., 159(R3), *238*
Robinson, R. F., 403(R4), *424*
Rochester, N., 337(B3), *366*
Rock, S. M., 343(B5), *366*
Rodebush, W. H., 34, *76*
Roebuck, D. S. P., 388(B5), 396(B5), *420*
Rohsenow, W. M., 13, 14, 15, 16, 20, 21, 22, 51, 55(R12), 59, *76*
Romie, F., 54(M6), 64(B16), *74*, *75*
Rose, A., 332(R2, R3), *367*
Rosen, J. B., 228, *238*
Rosenblum, C., *424*
Rosenhead, L., 162, *237*, *238*
Roseveare, W. E., 173(P6), 196(P6), *238*
Roshko, A., 267, *285*
Roth, W., 147(R6), *152*
Rothschild, W. G., 391(B9), *420*
Rouse, H., 109(R7), *152*
Rubin, L. C., 276(B8, B9), *285*
Rumpel, W. F., 206(B19, R6), *234*, *238*
Rushton, J. H., 120, 121, *152*, 295(R4, R5, R6), 296, 297(R6), 308, 312, 315, *328*, *329*
Rust, F. F., 385(B8), *420*
Rutherford, W. M., 280(R5), *287*
Ryan, N. W., 83(R10), 98(R10), 106 (R10), *152*, *153*

S

Sachs, J. P., 296(S1), *329*
Sage, B. H., 179, *238*, 248(P3), 249(P2, S5), 250(S5), 252(C3, P1, P3, S7), 253(C5), 256(C3, P1, P3), 258(C3), 259(C3, P1, P3, S7), 260(J6, S6), 261(S7), 262(S7), 273(C2), 278(S1), 282(P2, P3), *285*, *286*, *287*, *288*
Salamone, J. J., 121(S1), 123(S1), 124, 128, 129, *153*
Salt, D. L., 97, 98(S3), 103(S2), *153*
Samuel, A. H., 379(S1), *424*

Santangelo, J. G., 4(W3), 7, 8(W3), 9 (W3), 10, *76*
Satterfield, C. N., 281, 282, *287*
Sauer, E. T., 7, *76*
Sawyer, R. L., 412(D5), *421*
Scarborough, J. B., 347, *367*
Schafer, A. C., 309(S2), *329*
Scheibel, E. G., 312(K2), 315(S3, S4), 316, *328*, *329*
Schilk, J. A., 332(R2, R3), *367*
Schlichting, H., 165, 179(S1), *238*, 248 (S4), 270(S4), 278(S4), 282, *287*
Schlinger, W. G., 179, *238*, 248(P3, S5), 249, 250(S5), 252(C3, P1, P3), 253 (C5), 256(C3, P1, P3), 258(C3), 259 (C3, P1, P3, S7), 260(S6), 261(S7), 262(S7), 282(P2, P3), *285*, *287*, *288*
Schmid, E. D., 344, *367*
Schmitz, L., 392(S7), *424*
Schnurmann, R., 103, *153*
Schottky, W. F., 158, *238*
Schrage, R. W., 181, *238*, 268, *288*, 335 (S2), *367*
Schubauer, G. B., 243, *288*
Schuler, R. H., 374(S4), 379, 382, *424*
Schuler, R. W., 266, *288*
Schulman, James H., 374(S5), *424*
Schultz-Grunow, F., 96, 119, 143, *153*
Schuurman, W., *237*
Schwartz, J., 386(H18), *422*
Schwarz, H., 374(W4), 379(S6), *424*, *425*
Schweigert, B. E., 410(S8), *424*
Schweppe, J. L., 56(S2), 62, 64(S2), *76*
Schwertz, F. A., 223, 228, *238*
Scorah, R. L., 51, 56(F1), *74*
Scott, E. J., 181, 182, 208, *238*
Scott Blair, G. W., 87(S7), 88, 97, *153*
Seaman, J. F., 405(S9), *424*
Seban, R. A., 267, *288*
Seder, L. A., 267(S17), *288*
Sege, G., 318(W2), 320(S6), 321(S6, W2), 322(S6, W2), *329*
Seifert, H. S., *286*
Seith, W., 159(S8), *238*
Seitzer, W. H., 391, 392(S10), *424*
Selke, W. A., 387(S11), 388(S11), 389 (S11), *422*, *424*
Seubold, F. H., 385(B8), *420*
Severs, E. T., 144, *153*

Shair, K. A., 391(B14), 392(B14), *420*
Shapiro, A. H., 255(S12), *288*
Shaver, R. G., 398(H5), *422*
Sheppard, C. W., 381, 389(S13), *424*
Sherwood, T. K., 159, 179, 180, 190(S9), 198(S9), 200, 201(S9), 203, 209, 210, 211, 214(S9), 215, 217(S9), 218(S9), *238*, 267, 268(G3), 271, 272, *286*, *287*, *288*, 290(S7), 304(S8), *329*
Simmonds, W. H. C., 23(S3), *76*
Sinfelt, J. H., 181, *238*
Sisko, A. W., 145(M2), *151*
Skinner, G., 249, 251, *288*
Slattery, J. C., 190, 193, 194, *238*
Smith, H. P., 387(F3), *421*
Smith, J. M., 266(S10), *288*
Smith, M. I., 306, *328*
Snider, R. F., 191, 206, *235*, *238*
Sobcov, H., 343, *368*
Sorenson, C. W., 398(H5), *422*
Soret, C., 280, 281(S19), *288*
Soroka, W. W., 334, *368*
Spalding, D. B., 228, *238*
Spalding, F. F., 391(B9), *420*
Sparrow, A. H., 412(D5, S14), *421*, *424*
Spencer, R. S., 107, 118, *150*, *153*
Spenser, G., 355, *368*
Spielman, M., 268, *288*
Spinks, J. W., 268(K12), *286*
Spotz, E. L., 185(H9, H10), 188(H8), 193 (B13), 227, *234*, *236*, *238*
Squires, L., 88(L2), *151*
Srivastava, B. N., 227(S15), *238*
Srivastava, R. C., 227(S15), *238*
Stalling, V. P., 266(S10), *288*
Stapleton, G. E., 400(S15, S16), 401 (H23), 404(H22, S17), *422*, *424*
Stearn, A. E., 173(S16, S17), 196(K6), 197(S16, S17), *237*, *238*
Stein, G., 383(R2), 389(L15, L16), 404, *423*, *424*
Stein, N., 206(B19), *234*
Steingrimsson, V. B., 206(B19), *234*
Stevens, R. F., 342, *368*
Stevens, W. E., 83(R10), 98(R10, S10), 103(S10), 106(R10), *152*, *153*
Stevenson, D. P., 197(H12), *236*
Stewart, D. W., 159(S18), *238*
Stoebe, G. W., 61, *76*

Stokes, R. H., 159(R3), 195, 205, *238*, 268(S21), *288*
Strehlow, R. A., 206(S20), *238*, 268, *288*
Strong, J. D., 334(S7), *368*
Summerfield, M., *75*, 266, *288*
Sun, K. H., 390, *424*
Sutherland, W., 187, 195, 196(S21), 204, *238*
Swallow, A. J., 386(S21), *424*
Swanson, C. P., 400(S22, S23), *424*
Sweeney, K. M., 346(B4), *366*
Symonds, G. H., 365, *368*

T

Tabaoda, O., 411(B1), *420*
Takagi, S., 32, *76*
Tarpley, W., 403(T1), 413(T1), *424*
Taub, F. B., 403(B12), *420*
Taylor, G. I., 242, *288*, 298, *329*
Taylor, P., 396(T2), *424*
Taylor, Sir G., 227, *238*
Thakar, M. S., 198, *237*
Thatcher, C. M., 267(B12), *285*
Thiele, E. W., 203, *238*
Thomaes, G., 280, *286*
Thomas, B. W., 148, *153*
Thomas, H. C., 227, *239*
Thompson, J. M., 83(A1), 84(A1), 86 (A1), 104(A1), *150*
Thompson, T., 377(T3), *425*
Thomson, Sir W. (Lord Kelvin), 26, *76*
Thornton, J. D., 317(T2), 318(T2), 319 (T2), 321(T2), 322(T2), *329*
Timmerhaus, H. C., 193, *239*
Tobias, C. W., 180, *239*
Tobolsky, A. V., 391, 392(S10), *424*
Tomlinson, N. P., 334(H8), *367*
Toms, B. A., 104(T2, T3), *153*
Torpey, W. A., *75*
Towle, W. L., 267(S17), *288*
Towsley, F. E., 118, *150*
Treybal, R. E., 197, 198(T6), *239*, 290 (T3), 291(T3), 298(F3), 301(T3), 304(T4), 308, *328*, *329*
Tsao, M-S., 159(F3), *235*
Tung, L. H., 181, 182, 208, *238*, *239*, 268 (T3, T4), 280(T2), *288*
Tupper, H. T., 319(G1), 321(G1), *328*

U

Uehling, E. A., 190, *236*, *239*
Uhl, W. C., 349(U1), *368*
Uhlenbeck, G. E., 170(W4), 183, 190, *239*

V

Van der Meer, W., 106, *153*
Van Der Veen, B., 268(K13), *286*
Van Dyck, W. J. D., 317(V1), 322(V1), *329*
van Selms, F. G., 84(K2), *151*
Vaughn, R. D., 109, 116, 132(V2), 140(V2), 141(V2), *153*
Vaughn, W. E., 385(B8), *420*
Vermeulen, T., 291(D1), 298, 299, 310, *328*, *329*
Vermijs, H. J. A., 314(V3), *329*
Verway, E. J. W., 87, *153*
Vogell, W., 193, *234*
Vogelpohl, G., 164(V1), *239*
Volmer, M., 23, 32, *76*
von Berg, R. L., 321(C1), 322(C1, W1), *328*, *329*
Vyazovov, V. V., 180(V2), 215, *239*

W

Waddel, H., *239*
Wainwright, L., 338, *368*
Walker, C. A., *421*
Walker, W. C., 88(L4), 143(L4), *151*
Walker, W. H., 174(W1), *239*
Wall, F. T., 159(W2), 196(W2), *239*
Walton, J. S., 198, *237*
Waly, A., 402(H26), *422*
Wamsley, W. W., 266, *288*
Wang, S. J., 121(B6), 124(B6), 131(B6), 132(B6), 134(B6), *150*
Wang Chang, C. S., 170(W4), 183, *239*
Warburton, P., 396(W1), *425*
Ward, A. F. H., 182, *239*
Ward, H. C., 106, 115, 146(W2), *153*
Washburn, H. W., 343(B5), *366*
Watson, K. M., 178, 194, *236*
Watson, W. F., 385(B7), *420*
Webb, G. B., 276(B8, B9), *285*
Weber, E. N., 382(S2), *424*

Weeks, B. W., 381(G4), *421*
Weiland, W., Jr., 267(L8), *287*
Weiss, J., 374(W3, W4), 383(R2), 385, 389(K2, L15, L16), 399, 403(T1), 404, 413(T1), *421*, *422*, *423*, *424*
Weiss, M. A., 228, *237*
Weisskopf, V. F., *239*
Welsh, D. G., 304(C4), *328*
Weltmann, R. N., 88(W3), 89(W4), 91(W4), 94(W4), 97, 99, 109, 114, 146, *153*, 164(W6), *239*
Wentworth, R. L., 281(S3), 282(S3), *287*
Westwater, J. W., 4(W3), 7, 8(W3), 9(W3), 10, 11(W2), 63, *76*
Wheeler, A., 182(W7), *23*
Whitehead, W. L., 381(S12), 405(W5), *424*, *425*
Whittaker, E. T., 347, *368*
Wiegandt, H. F., 321(C1), 322(C1, W1), *328*, *329*
Wilcox, D. E., 343(B5), *366*
Wild, W., 397, *425*
Wilhelm, R. H., 95, 103(W6), 106(W6), *153*, 219, *234*
Wilke, C. R., 61(M9), *75*, 159, 178, 180, 188, 197, 200, 223, *235*, *237*, *239*, 268, 271, *287*
Williams, G. M., 298(V2), 299(V2), *329*
Williams, R. R., Jr., 382(G6), *421*
Wilson, H. A., 218(W3), *239*
Wilson, K. B., 360, *366*
Winchester, C. M., 117, *151*
Winding, C. C., 106(W7), 121(W8), 124, 125, 127(W7), 128, 129(W8), *153*
Winning, M. D., 91, 93, 102, 103(G1), 110(G1), *151*
Winter, E. R. S., 188(W14), *239*
Wismer, K. L., 28(K2), 33, 46, 54(K2), 65(K2), *75*, *76*
Woertz, B. B., 267(S15, S16), 271(S16), 272, *288*
Wood, G. F., 115(W9), *153*
Wood, T. H., 400, 401(W8), 404(W8), *425*
Woodfield, F. W., 318(W2), 320(S6), 321(S6, W2), 322(S6, W2), *329*
Woodward, J. G., 147(W10), *153*
Wourtzel, E. E., 387(W9), *425*
Wright, J., *425*

Wright, N., 344, *366*
Wroughton, D. M., 95(W6), 103(W6), 106(W6), *153*
Wüst, O., 397(W10), *425*
Wynant, D. E., 411(B1), *420*

Y

Yong, T. F., 159(Y1), *239*
Yoshida, T., 5(Y1), 61, *75, 76*
Yost, H. T., 400(S23), *424*
Yudis, M., 403(T1), 413(T1), *424*

Z

Zabroski, E. L., 309(C5), *328*
Zemany, P. D., 344, *368*
Zettlemoyer, A. C., 88(L4), 89(Z1), 143(L4), *151, 153*
Ziegenhorn, G. J., 325(Z1), *329*
Zigan, F., 193, *234*
Zirkle, R. E., 402, *425*
Zuber, N., 13, 16, 17, 18, 19, 20, 21, 22, 59, 61, 67, 68, 69, 70, 71, *74, 75*
Zwanzig, R. W., *236*
Zwick, S. A., 21, 67, 68, 69(Z1), 70, 71, *76*

Subject Index

A

Acrylamide,
  radiation-induced polymerization, 392
Air stream,
  two-dimensional, schematic diagram of, 263
  temperature distribution in, 262

B

Beer,
  effect of low-dose radiation on, 411
Bingham plastics, 83–84
  friction factor—Reynolds number diagram for, 93
  heat transfer data, 132–134, 137
  mechanics of flow in round pipes, 90–95
Boiling,
  film, 4, 10
  growth of bubbles, 67–71
  of liquids, 1–76, see also Liquids, boiling
  nomenclature, 71–73
  nucleate, 4, 8
    and critical-temperature difference, 12–13, 42–50
    effect of agitation on, 62–64
      of geometric arrangement on, 58–59
      of impurities on, 64–66
      of pressure on, 60
      of short-wave irradiation on, 66–67
      of surface tension on, 60–62
      of surface texture on, 56–58
      of type of hot solid on, 54–56
      of type of liquid on, 51–53
    experimental values for, 50–67
    theoretical treatment of, 13–22, see also Nucleation
      empirical correlations, 21–22
      Forster-Zuber equation, 16–21
      Rohsenow equation, 13–16
  sound of, 11–12
  transition, 4, 9
  types of, 4–6
Buckingham (6-exp) potential, 184–186, 187
n-Butylamine,
  continuous extraction from kerosene into water, 307

C

Calculators,
  punched card, 334–336
  stored program, 336–338
Carbohydrates,
  effect of ionizing radiation on, 405
Cellulose,
  effect of radiation on, 394
Chemical engineering,
  solutions of diffusion equations of interest in, 198–228
Chemical reactions,
  initiation by ionizing radiation,
    experimental work on, 374–397
    fundamentals, 375–386
    investigations on practical processes, 386–397
Chlorination,
  induced by ionizing radiation, 389–390
Columns,
  agitated, multistage,
    comparison of the extractors, 311–316
    countercurrent, 310–316
    field of usefulness, 311
  pulsed, 317–323
    fields of usefulness, 318–319
    internal design for towers, 319–322
    packed towers, 322
    perforated plate towers, 319–322
    power for pulsing, 319
    pulsed mixer settlers, 322–323
    pulsing methods, 317–318

Computers,
  analog, 332–334
  automatic, in control and planning of manufacturing operations, 331–368
    control of manufacturing operations and, 341–345
    equipment, 332–341
    mathematical models, 348–356
    optimization studies, 356–366
    statistical and numerical analysis, 345–347
Condensations,
  radiation-induced, 389, 393
Conductivity,
  eddy, 256–258
    effect of distance parameters on, 257
    of Reynolds number and position on, 256, 257
Critical temperature difference, 42–50
  burnout problem, 42–43
  maximum efficiency, 42
  and nucleate boiling, 12–13, 43
  physical interpretations of, 43–50
    kinetic viewpoint, 48–50
    thermodynamic equation of state, 44–48

D

Dairy products,
  effect of ionizing radiation on, 408, 409
Dextran,
  effect of radiation on, 394–395
Diffusion,
  in binary systems, 170–177
    definitions of concentrations, velocities and fluxes, 170–171
    ordinary, 171–176
    thermal, 176–177
  calculation and estimation of diffusion coefficients, 182–198
  in complex systems, 220–228
  and correction in tube flow, analysis of, 227
  fluid mechanical basis for, 159–182
    vector and tensor operations, 229–231
  by forced and free convection, 179–180
  free double, separation of gases, 228
  in multicomponent systems, 177–178
  in multiphase systems, 180–182
  from a source in a skew velocity field, 228
  into and out of spherical particles, 228
  steady state, see Steady state diffusion
  theory of, 155–239
    nomenclature, 231
  in turbulent systems, 178–179
  unsteady state, see Unsteady state diffusion
Diffusion coefficients,
  calculation and estimation of, 182–198
  in dense gases,
    Enskog's rigid-sphere theory, 191–192
      applications to real gases, 192 ff.
  in dilute gases, 182–191
    Buckingham (6-exp) potential, 184–186, 187
    calculations of,
      classical calculations, 186–190
      empirical formulas, 190–191
      quantum calculations, 190
    intermolecular forces, 183–186
    Lennard-Jones (6-12) potential, 184, 186, 187
  in liquids, 195–198
    activated state theories (G6), 196–197
    empirical relations, 197–198
    hydrodynamical theories (F13), 195–196
  from sorption data, 227
Diffusion column, thermal,
  theory of, 223–227
Diffusion equations,
  of interest in chemical engineering, solutions of, 198–228
Drugs,
  radiation sterilization of, 413–414
  of containers, 414–415
  effective doses, 403
  engineering design, 415–419

E

Enskog's rigid-sphere theory, 191–192
  application to real gases, 192 ff.

Ethanol,
  nucleate boiling, effect of type of metal on, 55
Ether,
  nucleate boiling, effect of surface texture on, 57–58
Ethylene,
  radiation-induced polymerization, 391, 392
Extraction,
  liquid, continuous, countercurrent, multistage, 303, 304
    inadequacy of conventional equipment, 290
    mechanical aids, 289–329
      agitated vessels, 291–310
      centrifugal extractors, 323–326
      multistage counter-current columns, agitated, 310–316
      nomenclature, 327
      pulsed columns, 317–323
Extractors,
  centrifugal, 323–326

F

Fats,
  effect of ionizing radiation on lipids and, 405–406
Film,
  nonisothermal, diffusion through, 201–202
Fish,
  effect of ionizing radiation on, 408, 409
Flame propagation, theory of, 227
Flow,
  boundary, velocity distribution, 247–250
  laminar, material transport in, 270
  shear, kinetic energy spectrum of turbulence in, 244
  turbulent, effect of position on longitudinal fluctuations in velocity, 246, 247
    fluctuating longitudinal velocities in, 243, 246, 247
    macroscopic aspects of steady uniform, 247–255
    simultaneous thermal and material transport in, 278–281

  eddy properties, 280–281
  thermal transport in, 255–267
Fluid(s),
  behavior, recent developments in engineering classification of, 88–89
  classifications, 79–90
    fundamental basis, 79–80
  dilatant, 86–87
  moving, diffusion from a point source into, 218–219
  Newtonian, 80–82
    definition and representation, 80–82
  non-Newtonian, 82–90
    see also Non-Newtonian systems, classification, classical method, 83–88
  consistency index, 89
  extrusion, 118
  flow of, in annular spaces, 115
    laminar, 110
    nonisothermal, 116–117
    in round pipes, 90–95
    through fittings and entrance losses, 113–115
    two-phase, 115–116
  flow-behavior, design procedures for investigation of, 110–112
    effect of surface behavior on, 108–109
    effect of temperature and concentration on, 109–110, 111
    index, 89
  fluid kinetic energies, 112–113, 118–119
  friction factor-Reynolds number diagram for, 93, 100, 101, 102
  mixing of, 119–121
  velocity profiles, 107
    effect on flow-behavior index, 108
  plastic, see Bingham plastic
  pseudoplastic, 84–86, 95–107
    heat transfer data, 134–136
  pure, equations of change for, 159–160
    of continuity for, 160–161
    of energy balance for, 162–165
    of motion for, 161–162
  response to an imposed shearing force, 79
  stagnant, diffusion through, 199–201
Fluid-flow curves, 80, 81

Fluid properties,
  measurements of,
    determination of the absence of thixotropy and rheopexy, 142–143
    determination of shear-stress-shear-rate relationships, 138–141
    determination of the relationship between $D\Delta P/4L$ and $8V/D$, 141–142
Fluorocarbons,
  radiation-induced polymerizations, 392
Food,
  delicatessen, effect of low-dose radiation on, 410–411
  radiation sterilization, 404–413
    chemical changes, 404–406
    control of infestation, 411–412
    effect on flavor and appearance, 407–408
    effect on storage properties, 409
    effective doses, 401
    loss of nutrients, 406–407
    low-dose irradiation, 409–410
    toxicity, 408–409
Forster-Zuber equation, 16–21
Fruit,
  effect of ionizing radiation on, 408, 409, 411

G

Gases,
  dense, diffusion coefficients in, 191–195
  dilute, diffusion coefficients in, 182–191
  effect of ionizing radiation on, 382
  separation by free double diffusion, 228
Grains,
  control of infestation by irradiation, 411–412

H

Heat transfer, 121–138
  to dilute suspensions, 121–131
    comparative economics of suspensions and liquids as heat transfer media, 129–130
    design equations for use with dilute suspensions, 125–129
    evaluation of thermal conductivity, 122–124, 130
    of viscosities, 124–125, 130–131
  to highly non-Newtonian systems, 131–138
    laminar-flow region, 131–132, 136–137
    turbulent-flow region, 132–136, 137

I

Iodine,
  extraction between water and CCl_4, 306
Ionizing radiation
  application to chemical processes, 369 ff.
    dosimetry, 373–374
    fundamental considerations, 371–373
    practical considerations, 373
  application to industrial processes, 386–397
  biological action,
    mechanism of, 398–399
      chemical interpretation, 399
      physical interpretation, 398–399
  chemical intermediates formed in, 377–383
  effect on gases, 382
    on polymers, 393–397
  initiation of chemical reaction by, 374–397
    conversion of units, 375
    effect on aqueous systems, 378–380
    on gases, 382
    on nonaqueous liquids, 380–382
  prevention of sprouting by, 412–413
  reaction of chemical intermediates to, 383–386
  synthetic action, 388

L

Leidenfrost phenomenon, 3
Lennard-Jones (6-12) potential, 184, 186, 187
Liesegang ring formation, 227
Liquids,
  aqueous, effect of ionizing radiation on, 378–380

SUBJECT INDEX 445

boiling of, 1–76, see also Boiling
  bulk of volume-heated, 6
  critical temperature difference, 12–13, 42–50
  effect of pressure on peak heat flue and critical ΔT of organic, 60
  experimental values for nucleate, 50–67
  nomenclature, 71–73
  nucleate, of immiscible mixtures, 66
  photographic studies on, 7–10
  surface of subcooled, 6
diffusion in, 195–198
nonaqueous, effect of ionizing radiation on, 380–382
Luwesta (Centriwesta) Extractor, 325–326

M

Manufacturing operations,
  automatic computers in control and planning of, 331–368
    cost considerations in computing, 338–341
  computing equipment, 332–341
    punched card calculators, 334–336
    stored program calculators, 336–338
  control of, 341–345
    calculation of laboratory analyses, 342–344
    computer and automation, 344–345
    operations accounting, 341–342
  mathematical models, for, 348–356
Material transport,
  turbulent, 267–278
    diffusion coefficients, 268–269
    effect of position in stream on concentration parameter, 277
    generalized temperature distribution in uniform, steady, 265
    macroscopic, 267–268
      effect of turbulence, 268
      resistance at interface, 268
    prediction of, 275–278
      analytical method, 276–278
      equations of state, 275–276
      turbulent streams, 276
    temperature distribution around a sphere with, 279

    variation with composition in fully developed, 273
    variation in eddy diffusivity with position, 274
Mathematical models,
  for manufacturing operations, 348–356
  blending operations and, 352–354
  combinations of, 351
  definition, 348–349
  incorporation of economic factors, 356
  Monte Carlo method, 354–356
  optimization studies, 356–366
    factorial designs, 358–360
    general problems, 356–358
    linear programming, 364–366
    problems involving restrictions, 363–364
    sequential designs, 360–363
  types of, 349–356
Meat,
  effect of ionizing radiation on, 407–408, 409, 410
Metals,
  boiling of liquid, curves for, 52, 53
  experimental values for, 53–54
Methanol,
  boiling of, 7, 8, 9, 10, 11
Methylmethacrylate,
  radiation-induced polymerization, 392–393, 395
Microorganisms,
  mean lethal radiation doses, 400
Mixing theory,
  applications of turbulent diffusion to, 228
Mixtures,
  multicomponent,
    equations of change for, 165–170
    comments concerning, 169–170
    fluxes in terms of transport coefficients, 166–169
    in terms of fluxes, 165–166

N

Newtonian fluids, see Fluids, Newtonian
Non-Newtonian systems, 82–83
  time-dependent, 87–88
  time-independent, 83–87

Non-Newtonian technology, 77–153
 fluid classifications, 79–90
 heat transfer, 121–138
 mechanics of flow in round pipes, 90–112
 miscellaneous flow problems, 112–119
 mixing of non-Newtonian fluids, 119–121
 nomenclature, 148–150
 rheology and viscometry, 138–148
Nucleation, see also Boiling, nucleate
 theoretical treatment of, 22–42
  classical rate theory, 23–39
  statistical fluctuation theory, 39–42
Nukiyama's boiling curve, 3–4
Numerical analysis, 347
Nusselt number,
 predicted influence of Reynolds number on, 265

O

Oldshue-Rushton column, 312–314
Onions,
 prevention of sprouting by irradiation, 412–413
Organic reactions,
 induced by ionizing radiation, 386–390
Oxidations,
 due to ionizing radiation, 383–386, 387
  inorganic, 386–387
  organic, 388–389

P

Pharmaceuticals, see Drugs
Plate,
 flat, mass transfer from, 228
Podbielniak extractor, 323–325
Polyethylene,
 effect of radiation on, 389–395
Polymerizations,
 induced by ionizing radiation, 390–397
Polymers,
 effect of radiation on, 393–397
  cross-linking, 395–396, 397
  degradation, 394–395
Polystyrene,
 effect of radiation on, 396

Potatoes,
 prevention of sprouting by irradiation, 412–413
Prandtl number, 258–259
 effect of molecular, on total, 260
 of Reynolds number on total, 260
Proteins,
 effect of ionizing radiation on, 404–405

Q

Quenching, 2–3
 curves of, 5

R

Radiation,
 ionizing, see Ionizing radiation
 particulate, range of, in water, 376
Radiation sterilization, 397–420
 of drugs and pharmaceuticals,
  engineering design, 415–419
  sterilization of containers, 414–415
  sterilization doses, 401
 factors in, 399–404
  concentration, 401
  culture age, 400
  environmental conditions during exposure, 401–402
  hydrogen ion concentration, 402
  induced radioactivity, 402–403
  induced resistance, 401
  ion density, 402
  oxygen pressure, 401
  postirradiation conditions, 404
  preirradiation treatment, 400
  protective chemicals, 401
  sterilization doses, 403
  temperature, 401
  type of organism, 399–400
 of food, 404–413
  sterilization doses, 401
Reductions,
 due to ionizing radiation, 386, 387–388
Reynolds number,
 effect on eddy conductivity, 256, 257
  on Prandtl number, 260
  on ratio of eddy properties, 259
Reynolds stresses, 245
Rohsenow equation, 13–16

Rotating-disc contactor-RDC, 314–315
Rubber,
  effect of radiation on, 389, 395–396

S

Scheibel column, 315–316
Schmidt number, 272–275
Sharples Super Centactor, 326
Solids,
  radiation effects on, 382 ff.
Sprouting,
  prevention of, by irradiation, 412–413
Statistical analysis, 345–347
Steady state diffusion,
  in chromatographic column, 227
  in flow systems, 211–220
    into a falling film, 211–215
    from a point source into a moving fluid, 218–219
    through circular tubes, 216–218
    through a tubular reactor, 219–220
  in nonflow systems, 199–205
    accompanied by chemical reaction, 202–203
    through nonisothermal film, 201–202
    through a stagnant fluid, 199–201
    with pressure diffusion, 204–205
    with thermal diffusion, 203–204
  into a moving sphere, 228
Sterilization, radiation,
  see Radiation sterilization
Styrene,
  radiation-induced polymerization, 391–392
Systems,
  binary, diffusion in, 170–177
    notation for concentrations, 172
    for mass-flux vectors in, 173
    for velocities, 172
  multicomponent, diffusion in, 177–178
  multiphase, diffusion in, 180–182
  turbulent, diffusion in, 178–179

T

Technology, non-Newtonian,
  see Non-Newtonian technology
Thermal diffusion column,
  theory of, 223–227

Thermal transport,
  turbulent, 255–267
    macroscopic, 266–267
    prediction of, 259–266
      analytical solutions, 263–266
      boundary conditions, 262–263
      general energy equation, 260–262
Tobacco,
  control of infestation by irradiation, 412
Turbulence,
  classic references to, 242–243
  correlation coefficient, 245
    effect of distance on, 246
  effect on macroscopic material transport in fluid systems, 268
  homogenous, physical manifestations of, 243
  kinetic-energy spectrum of, in shear flow, 244
  nature of, 242–247
    statistical, 243–244
  in thermal and material transport, 241–288
  nomenclature, 283–285
Turbulent core,
  velocity deficiency, 250–251

U

Unsteady state diffusion,
  accompanied by chemical reactions, 209–211
  in evaporation, 220–223
  in nonflow systems, 205–211
    equimolal counterdiffusion,
      bounded, 206–207
      partially bounded, 207
      unbounded, 205
    two-phase system, 208–209

V

Vegetables,
  effect of ionizing radiation on, 408, 409
Velocities,
  fluctuating longitudinal, in turbulent flow, 243

Vessels,
  agitated,
    batch operations, 294–295, 300–302
      rate and efficiency relationships, 305
    bench-scale equipment, 309–310
      pump-mix extractor, 309–310
    characteristics, 292–309
      emulsions, 308
      interfacial area, 298–299
      mass transfer rates, transfer units, and stage efficiencies, 299–308
      mixing effectiveness, 293
      power, 296–298
      scale up, 308–309
    continuous operations, 295–296, 302–306
      efficiency relationships for countercurrent, 305
    design, 294–296
    field of usefulness, 292
    in liquid extraction, 291–310
      extraction rates, 306–308
  power characteristics of a mixing impeller, single-liquid batch system in open, 297

N-Vinylpyrrollidone,
  irradiation of, 393
Viscometers, 147–148
  capillary-tube, 138–139, 141, 142
    design of, 143–146
  Orr-Dalla Valle suspension, 145
  rotational, 139–142, 143
    design of, 147
Viscosity,
  eddy, 247, 251–257
    definition, 247
    in material transport, 271
  relative, 252–255
Vitamins,
  effect of ionizing radiation on, 406–407

W

Water,
  boiling, experimental values for nucleate, 51–52
    effect of impurities on maximum temperature for superheated, 65
    of pressure on peak heat flux and critical ΔT, 60
Wetted-wall tower,
  total diffusivities in, 272

DOES NOT CIRCULATE